Color for the Sciences

For information about special quantity discounts, please email special_sales@mitpress.mit.edu

Printed and bound in the United States of America.

Library of Congress Cataloging–in–Publication Data

Koenderink, Jan J.
Color for the sciences / Jan J. Koenderink.
    p. cm.
Includes bibliographical references and index.
ISBN 978–0–262–01428–1 (hardcover : alk. paper)
1. Colorimetry-Textbooks. 2. Colorimetry–History. I. Title.
QD113.K58 2010
535.6—dc22

                                              2009043838

10 9 8 7 6 5 4 3 2 1

Color for the Sciences

Jan J. Koenderink

The MIT Press
Cambridge, Massachusetts
London, England

Contents

CONTENTS

Preface

This book started out as a hobby. For some years I experimented with a formal course in colorimetry at Utrecht University, in the department of Physics and Astronomy. Since then I have taught similar courses to a variety of students and scientific audiences. The topic is a very attractive one and of considerable intrinsic interest to students of the sciences. Unfortunately, "colorimetry" as a discipline has become an engineering topic, focused on applications, dominated by conventional praxis and committee definitions and standards. As a casual look at any generic text will show, this is quite unlike the way a topic in the sciences is typically taught. There exists no text on colorimetry that comes even close to one of the familiar introductory texts in mechanics, electromagnetism, or optics, for instance. It is not just a matter of level, for the level of no "introductory" text can be very high, but mainly of *perspective*. Colorimetry is frequently regarded as a minor branch of engineering rather than as a science.

I aim at students from the sciences who desire an understanding of the topic of "colorimetry." It is not a highway to engineering applications or praxis, there exist many other texts with that aim. There is no royal road to any scientific topic. A scientist has to rethink the concepts from the roots. Ideally, the student should come to disagree with me on various topics, then the way to individual development lies open.

Although my aims are clear enough, it proved to be surprisingly hard to write such a text. The topic diverges from the standard strands in the physics curriculum, thus it is necessary to explain why it is of interest to delve into certain investigations at all. I also need to allude to the history of the subject, which otherwise might appear totally unrelated.

Fortunately, the history of colorimetry involves many well-known scientists and humanists with key contributions in more familiar fields of endeavor. In *physics* one has Newton, Maxwell, and Schrödinger; in *mathematics* one has Grassmann, and Riemann; in *chemistry* there is Ostwald; in *medicine* one has Helmholtz (who could just as well be counted with physics or mathematics); in *philosophy* there is Schopenhauer; among the *writers* Goethe; and, finally, among the *painters*

Runge. Of course this list is by no means exhaustive, moreover it depends on where one draws the line between history and the present.

It is perhaps somewhat of a problem that the text remains rather distant from the standard accounts. This is necessary because the standard accounts are not open-ended but present the topic in frozen form, as praxis. Most texts are little more than recipe books. If colorimetry is to be reckoned among the sciences, its methods and definitions should be continually questioned, not fixed by committee definitions. After reading this book, the reader should find it easy enough to pick up the standard accounts and accept them for what they are worth.

This book is written as a textbook in casual style. I present exercises at the end of each chapter. Many of these are open-ended, and some can be developed into research projects. As is common in introductory textbooks of the sciences, I do not even try to come up with an exhaustive bibliography. The reader will find many useful references in the notes to the chapters.

The formal apparatus of colorimetry is linear algebra over the real numbers of infinite dimensional spaces, mostly without the convenience of a metric. In the majority of cases one is mainly interested in certain convex subsets of these spaces; thus, the theory of convex sets also plays a key role. I have tried to keep the formalism as simple as possible, and I always consider the linear algebra from a geometrical viewpoint. When necessary, I explain the formal methods in appendixes; thus the book is self-contained. Some prior exposure to linear algebra and analysis should be a sufficient basis. No prior exposure to colorimetry is expected, whereas a certain innocence is perhaps desirable.

Although the math is actually very simple, especially readers with formal training in the sciences may find it confusing at places. For instance, the trained physicist is used to linear spaces of infinite dimensions, but in the setting of quantum mechanics where the spaces of interest are Hilbert spaces and the metric (in the guise of the scalar product) is understood. In colorimetry, the linear spaces of primary interest fail to be Hilbert spaces because no scalar product is available. Thus mappings have no adjoints, it makes no sense to consider orthogonal projections, and so forth. Although colorimetry is actually (much) simpler than quantum mechanics, one is apt to commit errors due to conventional habits that do not apply.

Although the book will be mainly used for self-study, it is well suited for an introductory, formal course. It is very suitable for a facultative, "different" addition to the standard sciences curriculum.

The main content of colorimetry is treated in chapters 3–8 and 12–13. Chapter 1 gives a short historical introduction and chapter 2 an inkling of what colorimetry is about in real life. They are best assigned as preliminary reading. Chapter 9 is a short one that introduces the simplest and most intuitive and generally useful implementation of colorimetry that I can think of. Chapter 10 explores extensions pioneered by Schrödinger and Helmholtz, it can be skipped on first reading. Chapter 14 should be of interest to computer graphics students but contains no additional material on colorimetry as such. The chapter stands by itself and can be skipped on first reading. Chapter 15 is a short one that explores an interesting and potentially important extension of basic colorimetry. Chapters 16 and 17 are about topics related to colorimetry that may be primarily of interest to "computer vision" students. Finally, chapter 17 gives a few short hints on various derivative topics. It is up to the teacher what to assign of the appendixes. I would at least treat one of the models in class, and assign others as problems.

The book should be suitable for a one (chapters 3–7) or two (add chapters 8 and 11–13) semester course on colorimetry. The text and notes contain numerous suggestions for demonstrations (much recommended, and many demonstrations are easily adapted for a large audience) and individual explorations, either with pencil and paper or using a computer (I have good experiences with Mathematica as an environment).

The "models" I introduce in this book are especially useful in sharpening the understanding and they can be explored readily with pencil and paper. Such models are not to be found in standard texts because they are irrelevant to applications and the standard texts are not tuned to building understanding, but to promote effective praxis. Though indeed, good for nothing, I highly recommend that the reader spend much time on the models and—even better—construct additional models particularly adapted to illustrate additional elements of the theory. It is also of interest (both useful and fun) to compute various entities for the "standard observer." Since the standard observer is defined as a table, this involves numerical linear algebra of high dimensional spaces (dimension of the order of one hundred), and thus is very tedious by hand. Such linear algebra problems are a breeze on even a small personal computer, of course. In order to obtain a feel for the theory it is almost a necessity to acquire some expertise in this type of calculations.

Many of the exercises are "generic" in the sense that the reader is expected to think of various ways to explore a topic. For a regular course the teacher will find it easy to clone many exercises (and exam topics) on the basis of these examples.

Any remarks that might help me to improve the text are very much appreciated. My email address is J.J.KOENDERINK@TUDELFT.NL.

———————

Because I wrote this book by way of a hobby (in the margin of my actual work), it took me a unusually long time to complete. Apart from home, much was written in various hotel rooms over Europe and the U.S. I also used several occasions of prolonged absence from home for (always still too short to get really substantial work done) spurts of activity, most notably at the Institute for Mathematics and Its Applications, Minneapolis and at the École Normale Supérieure, Paris. Some final touches (a contradiction in terms) were applied while at The Flemish Academic Centre for Science and the Arts (VLAC).

Several people read versions of the manuscript in various stages of completion. I am especially indebted to Michael Brill, whose understanding of the topic far exceeds mine.

The suggestions by the proof reader of MIT Press taught me humility and were much appreciated. For all the errors left in the text, I of course assume full responsibility.

Writing a book is fun for the author but perhaps less enjoyable for his family. My wife Andrea took the brunt of the enterprise. I am thankful for her patience and moral support. Among women she is as close to perfect as they come.

Utrecht, May 8, 2010

Part I

Introduction

Chapter 1

About Colorimetry

For any book on "color," the question "What *is* color?" appears appropriate.[†] I skip this question, though. Why? Because I think it doesn't carry you much farther.

In order to get a feel for why this might be so, consider a few possible answers:

Physicist: "Color is nothing but the wavelength of electromagnetic radiation in the 470–700 nm[1] band, as Newton has already shown. This topic is old hat. Who cares?"

Physiologist: "Color is the (electrochemical) activation of the neural system as modulated by the three retinal cone types.[2] Nowadays the photopigment genetics has been fully cleared up, and color is well understood."

Psychologist: "Color is a basic sensation of the visual modality that is most easily measured by way of the Munsell atlas.[3] There are still some minor problems to be cleared up."

Philosopher: "Color is a *quale*.[4] Either you see color by wavelength or color is mere mental paint, we haven't decided yet."

[†]Various notes appear throughout the text, but this is the only footnote. Notes are collected at the end of each chapter. There is no need to consult them at first reading, or at any time at all. You might find the notes occasionally useful as a guide to additional reading and so forth. Best to simply ignore them right now.

Colorimetrist: "Colors are the equivalence classes of mutually indiscriminable physical beams. Colorimetry is a closed subject, though very useful in many applications. Simply follow CIE[5] recommendations and you can't go wrong."

Engineer: "Color is simply a transformation from radiant spectral densities to Lab[6] coordinates. Just refer to the manuals."

Artist: "Colors are what make the picture surface *vibrate*.[7] You can easily spend a lifetime and still not understand color."

Salesman: "Colors are what make the product sell, as simple as that."

Computer scientist: "Colors are 3-item, ordered data structures, typically represented as 24 bit (3-byte) structs.[8] What else did you want to know about it?"

Man in the street: "Color is what makes the red fire truck stand out from the green bushes. Any fool understands as much."

And so forth.

Each answer is representative of many answers that I have actually heard. Most are phrased in some kind of technical lingo that would make the target in-crowd nod in recognition. Did you *learn* anything in reading these answers? Did you notice that some answers are just plain *wrong*, while others involve a *confusion of levels*, show an overly *limited understanding* of the question, or are simply sophisticated ways to *dodge it*? You are probably confused, and rightly so. For "color" means many different things to many different people for many different purposes.

The question, "What is color?" should not be posed at the beginning, but rather at the end of a book on color. Or, even better still, the reader should be liberated of the urge to ask the very question by the time the book has been read, and not simply because the question was answered between the covers. It is possible to know much about color without really knowing what it is (or rather, can be). Instead of considering so-called deep questions ("Is the RED[9] I see the same as the RED you see?" is perhaps the most commonly heard one), it is more useful to discuss pertinent *facts*.[10] This is what I will do in this book.

If you are mainly interested in the deep questions you should consult the philosophical literature on color.

It is natural enough to be fascinated by "color," for color is part of the stuff that makes up your conscious experiences. Color is both emotion (thus "you") and is in things. For all of us, color is an indelible part of reality, despite the fact that many philosophers deny its very existence except (perhaps) as "mental paint". Artists tell us that the world of color is so rich and complicated that it forever evades the human power of description or measurement. Whether true or not, it is not a very helpful point of departure for a *science* of color, and you may actually wonder whether such a science is possible at all.

Indeed, color is one of the immediate and ever-present parts of experience that defied measurement for most of written history. Spatial extents, weights, and stretches of time were quantitatively measured by most ancient civilizations. Color was not. In order to specify a color, one simply had to keep a specimen of that fiducial color in stock. A "science" of color is first of all a method of *measuring* color in an objective manner. In this introductory chapter I sketch (very summarily) how such a science of "colorimetry" developed as part of the scientific development since the renaissance.

Color is light. You cannot have luminous experiences without having color experiences and vice versa. Light became measurable in the seventeenth century, when the science of "photometry" ("light measurement") was developed.[11] This became a true science by the end of the eighteenth century. It was from photometry that a science of "colorimetry" ("color measurement") eventually arose.

"Colorimetry" is literally "color measurement," as "geometry" is literally "land measurement." Geometry became a practical endeavor as people came up with tools that enabled quantitative measurement. The idea is simple enough. You take a stick (or any rigid object) and place it next to some arbitrary stretch. If the endpoints of the stick are in coincidence with the end points of the stretch, then that stretch is declared to be "one stick long." The stick may be arbitrarily moved (displaced and rotated) in order to bring points into coincidence. The stick may not be broken, bended, elongated, or compressed, though (thus, rubber sticks are tricky to use). Two different stretches that are each one stick long are said to be of the same length, no matter what their locations or spatial orientations might be. With some ingenuity you define lengths of several sticks (by suitably displacing the stick) or less than a stick (by subdividing the stick, *bisecting it, and so forth*). In this way you develop "geometry." The basic operation is

the judgment of coincidence of pairs of points.[12]

Measures for area, angle, and so forth are added to the toolbox in a fairly obvious fashion. Experience with measurement results leads to a science of geometrical measurement that allows the outcome of certain virtual (that is to say, not actually performed) measurements to be predicted from the value of other (actual) measurements. The generic example is the "Pythagorean theorem." For instance, if a point can be reached from a given point by moving three sticks to the right and then four sticks forward, then the two points will be found to be exactly(!) five sticks apart. Thus, using a rope tied in a loop with knots defining stretches of three, four, and five sticks, you can lay out a right angle. Perhaps sadly, most modern people have lost the ability to see the *magic* in this any more.

The basic operation of comparing two entities through a judgment of "coincidence" or "indiscriminability" is very basic and powerful. It can be used equally well in many different domains.[13] For instance, the measurement of weight was built upon a similar foundation from the earliest times on using scales. The equilibrium condition is again one of coincidence. In the seventeenth century *photometry* was developed along these lines, luminous intensities being matched via standard candles. The matching of luminous intensity is very similar to the use of scales. It is again a condition of coincidence or indiscriminability. In the nineteenth century physicists developed numerous systems of measurement using this principle.

A measurement of color can also be developed using the comparison idea. This is a very natural notion indeed, especially when you already have some familiarity with photometric measurements.

The start of colorimetry is conventionally considered to be the work of Isaac Newton in the seventeenth century, Newton studied the spectrum, did some experiments on color mixture, and provided many speculative thoughts. His fame was such that even today many scientists and (especially) philosophers believe that you "see color by

---

What is a "quale"?

Q: What is it like to experience REDNESS?

A: If you have to ask you'll never know!

---

wavelength," an unfortunate misconception due to Newton. It was only around the turn of the eighteenth to the nineteenth century that an "explanation of color" was sought not just in the "light," but also in the physiology of the observer. The key idea is conventionally attributed to Thomas Young.[14]

The true development of colorimetry took place during the course of the nineteenth century, mainly due to the empirical and conceptual work of James Clerk Maxwell and Hermann von Helmholtz, with notable formal insights due to Hermann Grassmann.

The conceptual crux that makes colorimetry into a true science is essentially due to Maxwell. It involves a giant conceptual leap, of a nature that even today is rarely appreciated even by the majority of scientists. Colorimetry has nothing to do with color as you experience it, color as *"quale"* as the philosopher puts it. Thus colorimetry is *not* about "what it is like to see red"; in a very real sense *colorimetry is not about color at all*, at least not about "color" as you know it.[15]

In photometry, you define "equal intensity" of two patches of light via their indiscriminability. A practical way to do this is to set up a simple apparatus such that an observer is presented with a luminous disk in an otherwise dark, featureless field of view.

The disk is split into two half-fields that sharply abut at the vertical diameter of the disk (think of the disk as presented in a frontoparallel plane). Each half-field is of absolutely featureless, uniform luminosity (see figure 1.1). The luminosity of the half-fields is brought about in some way that is not particularly relevant here.[16] I will assume that each half-field is due to some illuminating *beam*. Both beams elicit a luminous *patch*, and the two patches appear adjacent in the visual field of the observer.

As the observer peruses the disk either of two things may turn out to be the case (see figure 1.1).

- The half-fields are apparent; thus, the beams appear *different* to the observer.

- The half-fields are not apparent, and the observer is only aware of a uniform, undivided luminous disk. In this case, the beams are declared to be "photometrically equal."

The system works much like scales for the comparison of weights, if you are unable to distinguish which way the scales tip, the weights at either side are said to be equal.

It was Maxwell's genius to notice that the only *objective* measure of color is the judgment of indiscriminability of beams of radiation, a straightforward extension of photometry (figure 1.2). If you consider two beams of radiation such that the patches of light that appear when the beams that hit your eye cannot be distinguished by you, then the beams can be said to "have the same color" for you. This can be operationalized in many ways; you merely have to make sure that the observer has no way to distinguish the beams except for their "quality," thus the spatial and temporal nature of the beams should be made as similar as possible. Conceptually, at least, you can test *all* beams (of course there exists an uncountable infinity of these) and divide them into subsets of mutually indiscriminable ones. Such subsets are "equivalence classes" with respect to the vision of the observer. These equivalence classes are defined as the "colors" of colorimetry. Colorimetry seeks to find relations between beams and (colorimetric) colors as well as the structure of the (abstract) "space of colors."

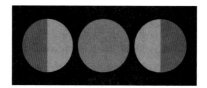

Figure 1.1 Three views in a typical split-field photometer. Equality obtains for the view in the middle.

Figure 1.2 Maxwell as a young man. The object in his hands is one of his color tops. Colorimetry was foremost in his mind at the time. The biography in which I found this photograph never mentions color at all. It is of no relevance to physicists today, no matter what Maxwell himself may have felt back then.

It is important to appreciate how *abstract* this notion is. "Colors" in colorimetry have no "color" in the sense of qualia; thus, a colorimetric description of a color gives no cue on "what it is like to see" (or "have in mind") that color. It is not so much that colorimetric colors are "colorless",[17] but rather that the concept of color as quale does not apply to them. Their ontology is different, they reside in different strata of existence. Such was Maxwell's genius that he saw that such an arcane definition is the very foundation on which a rich and quantitative science can be erected. He proceeded to develop much of that science (both empirically and conceptually) from scratch and on his own steam. Maxwell should count among your heroes.

I consider this most remarkable because even today, when I deliver a lecture on colorimetry to some learned audience (I have lectured to audiences of physicists, mathematicians, computer scientists, philosophers, and psychologists thus far), obviously stressing the all-important conceptual issues, then invariably, at the discussion period after the lecture someone will ask a question that shows that he or she completely missed the crux of the matter. The question will always be about color as quale. It is apparently unpalatable to the audience that colorimetry has nothing whatsoever to do with what it is like to see REDLY (or have REDNESS in mind).

This is a book about colorimetry, or, as I prefer, about the "geometry of color space." I will mainly deal with conceptual, formal (expressed in terms of mathematics) issues. Thus this book might have been called "theoretical colorimetry." Of course, I will mention empirical matters too, because the formal structure is *about* real phenomena. I will not discuss experimental methods, though, or quote empirical results extensively.

"Color science" is not a "science" any more than Christian Science is. Colorimetry *is* a science. Since "color science" should at least contain colorimetry, color science is a *much* broader topic than colorimetry. Color science means different things to different people. This runs from engineering issues in computer science and ergonomics to aesthetic issues in the arts and color therapies in offbeat medicine. This book is not about these (many of them important) fields of endeavor. There exist many books that focus on such issues, and the reader is invited to consult them.[18] Think of the present book as about a very tiny corner of color science, "theoretical colorimetry," which is again a tiny part of "colorimetry" as practiced on a daily basis (which is mainly about engineering and convention).

Theoretical colorimetry, or the geometry of color space, is a very attractive topic if you happen to enjoy the study of the formal structures of our scientific descriptions of the world. It is rather useless from the applications perspective, but has its own attractiveness and even beauty to the enquiring mind. Moreover, it is far from being a fully developed area, so the interested student will find plenty left to do.

The level of exposition is that of the typical undergraduate physics course. A basic understanding of linear algebra and some analysis will get you on the way.

1.1 Color in Science, Business, and Industry

Colorimetry is a true science,[19] but it is rarely pursued for its own sake. Most people would probably consider colorimetry a "dead science" in the sense that nothing *new* remains to be done. (Of course, I disagree with this view!) The majority of people then are *users* of colorimetry. Small wonder that their understanding of what colorimetry is about and what is important in it is strongly colored by their own discipline and interest.

For those with an engineering interest, the case is simple enough. They look for the state-of-the-art methods and apply these without questioning their origin. For those with an interest in the sciences, the case is rather less simple. They reinterpret the foundations of colorimetry to suit their own disciplines.

The sciences that use colorimetry on a conceptual level are mainly physiology and psychology. It tends to be rather obvious whether a writer on colorimetric topics originates from one of these sciences.

The physiologist is interested in *mechanism*. For such a person, colorimetry is all about retinal cones and photo-pigments. A friend once remarked to me some years ago, "Color vision is finally solved!" and referred implicitly to recent findings in the genetic code for retinal photo-pigments. This is evidently a rather *limited* view. Moreover, colorimetry proper has really nothing to say concerning retinal photo-pigments. All you can do is determine "human fundamental space,"[20] thus the "fundamental response curves" up to arbitrary linear transformations.[21] For the physiologists, this simply won't do, and they reinterpret and redefine colorimetry to suit their wishes. Something similar applies to the psychologist. As I explained earlier, colorimetry has only the most tenuous link with psychology. Again, for the psychologist this simply won't do. So the psychologist redefines colorimetry such as to deal with *color appearances*.

Since colorimetry is not about physiological mechanisms nor about color appearances, both the physiologist and the psychologist really redefine and extend colorimetry. This is not a problem as such, but the problem starts when this is forgotten or not understood to begin with. Unfortunately, this appears to be the generic case. It frequently leads to unfortunate misunderstandings, even in the modern literature.

In business and industry, people are mainly interested in standard-
ization. They need to specify the color of products and monitor variations of color. Since it is a nuisance to have to deal with human observers they need an "artificial observer," that is to say, some algorithm
or machine. But this works only if everyone agrees on the *same* standard. This is far more important than the minor question of whether
the standard is indeed anything *like* the generic human observer. Thus,
for business and industry, colorimetry is essentially *convention*. Large
groups of people are involved in polishing rules and recommendations,
other groups in pushing or even enforcing them. There is rather less
science in all of this than apparently meets the eye.

The engineer who designs or produces paints, car bodies, computer
screens, TV sets, photographic films, inkjet printers, and so forth needs
to be aware of the standards, but in addition has a certain interest in
getting things right. For such people, colorimetry is applied science,
to be sure, but they do have a genuine interest in effective procedures
that yield results that indeed apply to generic human observers. Some
people from this group are actually involved in the further development
of colorimetry as a science, although the larger fraction is quite happy
with "hacks." If you are involved with practical things, then something
is right if it works for your purpose. Issues of scientific "correctness"
simply don't arise.

Of course, all of this only makes sense, and there is nothing wrong
with it. It is somewhat to be regretted, though, that colorimetry as a
science doesn't fare very well. Indeed, it is hardly to be found amidst
the numerous hacks and arbitrary conventions. Strange to say, but if
you happen to be interested in colorimetry as a *science*, you may safely
skip much of the literature that purportedly deals with the topic.

1.2   A Short History of the Geometry of Colors

On June 10, 1854, Bernhard Riemann delivered his famous colloquium
"Über die Hypothesen, welche der Geometrie zu Grunde liegen" (On
the Hypotheses that Compose the Roots of Geometry[22]) at the occasion
of his "Habilitation". In the audience was *Herr Geheimer Hofrath*
Gauss, who may have been the only one to have fully appreciated the

content of the lecture. In his introduction "Notion of an n-ply extended magnitude" Riemann remarks,

> *so few and far between are the occasions for forming notions whose specializations make up a continuous manifoldness, that the only simple notions whose specializations form a multiply extended manifoldness are the positions of perceived objects and colours. More frequent occasions for the creation and development of these notions occur first in the higher mathematic.*[23]

That the "positions of perceived objects" can be described formally by way of the three-dimensional Euclidean geometry (and nowadays more general geometries) is a heritage from the ancient world. This "multiply extended manifoldness" simply *embodies* geometry for us. That Riemann saw fit to mention the colors as the only other example known to us from intuition is nothing less but remarkable. It it probably due to his familiarity with the works of Isaac Newton, James Clerk Maxwell, and (especially) Hermann Grassmann. Each of these scientists contributed crucially to the formal geometrization of the multiply extended manifoldness of colors.[24]

In the seventeenth century Isaac Newton studied the phenomenon of colors in the context of optics.[26] He described the spectrum and noticed a correlation between the wavelength of a monochromatic beam (to Newton the "refrangibility" of a "homogeneous light") and HUE.[27] Newton concluded (falsely) that all HUES are necessarily correlated with wavelengths of monochromatic beams.[28] He noticed phenomena of color mixture and drew up a "chromaticity diagram"[29] that consisted of a circular area, the circumference being populated with "homogeneous lights," the interior with mixtures of such lights. He suggested a barycentric construction for the result of such mixtures,[30] Thus, Newton ventured something like a topological color space with nontrivial geometrical (affine) structure. There are many problems with Newton's ideas (and they were offered frankly as guesses, except for the spectral composition), but Newton's authority was so great that even people today believe that "you see colors by wavelength."

Remarkable and admirable as Newton's conceptualizations were, they were not without glitches. Because of Newton's authority, some

Bernhard Riemann (1826–66)

Bernhard Riemann was one of the most important mathematicians of the nineteenth century. His influence on (mathematical) physics was considerable. Here I am mainly interested in his work on the structure of spaces. His work contains a broad generalization of Gauss's work on curved surfaces. It opened new perspectives on the classical homogeneous spaces (Euclidean space and Lobatchevskian space, to which Riemann added elliptical space). Riemann went further and developed the metrical description of inhomogeneous spaces (the metric changing from point to point). These ideas later obtained physical reality in the hands of Einstein. Helmholtz[25] and (later) Schrödinger were the first to use Riemann's ideas to put a metrical structure on color space. The issue of the metrical structure of color space (if there is any) is still not settled.

of these even corrupt our thinking today. One of these (especially important in color vision) is the notion that "white light is a confused mixture of homogeneous lights" or (as one would phrase it today) "general beams ("daylight") are incoherent superpositions of monochromatic beams," each "monochromatic beam" being characterized by a unique "wavelength" (*in vacuo*, one should add). Newton believed that he had shown this beyond any reasonable doubt through the (rightfully famous) *experimentum crucis*. He also believed that HUES correspond in a one-to-one fashion to refrangibility (wavelength). Even today, many of my physics students, and even my colleagues, believe this to be *literally true.* It is no doubt the basis for the (fallaceous) belief that "you see color by wavelength." Yet Newton's arguments contain a basic flaw. He indeed showed that you may *analyze* and *synthesize* daylight from "homogeneous lights," but to say that daylight is *composed* of homogeneous lights is an unwarranted generalization. It is like saying that a sausage is composed of slices because it is admittedly possible to slice a sausage.

Isaac Newton (1642–1727)

Isaac Newton became famous both as a physicist and as a mathematician. He went to school at Woolsthorpe in Lincolnshire, entered Cambridge University in 1661, and was elected Fellow of Trinity College in 1667, and Lucasian Professor of Mathematics in 1669. Newton was elected a member of Parliament in 1689, and he became Master of the Mint (at London) in 1699. Elected a Fellow of the Royal Society of London in 1671, he became its president in 1703 and was annually reelected for the remainder of his life. His best known work (the *Principia*) dates from the plague years 1665–66, when he fled Cambridge for the (supposed) safety of the country. His *Opticks*[31] appeared in 1704, just before he was knighted in 1705. The experiments and thoughts on which the *Opticks* is based date from 1668. From the early eighteenth century on the *Opticks* has been considered a paradigm of the "scientific method."

It was Johann von Goethe who—as the first—saw *flaws* in Newton's arguments. Due to Newton's status, nobody even tried to follow his reasonings, though. Newton's word was LAW. Unfortunately, Goethe was not a scientist, and his *Farbenlehre*[32] was largely ignored by scientists. Goethe's (indeed largely unfair) polemics against Newton only made things worse. What Goethe discovered empirically is known in optics as "Babinet's Principle"[33] (see p. 228), the fact that (in both geometrical and wave optics) complementary apertures yield complementary images.

Since a "spectrum" is simply the image of the "entrance slit" of a spectroscope, you can produce an "inverted spectrum" by changing to a "complementary slit." This is what Goethe discovered with his

**Johann Wolfgang von Goethe
(1749–1832)**

Johann Wolfgang von Goethe is often referred to as Germany's major poet. He was a universal genius both in the humanities as well as in the sciences. He was a social being, with many relations among his contemporaries, a great teacher and enthusiastic lover of women. From 1775 on he lived as minister at the Residenz of count Karl-August von Weimar. Of interest here is his work on color science. The larger part of this is laid down in the *Farbenlehre* (Color Theory), which is available in many translations. The book deserves reading by anyone interested in color.

continues next page . . .

> *... continued from last page*
> Goethe was a remarkable observer and has a special way of turning observations into science. He despises formal (mathematical methods) and sticks close to the phenomena. Ideally, "the phenomena themselves are the theory," but then you have to abstract their essentials, the "*Urphenomän.*" Different than for Newton for whom sunlight was "a confused mixture," for Goethe sunlight is the simplest, integral entity. Colors are a kind of incomplete sunlight; they are "shadow-like." The spectrum colors of Newton are nothing special. Moreover, they are not complete, for they lack purples. Goethe introduced the inverted spectrum and shows how the spectrum and the inverted spectrum together account for all colors on the color circle.

Kantenfarben (see p. 215). Newton's *experimentum crucis* (see p. 217) can be repeated with the inverted spectrum, and Babinet's Principle ensures that it works just as well. The resulting "homogeneous lights" are far from "monochromatic beams," though, showing that Newton's reasoning contains a flaw. Indeed,

> *daylight can be analyzed in terms of monochromatic beams, but it is in no way "composed" of it.*

There are infinitely many ways to describe daylight that are mathematically fully equivalent. For many purposes of physics, the monochromatic basis is most convenient. For the purposes of colorimetry, the monochromatic basis is largely irrelevant, however. Goethe's *Kantenfarben* ("edge colors") do as well and have quite a few advantages to boot.

The notion that the dimension of the space of colors is only three, and thus that not every homogeneous light contributes an independent degree of freedom, was around (albeit in vague form) from (at least) Newton's times. For our purposes, it suffices that Thomas Young clearly articulated this point in the early nineteenth century. The first one to see the consequences with amazing clarity was James Clerk Maxwell. It is not overstating the case to say that Maxwell developed color geometry single-handedly from a primitive state to (almost) maturity. Maxwell started serious work on colorimetry in 1849 using a simple "color top" and in 1858 first "gauged the spectrum." The

Thomas Young (1773–1829)

Thomas Young was a gentleman of independent means with a keen interest in science. Although he decided on a career in medicine, he did not really practice but continued scholarly studies at Emmanuel College Cambridge. In 1801, Young was appointed professor of natural philosophy at the Royal Institution, and in 1802 he became foreign secretary to the Royal Society. Young is well known for many reasons more important than his notions on color. In 1802, Young proposed that there might exist three types of retinal particles, each associated with one of the principal colors red, yellow, and blue, in his 1803 paper changed to red, green, and violet. Although Young's primacy is debated, his was clearly one of the earliest and clearest statements of the fundamental property of "trichromacy" that characterizes human spectral discrimination.

first series of experiments established the linear structure of colorimetric equations,[34] and the second series made it possible to predict the outcome of such equations from the structure of the spectra of beams.

Maxwell was twenty-four years of age when he wrote his first paper on colorimetry. By 1852 he had read Helmholtz's paper, which so infuriated Grassmann (see below). Where Helmholtz was at first reluctant to adopt Young's hypothesis, accepting it only in 1858, Maxwell adopted it from the start. He showed that studies of color blindness can in principle reveal the nature of Young's fundamentals. This was actually done much later; in 1886, König and Dieterici[36] provided the first good empirical estimates of the chromaticity of Young's fundamentals, indeed using color blindness.

Maxwell's "color top" is nowadays a mere plaything for kids that you buy in science museum shops. Maxwell used it to demonstrate the linearity of additive color mixture. One of his most important contributions was methodological. He did not rely on absolute color judgments, as everyone else did (and many people would like to do today), because he clearly perceived these to be subjective in the sense of idiosyncratic. People tend to fight forever over such silly questions as whether turquoise is "green" or "blue," and there is no end to this. But most people ("normal trichromats") will easily agree over judgments

James Clerk Maxwell (1831–79)

James Clerk Maxwell is best known as the theoretical physicist who formulated the fundamental structure of electromagnetic theory. He has many other accomplishments in physics. His work on colorimetry[35] is much less commonly known. In one biography, the photograph next to the frontispiece shows Maxwell as a young man holding a color top (figure 1.2). The legend fails to mention the nature of this odd atttribute (which must look mysterious to most readers), nor is there any mention of colors in the book. This is fairly typical. Yet his work on color was very important to Maxwell himself and is indeed of monumental importance to science. Maxwell's work led to Grassmann's formal color geometry.

of equality or discriminability, that is to say, whether two colors are or are not distinguishable. In his experiments with the color top, all the observers had to do (indeed, were permitted to do; Maxwell would simply ignore their opinions as to "the" colors they saw) was to decide whether a colored disk was qua color distinguishable from a surrounding annulus. Such judgments are very easy to arrive at, and they are the very crux of the science (indeed, it became a science in the hands of Maxwell) of "colorimetry".[37]

Later Maxwell constructed his "color box" (see p. 531), which is essentially an triple-barreled, inverse spectroscope. With the color top Maxwell could only compare the colors of colored papers, whereas with the color box he was able to study the discriminability of spectral distributions. Maxwell exploited the linear structure of colorimetry as he expressed the nature of human color vision in terms of a *basis* of the space of physical beams. He used the spectral basis, and, as we would say today, was the first to "gauge the spectrum." Once this is done, the (colorimetric) color of *any* beam can be *calculated*, so that you don't have to *measure* it. Once you have gauged the spectrum you are done with observations. Clearly this was a very important, even spectactular, development. This is the modern situation. If you want to know the (colorimetric) color of a beam, you load the specifications of the "normal trichromatic observer" from the Internet[38] and simply *calculate* it.

In the mid nineteenth century, Hermann von Helmholtz discovered empirically that homogeneous lights from the middle of the visual spectrum ("GREENS") have no "complementaries," that is to say, there exist no homogeneous lights that can be mixed with them such as to produce an ACHROMATIC ("WHITE") mixture. This caused a shock in the scientific world, for according to Newton's scheme, every homogeneous color has an "antipode" on the circumference of the color disk, and an equal mixture of them thus should get you at the (colorless) center of the disk. Helmholtz used methods that were fully acceptable from a technical point of view, though. Perhaps the only flaw in his work was that the definition of "white light" was informal. But no matter how you define "white light" (sunlight, average daylight, candlelight, and so forth, are all likely candidates), the greens (monochromatic beams from the "middle" of the visual spectrum) turn out to have no complementaries. The exact location of this "gap" in the spectrum does depend on the precise definition of the "white light," though.

Helmholtz continued (although he was active in many other fields) to contribute to the science of visual perception till the end of his life in the 1890's. Part of his work on color goes beyond colorimetry and runs into experimental psychology. Helmholtz had a particular gift in finding approaches that would fit into the exact sciences. Whenever possible, he would frame physical or physiological explanations for his findings, and whenever possible he would come up with (sometimes surprising) formal, mathematical descriptions. He was the first to attempt to go beyond the colorimetric framework due to Maxwell,

**Hermann von Helmholtz
(1821–1894)**

Hermann von Helmholtz received his academic training at the Medical Institute of Berlin because his parents were unable to pay for a regular education. He received his degree in 1842, then had to spend some time as an army surgeon at Potsdam to pay for his education. He could devote time to physiological studies, though. In 1848 he was offered a position as associate professor of physiology at Königsberg. Here he published his influential paper on the conservation of energy. He also did his first experiments on color vision and corrected Newton. In 1855 Helmholtz moved to the University of Bonn, and in 1858 to the University of Heidelberg, where he did much work on sensory physiology.

continues next page . . .

... continued from last page

In 1878, Helmholtz accepted a chair in physics at the university of Berlin. Most of the remainder of his career was dedicated to physics. For our purposes his "Handbuch der physiologischen Optik" (Handbook of Physiological Optics), first published in 1856, is the major source for his work on color.

Much of Helmholtz's innovative work on color was done very early in his career, though he remained interested all his life (for instance his attempt to generalize Fechner's law for color space dates from 1890). In his early work he was the first to correct Newton on the color circle. He showed that greens admit of no complementary spectrum colors and that the purples do not occur in the Newtonian spectrum. He also showed (like Maxwell) that the spectrum locus is a general, horseshoe-shaped curve, not a true circle.

Grassmann, and himself and try approaches that were only followed up in Schrödinger's work of the 1920s. Helmholtz was the first to attempt a Riemann metric for color space (in 1890), a metric changing from point to point, with nontrivial geodesics.[39] It is described in the second edition of his handbook, the last edition to be edited by himself. It was purged from the handbook's later editions and the English translation by his former pupil von Kries.

Helmholtz's publication so scandalized Hermann Grassmann that he set out to write a paper against Helmholtz in which he purportedly proved Newton "mathematically" right. Although Helmholtz eventually found himself on the right side of this dispute, Grassmann's paper "On the theory of color mixing"[40] became deservedly famous because it gives the formal (even axiomatic) structure of the geometry of colors. Maxwell's experiments were important in formulating these axioms (now fondly known as "Grassmann's Laws").

"Grassmann's laws" (see page 62) axiomatically define color space as a three-dimensional linear vector space, a projection of the infinite-dimensional linear vector space of physical beams of electromagnetic radiation (called "light" when you see it). That is to say, all except Grassmann's fourth law, which is more of a metrical nature. The fourth law defines "brightness" as a linear functional on color space

Hermann Grassmann (1809–77)

Hermann Grassmann was a top-rank mathematician, although he spent most of his life as a high school teacher. His work on linear algebra and geometrical calculus (*Ausdehnungslehre*) set the scene for modern developments and has not even be completely mined today. He started work on color on reading Helmholtz work which clashed with Newton's. In setting out to prove Helmholtz mathematically wrong he laid the foundations for formal colorimetry that remain valid today. Grassmann ended his career with important work on Sanskrit. He always had been remarkably broad, teaching Chinese to missionaries in his spare time.

(and thus on the space of beams). This "law" was thrown in by Grassmann for good measure, for he clearly saw that it stood apart from the other laws and considered it "less certain" and perhaps more of a convenience. This was good foresight on his part.

The topic of "heterochromatic brightness" is a particularly hairy one. The reason is that the notion does not fit into the colorimetric framework because it cannot be defined in terms of judgments of equality. When two beams are "equally bright," they still *look different* because of their different hues (one reddish, the other bluish, say).

The problem has never been solved from a scientific point of view. It was "solved" by committee when (in 1924) Grassmann's fourth law was proclaimed (exactly) true by *definition*. This definition is the basis for the international unit of luminous intensity, the lumen.[41] Needless to

Philip Otto Runge (1777–1810)

Philip Otto Runge was an important Danish painter of German romanticism. Being very religious, most of his works deals with beauty as divine revelation. As was the rule in romanticism, he often used intent vision into nature as a way to show a synthesis of man and world. Runge also understood colors in terms of religious symbolism. For him the three principal colors BLUE, RED, and YELLOW, were the symbol of the Trinity. Black and white are not colors because light is Good whereas black is Evil. Runge designed an order of all possible colors that is essentially modern and indeed similar to Schrödinger's color solid of 1920. It is described in his booklet *Farbenkugel: Construction des Verhältnisses aller Mischungen der Farben zueinander und ihrer vollständigen Affinität* of 1810. Goethe printed a letter by the painter as an appendix to the *Farbenlehre.*

say, there is little science here. In retrospect this forced solution turned out to be an unfortunate one that led to inconsistencies. However, for the industry any solution is better than none.

The seminal work of Newton, Young, Maxwell, Grassmann and Helmholtz led to a fairly complete science of "colorimetry" by the close of the nineteenth century. There existed a parallel wave of "color science" that was (very roughly!) launched by Goethe's *Farbenlehre.* Goethe had quite a different take on the space of colors. He was more interested in the colors of *things* than in the colors of "homogeneous

**Arthur Schopenhauer
(1788–1860)**

Arthur Schopenhauer was a German philosopher, one of the more important of the nineteenth century, perhaps best known for his *Die Welt als Wille und Vorstellung* (The World as Will and Appearance). In the pop-philosophy literature Schopenhauer is known as the "philosopher of pessimism" and misogynist attitude (due to his essay "On women"). Schopenhauer valued aesthetics, sympathy for others and ascetic living. His ideas on perception and art are best understood from his Buddhist's perspective and have close affinity to Goethe's mystical notions. Schopenhauer knew Goethe and may indeed be seen as Goethe's "pupil" on the topic of color theory. Goethe saw Schopenhauer as someone who could "sell" the *Farbenlehre* to the (scientific) world. In 1816, Schopenhauer wrote his book *Über das Sehn und die Farben* (On Sight and Colors) on color. It is of interest because it describes colors as "parts of daylight," a perspective that has been largely lost in modern science, but has its merits.

lights" or in "spectra," and to him the simplest possible color was that of *natural sunlight*. Sunlight was not a "confused mixture of homogeneous lights," as Newton would have it; it was a fundamental entity, complete in itself and not in need of any analysis that would pull it to shreds. Goethe was more of a poet than a scientist, perhaps, but he was well read in the natural sciences. In this parallel stream we also find the Danish painter Philip Otto Runge (more of a mystic than a scientist) who published a booklet, *Farbenkugel* (Color sphere) in which he describes the first essentially modern geometrical structure of the space of object colors (Goethe's book contains a letter by Runge as appendix); the philosopher Arthur Schopenhauer (a pupil of Goethe) who wrote *Über das Sehn und die Farben* (On Vision and Colors, 1815) in which he treats colors as (large) parts of daylight; and the psychologist Ewald Hering, who, like Goethe thought of colors in terms of

Wilhelm Ostwald (1853–1932)

Wilhelm Ostwald was born 1853 in Riga, Latvia, and died near Leipzig, Germany in 1932. Ostwald studied chemistry and was extremely successful. In 1909 he was awarded the Nobel Prize for his work on catalysis (1909), what he regarded as one of his minor topics. Ostwald was very successful both as a scientist and as a teacher: Among his pupils were Arrhenius (Nobel Prize 1920), Van't Hoff (Nobel Prize 1901) and Nernst (Nobel Prize 1920).

Ostwald was very active, both professionally and socially. He was active in philosophy, the history of science, and the popularization of science. He founded societies, journals, and book series. Among much he pursued the acceptance of rational paper formats (the A4 format on which I am writing) and of Esperanto as the standard language for international interactions in science. His busy life was quite interesting, and his autobiography makes very entertaining reading.

continues next page . . .

> *. . . continued from last page*
>
> Ostwald started serious work on color science only after his retirement. Remarkably, he managed to publish dozens of books and hundreds of papers on the subject. Most of these make interesting reading, though obviously there is some redundancy. (Ostwald used to teach his students: "After you have told them, tell them again!" an approach that is indeed very effective if you don't tire of it.) However, the absolute measure of quite novel ideas is highly remarkable for a single person. Ostwald was (as always) very successful with his color science. His books were read from high school to university and companies, his materials (the monumental color atlas for science, business, and industry, boxes of crayons for the kindergarten, and so forth) widely sold. History has not been kind to Ostwald's color science, though. You find his name in very few modern handbooks on color, and usually he gets only a few lines if he gets any. Moreover, when he is quoted, it is typically with blatant misunderstanding. Probably World War II and the fact that his works are written in German have something to do with this, but it is very clear to me that this has been to the detriment of the subject. Ostwald's ideas on various topics were in many respects more advanced than even the current "state of the art."

polarities. This parallel stream never established much contact with colorimetry proper, with the exception of the work by Wilhelm Ostwald in the early twentieth century, and psychophysical work on "opponent color theory" of the mid twentieth century. It is certainly a reasonable summary to say that this never counted for much in main stream colorimetry though. This is much to be regretted, since, as I will argue later, any coherent account of color science necessarily involves recognition of either strand.

Wilhelm Ostwald was particularly interested in the colors of *things*, less so in the colors of *beams*. This is an important difference of emphasis, because colorimetry proper deals exclusively with the colors (in the sense of their discriminability) of beams. For one thing, this forces you to go beyond monochromatic beams *in principle*.

Ostwald was a very effective communicator and I am not certain whether the stories in his many books reflect actual occurrences or are didactical tricks. Anyway, this one conveys one of his messages in

an especially clear way. In his 1936 book *Er und Ich* (He and me[42])
Ostwald asks a physicist:

> Q: *"What is the reflection spectrum of the best yellow
> paint?"*

The physicist (the sucker) answers:

> A: *"Zero throughout the spectrum except for the wave-
> length of 580 nm"* (assuming that a monochromatic beam of
> 580 nm looks "yellow"—it does to most people).

Ostwald then notices that a reflection spectrum that is such a "blip-
function" (everywere zero except at one point) must be the reflection
spectrum of a *black* paint because such a spectrum will yield reflected
beams of vanishing radiant power. Being rightly dissatisfied with the
physicist's answer, Ostwald put the best yellow pigments he could find
in front of a spectroscope and noticed that all these spectra agreed.
The pigments reflected all of the spectrum except for the blue (short
wavelengths) part. Ostwald found empirically that all highly colored
pigments reflect about half of the spectrum; thus they were invariably
very unlike "monochromatic beams." Based on these observations,
Ostwald developed a formal theory of object colors. As a particularly
nice byproduct, he was the first to be able to clear up the relation
between the spectrum and the color circle.

Erwin Schrödinger, a theoretical physicist, devoted much effort to
physiological optics, and especially the theory of colors. His papers
of the 1920s in the *Annalen der Physik* are fundamental to colorimetry.[43] Schrödinger managed to show that the bitter arguments between
Hering's and Helmholtz's followers on the "four color" versus "three
color" color theories were vacuous from the perspective of colorimetry
because they boiled down to a mere choice of basis in color space. He
also made progress on the difficult topic of heterochromatic brightness
(thus extending colorimetry fundamentally) by suggesting the only reasonable generalization of Maxwell's criterion of the judgment of equality known today (chapter 10). Schrödinger explored the novel principle
with very elegant mathematics, applying Riemann's geometry to color
space and going far beyond Helmholtz's initial attempt of the early
1890s. Schrödinger's research in colorimetry, especially its geometrical

structure, is exemplary. Unfortunately, it has hardly been followed up in modern times.

Erwin Schrödinger (1887–1961)

Erwin Rudolf Josef Alexander Schrödinger was born in 1887 in Vienna, Austria, and died 1961 in the same city. Schrödinger led a very lively (some would say chaotic) life. He was a very bright pupil and student and obtained a doctorate in physics in 1910. Although his main inclination was theoretical physics, he had a remarkable affinity for the experimental side, a fact that shows up clearly in his work. Although Schrödinger is perhaps best known for the wave equation of quantum mechanics, an elegant generalization of the Hamilton-Jacobi equation of classical physics, his body of work is quite broad. Of special interest here is the fact that he spent much effort on optics and especially physiological optics including color vision. Judging from the relative volume this part of his endeavor must have been important to him. As I remarked earlier, Schrödinger led an interesting life. When he arrived at Magdalen College, Oxford, he was accompanied by his wife and a female friend (the wife of an assistant) who carried his child. There must have been frowns at Oxford, but the Nobel prize for the work on quantum mechanics no doubt helped to smooth things out. Schrödinger apparently was something of a ladies' man, for in Dublin during the war he had two further daughters by different Irish women.

continues next page . . .

> *... continued from last page*
>
> Of interest for this book are Schrödinger's insights on the nature of life
> ("What is life," 1944), on physiological optics and—of course—color vi-
> sion. Especially his papers in the *Annalen der Physik* of the early 1920s
> are marvels of clarity and insight, and it is much to be regretted that
> they have been less influential than they have been. Schrödinger saw
> clearly that there is no problem whatsoever in the reconciliation of the
> "three-color" (Young-Helmholtz) and "four-color" (Hering) theories. His
> reformulation of Grassmann's work is the definitive paper on "austere col-
> orimetry" and his analysis of the heterochromatic photometry problem is
> exemplary. Pity that nobody listened.

Much later, in the 1970s, Jozef Cohen managed to make new and
important headway in the theory of colorimetry. His work has found
important applications in engineering because it allows one to do many
calculations that used to be very difficult or impossible with surprising
ease. Although this work is usually approached from a purely algebraic
perspective, Cohen developed the geometry of the space of colors in
novel ways, going beyond Grassmann's fundamental work.

Grassmann's laws make color space into a three-dimensional linear vector space. The space of physical beams is also a linear vector
space, though it is infinitely dimensional. The colorimetric colors are
defined as equivalence classes of mutually indiscriminable beams (using Maxwell's criterion of judgment of equality). Thus, from a formal
perspective, color space is a three-dimensional subspace of the space of
physical beams,[44] and Maxwell's gauging of the spectrum defines this
subspace through a projection operator. This allows you to find "the
color" of any beam, but the reverse is not possible: there is simply no
way to assign a unique beam to a given color. Typically, there exist infinitely many different beams that all would yield that same color. The
best you can do is to specify the subspace of beams that are "black,"
that is to say, are in the same equivalence class as the "empty beam"
(absence of a beam). This "black space" is the kernel of the projection
operator. (See chapter 8.)

Jozef Cohen took seriously the notion that any physical beam can
be split into a unique way into two components, one that is responsible

for the color, and the other "invisible" or "black." (This is the so-called "Wyszecki's hypothesis" of the 1950s.) In that case, any color would specify a unique beam that would be its "cause." The (infinitely many) different beams that yield a single color would then differ only in their "black components," which were assumed to be "causally ineffective." From a geometrical point of view, this means that you define color space as an *orthogonal* projection of the space of beams and split the space of beams as the direct sum of a "fundamental visual space" and a

Jozef Cohen (1921–1995)

Jozef Cohen was a psychologist with an interest in visual perception, especially phenomena of color vision. He started from vague ideas (which had been around for some time) that "colors" where caused by the "visual parts" of physical beams, whereas the human observer would be totally blind to the "black" (causally ineffective) parts of physical beams. Although not a mathematician, he managed to put these ideas on a formal basis. His famous "Cohen's matrix R" has become a key tool in colorimetry and is at the heart of many modern algorithms that let you compute things not easily possible before. Unfortunately, Cohen's approach was largely algebraic. He dealt with colorimetric objects in terms of matrixes, which he regarded as arrays of numbers. Thus, the geometrical content of his work remains obscure. For instance, he apparently did not notice that he implicitly introduced a metric that actually requires physical justification.

"metametric black space." In order for this to be possible, you require a scalar (or inner) product defined on the space of beams. Granted the inner product, orthogonality is defined and you can construct the required orthogonal projection. The "inner product" is a bilinear, symmetric functional on the space of beams that assigns a number to any pair of beams irrespective of their order. The inner product evidently has to be introduced on the basis of considerations of *physics*.

Unfortunately, Cohen approached the problem from a purely algebraic perspective, largely ignoring aspects of geometry and physics. He manipulated matrixes, that is, large (theoretically infinite) arrays of numbers. Eventually, he hit upon his "Matrix R." The matrix R, when applied to a spectrum (considered as a vector), yields the "fundamental component" of the spectrum and thus strips the spectrum of its (causally ineffective) "black component." The matrix R is thus a coordinate representation (in the monochromatic basis) of the orthogonal projection operator on the orthogonal complement of the kernel of Maxwell's projection. Cohen failed to see that this only works through the (implicitly assumed) scalar product and thus involves nontrivial physics.

Thus, Cohen's work is interesting because it puts an inner product structure on the space of beams, turning it into a Hilbert space,[45] and defines color space as a three-dimensional subspace through an orthogonal projection operator. This allows you to associate a unique "fundamental spectrum" with any (colorimetric) color. It defines a *metrical* colorimetry (see chapter 8) that extends the basic Grassmann structure and opens up many new possiblities (as well as problems).

The metric remains a major problem. I am at a loss to see how one could defend it on the basis of physics.

1.3   Color Appearances

Colorimetry, as the science originated by Maxwell and perfected by many all the way to Cohen's work, has *nothing whatsoever* to say about "color appearances," or color as "quale" as the philosopher would say. You can establish whether two beams of radiation will be distinguishable or not, but you can say nothing about how it "feels like" to have a

**Michel Eugène Chevreul
(1786–1889)**

Michel Eugène Chevreul was born at Angers in a family of mainly medical people. He grew up at Angers during the Revolution and the Terror. In 1803 he took up the study of chemistry. In 1824 Louis XVIII appointed him director of the dying department at the Manufacture Royale des Gobelins. He was to spend the next sixty-one years of his life at this post. He made important discoveries in both chemistry and the psychophysics of color. It is this latter fact that makes him important for color theory. His *De la loi du contraste simultané des couleurs et de l'assortiment des objets colorés* (published in Paris in 1839) is a monumental work that focusses on the mutual interaction colors in perception. The book was influential in the art world and was instrumental in the rise of "pointillism" and "divisionism" (think of Georges Seurat and Paul Signac). Chevreul was elected a member of the Académie des Sciences in 1826, and president in 1839 and 1867.

certain beam of radiation strike the eye. This is exactly why colorimetry manages to be an objective science and is also precisely why many people deem it insufficient or even useless. What good is a color science if it cannot even predict that a fire truck seen under broad daylight is certain (well, very likely) to look RED? I mean, the color of experience that we call "red"?

Early work on color appearance that might be called "(pseudo)-scientific" is due to the Goethe, Runge, and Schopenhauer branch of color science, and perhaps especially to the work of Chevreul and Hering. Of more modern work, I mention Munsell (because—more or less by way of historical accident—his color atlas became canonized) and (even later still) Edwin Land, who actually innovated visual psychophysics on the topic.

Chevreul was a scientist with a background in chemistry who had occasion to pursue the psychophysics of color perception when he noticed that some complaints of customers were not due to the quality of his dyes but to their visual systems. His book remains a classic and is still very much worth reading.[46] Chevreul managed to investigate much of the phenomenological "laws" of contrast and assimilation in fact, most of what has been done in later eras can be understood as footnotes to Chevreul.

Munsell devised a "color denotation system" that might be of use to painters, designers, and the like, that has become the de facto standard "color atlas" of today. Munsell was a painter and teacher. Conceptually, his color atlas is more primitive than Runge's "color sphere," the value of Munsell's system lies in the extensive psychometrical scaling

**Alfred Munsell
(1858–1918)**

Alfred Munsell was a Bostonian painter and teacher of art. He wrote "*A Color Notation*" in 1905 and published his famous color atlas in 1915. In 1917 he founded the Munsell Color Company. His atlas is still for sale.

(by eye measure) that puts a "well tempered" scale over the gamut of object colors.

Land may be seen as a "modern Chevreul" and—in my opinion—the person that set the first steps firmly outside of the realm of phenomena considered by Chevreul. I vividly remember demonstrations by Land in the 1970s that made a lasting impression on me.

Attempts to put the study of color appearances on a quantitative basis start with the seminal work by Fechner. His Psychophysical Law served as the paradigm for early speculations by Helmholtz and later work by Schrödinger. This type of formal analysis is still alive in current research.

A singular thread was introduced by Ewald Hering with his notion of "opponent colors" (*Gegenfarben*). Although there exist obvious tangencies to ideas of Goethe and Schopenhauer, Hering's ideas are best understood on their own. Hering published his ideas on black and opponency in 1874 in *On the Theory of the Light Sense*. He held

Edwin Land (1909–1991)

Edwin H. Land was a scientist, an inventor, and a (multi-)millionaire. His inventions include the sheet polarizer and the instant photography process. He became interested in color when working on the problem of instant color photography. His work on perceived hue in scenes was groundbreaking and went far beyond what was current in perceptual studies at the time.[47]

Gustaf Theodor Fechner
(1801–1887)

Gustaf Theodor Fechner was a German experimental physicist and psychologist. He spent most of his life at the Leipzig university, being appointed professor of physics in 1834. In 1839 he contracted an eye disease (probably due to his experimenting in vision) and resigned. He spent most of his life studying the relations between body and mind. His interests become clear when you see titles of his works: *Das Büchlein vom Leben nach dem Tode* (1836); *Nanna, oder über das Seelenleben der Pflanzen* (1848); *Zend-Avesta oder über die Dinge des Himmels und des Jenseits* (1851). His is a dual-aspect, monistic, pan-psychical mind/body view. For the present occasion the most important work is the *Elemente der Psychophysik* (Elements of Psychophysics) of 1860. The relevant concept for this book is Fechner's law.

that black does not differ essentially from white, red, or green. He introduced the color-white-black triangle. This is intriguing, since such ideas are by no way generally accepted even today. They were one of the obstacles in the acceptation of Ostwald's work for instance. In 1889 Helmholtz, writing in the second edition of the *Handbuch der physiologischen Optik*, argued, "Herr Hering's theory, . . . is able to explain all the facts of color mixing just as well, but also no better, than Thomas Young's theory," the difference being a mere coordinate transformation. This has been the view of many physics-oriented colorimetrists and has been a reason to generally ignore Hering. Hering was an excellent observer and psychophysicist, and his ideas are very original. They triggered a rather independent line of psychophysical color research that can be traced throughout the twentieth century. It obtained a boost when electrophysiological results apparently put Hering in the right. In the meantime, much more data has become available and the picture has become sufficiently complicated that the initial euphoria had died. However, Hering's notion of opponent colors is definitely here to stay, and I will return to it at a number of places in this book.

Ewald Hering (1834–1918)

Ewald Hering was born in Altgersdorf (Saxony) as the son of a Lutheran pastor. He entered Leipzig University in 1853, and studied medicine. He practiced as a private physician in Leipzig between 1860 and 1865, then became a tutor in 1862. His first scientific work was the *Beiträge zur Physiologie*, on binocular vision and depth perception, written between 1861 and 1864. He was appointed professor of physiology at the Josephinum in Vienna in 1865. In 1870, he succeeded Jan Evangelista Purkinje at the University of Prague. Here he stayed on for twenty-five years and here he wrote his *Theory of the Light Sense*, which introduces the concept of opponent color vision. In 1895, Hering succeeded Karl Ludwig at the University of Leipzig. He retired in 1915 and died in 1918.

continues next page . . .

... continued from last page

Hering was a very outspoken figure, both in social life and in science. He sought confrontation and was a master of sometimes pugnacious polemics. He was very successful as a teacher and had a broad circle of followers during his lifetime. As it happened, he became known as the natural opponent of Hermann von Helmholtz, the polemics between these two figures (or, especially, their camps of followers) being a striking part of the scientific scene in Germany of the end of the nineteenth century.

In color vision, he must be credited for his concept of *black* as just another color (instead of "nothing") and for the concept of opponent colors. Whereas Helmholtz had a typical physicist's view of nature, including the human mind, Hering thought in terms of a mixture of Lamarckian and Darwinian evolutionary theory. He was keenly aware of the fact that the body contains huge amounts of information concerning the world and thought that there is a kind of continuous spectrum between, for instance, the optics of the eye and such things as motor programs or even conscious perception and thought. According to Hering, "Every organized being of our present time is the product of the unconscious memory of organized matter."

One interesting notion of Hering is that the brightness of a pure color cannot even be *defined* apart from the achromatic component that accompanies it. All purely chromatic sensations (or mixtures of them) have no intrinsic brightness. The idea has generally been ignored, though for no obvious reasons.

Two additional approaches are of more modern origin. One is the study of ecological matters, the idea being that color vision evolved as a tool to get on in the world and that as a consequence the visual system is likely to be "tuned" to the structure of the generic human (or primate) biotope. Thus the study of the world doubles as the study of the mind. The other approach is closely connected to this but focuses more directly on various "image processing algorithms." This is essentially an engineering approach. One considers "how the visual system should have been designed" given various boundary conditions that derive from the ecological approach and an understanding of physiology.

Apart from the (semi)scientific approaches mentioned so far, color experiences are the subject of aesthetics and a variety of practices.

Much of interest is available from such sources, though these data are perhaps even less easily integrated in an overall scheme that would still be "scientific." But, of course, not everything that might or should interest you is necessarily science.

Important as the topic of color appearances is, this book is not about them. However, I firmly believe that the study of color appearances cannot commence without firm roots in colorimetry. Thus the book may be considered a prolegomena to the study of color as experience. There exist only few texts with this orientation, some of my favorites[48] being already quite dated (though by no means outdated!).

Exercises

1. Read sources of colorimetry   [As extensive as you want] Find some of the key early sources on colorimetry, such as Newton, Maxwell, Grassmann, and Helmholtz. These are not difficult to read, and you will find them very informative.[49] Which part of the observations and conclusions of these authors still stands? You may want to return to these sources (and probably enjoy them) as you progress through this book.

2. Read some sources on "Color Science"   [As extensive as you want] Interesting works are by Goethe, and Chevreul, and (later) Hering, Ostwald, and Land. Notice how these give a completely different account of colors than found in the literature of colorimetry.

3. Read some sources on color in art and design   [As extensive as you want] Interesting and easy to find are Wassily Kandinsky's *Über das Geistige in der Kunst* (On the spiritual in art) and Josef Albers.[50] Try to relate these accounts to the scientific concepts.

4. Consider Riemann's proposition   [Conceptual] In his famous address, Riemann mentions the space we move in and the space of colors as the only multiply extended continua known to us by experience. Consider differences between these spaces. For instance, you probably know that "colors can be mixed," but on the other hand it is not clear what a "mixture of locations" might mean. Also consider relations between these spaces; for instance, although it is clearly possible for a color to be present at two distinct locations, it makes no sense to consider two distinct colors at a given location (or does it?). Discuss such problems in detail. Are these two spaces indeed on a par?

5. Analyze technical sources   [Difficult] For this exercise you already need to know about colorimetry, so you may want to return to it after having read most of this book. Analyze some of the key technical sources (Newton, Maxwell, Helmholtz, Grassmann, Ostwald, and Schrödinger) in the light of your superior (modern, I mean) knowledge.

Did they commit errors? Serious ones? Were they to blame (consider the data at their disposal at the time)? Do the errors reveal erroneous preconceptions current at the time? Where did these come from? Did they "stand on the shoulders of giants"?

Chapter Notes

1. The unit "nm", that is "nanometer", is 10^{-9} m (meter), that is a really small spatial extent (going by body scale) when you remember that a hairbreadth is about 100μm, or 10^{-4}m.

2. The "cones" are specialized retinal cells containing a photopigment. These specialized neurons convert absorbed radiation into electrochemical (neural) "signals."

3. The "Munsell atlas" is a collection of colored pieces of paper, all neatly organized and labeled. It allows you to specify color by means of the colored paper in the atlas that comes closest.

4. A "quale" is an aspect of your *awareness* (such as "coloredness"), that is, it is not anything *physical*. Sadness (in faces for instance) is a quale.

5. CIE stands for Commission Internationale d'Éclairage.

6. The "CIE-Lab coordinates" are the currently most popular color classification system that also purports to be a subjective metric. That is to say, the more different the CIE-Lab coordinates (a triple of numbers) are, the more different the corresponding colors look. A certain minimum difference guarantees that two colors are "just noticeably different."

   All this is only approximate (and not a little pretentious), but the industry loves it since it obviates the need of having people actually *look* at color. You do a simple measurement instead. This is color without an observer!

7. Artists, of course, do not explain what they mean by this. They just feel the vibes. It is definitely true that certain colored patterns may seem to "vibrate" (as if there were a temporal modulation, which there isn't) though. Crude examples abound in the op art popular in the 1960s and 1970s.

8. This is all very technical. A "struct" is a format (think of a form to fill in) that accommodates a bunch of named values (like "height" and "weight" of persons), a "byte" is an ordered sequence of eight "bits", a "bit" is a slot that can take either one of two values.

9. I will occasionally use SMALL CAPITALS to mark words that refer to "subjective" entities such as feelings, emotion, and impressions. When I use "red" it will (at least in most chapters of this book) refer to a technical, formal concept, whereas when I use "RED" it refers to your experiences or feeling of the "red" quality. For the sake of sanity, I promise not to overdo this. You will need to exercise your own judgment. I count on it.

10. "Facts," of course, are *objective* facts, not simply anything you yourself happen to be convinced of.

11. Photometry was developed almost simultaneously by Lambert and by Bouguer in the eighteenth century.

12. Remember that Eddington defines physics as the prediction of the coincidence of pointers with marks on scales.

13. For example, physics, according to A. S. Eddington, *Nature of the Physical World (Gifford Lectures 1927)*, (Cambridge: Cambridge U.P., 1928).

14. The history is actually somewhat more involved and not at all undisputed. There is an ample literature on the topic.

15. Although this is not generally acknowledged, the same holds for space. A perceived stretch is as much of a quale as a color is. Perceived stretches are ontologically different from physical stretches and the psychophysics of perceived geometrical relations (e.g., stretches) is very complicated and far from being fully understood.

16. Some pretty sophisticated devices have been developed during the course of (visual) photometry, one of the best being the "Lummer-Brodhun cube," which is discussed in many old-fashioned books on technical optics.

17. Notice that the very notion of a "colorless object" is self-contradictory.

18. J. Gage, *Color and Culture: Practice and Meaning from Antiquity to Abstraction*, (Singapore: Thames and Hudson, 1993); —, *Color and Meaning. Art, Science and Symbolism*, (Singapore: Thames and Hudson, 1999).

19. This is the title of a well known book, D. B. Judd, and G. Wyszecki, *Color in Business, Science, and Industry*, (New York: John Wiley, 1975).

20. Its complement, "human black space," is actually more fundamental because it can be defined without taking recourse to a metric.

21. The "fundamental response curves" are indeed less precisely known than "fundamental response curves up to arbitrary linear transformations." The reason is that the former are *physiological* entities, the latter *psychophysical* entities.

22. The lecture has been printed in the thirteenth volume of the *Abhandlungen der Königlichen Gesellschaft der Wissenschaften zu Göttingen*. Several English translations exist, among which one by Clifford. A good translation with explanatory notes is contained in Spivak's book (M. Spivak, *A comprehensive Introduction to Differential Geometry (5 vols)*, (Berkeley, CA: Publish or Perish, 1970)).

23. Translation by William Kingdom Clifford: B. Riemann, "On the Hypotheses which lie at the Bases of Geometry," *Nature* 8, pp. 14–17, 36, 37 (1873).

24. More extensive accounts can be found in K. T. A. Halbertsma, *A History of the Theory of Color*, (Amsterdam: Swets and Zetlinger, 1949); W. S. Stiles, "Colour vision: A retrospect," *Endeavour* XI, pp. 33–40 (1952); J. D. Mollon, "The origins of modern color science," in

The Science of Color, edited by S. K. Shevell (Washington, DC: Opt.Soc.Am., 2003), Chap. 1.

25. Helmholtz actually preceded Riemann in the development of geometry of general "multiply extended continua."

26. The easiest source is Newton's *Opticks* which has been reprinted numerous times. I. Newton, *Opticks: or, a Treatise of the Reflections, Refractions, Inflections and Colours of Light*, (London: W. and J. Innys, 1718).

27. Notice that I use small capitals to indicate *subjective* terms such as HUE. These are called "subjective" because—in principle—not amenable to objective verification. People need not agree on HUE; philosophers would say that HUE is a *quale*. More on that later.

28. I use modern language here for clarity. Newton would speak of "refrangibility," that is, how much the rays were deviated by his prism.

29. "Chromaticity diagram" is a modern concept with a formal, technical meaning. Of course, Newton's meaning was quite different. It is hard, today, to guess what Newton exactly meant. Most probably he didn't know himself. I would say that Newton (at best) entertained some form of "phenomenological model," and that he would not have defended it as "mathematically certain."

30. Newton had no empirical evidence whatsoever as to the linearity (in color space) of the additive mixture of beams. These were mere guesses on Newton's part.

31. Newton, *Opticks* (ibid). Newton's first publication on color is from 1672.

32. Goethe's *Farbenlehre* was published between 1808 and 1810 in three volumes. Many editions and translations are readily available.

33. Most books on optics will have some remarks on Babinet's principle which is often considered trivial. See, for instance, M. Born, and E. Wolf, *Principles of Optics: Electromagnetic Theory of Propagation, Interference and Diffraction of Light*, (New York: Cambridge U.P., 1959).

34. The concept of "colorimetric equation" is the crux of colorimetry. I will explain it in detail later.

35. J. C. Maxwell, "Experiments on colour, as perceived by the eye, with remarks on colour-blindness," *Trans.Roy.Soc.Edinburgh* 21, pp. 275–298 (1855); —, "On the theory of compound colours, and the relations of the colours of the spectrum," *Phil.Trans.* 150, pp. 57–84 (1860).

36. A. König, and C. Dieterici, "Die Grundempfindungen und ihre Intensitäts-Vertheilung im Spectrum," *Sitz.ber.Akad.d.Wiss.*, Berlin 2, pp. 805–829 (1886).

37. Maxwell, Ibid.

38. The CIE (*Commission Internationale d'Éclairage*) is the GUARDIAN of this information.

39. Helmholtz's metric was based on Fechner's Psychophysical law."

40. H. Grassmann, "Zur Theorie der Farbenmischung," *Ann.Phys.Chem.* 89, pp. 69–84 (1853). It is reprinted in *Gesammelte Werke*, Vol. 11, 2. pp. 161–173.

41. According to the International System of Units, *One lumen is the luminous flux emitted within a solid angle of 1 steradian by a point source with an intensity of 1 candela*, whereas the candela is defined as *The luminous intensity, in a given direction, of a source that emits monochromatic radiation of frequency 540 terahertz and that has a radiant intensity in that direction of 1/683watt per steradian*". Notice the "magical numbers" that render this definition perhaps less than elegant. A typical use for the lumen is to describe how much light ordinary household light bulbs give off. The packaging usually gives the bulb's light output in lumens; a new 75W incandescent bulb puts out about a thousand lumens.

42. W. Ostwald, *Er und Ich*, (Leipzig: Theodor Martins Textilverlag, 1936).

43. In a first paper Schrödinger develops the theory of object color: E. Schrödinger, "Theorie der Pigmente von größter Leuchtkraft," *Ann.Phys.* 62, pp. 603–622 (1920); in a second he extends the colorimetric basis (equality of beams) and ventures a theory on maximum similarity: —, "Grundlinien einer Theorie der Farbenmetrik im Tagessehen," *Ann.Phys.* 63, pp. 397–426, 427–456, 481–520 (1920); —, "Farbenmetrik," *Z.f.Physik* 1, pp. 459–466 (1920).

44. That color space is a three-dimensional subspace of the space of physical beams is indeed formally true, though many treatments of colorimetry complicate matters by introducing an—unnecessary—level of indirection. I will explain details later.

45. "Hilbert space" is a technical notion from mathematics. A Hilbert space is a separable space (given any two distinct points, there two exist disjunct open sets such that each set contains one of the points) that is complete (every Cauchy sequence has a limit) and has an inner product. Hilbert spaces are *nice*; everyone likes them. They are the arena for much of mathematical physics.

46. Of course, the book is not written in the modern terse style scientists have come to prefer; thus, Chevreul spends many pages giving advice to women of color choices in dress based upon their complexion and hair color and so forth. I enjoy reading such things, but many of my friends in physics would be irritated to the extreme.

47. Important papers on color experiences by Land are E. H. Land, "Color vision and the natural image. Part I," *Proc.Natl.Acad.Sci. U.S.A.* 45, pp. 115–129 (1959); —, "Experiments in color vision," *Sci.Am.* 200, pp. 84–94, 96–99 (1959); a theoretical account by

Land is found in —, "The retinex," *Am.Sci.* 52, pp. 247–264 (1964); —, and J. J. McCann, "Lightness and retinex theory," *J.Opt.Soc.Am.* 61, pp. 1–11 (1971); —, "The retinex theory of colour vision," *Proc.R.Inst.Gt.Brit.* 47, pp. 23–58 (1974); —, "The retinex theory of color vision," *Sci.Am.* 237, pp. 108–128 (1977).

48. Some of my favorite texts are O. M. Rood, *Modern Chromatics*, (New York: van Rostrand Reinhold, 1973 (orig. 1879)); R. M. Evans, *An Introduction to Color*, (New York: Wiley, 1948); P. J. Bouma, *Physical Aspects of Colour*, (Eindhoven: N.V. Philips Gloeilampenfabrieken, 1947); R. S. Berns, *Billmeyer and Saltzman's Principles of Color Technology*, (New York: Wiley, 2000).

49. Unfortunately, many of the sources are not in English. You may find it useful to find D. L. MacAdam, *Sources of Color Science*, (Boston, MA: MIT Press, 1970); L. M. Hurvich, and D. Jameson, *E. Hering: Outlines of a Theory of the Light Sense*, (Cambridge, MA: Harvard U.P., 1964).

50. J. Albers, *Interaction of Color*, (New Haven, CT: Yale U.P., 1975); W. Kandinsky, *Über das Geistige in der Kunst*, (Bern-Bümplitz: Benteli-Verlag, 1959); S. Quiller, *Color choices*, (New York: Watson-Guptill, 1989).

Chapter 2

Colorimetry for Dummies

In this chapter I describe the colorimetric procedures that are conventionally used to arrive at *standard results*. I will not assume any prior exposure to the topic, hence the reference to "dummies" or "complete idiots." No disrespect intended—really! The remainder of this book will serve to *demystify colorimetry* for you. There are many books on the topic, rather repetitious of course, and almost any of these will do if you want a somewhat less terse exposition.[1] Don't expect any scientific or conceptual insights from this (fortunately very short) chapter. You may find it useful,[2] though, and it will provide a good basis from which to appreciate my exposition of colorimetry in the other chapters. If you have ever done standard colorimetric calculations before, you may safely skip this chapter.

2.1   Colorimetric Problems

It is not hard to think of many generic problems that you might expect to be able solve given some basic expertise in colorimetry:

1. Given the spectrum of a light beam, find its color,

2. Given a color, find a light beam that will evoke it,

3. Given an object and an illuminant, find the color of the object,

4. Given two illuminants and a color, find a paint that evokes that color under either illuminant,

5. Given an object, find the color shift for a certain change of illuminants,

6. Given two objects, decide whether they can be distinguished (by color) under some illuminant,

7. Given two beams, decide whether they will look different.

It is to be expected that various types of problems might be distinguished and that some problems may be more complicated than others. Such is indeed the case. The *first* problem is the simplest, generic colorimetric problem. The *second* problem is of a different nature. It has infinitely many solutions, even though the beams that solve the problem are rare (a beam picked at random has probability zero to be a solution). The *third* problem is just as simple as the first one, for by calculating the beam scattered to the eye by the object—involving mere physics—you have converted it into an instance of the first problem. The *fourth* problem is not simple, even though there exist infinitely many solutions. This is the type of problem that the majority of users of colorimetry will find it next to impossible to solve. The *fifth* problem is again a simple and common one. Using the method of solution for the third problem, you find two colors. The *sixth* problem is different. Using the method of solution for the third problem, you easily find the colors of the objects under the various illuminants. The problem thus boils down to the decision of whether such a pair of colors is discriminable. You need some metric here. It is readily available though, thus rendering the problem a simple one. The *seventh* problem is essentially equivalent to the sixth.

2.2   Solving a Colorimetric Problem

I pick a generic problem in order to illustrate the use of various generic colorimetric calculations.

Suppose you are in the cosmetics business and propose to sell a package containing a lipstick and nail polish of matching colors. The lipstick will be a grease product with some organic dye. It is important that it not irritate the skin and have good hiding power. The nail polish is some lacquer that contains some pigment (the organic dye used in the lipstick being less suitable). It is important that it have good mechanical properties, and its hiding power should be low, yielding a pretty translucent effect. As a consequence, the reflectance spectra of these products will almost certainly be different from each other. Yet they should evoke matched colors, both in daylight and under artificial illumination (say incandescent light, though fluorescent tubes may be important for use in office environments). These illuminants have markedly different spectra, thus creating a potential problem. It is likely that the lips will fail to match the nails under exotic bar illuminations, and it will be largely impossible to prevent such unfortunate events completely.

2.2.1 Finding and preparing the tools

In order to solve colorimetric problems, you will need a number of tools. The toolbox is small enough, you can obtain all you need from the Internet. Type "color vision database" in Google. You will find many search results, many of them useful. Pick "CVRL Color & Vision database" (`http://cvision.ucsd.edu`) maintained by the Color Vision Research Laboratories of the University of California at San Diego. Look for "CMF's" ("Color Matching Functions") and load the "CIE 1964 10-deg, XYZ Color Matching Functions." These are tables at 1nm or 5nm intervals (your choice) from 360 to 830nm. At each wavelength you get a triple of real numbers at (low) machine precision. You will not need more than this.

This is what the table of color matching functions looks like:

...	...	...	...
560	$1.5574E + 00$	$9.1620E - 01$	$-9.3000E - 03$
565	$1.8465E + 00$	$8.5710E - 01$	$-8.7000E - 03$
570	$2.1511E + 00$	$7.8230E - 01$	$-8.0000E - 03$
575	$2.4250E + 00$	$6.9530E - 01$	$-7.3000E - 03$
580	$2.6574E + 00$	$5.9660E - 01$	$-6.3000E - 03$
...	...	...	...

The first column contains the wavelength in nanometers. This column is largely superfluous, since it is the same in all tables you are going to use. You will typically ignore this column. The other columns specify a triple of real numbers, one triple for each wavelength.

Obtaining the data

You need spectral descriptions of your products. This involves physical measurements of radiant power spectra and spectral reflectance factors. Thus you should either buy or borrow some standard spectrometer[3] and apply it to your samples. You will obtain a table of reflectance as a function of wavelength.

Here is a typical sample:

```
     ...
    0.1307
    0.1285
    0.1289
    0.1274
    0.1260
    0.1247
     ...
```

The values are for certain wavelengths; the instrument does not add a "first column" of wavelengths, because it is the same in every instance. You should refer to the manual.

You can of course measure daylight and incandescent light spectra (and for exotic bar illuminants there may be no other option), but it is far more convenient to load tables of standard spectra from the internet. You will need "CIE illuminant D65" (average daylight of color temperature 6000°K) and "CIE illuminant A" (incandescent light). These come tabulated at 1nm or 5nm intervals from 300 to 830nm. Here is a piece of the CIE illuminant A table:

...	...
560	100.000000
561	100.715000
562	101.430000
563	102.146000
564	102.864000
565	103.582000
...	...

The average daylight table has a similar structure. The first column is again the wavelength in nanometers. You will generally ignore it.

The final preparation is to make sure that all your tables are in sync, that is to say, sample the same region of the spectrum at the same (constant) intervals. A range of 360 to 750nm at 10nm intervals is ample; anything more elaborate is fine (but useless), and you could easily make do with less. This may well turn out to be an annoying chore of interpolation. Fortunately, modern packages like Mathematica make this a relatively painless operation.[4]

2.3   Calculation of Color Coordinates

Your first colorimetric calculation is that of the colors of the lipstick and the nail polish under the two illuminants. You start by finding the spectra of the beams scattered to the eye by multiplying the spectral reflectance tables with the illuminant tables,[5] entry by entry. Next you multiply these spectra with the "Color Matching Functions" table. Each entry of the former is a single number, each entry of the latter an ordered triple of numbers. You multiply each of the three numbers of the triple with the corresponding entry of the scattered beam. Finally, you compute the sum of all triples in the resulting table, column by column (each row being a triple) and multiply these with the wavelength interval of your tables.[6] You end up with a single ordered triple of numbers. This is called the "color". The numbers are the "color coordinates" or, more specifically, the

"CIE XYZ color coordinates."[7] You end up with four such "colors," namely both the lipstick and the nail polish under both average daylight (CIE illuminant D65) and incandescent light (CIE illuminant A). The meaning of this is that two beams are not distinguishable by the generic human observer if the colors of the beam are identical. It is very unlikely that you will ever encounter such (exact) identities, though.

2.3.1 Calculation of chromaticity coordinates

What if you illuminate with two light bulbs instead of a single one? Then all three color coordinates become twice as large. This suggests that the absolute values of the coordinates are not particularly useful. Perhaps you should somehow "normalize" the colors. The standard way to do this is to divide each color by the sum of its coordinates. Then the outcome is independent of the number of light bulbs used to illuminate the samples. The resulting triple of numbers is called the chromaticity and the individual numbers chromaticity coordinates, or, more specifically, CIE xyz chromaticity coordinates.

Since the sum of the three chromaticity coordinates will always equal one, you may as well drop one of the chromaticity coordinates. The lacking one is computed easily enough if you ever want it. Conventionally, you drop the last coordinate, ending up with a pair of numbers. This pair of numbers is typically called "the" chromaticity coordinates, or, more specifically, the CIE xy chromaticity coordinates. You can use these to plot the chromaticity as a point in the CIE xy-chromaticity diagram, more usually just called the chromaticity diagram (figure 2.1). When two colors end up at the same point in the chromaticity diagram, they are the "same color," differing only in "intensity," or, technically more correct, "luminance". The luminance of a beam is a familiar enough photometric quantity. It is suggested that if two colors end up at nearby points in the chromaticity diagram they will look similar. It is even often suggested that the location in the chromaticity diagram indicates the "color" (as an experienced entity) of a beam.[8] Both suggestions are misleading. When two colors have the same chromaticity they may look entirely different. Even if the nice color plate of your textbook suggests that the color should look orange, it may actually look brown or even black.

The colors in the chromaticity diagram fill only the interior of a convex figure, bounded by a certain convex arc and a straight segment. The points on the convex arc are labeled by wavelength. They represent the chromaticities of the "spectral colors" or "monochromatic colors." The points on the straight segment are various hues of purple, running between red and blue. The hues associated with the points on the boundary of the convex region change continuously and do not repeat. They form a periodic sequence suggestive of the "color circle" as used by artists. It is even often suggested that the boundary of the convex region is the scientifically correct correlate of the color circle, the latter being a mere

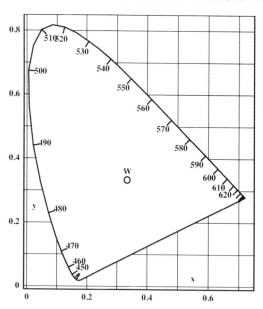

Figure 2.1 The CIE xy chromaticity diagram with "white point" **w** and the "spectrum locus" labeled with wavelengths in nanometers. Coordinate x on the abscissa, and y on the ordinate.

subjective construct (artists not being scientists, this is understandable). In many books the chromaticity diagram is printed in vivid colors. Since the hue changes continuously along the boundary, there has to be a point inside the region that is hueless. This point is called "achromatic," or, more often, "white." The CIE has conveniently placed the white point (for a certain standard illuminant of course) at $x = 0.333\ldots$, $y = 0.333\ldots$ (thus chromaticity $\{\frac{1}{3}, \frac{1}{3}, \frac{1}{3}\}$, the coordinates summing to 1).

Points other than the white point lie on a radius from white to some point on the boundary. If the latter point lies on the curved part of the boundary, you find a corresponding "spectral color"; thus the point can be labeled with a "dominant wavelength." The distance to the white point evidently has to do with the vividness of the color. One defines a suitable "saturation" to quantify this.

For the majority of users the chromaticity diagram is the representation of colors. All reasoning takes place with respect to the chromaticity diagram. Sophisticated users use it for various simple computations or geometrical constructions. Few general users understand the meaning of chromaticity on a deeper level. It is often suggested that the chromaticity diagram is essentially due to Newton, who—as everyone knows—discovered the spectrum and the relation between color and wavelength.

2.4 Conversion to the CIE-Lab Space

By now you have the lipstick and nail polish colors as seen under daylight and incandescent light plotted in the chromaticity diagram. So far, so good. What next?

In order to judge color differences you need some metric. The distances in the chromaticity diagram cannot serve as a metric. Fortunately, the indefatigable CIE again provides exactly what is needed here.

There exists a set of formulas that you can use to convert the XYZ-color coordinates into the so-called CIE-Lab coordinates. These formulas are complicated and look like a formidable theory might be hiding behind them.[9] The formulas contain many mysterious numbers and nonlinear transformations (cube roots and the like). Fortunately, you don't have to understand all this; you simply plug the XYZ-coordinates in the formulas and you get the Lab-coordinates out. Many computer programs (such as Adobe Photoshop) will do this for you, so you needn't even bother with the formulas if you don't feel like it.

The CIE-Lab coordinates of a color are a set of three numbers. The first number (the "L") is the "intensity," whereas the others (the "ab") specify the chromaticity. The intensity is actually a *relative* measure, thus you need to specify a "reference" if you do the XYZ to Lab transformation. The reference is the "white" color, for which you can substitute the color of the illuminant.

The ab-coordinates represent the color through its "red-greenness" (the "a") and its "yellow-blueness" (the "b"), the so called (Hering) opponent signals.

2.4.1 The CIE-Lab metric

It is suggested that the Lab-coordinates are mutually independent and that equal steps in these coordinates look equally spaced. The Euclidian distance in Lab space is taken as a measure of how different two colors look. A distance of one Lab-unit is supposed to be a just noticeable step or JND.[10]

This puts you in the position to put a precise measure on the difference between the colors of the lipstick and the nail polish. It solves your problem, at least the colorimetric aspect of it. Having come this far, you may call yourself a power-user of the colorimetric routine. Few professionals know more about it, or need to know more about it.

2.5 Colors of Colored Things

Notice that you can simply *calculate* whether the colors of the lipstick and the nail polish are distinguishable, there is no need to *look*. You might have thought that the problem would call for some psychophysics, to have, for instance, a hundred persons look and venture an opinion on it and perform some statistical trick on the answers. Of course, you would probably have to select a suitable group; at least the observers should have no medical record of colorblindness or anomaly, and preferably they should be people who would notice a mismatch between lips and nails in daily life (thus probably a strong bias on the female gender). In order to keep the procedure objective you could provide a questionnaire and reject people who failed a minimum predefined profile. Such a test would be a major undertaking, requiring professional help, for even trained scientists are certain to commit numerous beginners errors when trying their hand on psychophysics. Your research would be subject to all kinds of doubt. How much more convenient to simply do the calculation! Now you cannot go wrong, for you can hide behind the CIE and come up with a report that is certified to be scientifically correct. Moreover, you save the hassle and expense with the observers. Small wonder that this is the industry standard in color science.

Few people ever wonder how much of the calculation is indeed science, how much "guestimation" or even "make-believe" (more academically correct terms would be "accepted convention" or "dogma"). And exactly what replaces the observer(s)? If you indeed care about understanding such differences, this book is for you.

Exercises

1. Set up shop [A REQUIRED CHORE] In order to do the exercises, you need to set up shop, for many exercises will be numerical and you need to be ready for that. Load the color matching functions, various illuminant spectra, and perhaps a database of object reflectance spectra from the Internet. Bring the various tables into the same format. For this you may have to use interpolation, to discard parts of tables from either end, or to use padding (e.g., with zeroes) in order to extend tables.[11] This will take effort, but it is worth it.

2. Find illuminant colors [EASY] Find the colors of some illuminants, daylight, incandescent light, and so forth. Plot the results in color space. Find the chromaticities and plot the results in the chromaticity diagram.

3. Draw the chomaticity diagram [A REQUIRED EXERCISE] Plot a somewhat more fancy chromaticity diagram by including the curve of chromaticities of monochromatic illuminants. Since you will probably use this type of plot often spend some effort in adding a wavelength scale, etc. It is useful to prepare a skeleton chromaticity diagram into which you can inject various chromaticity data.

4. Find object colors [SIMPLE TO ELABORATE] Compute the colors of some objects (you need spectral reflectance data, and there is plenty to find on the Internet) and plot them in both color space and the chromaticity diagram. Study the color changes as you change the illuminant. Do the same for some artificial reflectance spectra, uniform, linear slope, partwise constant, sinusoidal, etc. If you are the adventurous type, you may enjoy finding the colors of the rainbow at noon or at sunset.

5. Color temperature [CONCEPTUAL] Compute the chromaticity of a Planckean radiator. Plot the "black body radiation" beam in the chromaticity diagram. Find the limit points ("reddish glow" and "white hot" source). How would you define color temperature for arbitrary beams? How does the CIE handle this?

6. CIE-Lab space [SIMPLE] Map the object colors you have found to Lab space. Compute distances and compare with the configuration in the chromaticity diagram. How do the distances change as you change the illuminant?

Chapter Notes

1. Simple and short: D. Malacara, *Color Vision and Colorimetry. Theory and Applications*, (Bellingham, WA: SPIE Press, 2001); more extensive and technical: H. R. Kang, *Computational Color Technology*, (Bellingham, WA: SPIE Press, 2006); particularly nice: R. S. Berns, *Billmeyer and Saltzman's Principles of Color Technology*, (New York: John Wiley, 2000); much more comprehensive: M. Richter, *Grundriss der Farbenlehre der Gegenwart*, (Dresden: Theodor Steinkopf, 1940); and M. Richter, *Einführung in die Farbmetrik*, (Berlin: Walter de Gruyter, 1976).

2. If you are going to do serious colorimetric work you *need* to consult a book on praxis. This book is *not* on praxis, but on *concept*. You need both, except when you are already adept in the conventional business of colorimetry. If not, then the book by Berns (ibid) is my primary recommendation. Be sure to read it. You will pick up many things that you will not get from my book and that will prove to be very useful to you. Don't reinvent the wheel yourself (at least when you have practical needs, I'm not talking concepts here) and make sure you are aware of the "state of the art."

3. Either you know what "spectrometers" are (thus you need no further advice) or you do not. In the latter case it is sufficient to know that there exist instruments of that name designed to be used by people who hardly know what "spectroscopy" is. Simply follow the manual. There are a great many errors you can make if you do not know what you are doing, but few users care as long as they get numbers out of the machine. For the moment, that is the right attitude. I will not explain to you how to use such instruments intelligently. You will have to ask an (old style) physicist or an optics engineer.

4. There exist dedicated packages for doing colorimetry. Since colorimetry is so simple, I think they are hardly worth your money. Using a generic math package, you can do whatever you want, and you can check whether things are actually correct (I hate to hand over responsibility to some package the innards of which I do not know). What you will use most is numerical linear algebra of the simplest kind.

5. This simplifies the actual physics to the point of being a caricature, but it is the generally accepted practice.

6. Thus the sampling frequency will have no influence on the result. You are actually approximating an integral over the wavelength domain by a Riemann sum here, but there is no need to know this.

7. You may well ask: "coordinates of what"? Well, of colors, of course. But then, here the colors are defined via the coordinates. I agree that this is perhaps less than satisfactory. People of a geometrical inclination think of these "colorimetric colors" as "points" in some "color space." This color space is apparently three-dimensional (since you have three "coordinates"). For virtually all users, "color space" remains a mysterious entity. It helps not to think about it too much.

8. Since the intensity has been disregarded, don't be surprised if a color that "is supposed to look RED" will actually look BROWN or BLACK, though.

9. What is actually behind these formulas are endless committee meetings. There is no such a thing as the basic theory of Lab-space. See K. McLaren, "The Development of the CIE 1976 ($L^\star a^\star b^\star$) Uniform Colour Space and Colour-difference Formula," *J.Soc.of Dyers and Colourists* 92, pp. 338–341 (1976).

10. JND stands for "Just Noticeable Difference."

11. No need to go out of your way in order to extent the wavelength scale far into the ultraviolet or infrared. These regions are invisible anyway, though people sometimes fight over the exact boundaries. This silly fight will never stop because there are no such fixed boundaries. Depending upon the data you have available, it will be fine if you standardize on a wavelength range from anything between 350-400 nm and anything between 700-750 nm. No doubt many will protest against your choice. If you need to regain your sanity you should for once go out of your way to extent the range (zero to infinite nanometers is the goal in this crazy enterprise) and notice the (tiny!) difference it makes.

Chapter 3

The Space of Beams

3.1   Patches of Light and Beams of Radiation

In the simplest possible setting, colorimetry deals with beams. A beam is short for an "incoherent beam of electromagnetic radiation in the wavelength band of *ca.* 350–800nm." Since I assume that you had a first-year course in physics, the explanation can be short:

Electromagnetic radiation is described by Maxwell's equations as a propagating wave of electric and magnetic disturbances. It propagates in vacuum with the "speed of light c."[1] The radiation transports radiant energy and may interact in various ways with (mainly the electronic structure of) matter.

Incoherent radiation is radiation that is essentially chaotic, because generated by random causes. The key example is thermal radiation, or "black body radiation." The typical beams you deal with in daily life (sunlight, incandescent light, fluorescent tubes, candles, and so forth) are thoroughly chaotic (or incoherent), so you rarely encounter the type of interference phenomena (e.g., "speckle") that you see in surfaces illuminated with (coherent) laser beams.

The wavelength of radiation is an entity that applies to periodic (coherent) disturbances. Such waves are characterized by a characteristic frequency ν and wavelength λ, related through $\lambda\nu = c$. Physicists often use the angular frequency $\omega = 2\pi\nu$ and wavenumber $k = 2\pi/\lambda$. Then obviously $\omega = ck$.

The wavelength band tells you the *type* of electromagnetic radiation. Electromagnetic radiation appears to you in many different guises. Radio waves and X-rays simply pass right through you, the former being harmless, the latter bad for you.[2] Only wavelengths in the visual region, *ca.* 350–800 nm, give rise to *luminous perceptions*[3] (figure 3.1). Such rays used to be denoted "light" in the nineteenth century physics texts.

The eye is sensitive only to a limited spectral region. It even discards much of the available daylight[4] (see figure 3.1). I will not pursue the reasons for this, interesting as they might be in their own right. I will simply take the fact for granted.

3.1.1   Patches of light

When a beam of radiation enters your eye, you typically become aware of a "patch of light." It is not that you may not have luminous perceptions without such a beams, for you have them in your dreams or when you are hit on the eye, and so forth. Neither is it the case that luminous perceptions will *necessarily* arise when a beam hits your eye. You may be temporarily blind or inattentive for instance. Nevertheless, "when the circumstances are appropriate" nobody will be surprised when

> *a beam of radiation of a suitable variety is likely to evoke the experience of a patch of light in typical human observers* (figure 3.2).

The shape and temporal behavior of the patch are correlated with the spatiotemporal structure of the beam. For instance, it is easy to set up things such that the beam will evoke the experience of a steady, uniform circular disk of a certain extent. I will completely abstract from the spatiotemporal properties here and assume that you will always see steady, uniform, circular patches "of light" in an otherwise dark surround. This typically happens

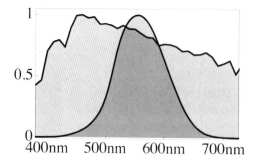

Figure 3.1 The sensitivity of the eye (smooth bell-shaped curve) compared with the spectrum of daylight (wriggly curve). The curves have been arbitrarily scaled in order to enable easy comparison.

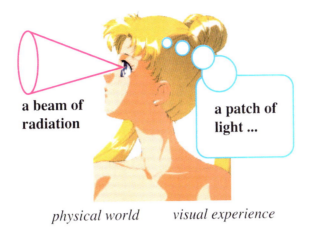

Figure 3.2 Beams of radiation and patches of light.

when you look into the ocular of some optical instrument. In this book I will assume that you have "beams" of the appropriate kind, which invoke "patches of light" in human observers of the appropriate kind (not blind, not asleep, attending to visual experiences, and so on). I assume that the patches are somehow "caused" by the radiation. The "causation" involved here is of a special kind, for though much of the process is just physics and/or physiology, the "patch of light" is a *mental* entity. "Light" and "radiation" have quite distinct ontologies; thus, any "causal connection" between them is highly problematic. It relates to the philosophical problem of the relation between the physical and the psychical.

Phenomenologically, the mental is evidently ontologically prior to the physical, although most scientists prefer to stick to the (irrational but generally respected) opposite view.

3.2 The Photometric Paradigm

One way to avoid psychology—for the moment, that is, for eventually you *have* to consider psychology when you discuss *mental* entities such as "patches of light"—is to use the human observer as a "null instrument." Here is the rationale. Suppose that an observer is unable to distinguish between two patches of light. This may happen (but this is a trivial example) when the patches are physically identical (that is, prepared in the same way). This is a fact that can be established objectively, just as objectively as establishing that a certain person is blind. In fact, inability to distinguish between two entities may be considered a kind of *selective blindness*, especially if the entities are *not* the same in a physical sense. Then you have established a nontrivial fact (e.g., two distinct physical beams cannot be distinguished by the observer) about vision that has nothing to do with psychology or the fact that patches of light are mental things. After all you don't ask for a report on what the patch looks like or what it feels like to experience the patch, and so forth. Specifically, you don't ask the observer to venture an opinion as to the "color" of the patch.[5] You establish mere indistinguishability of patches, not whether they look red or blue.[6] This general rationale is behind the discipline of "photometry," and because of the extreme "response reduction" ("equal" or "not equal"), photometry is an exact science and not psychology.[7]

3.2.1 Response reduction

Sound photometric practice insists on extreme *stimulus reduction* as well as extreme *response reduction*. Of these, the latter is *essential*, the former a mere convenience. The stimulus reduction is indeed convenient because it renders the observer's task unambiguous. Instead of telling the observer to attend to *this* but disregard *that*, you try to make sure that there is only *one* feature to be attended to and nothing else in sight. A generic implementation is the "split-field" or "half-fields" configuration. When the observer looks into the photometer a circular disk, possibly split into two hemidisks by a vertical boundary, is the only thing visible. The task of the observer is to decide whether a single, undivided disk is visible, or whether the vertical separation can be detected. I will assume this configuration and speak of the "left" and "right" half-field as a matter of course. The two half-fields appear as steady, uniform areas of "light." They are evoked by (physical) beams **a** and **b** (say), and I will use the notation

$$[\mathbf{a} \mid \mathbf{b}],$$

for this configuration. If the observer cannot see the division, a "*photometric equality*" is registered as

$$\mathbf{a} \,\square\, \mathbf{b}.$$

More usually, the division will be apparent, and you have a *photometric inequality* registered as

$$\mathbf{a} \,\boxtimes\, \mathbf{b}.$$

Of course *inequalities are the rule*, for it is very special for two randomly chosen beams to be indistinguishable.

In photometry proper, people deal only with the "intensity of light," by which they really mean the "luminance of beams."[8] You use the fact that beams can be attenuated in various ways. Suppose I have the configuration $[\mathbf{a} \mid \mathbf{b}]$ and **b** "looks brighter" than **a**.

(Here I assume that the patches are not differently "colored," which is very special. You may think of **a** as an attenuated copy of **b**. I'll consider the general case later.) Obviously, "looking brighter" is a psychological fact that doesn't make sense in photometry proper. However, you may take it as a hint that a suitable attenuation factor μ (say) might render **b** "equally bright" as **a**, and here you really mean "photometrically equal." Thus you look for the condition

$$\mathbf{a} \,\square\, \mu \mathbf{b}.$$

A convenient way to do this is to give the observer control over the value of the attenuation μ (e.g., using a knob) and instruct the observer to bring about the equality $\mathbf{a} \,\square\, \mu \mathbf{b}$ by turning the control. The attenuation factor and the beams are physical entities; thus, $\mathbf{a} \,\square\, \mu \mathbf{b}$ is a relation between physical entities. The observer appears only *implicitly* as instrumental in bringing about the "$\bullet \,\square\, \bullet$" (two open slots) relation. It is in principle not different from the judgment of equilibrium of scales in comparing weights.

3.2.2   Photometry and radiometry

This simple type of photometry is very much a nineteenth century thing because (in physics) it has been replaced with radiometry. I will not discuss the technical side of radiometry here. Conceptually, radiometry is the comparison of a beam with a suitably attenuated "standard beam" (in practice, the "standard" is probably implicit in the calibration of the instrument). Then the only difference between photometry and radiometry is the spectral sensitivity of the human observer, which can be mimicked by a selective filter in front of the radiometer. In cases where the human selectivity is implied (as in much of photography) such filters are standard (any photographic exposure meter offers an—approximate—example). Thus the practice of photometry involving a human observer has become a thing of the past.

3.2.3   Colorimetry

Although "photometry" is indeed a thing of the past, "colorimetry" is not (well ..., not quite yet[9]). As the discipline of photometry developed in the eighteenth century, people noticed that a "single knob" often did not suffice to bring about photometric equivalence. To put it in psychological terms, two patches of light may look "equally bright," yet look *different*. For instance, one might look "greenish" whereas the other might look "reddish." In such cases photometric equivalence cannot be obtained (figure 3.3).

Clearly you need additional knobs. The first one to suggest that three knobs might do *in all cases* was Thomas Young, at the beginning of the nineteenth century. Around the mid nineteenth century James Clerk Maxwell pioneered colorimetry, essentially a variety of

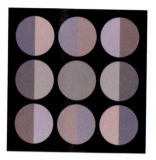

Figure 3.3 Sometimes photometry becomes very hard. This happens when you cannot obtain true equality by turning the (single) knob. In such cases you need additional degrees of freedom (more knobs). Then photometry turns into colorimetry. Here you see the (in this case two-dimensional) "environment" of an equality setting.

photometry, but with *three knobs* (figure 3.4). Maxwell designed the necessary apparatus, performed the necessary experiments, and worked out much of the necessary theory, all by himself. If you are interested in colorimetry he should be among your heroes. Almost exactly during the mid nineteenth century the mathematician Hermann Grassmann came up with (almost) the complete theory of colorimetry (see figure 3.5), and Hermann von Helmholtz came up with both empirical and theoretical studies that turned the field into a mature science. Both of these latter nineteenth century giants of science deserve to be among your heroes, too.

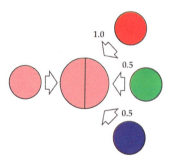

Figure 3.4 The colorimetric procedure. Here some color is being produced as the sum of fractions of three "primaries." In terms of this basis the color is {1, 0.5, 0.5}. Although the *colors* of the two half-fields look the same, the *beams* that fill the half-fields are (in all likelihood) *very* different. Thus colorimetry is all but trivial.

3.2.4 The linear structure of colorimetry

The basic idea (due to Maxwell) is simple enough (see figure 3.4). Select (almost any) three beams **u**, **v**, and **w**, and for any beam **a** establish the "colorimetric equality"

$$\mathbf{a} \,\square\, (\alpha\mathbf{u} + \beta\mathbf{v} + \gamma\mathbf{w}).$$

The "$\square$" operator stands for "the beam denoted on the left side is not discriminable from the beam denoted on the right side." Then the triple $\{\alpha, \beta, \gamma\}$ are called the "colorimetric coordinates" of the beam **a** relative to the triple of "primaries" $\{\mathbf{u}, \mathbf{v}, \mathbf{w}\}$. It turns out that the whole system is *linear*, that is to say, the colorimetric coordinates depend linearly on the beam and on the primaries. The linearity is perhaps surprising (because so little of physiology or psychology leads to linear relations), but enormously useful because it turns colorimetry into a very simple and useful science. It was Grassmann who noticed this explicitly and who managed to show how the formal properties derived from simple empirical observations, usually known as "Grassmann's laws". Grassmann's laws are generalizations from empirical observations (figure 3.5), if you accept them as "axioms" you obtain the for-

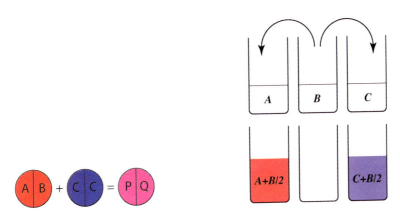

Figure 3.5 The figure at left illustrates the Grassmann Law I will perhaps most often use, namely $\mathbf{A}\,\square\,\mathbf{B} \to \mathbf{P}\,\square\,\mathbf{Q}$ for any beam **C** such that $\mathbf{P} = \mathbf{A} + \mathbf{C}$, $\mathbf{Q} = \mathbf{B} + \mathbf{C}$. Notice that though **A** and **B** *look* the same, they need not be identical beams (in fact, for the two half-fields the beam **C** need not be identical either as long as they only *look* the same). Such a state of affairs is by no means logically necessary. For instance, it does not apply to chemical mixtures (figure at right), a possible demonstration from the chemistry lab: Start with three vials containing colorless liquids (A, B, and C). Pour half of B into A and the other half into C. It is possible that you obtain a red and a blue liquid! Such a demonstration will never work for beams; that would indeed be a direct violation of Grassmann's laws.

Grassmann's laws

Grassmann's Laws were formulated by Grassmann[10] in a famous paper of the 1850s, but not formulated concisely as a set of axioms. However, it is not hard to distill a set of axioms from the paper and various authors have done so, with somewhat different results.[11]   Here is a set according to Bouma[12] (I quote verbatim):

(i) "If we select three suitable spectral distributions (three kinds of light) we can reproduce each color completely by additive mixing of these three basic colors (also called *primary* colors). The desired result can only be attained by one particular proportion of the quantities of these primary colors."

(ii) "If two different light spots give the same color sensation they continue to do so if we increase or decrease the brightness of both by the same factor (thereby leaving the relative spectral distributions unchanged)."

(iii) "Two light mixtures which, when displayed next to each other, produce the same color sensation, act in exactly the same manner when mixed with other lights. They may therefore be substituted one for the other in any mixture."

Bouma adds to this: "The contents of Grassmann's laws mentioned here indeed originate from Grassmann, but the formulation in his original publication is quite different from ours. Moreover, Grassmann adds two other laws, one on the continuity of colors (which will be taken by many people as self-evident) and another on the additivity of brightnesses."

That Grassmann's laws are far from self-evident is illustrated in figure 3.5. Many of the alternative framings of Grassmann's laws are useful or instructive, I will illustrate a few in a number of boxes on the following pages.

mal theory of colorimetry. Maxwell already understood much of this, and his experiments and theoretical exercises remain a model to any scientist.

If you want to review your understanding of linear geometry read appendix C.2. The formalism introduced there ("Dirac notation") will be used throughout the text. The physical implementation of linear geometry in colorimetry is largely the linear combination of beams. Throughout most of the book, this will mean the "addition" or "superposition" of incoherent beams of radiation. However, you need to be aware of other methods of "color mixture", not all of them respect linearity. (See figure 3.6 and appendix D.3.)

3.2.5   The ontological status of colorimetry

In retrospect, it is not difficult to see the reason for the linear structure of colorimetry. The colorimetric equality is due to equality of absorption of radiation in the three types of

Figure 3.6 Three different mixtures, from left to right: additive, partitive, and multiplicative (or subtractive). The latter implements the kindergarten rule "blue and yellow make green" (but notice that the blue is really turquoise).

spectrally selective retinal cones.[13] In the final analysis it is only these absorptions, which are due to the photopigments, that define the relevant input to the brain.

> *A colorimetric equality is the result of identical inputs to the brain; thus, subsequent brain processes are irrelevant in the sense that they cannot turn the equality into an inequality.*

Grassmann's "Symmetry Law"

The "symmetry law" is almost trivial:

$$\{\forall_{\mathbf{A},\mathbf{B}} \mid \mathbf{A} \,\square\, \mathbf{B} \Longrightarrow \mathbf{B} \,\square\, \mathbf{A}\}.$$

Of course the colorimetric equality is not affected by interchanging the two half-fields, only a pure mathematician (as Grassmann certainly was, although this "law" was pushed not so much by him as by others) would make a point of it. Apparently "Grassmann's laws" are abstractions that would not occur to the down-to-earth scientist.

I can illustrate such laws with Maxwell's color disks by showing the composition of the disk and what the disk looks like when spun, thus averaging over the spatial composition in the angular direction. Because spinning a disk does not affect the radial distribution I can also illustrate comparisons. This illustrates the symmetry law:

Left and right the two disks, at the center either disk as spun. The result is not supposed to surprise you! It is the full content of the "symmetry law."

Grassmann's "Transitivity Law"

The "transitivity law" is again almost trivial:

$$\{\forall_{\mathbf{A},\mathbf{B},\mathbf{C}} \,|\, (\mathbf{A} \,\square\, \mathbf{B}) \wedge (\mathbf{B} \,\square\, \mathbf{C}) \implies \mathbf{A} \,\square\, \mathbf{C}\}\,.$$

Here is the transitivity law demonstrated:

At far right either of the three disks as spun. This is the full content of the "transitivity law." Again, unsurprising!

Grassmann's Proportionality Law

The "proportionality law" seems again almost trivial:

$$\{\forall_{\alpha,\mathbf{A},\mathbf{B}} \,|\, \mathbf{A} \,\square\, \mathbf{B} \implies \alpha\mathbf{A} \,\square\, \alpha\mathbf{B}\}\,.$$

Here is the proportionality law demonstrated:

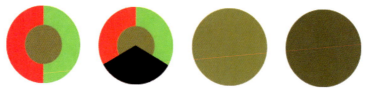

The two disks at right show the appearence of the two disks at left when spun. The result may not surprise you, but this "law" is far from being trivial once you come to think of it in terms of physiological mechanisms. What is at stake here is *linearity*, a very important property because of the formal implications, but a property that is rarely found in nature except in the "pure" theories of basic physics and (but only approximately) for generally very minor deviations from some set value.

It is a most important empirical fact.

Grassmann's Additivity Law

The additivity law has real content. It is far from trivial:

$$\{\forall_{\mathbf{A,B,C,D}}\,|\,(\mathbf{A}\,\square\,\mathbf{B}) \wedge (\mathbf{C}\,\square\,\mathbf{D}) \implies (\mathbf{A}+\mathbf{C})\,\square\,(\mathbf{B}+\mathbf{D})\}\,.$$

Here is the additivity law demonstrated:

The disk at right shows what the three disks look like when spun. Again, what is at stake here is *linearity*. Many people are not surprised, but you should carefully think about it. Could you *imagine* the world to be different? This is perhaps the most commonly invoked law in the whole of colorimetry!

Grassmann's "Continuity Law"

The continuity law it is considered trivial by many authors, and is indeed often omitted in modern texts. Here it goes:

If a beam is varied continuously, the patch varies continuously.

Here is the continuity law demonstrated:

Left is the structure of the disks, right their appearence when spun. Of course, one could add infinitely many intermediate cases, and evidently I would obtain a continuous progression from green to red. This is the property singled out by Riemann: The colors form a *continuously extended manifold.*

Grassmann's Fourth Law

Grassmann considered the fourth law as less certain and he apparently saw it more as convenience than fact. This showed keen judgment on his part. Here it goes:

$$\left\{ \forall_{\mathbf{A},\mathbf{B}} \mid (\|\mathbf{A}\| \equiv \|\mathbf{B}\|) \implies (\|\frac{\mathbf{A}+\mathbf{B}}{2}\| \equiv \|\mathbf{A}\| \equiv \|\mathbf{B}\|) \right\},$$

where $\|\mathbf{A}\|$ denotes the "brightness" of $\mathbf{A}$ and $\equiv$ stands for "equally bright." Here the failure of the fourth law is demonstrated:

Brown (the center triple), not yellow (the right triple), is in between red and green as is proven when the leftmost disk is spun. The brown looks darker than either the red or the green to most people, thus violating the fourth law. That the fourth law is indeed seen to be (very) wrong even on cursory inspection is perhaps surprising in view of the fact that so many able scientists even a century after Grassmann took it for law.

This is why psychology has nothing whatsoever to do with colorimetry. This absolutely essential point is often misunderstood. Yet this is the crux of colorimetry. To put it negatively, colorimetry has nothing to say regarding color as you know it. Colorimetry is quite unable to predict whether a beam will look reddish or greenish to you. On the positive side, colorimetry lets you decide whereas two *physically distinct* beams will look the same with uncanny precision (that is, a few decimal places, unheard of in psychology). Thus colorimetry is both *trivial* (to the psychologist) and *useful* (to the engineer).

As I said already, these facts typically go unappreciated. When I deliver a lecture on colorimetry I often see people nod, making me feel good. Then, after the lecture, in the discussion period, invariably someone asks a question showing that they confuse colorimetry with psychology (i.e., "color science," which is about "what is is like to experience red"[14]). This makes me feel rather depressed, though I soon get over it because I became used to it. If color science[15] is *anything* at all, it is not a *science*. Colorimetry definitely *is* a science, but it is not about color in the sense intended by the general audience. Later on I will define precisely what colorimetric color is. But be warned, it will not be what you expect.

3.3 The Formal Structure of the Space of Beams

3.3.1 The topological structure of the space of beams

From the physics you have no other constraints on the radiant power spectra than (firstly) the nonnegativity of the spectral radiant power throughout the spectrum and (secondly) the finite total radiant power. Thus any function of photon energy that is nonnegative throughout and is integrable, is (in principle) an acceptable (that is to say, *physically possible*) power spectrum. In practice the measurement accuracy is necessarily finite, and so is the spectral radiant density. Thus, it makes sense to limit the possibilities somewhat.

Let me augment the set of real numbers with the "nil-squared numbers."[16] A "nil-squared" number is a number whose square is identically zero, though it is itself different from zero. The nil-squared numbers are the nontrivial solutions of the equation $x^2 = 0$. I write composite "complex" numbers as $x + \varepsilon y$, where x and y are real numbers and $\varepsilon^2 = 0$, $\varepsilon \neq 0$. These complex numbers are called "dual numbers." Addition is simple,

$$(x_1 + \varepsilon y_1) + (x_2 + \varepsilon y_2) = (x_1 + x_2) + \varepsilon(y_1 + y_2),$$

whereas multiplication is explained as

$$(x_1 + \varepsilon y_1) * (x_2 + \varepsilon y_2) = (x_1 x_2) + \varepsilon(x_1 y_2 + x_2 y_1).$$

I think of any real number x (say) as being at the core of an environment $x + \varepsilon y$ of dual numbers, an "infinitesimal environment". As a hard-nosed physicist, I don't expect anything to change qualitatively as I stay in such an infinitesimal place. Since in these environments little happens, I will only accept functions $f(x)$ as possible power spectra that satisfy (apart from nonnegativity)

$$f(x + \varepsilon y) = f(x) + \varepsilon H y,$$

for certain $|H| < \infty$, that is to say, functions that are "locally affine". Such functions are "smooth", which is what the physicist takes for granted. This almost instantly removes much of the complexity that causes the mathematician so many headaches. It respects the integrity and coherence of the continuum in a way that is intuitive to the classical physicist.[17]

Thus a "purely monochromatic" spectrum like

$$f(x_0) = 1, \ f(x) = 0 \text{ if } x \neq x_0,$$

(the "blip function") is ruled out. Such monstrosities do not exist if you handle the continuum the way it should be, which is no doubt the physicist's way.

Any function possesses derivatives of arbitrary order everywhere. Of course, this is extremely convenient. Most of the zoo of badly behaved examples that you—no doubt—have been pestered with in your math classes can simply be forgotten about. Thus you have

$$f(x + \varepsilon y) = f(x) + \varepsilon f'(x)\, y = f(x) + \varepsilon H y,$$

which is indeed the *full* Taylor expansion. Thus

$$
\begin{aligned}
\cos \varepsilon x &= 1, \\
\sin \varepsilon x &= \varepsilon x, \\
\mathrm{e}^{\varepsilon x} &= 1 + \varepsilon x,
\end{aligned}
$$

for example.

Formally, a beam will be a density (radiant power spectral density as function of wavelength say) $S(\lambda)$, where the λ domain is understood as the dual number axis. Thus the spectra are smooth functions. The "space of beams" is an infinitely dimensional Hausdorff space. Due to the local affinity, it does not contain such infamous unpleasantnesses like the characteristic function of the rational numbers and so forth.

3.3.2   Linearity and convexity

I will consider the totality of beams and call it the space of beams $\mathbb{S}$. Is it a *linear* space? In a certain sense, yes. For instance, beams can be added together, and a beam can be multiplied with a scalar.

In order to add beams I use one of many possible optical tricks. For instance, I may produce a patch of light by shining a beam on a white screen (I typically use a projector or beamer). I may superimpose two such patches (see figure 3.7) and thus add (superimpose) the beams. Other methods use semisilvered mirrors, beam splitters, and so forth.

Multiplication with a scalar is easy when the factor is less then one, then I use a neutral density filter (sunglasses that is) that absorbs some of the beam.[18] Factors exceeding one can be reached through addition, for instance adding a beam to a similar beam produces a factor of two. There clearly exists a null element, the "empty beam" **o**, which is particularly easy to produce because it is no beam at all. Add the empty beam to any beam and nothing changes, multiply any beam by zero (just block it) and you get the empty beam. The optics of incoherent beams is such that in all such cases linearity obtains. It is not that there are no problems, though. Suppose I have some beam **a** and I *subtract* it from the empty beam. I should get the "additive inverse" $-\mathbf{a}$, for

$$\mathbf{a} + (-\mathbf{a}) = \mathbf{o}.$$

Figure 3.7 This figure illustrates the addition of two beams. You simply shine two beams on a white projection screen, such that the two patches of irradiated surface overlap. The addition is of the beams scattered to the eye. If the beams are incoherent ("natural light") you obtain a true linear addition. This is known as the "additive mixture" of beams. It is paradigmatic for the operations on beams commonly done in a colorimetric context.

But *such beams do not exist.* I cannot add a beam to another beam so as to extinguish the former; I invariably get more. Here you encounter a problem, because in a linear space, all linear combinations of elements of the space should be in the space. Linear spaces are complete and convex.

Convexity is okay, for I can obviously interpolate linearly between any two beams.[19] But completeness is not okay, because beams have no additive inverses. (See figure 3.8.) Thus the beams merely form a *convex set*, but not a *linear space*. This set has to be a *solid cone* with apex at the empty beam, because the existence of a beam implies the existence of a half-line of beams that contains both the empty beam and the given beam.

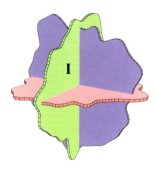

Figure 3.8 The "positive part" of a Cartesian coordinate system is the part where all coordinates are positive (or at least nonnegative). Here the positive part of a three-dimensional part is shown, it is often called the "first octant". In the space of beams the positive part is "the first orthant". "Positive part" is the most convenient term. Only beams in the positive part are real, all others virtual.

3.3.3   Formal definition of beams

There exists a way out of this dilemma that is due to the peculiar structure of the photometric method. Suppose you have an equality

$$\mathbf{a} \,\square\, \mathbf{b}.$$

Experience shows that such an equality cannot be destroyed by adding the same beam $\mathbf{c}$ to either side, thus

$$\mathbf{a} \,\square\, \mathbf{b} \rightarrow (\mathbf{a} + \mathbf{c}) \,\square\, (\mathbf{b} + \mathbf{c}).$$

This is actually one of Grassmann's laws. But this lets you define the additive inverse! Suppose I have a beam $\mathbf{a}$ and I put it in the left side of the photometer half-fields, leaving the right half empty; thus, I have

$$[\mathbf{a} \,|\, \mathbf{o}].$$

If the additive inverse $\mathbf{a}$ existed, I could likewise show it as

$$[\mathbf{o} \,|\, -\mathbf{a}].$$

But this is clearly impossible. If I add $\mathbf{a}$ to either side, I obtain

$$[\mathbf{a} \,|\, \mathbf{0}],$$

again. Consequently

> *the additive inverse of* $\mathbf{a}$ *on the left side is simply* $\mathbf{a}$ *on the right side or vice versa.*

If you use *pairs* of beams, you can construct a complete linear space of beams (figure 3.9).

In retrospect, this makes sense because in colorimetric practice you *never* venture an opinion as to the properties of any *single* beam. You only judge equality/inequality for *pairs* of beams. Thus

> *the basic entities for colorimetric purposes are not* beams, *but* ordered pairs of beams *such as* $[\mathbf{a} \,|\, \mathbf{b}]$.

Using Grassmann's laws you can also turn $[\mathbf{a} \,|\, \mathbf{b}]$ into

$$[\mathbf{a} - \mathbf{b} \,|\, \mathbf{o}].$$

If you "normalize" expressions such as $[\mathbf{a} \,|\, \mathbf{b}]$ to the form $[\mathbf{a} - \mathbf{b} \,|\, \mathbf{o}]$ (right half-field the empty beam), you may as well drop the second half (it is always the empty beam anyway).

Thus you see that

> *the photometric configuration* [**a** | **b**] *is equivalent to a (virtual) beam* **a** − **b**.

(See figure 3.9.)

If **a** □ **b** then **a** − **b** □ **o**, and I call the beam **a** − **b** a "black beam" because it looks like the empty beam (like nothing!). As a beam it is different from the empty beam, though (not "nothing" but *something*, though not something *real*). Of course, such beams need not exist as physical entities (for instance, if **a** exists −**a** clearly does not). But they *can* be exemplified as the physical configuration

[**a** | **b**].

I will denote such beams virtual beams or (if the problem of existence doesn't come up) just beams. Actually existing beams I will refer to as physical beams, or real beams.

In colorimetry, you deal with virtual beams throughout. Maxwell was perhaps the first to understand and exploit this. In a colorimetric equality

a □ (α**u** + β**v** + γ**w**),

equality can often be obtained only for cases where some of the colorimetric coefficients are *negative*. Maxwell's solution was simple: Swap the negative beam over to the other side of the equation and it becomes real.

Given any pair of beams in a colorimetric equation **a** □ **b** I can find the beam **c** such that $c(\lambda) = \min\{a(\lambda), b(\lambda)\}$ for all wavelengths and subtract it from both beams. This does not spoil the equation of course. Then the reduced beams have mutually nonoverlapping support and are thus orthogonal. (Example in figure 3.10.) I will refer to this procedure as photometric reduction. (Orthogonality is well defined despite the lack of a metric, this is possible because orthogonality is defined as the lack of spectral overlap here.)

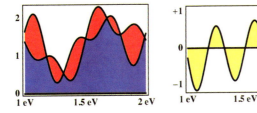

Figure 3.9 Two spectra (left) and their difference spectrum (right). The difference spectrum is virtual, it contains both positive and negative radiant spectral densities. Here I use photon energy instead of wavelength. I will use either label on monochromatic beams freely in this book.

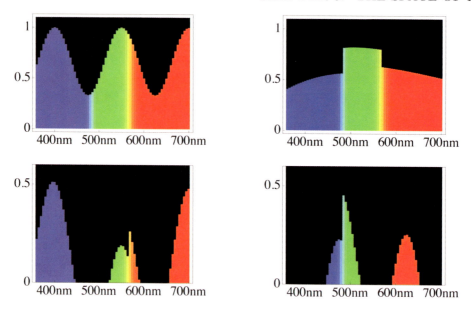

Figure 3.10 The notion of photometric reduction. On the top two spectra of beams that yield a colorimetric equation (yes, really). Different as these spectra are, their patches cannot be distinguished. On the bottom row are the reduced spectra. The reduced beams are real (by construction) and orthogonal because they lack any spectral overlap (again: by construction). The reduced beams also yield a colorimetric equation, though they yield different colors from the original beams. If you subtract either the original or the reduced beams from each other, you obtain a virtual spectrum that is black, that is to say, that makes a colorimetric equation with the empty beam.

Here I have reached a point where the formal structure of colorimetry becomes transparent. Physical beams are points of the positive part of the space of spectra, say $\mathbb{S}^+ \subset \mathbb{S}$. The actual entities that occur in colorimetry are ordered *pairs* of beams, thus oriented line segments in $\mathbb{S}^+$. Photometric reduction imposes an equivalence relation, for all oriented line segments that differ only by a shift are equivalent. Shifting one endpoint (say the second one, the right side of the splitfield) to the origin yields a convenient unique representative of the equivalence class. Such geometrical entities can be interpreted as points of $\mathbb{S}$ though. Thus the points of $\mathbb{S}$ can be taken to stand for equivalence classes of splitfield presentations. These are the entities colorimetry deals with. This takes away the mystery of virtual beams and it makes the space of beams $\mathbb{S}$ into a true linear space, not just a convex part. In this representation, Grassmann's laws simply boil down to the statement that there exists a linear subspace $\mathbb{K} \subset \mathbb{S}$ of codimension 3 whose elements cannot be (visually) distinguished from the empty beam (the origin of $\mathbb{S}$). "Colors" are affine hyperspaces parallel to the "black space" $\mathbb{K}$. This is the simplest geometrical account. The black space $\mathbb{K}$ appears as the essential invariant of color vision. Notice that it in no way depends upon any arbitrary convention such as the choice of colored papers in Maxwell's experiments.

3.3.4 The structure of beams

Thus far I have spoken of beams as integral quantities and I have not inquired into their *structure*. I do not mean the spatiotemporal structure here, at least not at the scale of the observer. Beams have certain "qualities" other than (spatiotemporal) geometrical ones. The first to study such qualities scientifically was Isaac Newton in his famous prism experiment (figures 3.11 and 3.12). Newton projected the "spectrum" of sunlight on a wall. Although sunlight is colorless, the spectrum (literally meaning "ghost") shows a continuous sequence of hues, namely Red, Yellow, Green, Cyan, and Blue. Newton believed that he had decomposed sunlight into simpler components (the "homogeneous lights", sunlight being a "confused mixture" of these) and that each homogeneous light was associated with a certain color (or hue) in your perception of it. Although Newton understood the colors to be *mental* ("the rays are not colored"), he believed in a strict and lawful correspondence between the type of homogeneous light and the hue. This is a bizarre notion indeed, but even today many scientists still entertain this silly idea, often denoted "seeing by wavelength" in the literature of philosophy of color.

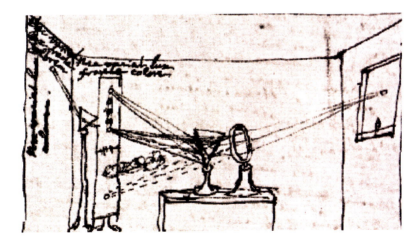

Figure 3.11 Newton's experiment showing the solar spectrum (right-hand part) and the atomic nature of homogeneous light (left-hand part). Notice that sunlight enters from the right via a hole in the window shutter.

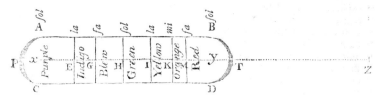

Figure 3.12 Newton's drawing of the solar spectrum. (Notice the "7 colors of the rainbow" and the musical scale.) The spectrum is a linear series of superimposed ("monochromatic") images of the sun. From this drawing you can estimate the spectral resolution of Newton's method. It is roughly a fifth of the extent of the visual region, thus pretty lousy by modern standards. Yet you may be ensured that the colors must have looked very vivid though.

The fact that Newton was wrong on both counts has had unfortunate consequences even to the present day. When philosophers say "color is either *seeing color by wavelength* or else they are (mere) *mental paint*" this echoes the notion that each hue corresponds to a specific homogeneous light. When the physicist says that "sunlight is actually composed of monochromatic beams," this is Newton's idea of sunlight as a "confused mixture". Both notions are easily demonstrated to be *false*.

Consider the "seeing by wavelength" notion. That it is indeed *false* can be appreciated easily if you consider the following simple experiment. Take a "red" and a "green" homogeneous light. There indeed are such beams to most human observers. Then the sum of the two may look "yellow," just as yellow as "the" yellow homogeneous light[20]. Indeed, if done right, the human observer cannot tell the two apart. Here you have that three "homogeneous lights" **r**, **y**, and **g** (say) such that **r** + **g** cannot be distinguished from **y**. But then **y** can have no individual existence, since I cannot uniquely point it out. Thus "hue" is not simply caused by "type of homogeneous light" (figure 3.13).

Next consider the "sunlight is a confused mixture" notion. That it is indeed *false* is obvious if you consider that:

> *The fact that sunlight can be **decomposed** into monochromatic components (indeed, that can be done) in no way implies that sunlight is **composed** of these same components.*

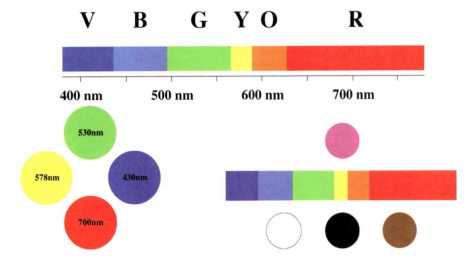

Figure 3.13 Top: If the circumstances are at all "reasonable," most observers agree pretty much on the wavelengths in the Newtonian spectrum at which they locate "unique hues" such as "green". Here are some values that most people would (at least roughly) agree on. Bottom-left: Some "unique hues", like yellow for instance, are quite sharply located for a given observer. A person may reproduce the wavelength of "yellow" within a few nanometers; Bottom-right: Many colors are never seen in the spectrum. Examples are white, black, gray, and brown. The purples are special in that they look very vivid (no obvious white or black content) just as the "spectral" colors do, yet are not among the "spectral hues".

By the same logic, a sausage would be composed of slices before you ever do the slicing. Sunlight can be decomposed in an infinite number of (very different) ways, and the spectral decomposition is only one possible instance. All such decompositions are equivalent in the sense that they are related through linear operations. No such decomposition is actual unless it is actually performed.[21]

A linear space has no atomic parts by its very construction. All infinitely many bases that may be used to describe it are fully equivalent. The elements of one basis can be "decomposed" in terms of the elements of any other basis and vice versa. A basis is simply a convenience, its elements are in no way "atomic parts". The spectral basis has obvious advantages in physics since Maxwell's equations are invariant with respect to translations in the time domain. These advantages count for little in colorimetry, and different choices of basis are often convenient. As an example, I will introduce the basis of Goethe's "edge colors" later in this book. Athough the spectral basis has its uses, it is perhaps slightly awkward because the basis vectors exist only in some limiting sense and cannot really be implemented. A spectral slice of daylight of zero spectral width is simply the empty beam. Monochromatic beams are nonentities and can be realized only approximately.

It is amazing and most unfortunate that these facts rarely, if ever, find mention in accounts of Newton's (indeed seminal and admirable) work on color. I have not encountered them in textbooks on colorimetry. It is no mere hairsplitting either. As I will show later, the difference has important consequences.

3.3.5   The spectral basis

The spectral decomposition is just one (though very useful) way to describe the qualities of a beam. The decomposition can be likened to the introduction of a coordinate system. Instead of pointing at an object, you may give its coordinates with respect to a fiducial origin. The object is located x steps in the X-direction, y steps in the Y-direction, z steps in the Z-direction,[22] or "the object is at $\{x, y, z\}$" in the XYZ-coordinate system. Specifying three numbers is often more convenient than pointing. Notice that the numbers depend on the choice of coordinate system though (e.g., X-direction toward the east, Y-direction towards the north, Z-direction vertically upward), many choices are open to you. It is the same with beams. The spectroscope defines the coordinate system, the spectrum is an infinite set of coordinates. Indeed, the dimension of the space of beams $\mathbb{S}$ is infinite.

The monochromatic beams of unit radiant power form a convenient basis of the space of beams $\mathbb{S}$. Notice that "monochromatic" does not imply that the spectrum of these beams is a blib function. The spectra are still smooth, though the spectral density is concentrated in such a narrow spectral region that is of no interest to inquire after their precise shape. For all practical purposes you can consider the spectral support as negligible and the spectrum sufficiently described by location and total spectral power.[23]

I will use the notation

$$\mathbf{m}_\lambda \text{ or } |\lambda\rangle,$$

(see the appendix C.2 for an explanation of the Dirac notation) for a monochromatic beam of unit radiant power and wavelength λ. The wavelength of the beam λ is used as a continuous index. In practice, it runs roughly from about 350 nm to 800 nm, the visual region. There exist machines called "monochromators" that yield the component of any given beam ($\mathbf{s}$ say) in the direction of any monochromatic beam. The radiant power of such a monochromatic component, then, is the coordinate s_λ of the beam for the index λ. If the physical dimension (of radiant power) is associated with the monochromatic beams $\mathbf{m}_\lambda$, then the coordinates are pure numbers. Notice that s_λ depends continuously upon

Dimension of the spectral basis

The wavelength is a positive, continuous index, thus $\lambda \in \mathbb{R}^+$. A spectral radiant power density is a nonnegative function $s(\lambda)$. Such functions are physical entities, usually operationally defined (for instance, via spectrographs). Thus they must have mathematically simple properties, for instance, a function taking on one value on the rational numbers and another on the irrational reals makes no physical sense. You may assume that such functions are only known to some finite resolution and thus just as good as smooth approximations. This again means that you can always discretize such functions without loss of precision. In almost any practical setting, a sampling distance of 5 nm or (rarely) 1 nm would amply suffice. Combine this with the extent of the visual spectrum and you will understand that the practical dimension of the space of beams is somewhere between fifty and five hundred, rather than infinity. Such dimensions are easy to handle in modern computer packages such as Mathematica.

the index λ, which is why I will often write $s(\lambda)$. A plot of $s(\lambda)$ against λ is called "the spectrum of the beam **s**". Expressed in the spectral basis, the beam can be written

$$\mathbf{s} = \sum_\lambda s_\lambda \mathbf{m}_\lambda,$$

or, equivalently,[24]

$$|s\rangle = \sum_\lambda s_\lambda |\lambda\rangle.$$

Here the summation over the continuous index λ has to be performed as an integration, of course.[25] In practice, the wavelength domain is typically *sampled* and you need to worry about the bin width $\Delta\lambda$ (say). All this is rather standard, you simply approximate the integrals over the wavelength domains through Riemann sums.

Notice that a monochromator tuned to the wavelength λ and followed by a detector of radiant power will produce a number when you feed it a beam and that it will do so in a linear way. Thus, such a machine is the physical implementation of the linear functional $\langle\lambda|$ associated with the spectral basis (see the appendix C.2). Contracted upon a beam, you get the coordinate of the beam in the direction λ, that is to say,

$$s_\lambda = \langle\lambda|s\rangle.$$

The procedure is known as spectroscopy, and it is routinely performed on a daily basis in numerous laboratories over the globe. Thus the construction

$$|s\rangle = \sum_\lambda \langle\lambda|s\rangle\,|\lambda\rangle,$$

can be understood in terms of physical operations as

$$|s\rangle = \sum_\lambda \underbrace{\langle\lambda|}_{\text{spectroscopy}} |s\rangle \overbrace{\underbrace{|\lambda\rangle}_{\text{monochromator}}}^{\text{synthesis}}.$$

The contraction of the ket $|s\rangle$ on the bra $\langle\lambda|$ is standard spectroscopy, the basis vectors $|\lambda\rangle$ can be produced by a monochromator and a black body source (for instance), the beam $|s\rangle$ can be "synthesized" by addition of suitably scaled monochromatic beams. After the synthesis the monochromatic beams are no longer "parts" or "components" of the beam (as Newton would have it); rather, the synthesized beam is physically identical to the original beam. Notice that[26]

$$|s\rangle = \sum_\lambda |\lambda\rangle\langle\lambda|s\rangle = \left(\sum_\lambda |\lambda\rangle\langle\lambda|\right)|s\rangle,$$

which implies that

$$\sum_\lambda |\lambda\rangle\langle\lambda| = \hat{1},$$

i.e., the *identity operator.* This expresses the basic fact that the spectral basis is "complete."

With respect to these issues, the psychophysical law of univariance is relevant. It succinctly expresses the empirical fact that the retinal photoreceptors are photon counters (appendix C.1), rather than caloric devices, and that once a photon has been "detected" (that is to say: converted into an electrochemical event that counts as a neural signal, that is, that which has causal influence on the activity of brain structures), the visual system has lost any clue as to the photon energy (or, equivalently, the wavelength) of the absorbed photon.[27]

This empirical fact has important consequences for colorimetry, though its fundamental importance is generally downplayed in the textbooks. In many texts it isn't even mentioned.

3.1 Wavelength versus photon energy

There exist two categorically different types of radiation detectors. One is a caloric device that exploits the fact that many substances absorb radiation by converting radiant into thermal energy, like a cat sleeping in the sun grows hot. The other type exploits events whereby a photon of radiation kicks an electron in a material to another state. Some of such devices can be used to measure the photon energy, but many only let you record the mere occurrence of such events. Most photosensitive devices in current use are of the latter type. The photoreceptors in the human retina convert photon related events into electrochemical (neural) events. Any record of the actual photon energy is lost, and only events (or average event rates) are recorded. The eye is sensitive enough that single photon events can be shown to have causal effects on behavior.[28]

For caloric devices it is convenient to express radiant flux in terms of radiant power per wavelength interval as a function of wavelength. For photon counters, it is more natural to express radiant flux in terms of photon number flux per photon energy interval as a function of photon energy.

For the case of human vision, the natural description is by photon number flux per photon energy interval as a function of photon energy. This reflects the way the retina functions, and, moreover, photon energy is what is biologically relevant, not wavelength. Only in a few cases does the wavelength of radiation relevant biological meaning, whereas photon energy is almost of universal biological importance.

For a beam of monoenergetic photons (or monochromatic radiation) you can simply convert from one representation to the other. For general, "broadband" radiation you cannot convert radiant power into photon number flux un-less you know the spectrum. Since it is only the latter type of beams that you encounter in the daily environment, the photon description is much to be preferred in almost all cases.

That so much of the literature still uses radiant power on wavelength basis is due to historical reasons dating from the time before the concept of "photon" was introduced and to the persistent but baseless idea that the wavelength somehow has to do with color. The latter idea was initiated by Newton and often explained through a (mistaken) analogy with sound. For instance, though it is often said that you "see about an octave," this is totally irrelevant in the case of vision. In vision, wavelength is simply a convenient numerical label on monoenergetic beams (figure 3.14), the absolute *wavelength* (as a geometrical measure) being quite irrelevant.

A convenient measure of photon energy is the "electronVolt" (or "eV") which is the kinetic energy an electron gains if it traverses a potential interval of one volt. In terms of standard units, 1eV equals roughly $1.6\ldots 10^{-19}$J. In order to obtain a feeling for the magnitude, at room temperature an air molecule has a kinetic energy of about 0.03eV. The visual range spans a range of photon energies from about 1.6eV to 3.3eV. This is a biologically most important range, because it is the range of chemical energies. Photons in the visual range often do things to substances without actually disrupting molecules. The UV

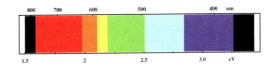

Figure 3.14 A schematic picture of the spectrum with wavelength and photon energy scales.

range is disruptive, and thus bad for life, and the IR range only shakes things a little bit, and thus is generally not informative to visual agents. In order to get a feeling for the translation, 500nm is about 2.5eV and 2eV is about 620nm. A convenient calculator is:

With some experience the photon energy scale becomes just as familiar as the old wavelength scale, trust me.

It is also useful to obtain a feeling for how many photons there are in the natural environment. It is easy enough to do a simple calculation for about the heaviest photon bombardment you may experience in daily life, that is looking at a white piece of paper on a sunny day on the beach. In order to do the calculation, you need a few simplifying and rather rough assumptions:
— the source is the sun, a black body at a temperature of ca. $5700°$K;
— the sun's diameter is about half a degree;
— the sun's height is $45°$ (say);
— the absorption by the atmosphere is neglected;
— the paper is a Lambertian surface of unit albedo;
— the pupil of the eye has 2mm diameter;
— the eye detects the photon energy range of 2–2.6eV (say);
— the conversion efficiency of the retina (including absorption by the eye media, and so forth), is about one percent;
— the diameter of a retinal cone is about $2\mu m$, back focal length of the eye is about 17mm;
— the integration time of a retinal cone is about 10ms.

Going through the standard radiometric calculations, I arrive at a photon count of about 400 events per integration time per receptor. Notice that this is about the maximum you'll ever get, whereas the eye function over a range of radiance of many decades. Thus there must be a large range over which the receptors are essentially photon counters.

This figure is far more informative than just the number of photons per meter cubed in the sun's beam. Figure 3.15 gives you the number of photons as femto mole[29] per unit photon energy increment in electron volt. The photon volume number density is useful in various radiometric calculations, but eventually the only thing that interests you will be the rate of photon absorptions in your retinal receptors, or "quantum catch," as some people say.

Because virtually all data you can draw from the web is in terms of radiant power per unit wavelength interval, you will have frequent occasion to do conversions. Though of course a pain in the neck, there is nothing hard about it, and you can simply prepare some standard conver-

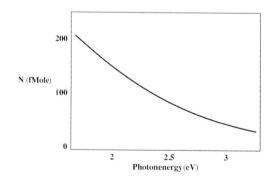

Figure 3.15 The spectrum of sunlight in terms of photon numerosity (femto moles) per unit energy increment (1eV) on photon energy basis. This was straightforwardly calculated from Planck's formula; thus, it illustrates a black body radiator at the temperature of the sun's photosphere, rather than the actual solar spectrum.

sion algorithms to take care of these chores more or less automatically.

You will have to convert the color matching functions (figure 3.16) as well as the various illuminant spectra (example, figure 3.17).

As an aside, a photon count of 400 events per integration time per receptor implies that on repeated samples you have to reckon with a random variation of the order of 20, that is 5 percent. The reason is that the photon absorptions occur randomly in time, the counts following Poisson statistics. A photon number flux N has an intrinsic uncertainty $\sqrt{N}$. This noisiness has no causal explanation, it simply happens, a fact that reflects one of the wonders of quantum theory. As the level of illumination is lowered the noise increases, the relative RMS noise level being inversely proportional to the level. At low radiation levels, "photon shot noise" limits the incremental detection threshold for the human observer. The only way to combat noise is to trade it against resolution. Pooling receptor outputs (thus decreasing spatial resolution)

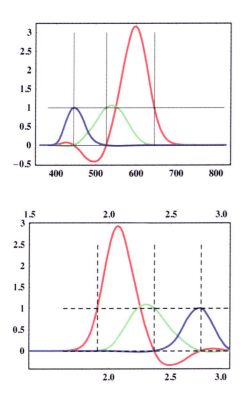

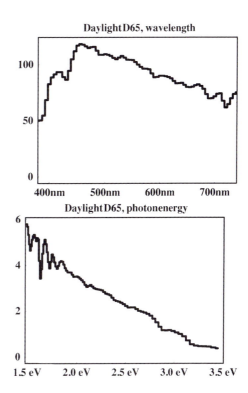

Figure 3.16 The CIE 1964 xyz data have been converted to RGB (top) and converted to the photon number density description (bottom).

Figure 3.17 Average daylight (D65) in the conventional representation (top) and converted to the photon number density description (bottom).

and/or lengthening the integration period (thus decreasing temporal resolution) will both help to decrease the noise level, or "combat noise", as people say. In this respect the eye is not different from a digital camera based on a CCD-chip.

The effect of this photon shot noise is illustrated in figure 3.18. For each subsequent image the radiation level was increased by a factor of ten. Since the shot noise is independent in the color channels you obtain a mixture of luminance noise and chrominance noise. Most people experience the latter as more objectionable than the former.

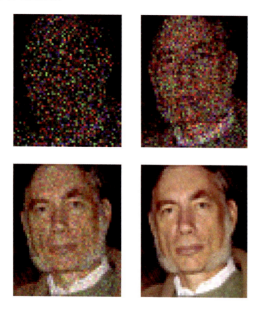

Figure 3.18 The effect of photon shot noise. The radiation levels of the four images are a factor 10 apart for each pair, a total range of 1,000.

3.2 Review of linear geometry

This appendix serves a dual purpose. One is to introduce the terminology I use in the book; the other is to introduce the Dirac notation. The appendix is too short to *teach* you linear algebra, I assume you already followed the standard course. You may not be familiar with the particular terminology I will adopt, though, and (especially if you are not a physicist) you may not be familiar with—the very convenient—Dirac notation. Skimming through this appendix may save you some perplexities later.

3.2.1 Linear Spaces

A simple space you may use to obtain an intuition for the structure of linear spaces is the surface of a blank page of paper. Since you will consider only spaces with a well defined *origin*, you may as well put a dot on the page and call it "origin". Write an "**o**" next to in order to remind you that this is the **o**rigin, or the "null element" **o**, or the ket $|o\rangle$ (figure 3.19 left). These are all notations for the same entity with somewhat different uses.

"Points" are indicated as dots on the paper and are named **p** ("point p") (figure 3.19 center) or $|p\rangle$ ("the ket p" or "the vector p"), or as arrows with tail at **o**, tip at **p** (figure 3.19 right).

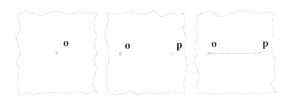

Figure 3.19 Left: The origin **o**; center: a point **p**; right the point **p** drawn as an arrow, or "the vector **p**" or $|p\rangle$.

> ## Linear space
>
> Remember the formal definition of a "linear space":
> A quadruple $(\mathbb{V}, +, ., \mathbb{R})$ is a *linear space* over $\mathbb{R}$ if $(\mathbb{V}, +)$ is an additive group with a multiplication $\mathbb{R} \times \mathbb{V}; (\alpha, \mathbf{x}) \mapsto \alpha.\mathbf{x}$, that satisfies the following properties:
>
> (i) $\alpha.(\mathbf{x} + \mathbf{y}) = \alpha.\mathbf{x} + \alpha.\mathbf{y}$;
> (ii) $(\alpha + \beta).\mathbf{x} = \alpha.\mathbf{x} + \beta.\mathbf{x}$;
> (iii) $\alpha.(\beta.\mathbf{x}) = \alpha\beta.\mathbf{x}$;
> (iv) $1.\mathbf{x} = \mathbf{x}$.
>
> If you are the formal type you probably much prefer this over the figures in this chapter (so much more succinct and precise). This appendix is evidently not for you then. You can safely skip it.

Clearly arrows and points can be identified, specifying the same type of geometrical entity. Arrows are generally preferable, because less confusing in your drawings. They explicitly show the relation to the origin. You can scale and add arrows via the familiar constructions, for instance the parallelogram construction for addition (figure 3.20 left). This should be familiar from the basic physics courses.

3.2.2 Vector bases

Since any point in the plane can be regarded as some linear combination of two given vectors, it is often convenient to choose such a basis of two vectors (figure 3.20 right). This works only if

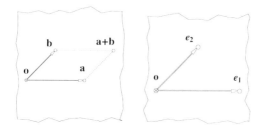

Figure 3.20 Left: Vector addition $\mathbf{a} + \mathbf{b}$ via the parallelogram construction; right: the vector basis $\{\mathbf{e_1}, \mathbf{e_2}\}$.

the pair of vectors is independent, that is to say, the vectors may not be each other's multiples. Otherwise the choice of basis is up to you. You may as well draw your basis right now and name the basis vectors $\mathbf{e_{1,2}}$, or (equivalently) $|e_{1,2}\rangle$. If you take some arbitrary vector $\mathbf{p}$, you can write it in terms of the basis as

$$|p\rangle = p_1 |e_1\rangle + p_2 |e_2\rangle.$$

Here the scalars (scalars are simply real numbers) $p_{1,2}$ are called the coordinates of the vector $\mathbf{p}$, where I often omit "with respect to the basis $\mathbf{e_{1,2}}$," although it is really necessary to do so. If you change the basis the coordinates of $\mathbf{p}$ change, although $\mathbf{p}$ is still the same *geometrical entity*. Thus coordinates are less general than the entity itself. Coordinates are often convenient, however. If you do numerical linear algebra by computer, they are *necessary*.

3.2.3 Linear functionals

You find the coordinates by way of a reverse parallelogram construction. A more convenient way to do so is to introduce another type of geometrical entities called "dual vectors," "bras," or "linear functionals" (all the same thing). A linear functional is simply a function that yields a real number for a vector argument. Think of a linear functional as a *slot machine* with a single

slot. Such machines are *very picky*. The slot only accepts vectors. The moment you put a vector in its slot the machine pops out a number. Consider the linear functional $\varphi(\bullet)$. If you fill its slot with a vector $\mathbf{a}$ out pops a number a (say), thus $\varphi(\mathbf{a}) = a$. The functional is called linear because for any pair of vectors $\mathbf{a}$, $\mathbf{b}$ and scalars p, q you have that

$$\varphi(p\mathbf{a} + q\mathbf{b}) = p\varphi(\mathbf{a}) + q\varphi(\mathbf{b}).$$

A linear functional, often called a "bra," and is written $\langle\varphi|$ instead of $\varphi(\bullet)$. Filling the slot is then written $\langle\varphi|a\rangle = a$. Thus a bra and a ket (linear functional and vector) together make up a bracket whose value is a scalar. In this notation you see that a linear functional and a vector eat each other. As the cool kids say, "a vector is *contracted* upon a dual vector so as to produce a scalar (number)" (see figure 3.21). If I leave open the vector slot in the bracket I see that $\langle\varphi|\bullet\rangle$ (not different from $\langle\varphi|$) is a single slot machine that takes vectors to produce scalars. Conversely, from $\langle\bullet|a\rangle$ (not different from $|a\rangle$) you see that vectors are linear slot machines that accept bras to produce scalars. Thus vectors are linear functionals on bras. The bras and kets are mutual duals, which is why the linear functionals are also known as dual vectors. Indeed, you easily show that they span a linear vector space, appropriately called the dual space.

3.2.4 Dual bases

Consider linear functionals $\varepsilon_{1,2}$, such that

$$\langle\varepsilon_1|e_1\rangle = 1, \quad \langle\varepsilon_1|e_2\rangle = 0,$$
$$\langle\varepsilon_2|e_1\rangle = 0, \quad \langle\varepsilon_2|e_2\rangle = 1,$$

which is quite special of course. Notice that

$$\langle\varepsilon_1|a\rangle = \langle\varepsilon_1|\,(a_1|e_1\rangle + a_2|e_2\rangle)\rangle = a_1,$$
$$\langle\varepsilon_2|a\rangle = \langle\varepsilon_2|\,(a_1|e_1\rangle + a_2|e_2\rangle)\rangle = a_2,$$

so you can write

$$|a\rangle = \langle\varepsilon_1|a\rangle|e_1\rangle + \langle\varepsilon_2|a\rangle|e_2\rangle,$$

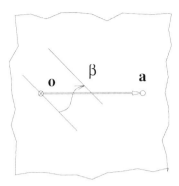

Figure 3.21 The contraction $\langle\beta|a\rangle$ of the vector $|a\rangle$ on the linear functional $\langle\beta|$. The linear functional is represented by its zero and unit level lines, the direction of increase is suggested by the wavy arrow. The vector $|a\rangle$ is estimated to reach towards the level lines for level two, thus $\langle\beta|a\rangle = 2$. (In most cases you will need to do such interpolations or (like here) extrapolations. Of course, you may draw as many additional level lines as you need.)

or, what amounts to the same

$$|a\rangle = (|e_1\rangle\langle\varepsilon_1| + |e_2\rangle\langle\varepsilon_2|)\,|a\rangle.$$

The "operator"

$$|e_1\rangle\langle\varepsilon_1| + |e_2\rangle\langle\varepsilon_2|,$$

is a slot machine that accepts vectors and produces a vector in return. In fact, it simply returns a clone of the vector you put into its slot, so it is an *identity operator* $\hat{I}$. It is a very important and useful geometrical machine as you will soon appreciate.

The linear functionals are best visualized as slopes or gradients that can be represented through the level lines ("isohypses" or "loci of constant height" as on a geographical map). Since all level lines are parallel and straight, and of constant spacing, a single pair will do. Moreover, since the zero level line must pass through

the origin anyway, you can actually make do with a *single* level line, most conveniently the one for unit height. But although this elegance and efficiency makes for easier drawing, you should always conceive of the linear functionals as slopes of infinite extent. The level lines (including the interpolated level lines for heights of fractional value) really *cover all of the space*.

How do you draw the dual basis $\varepsilon_{1,2}$ on your paper? You draw it by drawing loci of constant functional value. It is clearest to draw both the zero and unit level lines and to indicate that they belong together, as done in figure 3.21.

Now it is easy enough to figure out how to draw the dual basis (see figure 3.22). Notice that you obtain a parallelogram on the basis vectors. It is this figure of basis vectors and linear functionals (the dual basis) that make up a complete "scaffold" for the space. Usually you

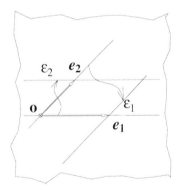

Figure 3.22 The all-important concept of the dual basis. The (vector-)basis is $\{|e_1\rangle, |e_2\rangle\}$, the dual basis consists of the linear functionals $\{\langle\varepsilon_1|, \langle\varepsilon_2|\}$. The construction is such that $\langle\varepsilon_1|e_1\rangle = 1$, $\langle\varepsilon_1|e_2\rangle = 0$, $\langle\varepsilon_2|e_1\rangle = 0$ and $\langle\varepsilon_2|e_2\rangle = 1$. This ensures that the linear functional ε_1 measures the $\mathbf{e_1}$ component of any vector, whereas the linear functional ε_2 measures the $\mathbf{e_2}$ component.

will need such a scaffold in order to be able to do numerical calculations with geometrical entity. You are quite free in your choice of scaffold, though. Usually the nature of your problem will suggest a convenient choice to you. If you do automatic number crunching, this doesn't matter at all since computers couldn't care less about conveniences. Then any old scaffold will do.

Using the level lines it is a simple matter to construct linear combinations of linear functionals (see figure 3.23). Suppose you need to draw

$$\langle\varphi| + \langle\psi|.$$

Then you find the point of intersection of the zero level line of $\langle\varphi|$ and the unit level line of $\langle\psi|$, after that the point of intersection of the unit level line of $\langle\varphi|$ and the zero level line of $\langle\psi|$. The line connecting these two points is the unit level line of

$$\langle\varphi| + \langle\psi|.$$

The zero level line of $\langle\varphi| + \langle\psi|$ is simply the line through **o** that is parallel to the unit level line.

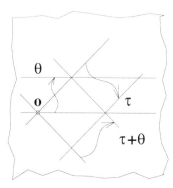

Figure 3.23 The addition of two dual vectors ϑ and τ is done easily via a dual parallelogram construction. It is very similar to the parallelogram construction of vector addition. Familiarity soon renders this obvious.

Done. Notice that this is a kind of dual parallelogram construction.

3.2.5 The "graph paper" concept

When you draw many level lines of the dual basis at equal intervals, you produce *graph paper*. Thus the dual basis trick to find the coordinates simply amounts to the use of graph paper. You have been using dual bases since you were a kid. The only progress is that you can easily work in arbitrary bases with oblique, differently scaled axes.

What if you change the basis? Say you have basis vectors $|g_{1,2}\rangle$ and dual basis $\langle\chi_{1,2}|$. Now you obtain coordinates $a'_{1,2}$ instead of $a_{1,2}$ (say). How are these coordinates related? Here you have the simple calculation

$$
\begin{aligned}
|a\rangle &= \sum_i a_i |e_i\rangle \\
&= \sum_i a_i \left(\sum_j |g_j\rangle\langle\chi_j|\right) |e_i\rangle \\
&= \sum_j \left(\sum_i t_{ji} a_i\right) |g_j\rangle \\
&= \sum_j a'_j |g_j\rangle,
\end{aligned}
$$

thus

$$a'_j = \sum_i t_{ji} a_i.$$

with

$$t_{ji} = \langle\chi_j|e_i\rangle.$$

Notice the simple trick: I inserted the identity operator

$$\sum_j |g_j\rangle\langle\chi_j|,$$

and expanded the expression. I carefully distinguished the two summations (i and j instead of two i's). The so called matrix elements t_{ji} appear as the entries in the code book needed for the translation between the two bases.

People say that picking a basis is to adopt a representation. The idea is that the *same* geometry can be expressed in many different representations. Clearly any representation is as good as any other because you know how to change representation or translate between representations using the code books defined by the pair of representations between which you translate.

A change of representation is merely a change of graph paper. The objects drawn on the graph paper are not affected by that at all, they exist quite independent of your choice of favorite graph paper.

3.2.6 Basis transformations

It is calculations like this that start to show some of the elegance and power of the Dirac (bra and ket) notation. The procedure is always the same: I replace the "|" with the identity of my choice. For instance

$$|a\rangle = \left(\sum_i |g_i\rangle\langle\chi_i| \right) |a\rangle = \sum_i a'_i|g_i\rangle.$$

Here the first expression is the vector **a** as geometrical entity, that is to say quite independent of any particular basis. Then I introduce a special identity and end up with **a** expressed in terms of the **g$_i$** basis. Notice that "||" is not different from "|"; in fact, "|" can be read as the identity and is merely a placeholder for some identity slot machine. Thus, the Dirac notation makes it very obvious how to proceed in algebraic calculations. They become no-brainers. (But you should refuse to succumb to the temptation to go algebraic, *intuition means geometry*!) In doing such calculations, you continually use these two properties of a basis:

$$\langle\chi_i|g_j\rangle = \delta_{ij}.$$
$$\sum_i |g_i\rangle\langle\chi_i| = \hat{1},$$

where the "Kronecker symbol" δ_{ij} equals 1 if $i = j$ and 0 if $i \neq j$, whereas $\hat{1}$ is the "identity operator" on kets. The first equation defines the relation of dual basis to the basis and the second equation expresses the fact that the basis is indeed a basis (a "complete" set of vectors, that is, a set that spans the space). The latter equation is generally known as the "completeness relation" for this reason. Apart from these, you use the fact that scalars commute, and so does the product of a scalar with a bra or ket. Although the summations in the example run only from 1 to 2, this works in any (finite) dimension without any special problem If you feel familiar with the two-dimensional case, you should have no problem to apply your intuition to the thousand-dimensional case (and you will have frequent occasion to do so in colorimetry). Colorimetry goes 1, 2, ..., 100(= ∞) or 1, 2, ..., 1000(= ∞) at worst.

3.2.7 The scalar product

Notice that though bras and kets *eat each other*, a pair of bras or a pair of kets can do little else together but simply *coexist*. A bra, understood as slot machine, is very picky and only accepts kets. Likewise, a ket, understood as a slot machine, is equally picky and accepts only bras. Thus there exists no simple way to combine two vectors such as to obtain a number (a scalar). It is important to notice this fact because most people in the sciences have at one time in their careers followed a course in vector calculus and when they see two vectors they produce a number at the drop of a hat, applying the so-called dot product (also known as inner product or scalar product).

The dot product is a *bilinear symmetric slot machine* whose two slots both accept a vector and deliver a number in return. I will not consider such geometrical entities here because they essentially introduce a *metric*. Using the dot product you find such things as the *length* of a vector or the *angle* subtended by a pair of vectors. Much of colorimetry is quite independent of the existence of a metric, so (at least for the time being) I will disregard such constructions. (But see chapter 8.) The reason is that the scalar product comes at a cost. You have to commit yourself to some kind of theory that makes it meaningful. In our context, that may come from physical optics of beams or from color science (psychology of color experiences). In either case, such a theory would come from a context outside of that of colorimetry proper.

In the absence of a dot product there is no way to uniquely identify a given ket with some bra (or vice versa). Thus I have to *insist* on a sharp distinction between kets and bras. Some colorimetric objects are intrinsically bras, others kets and *the difference matters*. It is indeed a sad fact that much of the colorimetric literature does not recognize this. It is why I seriously considered calling this book *Colorimetry done Right*. That I refrained from doing so is not because I do not hold this to be correct.

3.2.8 Linear operators

I will have occasion to consider linear operators (see figure 3.24). A linear operator $\hat{\mathrm{T}}$ is a single slot machine that accepts a vector and yields a vector in return. The operator is called linear because

$$\hat{\mathrm{T}}(p\mathbf{a} + q\mathbf{b}) = p\hat{\mathrm{T}}\mathbf{a} + q\hat{\mathrm{T}}\mathbf{b}.$$

Consider the coordinate representation of an operator. You have $\hat{\mathrm{T}}|a\rangle = |b\rangle$ say. Expanding this in the usual way, you have

$$
\begin{aligned}
b_i &= \langle \varepsilon_i | b \rangle \\
&= \langle \varepsilon_i | \hat{\mathrm{T}} | a \rangle \\
&= \sum_j a_j \langle \varepsilon_i | \hat{\mathrm{T}} | e_j \rangle \\
&= \sum_j T_{ij} a_j,
\end{aligned}
$$

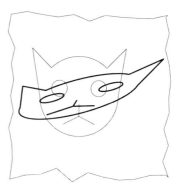

Figure 3.24 Example of a linear transformation. In this case the "cat" configuration was subjected to a linear transform.

thus $b_i = \sum_j T_{ij} a_j$ with

$$T_{ij} = \langle \varepsilon_i | \hat{\mathrm{T}} | e_j \rangle.$$

The coefficients T_{ij} are called the "matrix elements of $\hat{\mathrm{T}}$ in the basis $\mathbf{e_i}$." The matrix T is an array of numbers that depend upon the choice of basis, whereas the operator $\hat{\mathrm{T}}$ is a geometrical object in its own right, and thus is the *more general concept*. Using the standard trick—introducing the identity of your choice—you easily derive expressions that let you find the matrix in one basis in terms of the matrix in another basis. For instance,

$$
\begin{aligned}
T_{ij} &= \langle \varepsilon_i | \hat{\mathrm{T}} | e_j \rangle \\
&= \sum_{kl} \langle \varepsilon_i | g_k \rangle \langle \chi_k | \hat{\mathrm{T}} | g_l \rangle \langle \chi_l | e_j \rangle \\
&= \sum_{kl} \langle \varepsilon_i | g_k \rangle \langle \chi_l | e_j \rangle \, T'_{kl}.
\end{aligned}
$$

Here the $\langle \varepsilon_i | g_k \rangle$ and $\langle \chi_l | e_j \rangle$ are the code books that translate between the bases in either direction.

In the colorimetric literature, the emphasis tends to be on the *matrixes* rather than on the *operators*. This is in (geometrically speaking) bad taste because it tends to mask the actual geometrical meaning. Most of the literature simply applies algebraic manipulations in an opaque manner and never gets around to a geometrical interpretation. Such a practice is to be strongly condemned if the aim is understanding and lucid description.

There exist many types of useful and interesting linear operators. The most important dichotomy has to do with operators that transform a basis into yet another basis, and those that do not. The former type induces isomorphisms and can be *inverted*. The latter type clearly cannot be inverted. If an operator cannot be inverted, there exists a linear subspace that is annihilated by the operator, the so-called kernel or null space of the operator. You will have plenty of occasion to deal with such "degenerate" operators in colorimetry.

3.2.9 The kernel, domain and range of a linear operator

The set of all points that are mapped to the origin by an operator $\hat{T}$ is called the "null space" or "kernel" of the operator, or ker$\hat{T}$. Both the null space ker$\hat{T}$ and the range range$\hat{T}$ are linear (sub-)spaces. The sum of their dimensions is the dimension of the domain of $\hat{T}$. In the example (figure 3.25) the transformation $\hat{T}$ maps the space on a one-dimensional subspace of the space (a line). Thus the plane is squashed into line. That is what is meant by a "degenerate" operator. Because the operator is *linear* you *must* have that

$$\hat{T}\mathbf{o} = \mathbf{o}.$$

But there exist many points different from $\mathbf{o}$ that are mapped on the origin too. All these points are members of the *null space* ker$\hat{T}$ of $\hat{T}$. Take any point like $\mathbf{a}$ that is *not* a member of the null space. It is mapped upon $\hat{T}\mathbf{a} \neq \mathbf{o}$. Because

$$\hat{T}\mathbf{a}' = \hat{T}(\mathbf{a} + \mathbf{k}) = \hat{T}\mathbf{a} + \mathbf{o} = \hat{T}\mathbf{a},$$

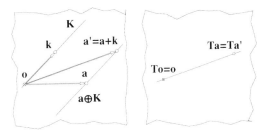

Figure 3.25 Let K denote the null space of an operator $\hat{T}$, and let $\mathbf{a}$ be a point outside of the null space. Then all points in the affine hyperplane $\mathbf{a} \oplus K$ have the same image as $\mathbf{a}$, and thus are "confused" by $\hat{T}$. The affine hyperplane $\mathbf{a} \oplus K$ has elements of the form $\mathbf{a} + \mathbf{k}$ for arbitrary elements $\mathbf{k} \in K$. Notice that range$\hat{T}$ is a single line, not the plane.

all points in the "affine line"

$$\mathbf{a} \oplus K,$$

(the $\oplus$ operator denotes the direct sum) are mapped on the *same point* $\hat{T}\mathbf{a}$. Thus the operator $\hat{T}$ does "confuse all points in the affine line $\mathbf{a} \oplus K$." In a sense, the action of the operator $\hat{T}$ is to map the *one-dimensional space* of affine lines $\mathbf{a} \oplus K$ on the *one dimensional space* range$\hat{T}$. You may say that the operator treats the domain dom $\hat{T}$ as "foliated" by the affine lines and regards only the affine lines as distinct elements. With such a perspective the two-dimensional domain appears as a one-dimensional space of folia. Another way to say this is that the affine lines are the equivalence classes under the action of the operator $\hat{T}$. This type of geometry will turn out to be the bread and butter of colorimetry.

3.2.10 Affine subspaces

An affine hyper(sub)space is *not* a linear space (for instance, it lacks a null element), but a linear subspace *translated* over a certain vector from the origin. Thus, affine hyperspaces are the direct sums of a (sub)space and a fixed vector. (Since a "fixed vector" is not a linear space, I use "direct sum" in a slightly restrictive sense here.) Such spaces become linear if you subtract the fixed vector from all of its elements. With some care, affine hyper(sub)spaces can be used as effectively as linear subspaces. You will encounter frequent occasions in colorimetry. (*Colors* are affine subspaces!)

3.2.11 Operators

In many important instances, you deal with transformations from one linear space to *another* linear space, even another space with a *different dimension*. This does not complicate matters very much. Transformations of a space into itself are called operators. This is just another name. It can be convenient, but it doesn't mean much.

Since operators can be complicated to deal with, people often try to break them down into manageable parts. One way to try to do this is to break down the space and study how the operator works on its subspaces. In order to explain this, you first need to understand the notion of direct sum.

3.2.12 The direct sum

In the theory of sets, you have the useful notions of set intersection and union that let you break down sets in parts or merge sets to larger sets. Something like that can be done in linear geometry, though the notions of set union and intersection tend to be less useful.

Consider two one-dimensional spaces, the first one with basis $\{\mathbf{e_1}\}$, the second with basis $\{\mathbf{e_2}\}$. The first space is a line in the $\mathbf{e_1}$ direction and the second a line in the $\mathbf{e_2}$ direction. Their union is the set of two intersecting lines, which is *not* a linear space, but an entity that we are unable to deal with using our toolset that applies to linear spaces. That is why set union is generally useless here.

A better way to deal is this is to insist on sets as *linear spaces*.

Thus I define the intersection of linear spaces $\mathbb{U}$, $\mathbb{V}$ as the largest linear space of elements that are also elements of both $\mathbb{U}$ and $\mathbb{V}$, and the "union" of $\mathbb{U}$, $\mathbb{V}$ as the smallest linear space of elements that are sums of elements in $\mathbb{U}$ and $\mathbb{V}$. This type of union is called the direct sum

$$\mathbb{U} \oplus \mathbb{V},$$

of $\mathbb{U}$ and $\mathbb{V}$. Thus any element $\mathbf{w}$ of $\mathbb{U} \oplus \mathbb{V}$ can be written $\mathbf{w} = \mathbf{u} + \mathbf{v}$ for certain $\mathbf{u} \in \mathbb{U}$ and certain $\mathbf{w} \in \mathbb{V}$. The direct sum is a linear space whose dimension is the sum of the dimensions of $\mathbb{U}$ and $\mathbb{V}$. In a similar way, I can write a linear space as the direct sum of a larger number of subspaces; in fact, a basis lets you express the space as the direct sum of the one-dimensional spaces spanned by the individual basis vectors.

3.2.13 Invariant subspaces of operators

An "invariant subspace" of an operator is a linear subspace that is mapped upon itself by the operator. If you can find such invariant subspaces, you express the space as a direct sum of these, with the result that the operator can be broken down into simpler components whose actions are confined to the invariant subspaces (figure 3.26). Ideally, the invariant subspaces would be the basis vectors of some special basis, such that the operator is broken down into operators on one dimensional subspaces, that is to say, simple scalings. If this is possible, the operator is said to be in its eigen representation (*Eigen* means "self" in German), the special basis is made up of the eigenvectors of the operator and the scaling factors are known as the "eigenvalues" of the operator. When possible, this is the preferred representation when dealing with the operator. Unfortunately, this is not always possible. For instance, consider a linear operator that looks like a rotation about the origin:

Figure 3.26 This transformation has two invariant subspaces. In one direction it expands by a factor 1.583... and in the other direction by a factor 0.41690....

It evidently has no invariant subspace of dimension one. It is possible to show that this is the worst case, though. The invariant subspaces are at most of dimension two. In colorimetric practice we will encounter *only* operators for which eigen representations exist.

3.2.14 Projections

A very important class of operators are the "projections." A projection $\hat{E}$ is an *idempotent* operator, that is to say,

$$\hat{E}^2 = \hat{E}.$$

Thus once you have applied a projection you are done, for repeated applications will never change anything. If $\hat{E}$ is a projection, then it is easy to show that

$$\hat{U} = \hat{I} - \hat{E},$$

is also a projection. For you have

$$
\begin{aligned}
\hat{U}^2 &= (\hat{I} - \hat{E})^2 \\
&= \hat{I}^2 - \hat{I}\hat{E} + \hat{E}\hat{I} - \hat{E}^2 \\
&= \hat{I} - \hat{E} + \hat{E} - \hat{E} = \hat{I} - \hat{E} \\
&= \hat{U}.
\end{aligned}
$$

Suppose that

$$|x\rangle = |y\rangle + |z\rangle,$$

such that

$$\hat{E}|y\rangle = |y\rangle \text{ and } \hat{E}|z\rangle = |0\rangle.$$

Then $\hat{U}|x\rangle = |z\rangle$, for

$$\hat{U}|x\rangle = (\hat{I} - \hat{E})|x\rangle = |y\rangle + |z\rangle - |y\rangle = |z\rangle.$$

Thus you have

$$|x\rangle = \hat{E}|x\rangle + \hat{U}|x\rangle.$$

The operators $\hat{E}$ and $\hat{U}$ split the space $\mathbb{X}$ into subspace spaces $\mathbb{Y}, \mathbb{Z}$ such that $\mathbb{X} = \mathbb{Y} \oplus \mathbb{Z}$. Here $\mathbb{X}$ is the space you're in. $\mathbb{Y}$ is the kernel of $\hat{U}$ and $\mathbb{Z}$ the kernel of the operator $\hat{E}$. The direct sum of the kernels is the space itself. Thus a projection $\hat{E}$ and its complementary operator $\hat{I} - \hat{E}$ generate a splitting of the space into the direct sum of their kernels. This is the type of structure that you will encounter over and over again in colorimetry (see p. 331). Notice that $\hat{E}$ projects along the $\mathbb{Z}$ direction(s) and operator $\hat{U}$ projects along the $\mathbb{Y}$ direction(s). These directions are completely general (figure 3.27); thus you should not confuse general projections with the (perhaps more familiar) orthogonal projections. Orthogonal projections can be defined only in spaces with a scalar product. You'll meet them later. For the moment, they don't even *exist*.

3.2.15 Duality

Notice that there is no reason to limit the notion of "linear operator" to kets, and you may apply the same notions to bras. What you *cannot* do is consider operators that accept kets and yield bras or vice versa, though, because that would imply the existence of a dot product (metric). You should make sure to distinguish clearly between operators operating on kets from those operating on bras. In order to make the distinction in the notation, I will sometimes (only if

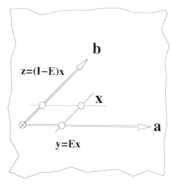

Figure 3.27 Here the plane of the paper is the direct sum $\mathbf{a} \oplus \mathbf{b}$. The projection $\hat{E}$ projects parallel to the $\mathbf{b}$ direction, whereas the complementary operator $\hat{I}-\hat{E}$ projects along the $\mathbf{a}$ direction. Notice that $\mathbf{b}$ is the kernel of the projection $\hat{E}$, whereas $\mathbf{a}$ is the kernel of the complementary projection $\hat{I}-\hat{E}$.

necessary) write $\hat{A}^\sharp$ for an operator that works on bras and $\hat{B}^\flat$ for an operator that works on kets. Thus you have

$$\langle\phi|\hat{A}^\sharp = \langle\psi|$$
$$\hat{B}^\flat|p\rangle = |q\rangle.$$

A product like $\langle\phi|\hat{B}^\flat$ is meaningless by itself. You cannot use it except (perhaps) in much larger expressions (of which it will not be a *proper* part). Notice that $\hat{A}^\sharp$ operates to the left, whereas $\hat{B}^\flat$ operates to the right. Thus

$$\langle\alpha|\hat{B}^\sharp|c\rangle,$$

is quite different from

$$\langle\alpha|\hat{B}^\flat|c\rangle.$$

There is simply no way to associate a unique left operator to a given right operator or vice versa. You need a scalar product (a metric) in order to be able to do that. (See chapter 8.)

3.2.16 Conclusion

This is about all the linear algebra needed in order to do colorimetric calculations in such a way that you may have an intuitive understanding of what the algebraic calculations *mean*. Here "meaning" is understood as "geometrical significance." The entities you encounter can invariably be visualized as vectors, dual vectors, linear spaces or subspaces, direct sums, linear operators such as scalings and invariant subspaces. The numbers you will meet are typically contractions of vectors on dual vectors or are matrix elements involving different bases or operators. In principle you can do all calculations through drawing, "more geometrico", though it will often be convenient to use numbers. But there is no need to become a mere symbol or number pusher. A clear geometrical understanding of what goes on is all-important. Perhaps unfortunately, even intelligent people routinely make stupid mistakes when they revert to mere symbol or number pushing, often out of sheer boredom. This is the human condition. Unfortunately, that is how the topic is usually taught, as matrix algebra. This is much to be regretted, since it would turn colorimetry into the dull and opaque discipline it is for so many people. There is no need for this at all.

Exercises

1. Grassmann's Laws   [FOR MATHEMATICIANS] Check whether Grassmann's laws indeed define colorimetry as being about linear algebra of pairs of beams. Could you come up with a more attractive set of axioms? How does this work out for purely physical beams (with nonnegative spectral density)?

2. Colorimetric reduction of beams   [EASY] Given a pair of beams, do (explicitly) a photometric reduction of them.

3. The spectrum cone   [REQUIRED] All real beams yield colors that lie in the convex spectrum cone. Plot it in three dimensions using the CIE 64 XYZ color matching functions loaded from the Internet.

4. Waves versus photons   [ELABORATE] Thus far your description has been in terms of radiant power on wavelength basis. Convert everything to photon number density on photon energy basis. You need to consider spectra of beams, spectral reflectance, color matching functions, and so on. It is always possible to convert some colorimetric quantity either way?

5. The photon number density   [FOR PHYSICISTS] Explicitly check my calculation regarding the number of photons being absorbed in your retinal receptor cells during normal daytime. Do you agree with my choice of the various parameter values? Does it matter?

6. Linear geometry   [HARD LABOR] In colorimetric calculations you work in very high dimensional linear spaces. It is a good exercise to try out the various relations discussed in the appendix. Use a random generator to generate high dimensional objects if need be. Check the various relations numerically. Make sure to handle problems with numerical tolerances intelligently. (For a start think of algorithms to decide the logical expression $a = b$ for arbitrary objects.) This is a good occasion to set up a small toolbox of algorithms for common problems. You'll need it later.

7. General primaries   [CONCEPTUAL] There is no reason why the primaries should be unit radiant power monochromatic beams, or general monochromatic beams, or general beams, they can even be *colors* (equivalence classes of beams). For each of these generalizations, go through computations and constructions once again. Is anything lost or gained?

Chapter Notes

1. The speed of light (in vacuum) is one of the fundamental constants of nature. Its value is $c = 299,792,458$ meters per second. This has been an *exact* value (not a measurement!) since October 21, 1983, when the fundamental SI unit distance (the meter) was defined in terms of the speed of light as the distance light travels in vacuum in $1/299,792,458$ of a second. Of course, these facts are of little importance to colorimetry. See J. D. Jackson, *Classical Electrodynamics*, (New York: Springer, 2004); M. Born, E. Wolf, *Principles of Optics*, (Cambridge: Cambridge U.P., 2000).

2. Of course, radiation that is "bad for you" apparently doesn't pass through your body *completely.*

3. A nanometer (nm) is 10^{-9} meters. Thus, wavelengths in the visual are about $5\,10^{-7}$ m. A human hair has a diameter of about 100 μm, or 10^{-4} m. Thus you have about 200 wavelengths in a hair's width. At reading distance you can resolve about one tenth of a millimeter (10^{-4}m, a hair's breadth). For most daily life purposes these wavelengths are so small that you may safely ignore them.

4. Here I really should have said "day*radiation*," because unseen radiation is not light. However, I will not be pedantic in this book.

5. This is, of course, the problem of "qualia."

6. Here I should really start a big show of writing RED (or some other typographic convention) when I mean the *quale* "red" (what it feels like to experience "red") instead of "red" (another typographic convention) when I mean "red" in the *colorimetric* sense (yet to be explained). I try to avoid this morass, but you have to help me out here and think for yourself. Remember that people are wont to confuse the issue. Although this may lead to confusion I consider it too pedantic to use this typographical convention consistently throughout the book.

7. It is not that there is anything particularly *wrong* with psychology (I have worked in both physics and psychology, and I enjoy either of them), it is simply that psychology is not among the "natural sciences." My friends in psychology will hate me for saying this, but it is simply true.

8. I will give the technical definition of luminance later.

9. In many fields, colorimetry actually already went the way of radiometry. Think of calibrating your printer, scanner, camera, monitor, and so forth. But in many respects colorimetry cannot yet be done without human observers, although the industry tries hard. Eventually much of colorimetry will go the way of radiometry, albeit with a great many essentially arbitrary conventions that are "good enough for government work" but rather suspect and in particularly bad taste from a scientist's (rather than engineer's) perspective.

10. H. Grassmann, "Zur Theorie der Farbenmischung," *Poggendorffs Ann.Phys.* 89, pp. 69–84 (1853). See also H. Frieser, "Die Graßmannschen Gesetze," *Die Farbe* 2, pp. 91–108 (1953).

11. A particularly clean formalization is due to Kranz. See D. H. Kranz, "Color measurement and color theory. I. Representation theorem for Grassmann structures," *J.Math.Psychol.* 12, pp. 283–303 (1975).

12. P. J. Bouma, *Physical Aspects of Colour*, (Eindhoven: Philips Industries, 1947).

13. The photosensitive cells in the retina that appear to be responsible for photopic color vision are the cones. There exist three distinct types whose action spectra peak at different spectral locations.

14. This is the problem of *qualia* again. Colorimetry has *nothing* to do with that. The colorimetric "colors" have no "colors" (*quale*, I mean). This is apparently too much for many people to stomach.

15. In fact, nothing that *calls* itself a science (Christian Science, and so forth) actually *is*. "Color science" is no exception.

16. Such "nil squared numbers" are a special case of the complex numbers originally studied by Cayley. A good introduction is that by I. M. Jaglom, *Complex Numbers and Their Application*, (Moscow: Fizmatgiz, 1963).

17. Of course, quantum theory has changed this. Moreover, don't think that this account is without its apparent paradoxes. For instance, it is not consistent with classical logic; you cannot retain the law of the excluded third. Consider the nontrivial solutions of the equation $x^2 = 0$ (thus $x \neq 0$), then the dichotomy "either $x > 0$ or $x < 0$" does not apply. But believe me, it is worth it to join "intuitionistic logic." See J. L. Bell, *A primer of Infinitesimal Analysis*, (Cambridge: Cambridge U.P., 1998).

18. A simple, but very precise method is to use the "episcopister." This is a rotating disc with a variable open sector. The rotation speed should be higher than anything the visual system can resolve. Then you can multiply the beam with a factor set by the (angular) width of opening of the sector.

19. Convex sets are very similar to linear spaces in many respects. They turn out to be the correct generalization of linear spaces for the purposes of colorimetry. Most of the spaces you will encounter are actually convex subsets of linear spaces. Although you have to use the tools of linear algebra with care (they might easily get you outside your fiducial convex set), you gain a number of very powerful additional tools that are often of considerable practical utility. Examples include linear programming and iterative projection on convex sets.

20. This is the so called Rayleigh match (L. Rayleigh, "Experiments on colour," *Nature* 25, pp. 64–66 (1881)). The observer sees a $2°$ split field with 590nm (yellow) on one side and a mixture of 679nm (red) and 545nm (green) on the other side. By changing the red/green ratio and the luminance of the mixture, a perfect match can be obtained.

21. T. Holtsmark, "Newton's *Experimentum Crucis* reconsidered," *Am.J.Phys.* 38, pp. 1229–1235 (1970); —, "Das Experimentum Crucis und die Theorie der Dispersion," *Opt. Act.* 18, pp. 867–873 (1971); —, and A. Valberg, "On complementary color transitions due to dispersion," *Am.J.Phys.* 39, pp. 201–204 (1971).

22. If x and y span the ground-plane, then the "steps in the z direction" will no doubt require you to use a ladder or staircase.

23. The formal theory is that of the "tempered distributions" of Schwarz. See L. M. J. Florack, *Image Structure*, (Dordrecht: Kluwer, 1997).

24. The funny notation "$|\bullet\rangle$" is Dirac notation (see P. A. M. Dirac, *The Principles of Quantum Mechanics*, (Oxford: Clarendon Press, 1930)), which I will explain shortly.

25. Although the wavelength is a continuous parameter, in real life you typically *sample* it (at 5 nm intervals for instance). Thus you would indeed perform a *summation* rather than an *integration*. The Dirac notation is usually clearer when you write summations throughout. In case an index happens to be continuous, you can either integrate or use a discrete approximation such as a Riemann sum. It really should make no difference whatsoever. Just watch your step.

26. Notice that I have reversed the sequence of $\langle\lambda|s\rangle$ and $|\lambda\rangle$. You may wonder whether the product is commutative? Indeed it is, for $\langle\lambda|s\rangle$ is a scalar quantitity, and $|\lambda\rangle$ a vector quantity. Scalars and vectors commute. It is important that you are aware of this since such transpositions of order are frequently necessary in almost any calculation.

27. This is by no means necessary; you can build physical detectors that let you record photon energy. But the fact is that photoreceptors in the eye only record events.

28. See B. Sakitt, "Counting every quantum," *J.Physiol.* 223, pp. 131–150 (1971).

29. A mole of photons means *a lot* of photons. Remember that Avogadro's number is approximately equal to 6.0221499×10^{23} and that "femto" stands for 10^{-15}.

Part II

Basic Colorimetry

Chapter 4

Basic Colorimetry

Basic Colorimetry is the colorimetry of beams, rather than illuminated surfaces (things), in the absence of a notion of illuminant, achromatic beam, or white, and in the absence of a metric in the space of beams. These various concepts will be discussed in later chapters. Here I discuss the absolute *rock bottom structure*. I will occasionally refer to it as "austere colorimetry." Indeed, for most applications you will want to augment it in various ways.

4.1 The Colorimetric Paradigm

Select a triple of "primaries"

$$\{|p_i\rangle\} \quad (p = 1, 2, 3).$$

Primaries are simply *fiducial beams of radiation*. Essentially any choice will do, as long the beams are independent. If you pick a random triplet of beams, chances are ("with probability one," as the cool kids say) that you have such an okay set.[1] When Maxwell started his colorimetric work, his primaries were simply colored papers that were mixed with his color top (figure 4.1). In the figure, the disk is set to mix gray from red, green, and blue primaries. Maxwell probably didn't know the spectra associated with these primaries, and he didn't need that knowledge either, a fact of which he was aware.

Given the primaries, any given beam $|a\rangle$ can (for colorimetric purposes) be represented as

$$\mathbf{a} \,\square\, \left(\sum_i a_i \mathbf{p_i} \right).$$

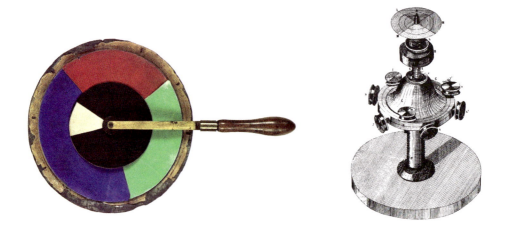

Figure 4.1 Maxwell's "color top." When the disc is spun very fast, you see an inner circle with an outer annulus. In this case the inner circle will be dark gray (a mixture of a small segment of white and a large segment of black paper). The annulus is a mixture of the primaries, which are red, green, and blue papers. Maxwell ingeniously arranged things such that the sector angles could easily be adjusted. As you can see at right, Maxwell went to great lengths to make his color top a technical marvel.

The representation $\sum_i a_i |p_i\rangle$ is (almost always) physically distinct from $\mathbf{a}$ though it *looks the same*; thus,

$$\mathbf{a} \, \Box \, \left(\sum_i a_i \mathbf{p_i} \right) \not\approx \mathbf{a} = \sum_i a_i \mathbf{p_i}.$$

This is exactly why colorimetry is of interest:

> *Physically different beams may look exactly the same.*

It is as if a beam contained some "invisible fluff" such that beams that *look* the same would actually *be* the same when stripped of their invisible fluff. This is evidently an attractive idea. It is known as Wyscecki's hypothesis.[2] Wyscecki's hypothesis is unfortunately *wrong* as it stands. However, with a little work the notion may be given some sense. I'll consider it later, in chapter 8.

The coefficients a_i are often called "tristimulus values." In the formalism introduced here, they are evidently dimensionless quantities.[3]

It is convenient to call the sets of all beams that look the same *colors*. Thus, colorimetric colors are defined as the equivalence classes induced by the equivalence relation $\bullet \, \Box \, \bullet$. This definition will take getting used to. For instance, a colorimetric color has no color in the conventional sense, that is to say, it cannot be said to be RED, GREEN, BLUE, and so forth.

Notice that though physical identity implies colorimetric equivalence, the inverse statement is not true; thus,

$$\mathbf{a} = \mathbf{b} \quad \rightarrow \quad \mathbf{a} \,\square\, \mathbf{b}$$
$$\mathbf{a} \,\square\, \mathbf{b} \quad \nrightarrow \quad \mathbf{a} = \mathbf{b}.$$

The reason is, of course, that colors are large subsets of the set of beams, rather than specific beams.

4.1.1 Black beams

Suppose you find another beam $\mathbf{b}$ that looks the same as $\mathbf{a}$, that is to say, $\mathbf{a} \,\square\, \mathbf{b}$, or $a_i = b_i$ for $i = 1, 2, 3$. Then the colorimetric coordinates of $\mathbf{a} - \mathbf{b}$ must be zero; thus $\mathbf{a} - \mathbf{b}$ looks the same as the empty beam. I will call it "black."[4]

$$\mathbf{a} \,\square\, \mathbf{b} \rightarrow (\mathbf{a} - \mathbf{b}) \,\square\, \mathbf{0} \rightarrow \mathbf{a} - \mathbf{b} \in \mathbb{K} \subset \mathbb{S}.$$

Of course, "black" is merely a catchy term that sticks easily to mind. Such black beams are not the *same* as the empty beam, though they certainly *look* the same (literally *like nothing* that is). I denote the space of all black beams the black space $\mathbb{K}$. It is obviously a linear vector space, that is, a subspace of the space of beams $\mathbb{S}$. For generic (trichromatic) observers it has codimension three. Thus, the black space is *very large*. That human vision has an infinitely dimensional black space is one tragedy of the human condition that cannot be remedied (e.g., with special spectacles[5]). All black beams with nonzero spectral radiant power in the visual range are virtual beams.

The triplet of primaries is in no way unique, for you may add arbitrary black beams to the primaries and you will still obtain the same colorimetric coordinates. Instead of specific beams you might equally well select (large) equivalent classes of identically looking beams for your primaries, that is to say *colors*.

Since beams of the same color differ by elements of the black space, colors are affine hypersubspaces in the space of beams, all parallel to the black space.

4.2 The Color Matching Functions

The basic geometry to keep in mind is shown in figure 4.2. Be certain to "mentally correct" for the low dimensions of the figure. This might indeed be confusing, but there is nothing I can do about it. There is simply no way to draw an intuitive and yet "correct" figure

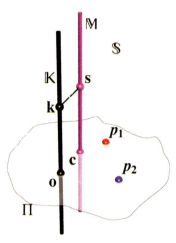

Figure 4.2 This is the basic geometry of colorimetry, scaled down to "intuitive" dimensions. The ∞ dimensional space of beams $\mathbb{S}$ factors into the direct sum of the black space $\mathbb{K}$ and one of its three-dimensional complements Π, the space spanned by the primaries $\mathbf{p}_i$. The beam $\mathbf{s}$ has representative $\mathbf{c}$ in Π (thus $\mathbf{s}\,\square\,\mathbf{c}$). The "color" is the metamer $\mathbb{M}$, which is the black space shifted by $\mathbf{s}$. Notice that $\mathbf{s} = \mathbf{c} + \mathbf{k}$, where $\mathbf{k}$ is a black beam that is fully invisible (it looks like the empty beam, that is, like nothing at all).

with the right dimensions; unfortunately, the paper of this book is only two-dimensional! The subspace $\Pi \subset \mathbb{S}$ spanned by the primaries is a complement of the black space $\mathbb{K}$, thus $\mathbb{S} = \mathbb{K} \oplus \Pi$. Of course, Π is quite arbitrary because the primaries are simply drawn out of a hat. The only requirement that they have to meet is to be independent. Any spectrum $\mathbf{s}$ is thus composed of a "black component" $\mathbf{k} \in \mathbb{K}$ and some linear combination of the primaries $\mathbf{c} \in \Pi$. The affine hyperspace $\mathbb{K} \oplus \mathbf{c}$ (same as $\mathbb{K} \oplus \mathbf{s}$) is the "metamer" $\mathbb{M}$, of which $\mathbf{s}$ is a member. This is the very backbone structure of colorimetry.

Since the colorimetric coordinates depend linearly on the beams, there has to be a triplet of linear functionals

$$\{\pi_i\}\ (i = 1, 2, 3) \text{ such that } \mathbf{a} = \sum_{i=1}^{3} a_i \mathbf{p_i} \text{ with } a_i = \langle \pi_i | a \rangle.$$

Of course, this triplet of linear functionals will depend upon your choice of primaries $\{\mathbf{p_i}\}$. You evidently have that

$$\mathbf{a}\,\square\, \left(\sum_{\mathbf{i}} |\mathbf{p_i}\rangle\langle\pi_\mathbf{i}| \right) |\mathbf{a}\rangle = \hat{\mathrm{P}}\mathbf{a},$$

introducing the operator $\hat{P}$ which is a projection in $\mathbb{S}$ because evidently $\hat{P}^2 = \hat{P}$ (prove it!). This colorimetric equivalence may also be written as an *equality* for beams, namely

$$\mathbf{a} + \mathbf{k} = \hat{P}\mathbf{a}, \text{ for some } \mathbf{k} \in \mathbb{K}.$$

The linear functionals satisfy the following (obvious) defining relations:

$$\langle \pi_i | \mathbb{K} \rangle = 0,$$
$$\langle \pi_i | p_j \rangle = \delta_{ij}.$$

Thus $\hat{P}$ is the identity when confined to the subspace spanned by the primaries.

Now suppose you select another triplet of primaries, say $|q_i\rangle$, which induce linear functionals $\langle \vartheta_i |$ (say). You have

$$\langle \vartheta_i | \mathbb{K} \rangle = 0,$$
$$\langle \vartheta_i | q_j \rangle = \delta_{ij},$$

and again $\mathbf{a} + \mathbf{k} = \hat{Q}\mathbf{a}$ for some $\mathbf{k}$ in $\mathbb{K}$. Here the operator

$$\hat{Q} = \sum_i |q_i\rangle\langle \vartheta_i|,$$

is again a projection in $\mathbb{S}$, though different from $\hat{P}$.

You easily check that $\hat{P}\hat{Q} = \hat{P}$ and $\hat{Q}\hat{P} = \hat{Q}$. The two projections project in the same direction, but on different subspaces. The operator $\hat{P}$ projects on the subspace spanned by the primaries $\{\mathbf{p_i}\}$, whereas the operator $\hat{Q}$ projects on the subspace spanned by $\{\mathbf{q_i}\}$, and these subspaces have little in common.

The two subspaces are in a one-to-one correspondence, because either projection is an isomorphism from one subspace to the other.

The operators $\hat{I} - \hat{P}$ and $\hat{I} - \hat{Q}$ apparently project on the black space. Thus any set of primaries yields a projector on the black space.

The difference $\hat{P} - \hat{Q}$ (which equals the commutator $[\hat{P}, \hat{Q}] = \hat{P}\hat{Q} - \hat{Q}\hat{P}$, prove it!) has to be a projector on the black space too, because

$$(\hat{P} - \hat{Q})\mathbf{a} = (\mathbf{a} + \mathbf{k}) - (\mathbf{a} + \mathbf{k}') = \mathbf{k} - \mathbf{k}' = \mathbf{k}''.$$

It projects on a subspace of the black space. This yields a method to construct black beams of specific types, which often comes in handy in colorimetric problems.

In order to keep the relevant geometry firmly in mind you may find it convenient to use figure 4.3 as an aid.

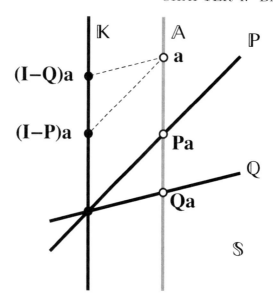

Figure 4.3 This drawing is only conceptual. Thus the $(\infty - 3)$-dimensional black space $\mathbb{K}$ is represented by the vertical line through the origin of the space of beams $\mathbb{S}$. The three-dimensional subspaces spanned by the primaries $\{\mathbf{p_i}\}$ and $\{\mathbf{q_j}\}$ are represented through the lines $\mathbb{P}$ and $\mathbb{Q}$ respectively. The arbitrary beam $\mathbf{a}$ projects on $\hat{P}\mathbf{a}$ and $\hat{Q}\mathbf{a}$. The affine hyperspace $\mathbb{A}$ through $\mathbf{a}$ parallel to $\mathbb{K}$ is the metamer of $\mathbf{a}$. The operator $\hat{P}$ also projects $\hat{Q}\mathbf{a}$ on $\hat{P}\mathbf{a}$, and the operator $\hat{Q}$ also projects $\hat{P}\mathbf{a}$ on $\hat{Q}\mathbf{a}$. The projections $(\hat{I} - \hat{P})\mathbf{a}$ and $(\hat{I} - \hat{Q})\mathbf{a}$ are the "black components" of $\mathbf{a}$. They are distinct; thus Wyscecki's hypothesis is apparently false. It helps to keep this drawing in mind, since it makes most relations immediately geometrically obvious.

4.3   Color Space

Since any beam can be assigned a triplet of colorimetric coordinates, people say that "human color space $\mathbb{C}$ is three-dimensional."[6] To make this more explicit, I pick a three-dimensional linear space with basis

$$\{\mathbf{e_1}, \mathbf{e_2}, \mathbf{e_3}\}.$$

Any old space and basis will do. The reason is that I will use this space merely as a canvas on which to draw (colorimetric) colors. If the colorimetric coordinates of a beam $\mathbf{a}$ are $\{a_1, a_2, a_3\}$, I will associate the point

$$a_1\mathbf{e_1} + a_2\mathbf{e_2} + a_3\mathbf{e_3},$$

with the beam.

Thus, for the primaries $\{\mathbf{p_i}\}$, I consider the linear map $\hat{A} : \mathbb{S} \mapsto \mathbb{C}$

$$\hat{A} = \mathbf{e_1}\langle\pi_1| + \mathbf{e_2}\langle\pi_2| + \mathbf{e_3}\langle\pi_3|.$$

This map then replaces the operator $\hat{P}$ constructed earlier. Notice that these maps are quite distinct, because $\hat{P} : \mathbb{S} \mapsto \mathbb{P}^3 \subset \mathbb{S}$, where $\mathbb{P}^3$ denotes the three-dimensional subspace spanned by the primaries $\{\mathbf{p_i}\}$. The transformations are connected by a linear map (an isomorphism) from $\mathbb{P}^3$ to $\mathbb{C}$. Of course this introduces an unnecessary and arbitrary indirection. However, this is the conventional setup.

When you change primaries, the "meaning" of the $\{\mathbf{e_i}\}$-space changes: Now the $\mathbf{e_i}$ represent the $\mathbf{q_i}$ (say) instead of the $\mathbf{p_i}$. Apparently the isometry between the spaces $\mathbb{P}^3$ and $\mathbb{Q}^3$ constructed earlier may double as the transformation needed to change the "meaning" of the $\{\mathbf{e_i}\}$-space. The unnecessary indirectness is certainly not very elegant.

Despite the lack of elegance, the concept of "color space $\mathbb{C}$" has its virtues. Each point of color space $\mathbb{C}$ represents an infinitely dimensional hypersubspace (e.g., $\mathbf{a} \oplus \mathbb{K}$) of the space of beams $\mathbb{S}$. This certainly aids the intuition when you deal with sets of colorimetric colors. Thus, the use of color space is essentially just this: *It allows you to think—and plot graphics—in 3D.*

The map $\hat{A}$ is what is conventionally known as the "color matching matrix". In terms of the spectral basis, you may load its coefficients from the Internet as the CIE-XYZ table. Of course, the color matching matrix depends upon the primaries and you will have frequent occasion to switch between representations. Fortunately this is easy enough. Let $\mathbf{c} = (\sum_i \mathbf{e_i}\langle\pi_i|)|s\rangle$ and $\mathbf{c}' = (\sum_j \mathbf{e_j}\langle\vartheta_j|)|s\rangle$ for any spectrum $|s\rangle$. There evidently has to be some isometry $\mathbf{c} = \hat{M}\mathbf{c}'$. Let's find its representation. You have $\sum_i \mathbf{e_i}\langle\pi_i| = \hat{M}(\sum_j \mathbf{e_j}\langle\vartheta_j|)$. Multiplying on the right with $|q_k\rangle$, you obtain $\sum_i \mathbf{e_i}\langle\pi_i|q_k\rangle = \hat{M}\mathbf{e_k}$. Finally, multiplying on the left with ε_l, you find $\langle\pi_l|q_k\rangle = \varepsilon_l\hat{M}\mathbf{e_k}$. Thus the matrix elements of $\hat{M}$ are $\langle\pi_l|q_k\rangle$, that are the colors of the q-primaries in terms of the p-primaries.

Because the choice of canvas is up to you, it makes no sense to pick anything but a thoroughly *nice* space (figures 4.4 and 4.5-left). Typically, I will use $\mathbb{R}^3$ for $\mathbb{C}$. A Cartesian orthonormal basis is definitely ideal to draw your picture of color space in. When you change primaries, this drawing will suffer a linear deformation (figure 4.5-right). Since the choice of primaries is essentially arbitrary, this means that the geometrical features to look for in color space should be invariant with respect to arbitrary linear deformations. Some writers on colorimetry draw the conclusion from this undeniable fact that it is best to choose as awkward a basis of color space as possible, just to remind you of the brute fact that color space is an affine space. To my mind, this is an uncalled-for form of masochism. Why use anything else but the canvas you like? Only don't become *attached* to it.

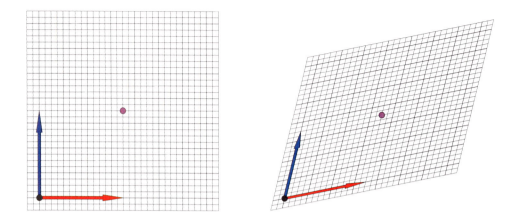

Figure 4.4 Left: A nice Cartesian cell in the plane. Standard graph paper is based on this ingenious idea; Right: An oblique cell in the plane. Although fully equivalent to the Cartesian choice, who in his right mind would pick on oblique axes unless there were a pressing reason? Yet authors of books on colorimetry often feel compelled to do so.

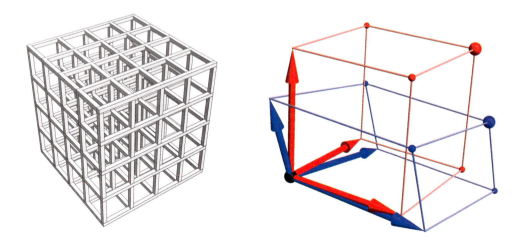

Figure 4.5 Left: This is my ideal place in 3D space, a "Cartesian jungle gym" $\mathbb{R}^3$. This is the preferred canvas upon which to draw color space. Right: A change of primaries. The color with coordinates $\{1, 1, 1\}$ is different in the red and the blue system. Of course the coordinates will not be $\{1, 1, 1\}$ when expressed in the other system. The vectors are the actual objects, the coordinates only their labels or names, which may change between languages.

Notice that the (conventional) terminology "color space is an affine space" is unfortunate, for $\mathbb{C}$ certainly has a natural origin.[7] No matter how you pick your primaries, the black beams will map on $\{0, 0, 0\}$. Notice that the primaries map on the basis vectors $\{\mathbf{e_1}, \mathbf{e_2}, \mathbf{e_3}\}$ of color space.

4.4 Colorimetry in the Spectral Basis

You can do quite a bit of interesting colorimetry without ever using a spectral representation at all. In fact, the spectral representation initially doesn't seem to be much of a help. It can be very useful though, as Maxwell was the first to point out. He invented the procedure of "gauging the spectrum," a topic I will consider next.

The basic idea is very simple. Colorimetric measurements are very time-consuming and thus a pain in the neck. Can you do with fewer measurements, or can you do all your measurements once and be done with? Yes, you can, because the colors depend linearly upon the beams. Thus, if you known the colors for any basis in the space of beams, then you need not do any measurements for a beam whose representation in that basis is known to you. The spectral basis is an instance of this. Suppose you have measured the colors of all (still infinitely many) the monochromatic beams of unit radiant power; then you can simply *calculate* the color of any beam from its spectrum. No need for any additional measurement. Finding the colors of all monochromatic beams was first done by Maxwell and is known as gauging the spectrum.

Although there is no need to use monochromatic primaries, it is convenient to do so, if for no other reason, than that such beams are easily specified and replicated from scratch in any laboratory. Thus, I will assume that the primaries are monochromatic beams of unit radiant power

$$\{|\lambda_1\rangle, |\lambda_2\rangle, |\lambda_3\rangle\} \quad \lambda_1 \neq \lambda_2 \neq \lambda_3.$$

Gauging the spectrum then yields the so called "color matching functions" $\langle\lambda_i|$ such that

$$|\lambda\rangle \,\square\, \left(\sum_{\mathbf{i}} \langle\lambda_\mathbf{i}|\lambda\rangle |\lambda_\mathbf{i}\rangle\right), \quad i = 1, 2, 3.$$

The triplet of color matching functions

$$\{\langle\lambda_1|, \langle\lambda_2|, \langle\lambda_3|\},$$

expressed in terms of the monochromatic basis, is conventionally known as the color matching matrix. The color matching functions are apparently the spectral representation (coordinate representation) of the dual spectral basis in terms of the spectral basis. However, in conventional colorimetric theory, people have no geometrical interpretation to speak of, and so they bluntly use matrixes as tables of numbers, and if someone mentions "vector" it only means a single column of numbers.[8] It will soon become evident how powerful the geometrical interpretation, is though.

The color matching matrix is convenient in colorimetric calculations. Notice, however, that it depends upon the choice of basis in the space of beams (the spectral basis) and also on the choice of primaries (the three wavelengths λ_1, λ_2, and λ_3). Thus, the color matching matrix in no way counts as an invariant descriptor of human color vision. You can find the color matching matrix for the generic human observer tabulated in many books. Nowadays you can also conveniently load it from the Internet.[9] Arguably the most useful to select are the "Stiles and Burch (1959) 10-deg, RGB Color Matching Functions." The primaries are

$$
\begin{aligned}
\lambda_1 &= 645.16 \text{ nm,} \\
\lambda_2 &= 526.32 \text{ nm,} \\
\lambda_3 &= 444.44 \text{ nm,}
\end{aligned}
$$

(Why these "magical numbers"? It is actually a shrewd choice, but if you want to understand you have to wait.)

The color matching functions are tabulated from 390 nm to 830 nm in 5 nm intervals in five significant digits (for the "sampling rate" see figures 4.6 through 4.8). When you load these data in a program like Mathematica, you are all set to start your own colorimetric calculations. This is exactly how I started out myself in order to prepare the illustrations used in this book.

4.5   The Spectrum Locus

Perhaps the first thing to try if you're the curious type is to plot the color matching matrix as a *curve*, the so-called spectrum locus, in color space (figure 4.9). Of course, you are not really plotting the color matching matrix, you have implicitly assumed a beam with a "uniform" (or "flat") spectrum here, and you plot the contributions of its spectral components to the color. You may just as well plot the spectrum locus for beams that do not have such a "flat" spectrum, and you would (of course) obtain a different result. I will consider such cases later. At this point of the discussion, a flat spectrum is simply a convenience.

This curve (for the "flat" spectrum) is the locus of the colors of monochromatic beams of unit radiant power (the "spectral basis" of $\mathbb{S}$), and thus indeed quite literally the spectrum locus. It is a curved and twisted space curve that issues forth from the origin into one direction and after frolicking around in color space returns to the origin from another

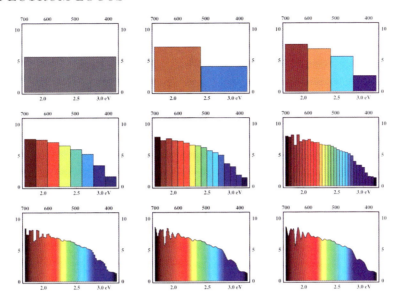

Figure 4.6 The spectrum of "average daylight" sampled at various resolutions. In practice any number of samples in excess of a dozen or so is fine for colorimetric purposes. With modern computers, working with hundreds of samples is no problem either, so I often find myself working in a hundred-dimensional space of beams. This is more realistic than "infinitely dimensional Hilbert space" and actually avoids a number of hairy problems. Your laptop is easily up to it.

direction. This general behavior is clear enough, for if the wavelength gets too small you enter the *ultraviolet* region, and when the wavelength gets too large you enter the *infrared*. Since both ultraviolet and infrared are not visible to the human observer, their colors are BLACK. Thus the curve must issue forth from the origin and eventually return to the origin again, and so it does.

4.5.1   The plane of purples

It is easy enough to find the limiting tangents to the spectrum locus near the origin from the numerical data. I will call these the "infrared limit" and the "ultraviolet limit." Although you cannot easily set wavelength limits to the IR (infrared) and UV (ultraviolet) regions, the limiting tangents are reasonably well defined.[10] Notice that the two tangent together span a plane, conventionally known as the "plane of purples" (figure 4.9 right). It is called plane of purples because it is made up of colors that are mixtures from monochromatic beams of very short and very long wavelengths. The former tend to look "BLUE," the latter

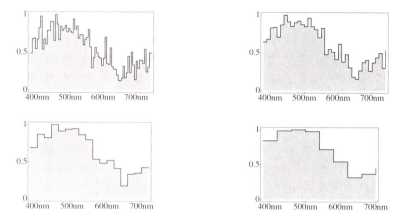

Figure 4.7 A "noisy" spectrum sampled at various bin sizes. The bin-width increases from left to right, top to bottom. With a narrow bin (top left) the spectrum looks rather noisy, while with a fairly wide bin (bottom right) the overall structure of the spectrum is easier to grasp. With respect to color the differences are very minor. You may just as well ignore busy wriggles in spectra.

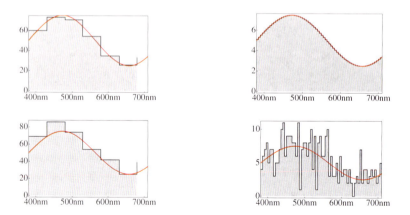

Figure 4.8 Upper row: Steadily narrowing the bin-width might be assumed to yield a progressively better approximation (in black) to the "real" spectrum (in red); Lower row: Actually, decreasing the bin-width also decreases the number of photon absorptions. The resulting increase in noisiness of the spectral density estimate destroys your chance to get an arbitrarily precise glimpse of the real thing. Only measured spectra exist; the so-called real thing is a non-entity.

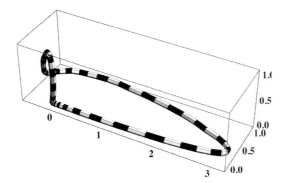

Figure 4.9 Left: The spectrum locus, straight from the shelf. The white/black stretches reveal the 5 nm intervals of the table. Right: The spectrum locus scaled so as to make its shape more apparent, with the plane of purples. Here I show only the sector of the plane of purples between the UV and IR limiting tangents.

"RED," and their mixture "PURPLE" or "MAGENTA". Of course, these are merely colorful terms, and you should not get the impression that colorimetry has anything to do with the quality of your luminous experiences.

Empirically (that is, by plotting the numerical data) you discover that the spectrum locus never *crosses* the plane of purples, it is confined to only *one side* of this plane. The spectrum locus lies almost in the plane of purples when it starts out in the direction of the UV limiting tangent, moves away from the plane, then return and runs again almost in the plane of purples at it progresses toward the origin along the UV limiting tangent. At some place it is apparently as far away from the plane of purples as can be. Since you are not in a metric space, it is not immediately obvious how to find the location of greatest remove from the plane of purples, though. Indeed, if there exists such a point at all, for if there is it should be invariant against the arbitrary affine deformations brought about by different choices of primaries.

4.5.2 The far green point

The simplest geometrical entity (after location) that you may derive from a curve is its local direction, or tangent. From the numerical data you simply take the difference of adjacent color coordinates to obtain an approximation of the tangents. It is of some interest to study the relation of the tangents along the curve to the plane of purples. Intuitively, there is at least one point on the spectrum locus where the tangent to the curve is *parallel* to the plane of purples. (You can easily find it by plotting the value of the volume of the crate subtended by the tangent and the two limiting tangents, for where it vanishes the tangent must be linearly dependent upon the limiting tangents, thus parallel [figure 4.10].) This is an important geometrical condition, because invariant with respect to affine deformations, hence independent of the choice of primaries. It is an invariant property of human vision. The parallelity occurs for a (unique) wavelength of 540.16 nm (roughly 540 nm). At this wavelength the spectrum locus is intuitively as far from the plane of purples as it ever gets.

Of course, you should remember that you implicitly assumed a beam with a uniform spectrum. The location of the far green point will depend on this. You can compute far green points for other spectral distributions, say average daylight, or the beam from a light bulb, and you are bound to find different results. The far green point for the uniform spectrum is evidently of some interest, though.

The wavelength of 540 nm may be called the location of extreme green, where "extreme" applies to the separation from the plane of purples, and "green" is merely a descriptive term. "Green" is experienced as the opposite of "purple." I have not seen the extreme green derived in the way I did here, although this wavelength often pops up as significant in colorimetric researches. Small wonder, it is perhaps the simplest example of an interesting affine invariant. Obviously, *all* geometrical properties of the spectrum locus that are invariant against the choice of primaries are of potential interest. You can find them though an analysis of the affine differential geometry of the curve.[11] I have never seen this done, though it is not particularly difficult to do so. I will not pursue the topic here because it would take me to far from the realm of conventional colorimetry. You may take this to be a worthwhile scientific endeavor.

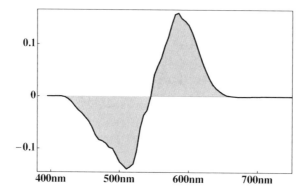

Figure 4.10 The oriented volume of the crate (arbitrary units) spanned by the tangents to the spectrum locus and the UV and IR limiting tangents. The zero crossing that defines the location of the extreme green point (the crate squashed flat) is clearly evident.

4.6 The Spectrum Cone

Since any real beam is necessarily a convex combination (that is, a linear combination with nonnegative coefficients) of monochromatic beams, the set of all real beams is a convex cone with apex at the black point and the spectrum colors as generators (figure 4.11). The sector of purples closes the gap in the spectral part of the cone mantel. That the spectral part of the boundary of the spectrum cone does not exhaust the boundary was not appreciated by Newton and led to an endless succession of misunderstandings throughout the nineteenth century. Eventually, it was Helmholtz who showed experimentally that the spectrum colors fail to exhaust the boundary of the spectrum cone, but the influence of Newton was so great that even Grassmann preferred to believe Newton, rather than Helmholtz's experiments.[12] But it is a *fact* as you can easily verify by plotting the numerical data. The colors in the sector of purples are not spectral colors, but mixtures of the extremities of the spectrum.

4.7 Chromaticities

One move that is invariably made in textbooks on colorimetry is to construct "*the* chromaticity diagram." This is somewhat of a misnomer, because you can construct such

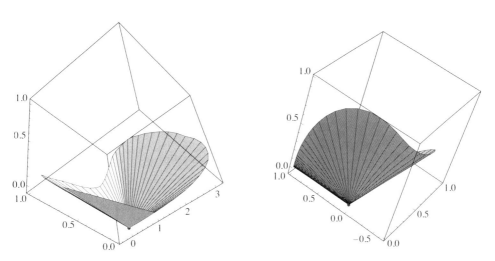

Figure 4.11 The spectrum cone viewed from two diametrically opposite directions. Remember that the plot refers to a uniform beam. The directions of the generators don't depend on that, but their lengths do.

diagrams in infinitely many ways. However, it has become customary to use a certain convention forcefully pushed by the CIE (Commission International d'Éclairage[13]) and most often referred to as "the" chromaticity diagram. Indeed, many people are not even aware that there exist (infinitely many) alternative choices. Usually the chromaticity diagram is introduced in a geometrically opaque manner. I am not particularly fond of chromaticity diagrams, because they are quite misleading to those who do not understand their structure and invite common errors. I know this only too well, having taught colorimetry to classes of physicists. However, because of their frequent mention, I will cover the topic here.

Chromaticity diagrams are popular because of the simple fact that they are *two*-dimensional, whereas color space is *three*-dimensional. It is their only claim to fame. It is much easier to draw in two dimensions than it is in three, and many people have trouble interpreting diagrams in three dimensions to begin with. The dimension that is most easily dropped is the intensity. The idea is that the COLOR of a beam does not change when you change the radiant power but not the spectral distribution of the beam. Thus, you may as well omit the intensity entirely. The problem is of course that there is simply *no way* in which this (that "the COLOR is independent of the intensity," I mean) can be understood in a formal or operational *colorimetric* manner. You cannot even check whether it is true, at least not within colorimetry. *Thus it isn't even false!* Apparently it is convenient to omit mention of this tricky point and to enter a mysterious region somewhere between colorimetry and psychology (whatever that may mean). Many authors even print their chromaticity diagrams in gaudy colors, which is probably the most certain way to confuse their readership as to the true nature of colorimetry. Nobody complains.

Here is the *geometrical way* of constructing chromaticity diagrams. The lines through the origin of color space are loci of equivalence classes of beams, such that all scalar multiples are in the locus. You may consider color space modulo such scalar multiples. You then obtain a space whose elements are the lines through the origin of color space. Such spaces have a familiar structure; they are *projective planes*. You can draw them on paper such that a point on your page stands for a line through the origin of color space. Any *point* of color space, except for the origin, is part of a line, thus can be mapped to a unique point on your paper. Thus you obtain a chromaticity diagram of color space. The only point that cannot be drawn in the chromaticity diagram is the origin, because the origin does not belong to a unique line. Thus the chromaticity diagram does not contain the "black point". Indeed, the chromaticity diagram does not have a natural origin at all, and thus *fails to be a linear space*. Projective spaces have a quite different structure than linear spaces. A simple and convenient way to obtain the mapping is to pick an affine plane in color space (not containing the origin) and to map any point (except the origin) of color space on the point of intersection of the plane and the line defined by the point and the origin (figure 4.12). The plane you use is conventionally called the chromaticity plane.

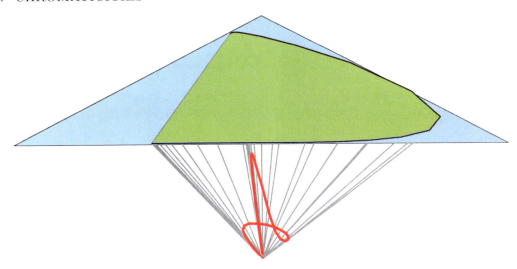

Figure 4.12 The chromaticity diagram generated by the color matching matrix of the CIE 1964 observer. The spectrum locus has been drawn in red. The bluish triangle has the CIE xy coordinates $\{1, 0\}$, $\{0, 1\}$, and $\{0, 0\}$. The greenish area contains all real chromaticities. It is the convex hull of the chromaticities of the spectrum locus. The straight part of its boundary is the line of purples.

There are a number of possible confusions here. Let $\mathcal{I}$ denote the equivalence relation that puts scaled copies of a color in the same equivalence class. Then $\mathbb{C}/\mathcal{I}$ (color space modulo the equivalence relation) is an abstract two-dimensional space with a projective structure. Because it has the dimension of a plane, it is often called "the projective plane $\mathbb{P}^2$." But notice that this is a purely formal matter. It has nothing to do with any plane in color space. You may draw a picture of $\mathbb{P}^2$ on paper and call it a chromaticity diagram. Indeed, you draw on a plane, but it has nothing to do with any plane in color space either. Thus, the concept of the chromaticity plane is ill defined.

If you use the construction mentioned earlier, then you have indeed singled out a special plane in color space and you may well call it the chromaticity plane. But notice that the *meaning* of the points of that plane has nothing to do with the corresponding points (in that plane) in color space. Such points are colors, but the points in the chromaticity diagram (no matter how you constructed it to begin with) are equivalence classes of colors. The meaning of the chromaticity diagram is invariant against arbitrary *projective transformations* (figures 4.13 and 4.14).

This makes the chromaticity diagram hard to interpret unless you have some geometrical expertise. If you have not, then the major benefit you can gain from the chromaticity

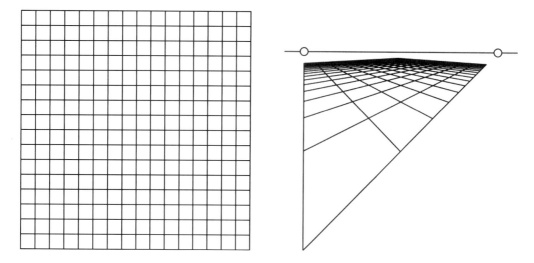

Figure 4.13 A perspective transformation. On the left is a Cartesian grid, and on the right its image under a projective transformation. I have indicated a "horizon" and two "vanishing points" of this transformation. It is simply the renaissance construction of the "pavimento" (a tiled city square in perspective rendering).

diagram is that it is easy to see whether a configuration in color space is coplanar with the origin, for then the configuration must be colinear in the diagram. Thus, if you spot three colinear points in the chromaticity diagram, then either color of the three can be mixed from the other two. Notice that you cannot do simple vector addition in the chromaticity diagram as you can in color space, though, for the chromaticity space fails to be linear. This is a major reason for common mistakes[14] and is why I recommend the use of color space over chromaticity diagrams. Only projective constructions in the chromaticity diagram make sense. Often they are very useful; thus it is perhaps surprising that the standard accounts ignore them altogether.

The spectrum locus can be mapped into the chromaticity diagram. The limiting tangents become points, and the locus a curved line that starts at the UV limiting tangent and runs toward the IR limiting tangent. Empirically, you find that the curve is convex[15] (no inflections) and looks a bit like a hook. The plane of purples becomes a straight line (through the limiting tangents) in the chromaticity diagram.

That the region of real beams is convex in figure 4.14 is accidental, for a projective transformation could undo this. However, it reflects the convexity of the region of real beams in color space, and that is a significant fact. Projective transformations that drag the regions of real beams over infinity are rarely of relevance to colorimetry.[16]

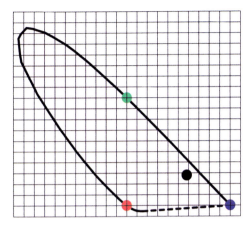 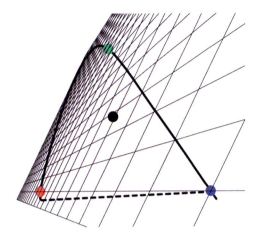

Figure 4.14 Example of a projective transformation (see appendix E.1) between two chromaticity diagrams. Although they look different to the innocent eye, the two diagrams are fully equivalent for colorimetric purposes. You are supposed to "see no difference".

Illustrations of chromaticity diagrams

In most texts you will find pictures of "the" chromaticity diagram in a fixed format, usually the CIE xy-diagram, plotted with the xy-coordinates on mutually orthogonal axes. Many people think of the representation as "the" chromaticity diagram. It is important to remember that it is only one of infinitely many possible chromaticity diagrams. In this book I feel free to use the representations that I feel are most informative given the discussion at hand, so you will find chromaticity diagrams of a variety of shapes. In many cases I will not even explain the precise definition used. Don't be bothered by this. It will be easy enough to get your bearings by searching out the familiar features-that is the spectrum locus, line of purples, spectrum limits and their complementaries, far green point, achromatic point, and so forth. Then apply your projectively invariant eye.

Notice that the points in the region of real chromaticities (corresponding to physical beams, I mean) may be labeled with chromaticity coordinates, but that these coordinates have no intrinsic meaning since a change of primaries would change them. You may wonder whether there might exist an *intrinsic* way to label the chromaticities?

4.8   Intrinsic Geometry

An intrinsic parameterization would have to relate points interior to the convex hull of the spectrum locus to the spectrum locus itself, for the only parameterization available to you right now is the wavelength scale along the locus. Here is a way to use this. Indicate any point except the purples by connecting it to the spectrum limits and finding the intersections of these lines with the spectrum locus. Thus you find two wavelengths $\{\lambda_1, \lambda_2\}$, say. For points on the spectrum locus you have $\lambda_1 = \lambda_2$, whereas for points on the line of purples these wavelengths are indeterminate. Thus you can indeed parameterize any chromaticity in an intrinsic manner, for the two wavelengths evidently do not depend upon the choice of primaries in any way.

Although this parameterization is both elegant and convenient (see figure 4.15), I have not seen it mentioned in any textbook on colorimetry, so I will not consider it further in this text.

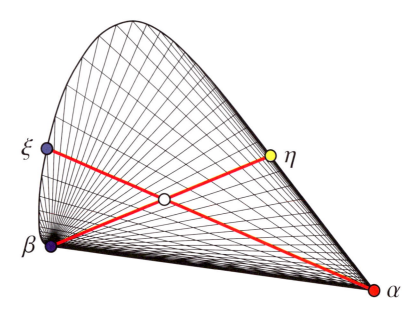

Figure 4.15 Any point in the interior of the spectrum locus of the chromaticity diagram can be specified uniquely through a pair of wavelengths as $\{\eta, \xi\}$. The spectral limits $\{\alpha, \beta\}$ provide a convenient reference.

4.9 Color Gamuts

"Color gamuts" are often important in applications, especially when colorimetry is used in the context of *images*, where naturally *many* colors are present simultaneously (see chapter 16). A color gamut is a set of colors selected according to some criterion, for example, all colors in a given image or a segment of an image (i.e., the "background gamut"). The spatial configuration is discarded, so the gamut is simply a flat list of colors.

In many (or even most) cases you need to describe the gamut in simpler terms than as a flat list of its members. The reason is that the actual members of a gamut are to some extent arbitrary in the sense that they would vary for various instances of an "image" due to noise factors, or change when you subject the image to minor transformations such as resampling.

One way to redefine a color gamut is to add all possible mixtures of the interpolation-type (that is $\mu\mathbf{a} + (1 - \mu)\mathbf{b}$ for $0 \leq \mu \leq 1$ for any pair of points $\mathbf{a}$, $\mathbf{b}$) to the set. Then the gamut is replaced with the *convex hull* of the set, and the convex hull is generated by its "extreme points."[17] Thus the addition of the interpolated colors actually leads to an enormously decreased set of colors that generate the gamut. This makes good sense in most contexts.

To find the set of colors that generate the convex hull from a flat list gamut requires some smart algorithms from computational geometry. It is easy enough to handle lists of a few thousand colors, and the best algorithms run in $n \log n$-time (n the length of the flat list).

In many (or even most) cases, the specification of the generating set is still *too specific*, and you would happily replace the convex hull for something even much simpler such as its bounding box, circumscribed ellipsoid, or perhaps even its covariance ellipsoid.

More to come...

When you are known to have an interest in color, people will ask you many questions, for most people have a keen interest in the topic of colors. No doubt you are eager to offer them your professional advice having read this book to this point. *It is wise to resist that temptation.* Few people are interested in the colorimetry of beams, most questions will concern the color of *things*. There is little of use you can tell them on the basis of what you may have learned thus far. Better wait until you've read some additional chapters. Patience does it.

In a later chapter I will discuss object colors (see chapter 11) and will show that all object colors form a gamut of finite volume, the so-called color solid (p. 245). On the other hand, in an image of a landscape a meadow might be described as "green," but in reality the colors in the "meadow area" of an image will form a gamut of finite volume instead of a mere point in color space. Such a fairly sharply located gamut is much like a color, except for its finite size. It is a "thick" point or color. These two examples span the whole range of gamuts, thus everything between colors proper and the set of all object colors. In a way, gamuts extend the notion of color, colors being mere singleton sets (gamuts). I will return to this important issue in chapter 15.

It is of some interest to consider the problem of the *union* or *intersection* of gamuts. If you specify a gamut as the convex hull, then finding the union merely involves finding the convex hull of the union of the sets of extreme points of both gamuts. Finding the intersection is rather more intricate. The convex hull of a flat list of finite length is a convex polytope, and the intersection of two convex polytopes is itself a polytope, though one that contains vertices not (necessarily) in either member of the intersecting pair. Although rather more coarse grained, the definition of the gamuts as bounding boxes renders such computations trivial.

Because gamuts are to be considered an extension of colors (as *points* in color space) that is of interest in many applications, the geometry of color space should really include the various operations on and relations between gamuts. Again, I will come back to the issue in a later chapter (see chapter 15).

4.1 The "Maxwell Model"

4.1.1 The role of models

In this section, I introduce the first instance of a model. Models are theoretical constructs that are designed to mimic some selected slice of experience to some desired degree of resemblance.

Because models are never designed to mimic *all* of your experience and do not have a *perfect* resemblance, there is a lot of freedom in the construction of a model. The only model that mimicked all your experience perfectly would be reality itself. It would be quite useless, like a geographical map on the scale of the landscape itself.

The *freedom* is what makes the model of conceptual interest. This is one way to bring reality within cognitive grasp. The freedom is used to make sure that the model has certain properties you deem desirable. Thus the class of models is very broad and *necessarily* ill defined. Some models are designed to mimic function, while others mimic form. Some models have a high degree of quantitative resemblance, while others a mere qualitative likeness. Some models are designed inside some formal mathematical framework, others are particularly nice to implement in a certain computational (computer) environment. Some models are designed to be constructed of wood, others to be assembled from empty beer cans. In short, models are fun.

The most important point—but unfortunately usually little understood—is that models are typically *not* in all respects and even quantitatively like the real thing. Models are not like theories. The use of a model lies as much in what it manages to ignore as in what (and how) it manages to model. By leaving out anything that is not essential, the model cuts down reality to pieces that you can easily chew.

I will use models as conceptual aids. Thus I will try to mimic certain selected functional or morphological properties of human vision. Usually I am interested only in the qualitative, structural aspect; thus, I am not interested in quantitative modeling of vision at all. In most cases I will try to construct models with the simplest possible formal structure that still yield the selected properties. Such models are an aid to your conceptual understanding. Often a number of different formalisms will do the job; then I will often take the viewpoint that any additional perspective on a topic can only serve to increase your understanding of it. Thus, I am not at all interested in unique models that account for whatever you might think of. Such models tend to be monstrosities full of magical numbers and arbitrary functions required to tweak things. Although unfortunately most common, such models are in the worst of taste, and a hindrance rather than a help to the understanding.

Notice that models need not be confined to mimic selected parts of experience. You may also design them to explore *extensions* of reality (then "mimic" is no longer an apt description). For instance, you might ask yourself: "What would colorimetry be like if human vision were hexachromatic instead of trichromatic?" Since no human being actually *is* hexachromatic such a model explores *extended reality*. Understanding hexachromatic (or tetrachromatic, and so forth) vision will no doubt increase your understanding of trichromatic vision, just as the study of higher-dimensional spaces has much deepened our understanding of three-dimensional space. Thus such models can be most rewarding.

It is conceivable that the genus *Homo* might develop tetrachromacy somewhere along its future line of evolution; thus, this example isn't even that weird.

4.1.2 The Maxwell model

Here is what I will try to understand with what I call the Maxwell model. The spectrum locus looks like the union of two roughly planar convex arcs in mutually "hinged" planes, almost like the outlines of a butterfly. The spectrum locus in the chromaticity diagram roughly looks like a hook, two roughly linear arcs meeting at an angle (in the "green"). This hook was empirically discovered by Maxwell and described by him as such. At first (his measurements were still pretty rough) he thought that the chromaticity spectrum locus really *was* a hook. As you know, in actuality it is somewhat rounded off. The Maxwell model is designed to capture this kind of structure in the simplest possible way.

In figure 4.16 I illustrate three spectrally selective mechanisms. There are several oddities to notice here. First of all, consider the wavelength region $(-1, +1)$. This is evidently nonsense, since wavelengths are of necessity positive. In this model *I do not even try* to model this as-

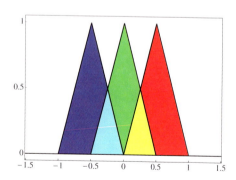

Figure 4.16 The spectral sensitivity curves of three spectrally selective mechanisms that make up the Maxwell model. Notice the odd (but convenient) wavelength scale. The three mechanisms are identical, though shifted, copies of each other.

pect of reality, though. If it worries you, you may shift the curves by a constant amount,[18] say to $(9, 11)$. This may buy you some mental rest, but it complicates the formalism unnecessarily, so I'll stick to the "impossible" region. It is not going to change the conclusions. The mechanisms I selected have unimodal spectral selectivity of finite support, and they are designed to mimic the relevant properties of biological photopigments. Notice that I have designed very (part-wise linear) schematic functional forms. You may want to use more complicated curves that look more like generic photopigment absorption. But why? It would unnecessarily complicate the formalism. I take three equal curves, equally staggered by convenient amounts. No doubt the actual photopigment curves are unequally staggered, have unequal heights and unequal shapes, but who cares? I will consider these curves to define the color matching matrix. It is easy to think of a convincing (fictional) story that explains the physics and physiology that brings this about. I won't even bother.

The spectrum locus is shown in figure 4.17, and the spectrum cone in 4.18. Notice that this simple model indeed captures exactly the butterfly structure that I believe I can see in the spectrum locus of human vision. The curve falls apart into branches, both planar, convex arcs. The main qualitative difference with the real locus is that in the real locus the curve is somewhat rounded off and the hinge between the two butterfly wings is less sharp. Such a structure is captured by a variety of models, an important one (which also has considerable utility) being the RGB color cube (figure 4.19; chapter 14). It is not a *necessary* feature, though. It is easily possible to conceive of otherwise quite "normal" models that fail to display this structure.

A way to understand the butterfly wings structure is as follows. As you move through the spectrum from -1 to $+1$, you meet the following, qualitatively different regions:

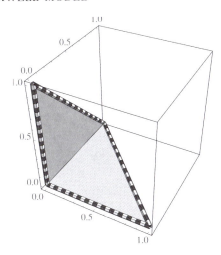

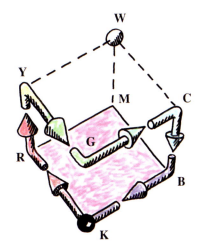

Figure 4.17 The spectrum locus of the Maxwell model. One way to understand this curve (really a polygonal arc) is as two planar, convex arcs, one running in the bottom panel, the other in the left-hand panel of the box. The planes are hinged at one of the axes, the one running in depth in the picture. The figure, understood in this manner, is like a pair of butterfly wings.

Figure 4.19 Many possible models yield the characteristic Maxwell butterfly wings structure. Here is the analog of the spectrum locus on the RGB color cube (chapter 14), also an excellent way to come to understand this basic feature. The letters stand for K (black), W (white), Y (yellow), G (green), C (cyan), B (blue), M (magenta), and R (red).

in the region −1 to −0.5 only the short wavelengths mechanism is active. Thus the system is effectively *monochromatic*, and the spectrum locus moves away from the origin along a straight line in the direction of the UV-limiting tangent;

in the region −0.5 to 0 the system is *dichromatic*. The spectrum locus runs from the "blue" to the "green";

in the region 0 to +0.5 the system is likewise *dichromatic*. The spectrum locus runs from the "green" to the "red". There isn't really any truly trichromatic region in this model, that is to say, a region where the spectral sensitivity of the three mechanisms overlaps. For that to happen the curves would have to be broadened;

Figure 4.18 The spectrum cone for the Maxwell model. Notice the sector of purples.

in the region +0.5 to +1 only the long wavelengths mechanism is active. Thus the system is again effectively *monochromatic* and the spectrum locus moves toward the origin along a straight line in the direction of the IR-limiting tangent.

Effectively monochromatic and dichromatic regions can also be identified in the human case.

Since the conical surface generated by the monochromatic beams does not fail to exhaust the boundary of the spectrum cone, a sector of purples is needed to close the gap (see figure 4.18). Again, this spectrum cone looks much like the actual one of human vision, the main difference being that it is a bit too angular, the real cone being somewhat rounded off.

Thus the Maxwell model, as simple as it is, already captures much of the actual structure of the human color system.

The Maxwell model may well be the clearest example, but many simple models achieve something similar. One model, which has also considerable practical utility, is the RGB model that I will discuss in detail later (chapter 14). However, the concept of the RGB cube should already be familiar. If you trace the "spectrum locus" on the RGB cube, you also obtain the characteristic creased structure.

4.2 The cone action spectra

The reason for trichromacy is to be found in the physiology of the retina. Since this is a topic that has nothing to do with colorimetry per se, I'll spend just a few words on it. There is a wealth of data on this, both in the regular literature and on the web.[19] Unfortunately, the fact that physiology is on an other ontological level than colorimetry appears to be lost on the majority of authors. Yet, once you travel this path, where is the end of "explanation"? Genes, molecules, atoms, quarks, maybe? Retinal physiology should be regarded an important and interesting field of study in its own right, but not a necessary prolegomena to colorimetry.

In figure 4.20 I show the current best shot at the retinal cone action spectra. How these curves were arrived at is a long story, whether there exist variations, how these are genetically determined, and so forth, is another long story. In retinal physiology, such issues as relative receptor densities and so forth are obviously of interest, just as in colorimetry these are irrelevant. The relation to colorimetry is, roughly, that the color matching matrix must linearly depend upon the cone action spectra. Otherwise the precise shapes of these action spectra are irrelevant.

Notice the curious layout of these absorption spectra. Whereas the curves tuned to the long and medium wavelength overlap very heavily, the one tuned to the short wavelengths lies rather apart from these although there is some overlap. Ecological relevance and evolutionary basis have generated a large and interesting literature.

It is perhaps most enlightening to consider the ratios of photon catches throughout the visual range of the spectrum (figure 4.21). This figure can of course obtained from the previous one, but it "reads" much easier for the purposes

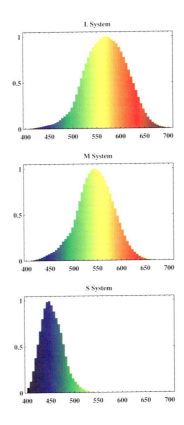

Figure 4.20 The human cone action spectra.

of colorimetry. After all, the spectral information is contained in the ratios, rather than the absolute magnitudes[20]

Notice that there is a range of short wavelengths at which the ratios are essentially constant, and another range at the long wavelengths where the same observation applies. In these ranges the system must be essentially color blind, these are the spectrum limits. There is a region beyond ca. 550nm where essentially only the medium and long wavelength sensitive cones

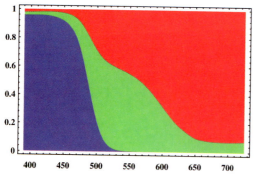

Figure 4.21 The distribution of the photon absorption ratios over the spectrum.

play a role, thus the system must be essentially dichromatic there. This is the straight part of the spectrum locus that you see in any chromaticity diagram. It is only a very short region, between ca. 470nm and 520nm where all three cone types change their sensitivity appreciably. In this region the system is really trichromatic; it accounts for the sharp bend in the spectrum locus seen in almost any chromaticity diagram, which is the property I tried to capture with Maxwell model.

Notice that this short appendix might well be expanded into a book. It would be a book about physiology, molecular biophysics and chemistry, genetics, ecological optics, the evolution of primates, and so forth, but could easily skip a chapter on colorimetry.

A topic on which I could find essentially no literature and yet seems not a little interesting to me is the following. You may wonder whether the brain knows about the manifold of photon energies at all (the visual region, an open linear segment) or whether it only knows the manifold of hues (the color circle with a quite different topology). Certainly my intuition never tells me that the hues stop at the BLUES and REDS and that the PURPLES are anything special. I am

aware only of a continuous, periodic sequence. Physiologically it boils down to the question: [21]

> *Is the fact that the "green" cones are located* between *the "red" and the "blue cones" (on the photon energy scale) in any way relevant to the anatomy or physiology?*

It seems to me that there exists no evidence that this is the case, except perhaps for the fact that the blue cones are qualitatively different from the red and the green cones. Colorimetry, of course, has nothing to say on the issue. All of the processing I have described thus far involves linear algebra, which remains perfectly intact under permutations of the coordinates. For Grassmann's color algebra it really is quite irrelevant whether the sensitivity peak of the green cone is in between those of the red and blue or not. If this is all granted, then it would follow that

> *spectra are quite irrelevant to physiology and psychology and only of use in the physical optics (what happens before the actual absorption of any photons in the photopigments).*

This should have consequences for much current work in physiology, psychology, and philosophy.

There is only one structural feature of the photon absorptions that I can think of that might conceivably enable the brain to figure out that the green receptor is in between the red and blue one. When you are confronted with a multitude of spectral distributions (as you continually are), the brain might notice that the photon catch in the green receptors correlates significantly with that in the red or the blue receptors but that the quantum catches in the red and blue receptors are much less correlated. Thus the green region (in whatever continuum) overlaps with the red and the blue, whereas the red and the green do not overlap. Hence the sequence must be RGB or BGR but certainly not RBG or GBR, at least if the brain may assume that the regions covered by the red, green and blue receptors are simply connected. Whether this is indeed a serious proposition can be judged when you regard the overlaps between the fundamental response curves. It turns out that the red and green fundamentals overlap almost completely (93%) whereas the blue overlaps *both* with the red (8%) and the green (14%).[22] Such overlaps are compatible with either a linear or a cyclical arrangement (figure 4.22). Thus, it seems to me that the mechanism will not work in the specific case of the human fundamentals.

Notice that this mechanism leaves the choice between an $\mathbb{I}^1$ ("spectrum") and $\mathbb{S}^1$ ("color circle") open. It doesn't let you pick either alternative.

This is perhaps a pity, for in principle this method might have been decisive (figure 4.23). Thus

there is no reason to assume that the open ended linear spectral sequence has any relevance to the brain.

Perhaps Occam's razor lets you pick the hue circle as the default topology. At least it avoids the (undecidable) choice of the spectral limits.

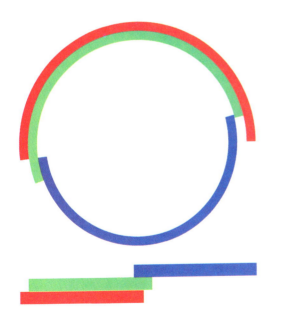
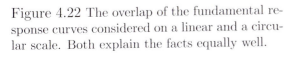
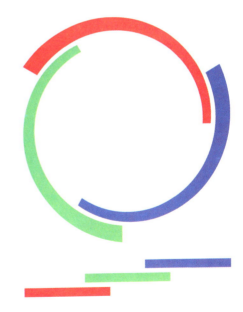

Figure 4.22 The overlap of the fundamental response curves considered on a linear and a circular scale. Both explain the facts equally well.

Figure 4.23 The overlap might well have been more informative. This is an "ideal case". Too bad.

4.3 Partitive mixtures

Color mixture is a topic that has led to numerous misunderstandings in the past and that even today does not fail to confuse people. The main obstacle is that the major tool of colorimetry, the addition of incoherent beams, is perhaps the least frequently encountered instance of color mixing. A method that closely resembles such mixture is the method used by Maxwell, that is partitive mixture, in Maxwell's case by way of the "color top." This method is well known because the color top is a familiar toy that is easily improvised with some colored papers and matchsticks or toothpicks. The difference with the addition of beams is only slight and does not pose any essential problems; the only problem is conceptual. This is in sharp contradistinction to the problems posed by mixing paints by stirring them together, and so forth, which leads to complicated problems of physical optics. Partitive mixture is not at all like that, it is a trivial modification of the addition of beams. The problem is entirely a conceptual one.

Although partitive mixture by way of the color top is of considerable historical interest, the major importance of partitive mixture is nowadays the optical mixture produced by various graphical processes, computer screens, painting in the "divisionist" style, and so forth.[23] An example is shown in figure 4.24.

In figure 4.24, you have yellow-blue checkerboard patterns in various degrees of coarseness. Depending on your eyesight and viewing distance you will not be able to spatially resolve the patterns in some cases. You then see the optical mixture, which happens to be gray. In fact, the pattern will be indiscriminable from the uniform gray pattern shown at bottom left.

A major polemic arose when Goethe noticed that you cannot obtain WHITE by mixing col-

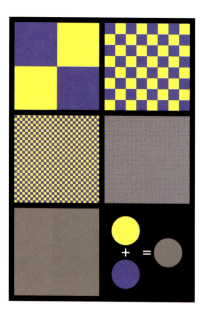

Figure 4.24 An example of partitive mixture.

ors on a color top. I guess this must have been noticed by various naive observers, but Goethe made a major point out of it, using it to attack the established physics professors. The latter invariably lectured that you could indeed mix WHITE from RED, GREEN, and BLUE—as taught by Newton—and "demonstrated" this with the color top, using colored papers. Goethe remarked that the best you could get was a dull GRAY, but never a true WHITE. The professors were stupid enough to fight back by defending Newton instead of simply giving in and substituting "achromatic" for "white." Of course, Goethe was perfectly right. You may want to try this.

The reason is obvious when you come to think of it. In the ideal situation of a demo for some audience, you would set up things such that three equal (120°) sectors of your color top would each reflect one third of the wavelength range. Then one sector might look RED, one GREEN, and one BLUE. Spinning the top would reproduce the spectrum of the incident beam, except for the fact that only one third of the total radiant power would be reflected as compared to the case that each sector would be white (100%) reflectance to begin with. Thus you obtain a (33.3...%) GRAY, not a WHITE. There is really no mystery involved; it is the immediate result from the fact that the three components must share the full surface of the top, that is to say, that the mixture is *partitive*, not *additive*. It is no more mysterious than the fact that a 120° sector reflects only one-third the radiant power of a full disk of the same reflectivity. Apparently neither Goethe nor the physics professors fully understood this point.

This is also the reason why Seurat's paintings look so disappointingly drab when you look at them from the distance that is needed to lose sight of the individual *touches*. They become a light gray. Of course, Seurat never intended them to be viewed in this way, that recommendation is due to art historians who misunderstood the very principle of divisionism. You should stand close enough to the painting that you can resolve the individual *touches*. Then you will get an impression of both the optical mixture and the way the color is divided. The surface of the painting becomes vibrant with color an the effect is very vivid, far from being dull gray.

How this works is only partly understood and largely mysterious. It has to do with the way your visual system handles its input, which is a process that runs on various spatial scales simultaneously.

Exercises

1. Change of primaries [OFTEN USEFUL] Given a set of beams find their colors. Now change the primaries and find the colors in this new representation. Find the transformation from one representation into the other. What happens when you change only a single primary?

2. The black space [CONCEPTUAL] Find a basis for the black space. In any good math environment this should be easy; for instance, in the Mathematica environment you use "NullSpace." Check that any linear combination of such blacks has indeed a "black" color. Plot spectra of random black beams. Can you construct a *smooth* black spectrum? Can you produce black spectra with few zero crossings? What is the minimum number?

3. Discrete representation [SIMPLE] Study the influence of spectral sample rate. What is the "practical" dimension of the space of beams (approximately, of course)?

4. The spectrum locus [REQUIRED] Plot the spectrum locus. Find the limiting UV and IR tangents and construct the plane of purples. Explicitly find the far green points for a number of illuminants. Plot these in the chromaticity diagram.

5. Projective transformations [REQUIRED] Experiment with various projective maps of your original (CIE) chromaticity diagram. Try some random projective transformations and see what happens. Construct a few "nice" ones. Find transformations from one diagram to another. (You may find appendix E.1 useful.)

6. Gamuts and convex hulls [INVOLVES COMPUTER HACKING] Generate a random gamut, then construct its convex hull and delete all but the gamut's extreme points. If you don't have access to an algorithm for generating convex hulls, you're in for some work. There are good algorithms to be found on the Internet. You'll have to implement one. It is worth the initial effort because it will find many uses in colorimetry.

7. Models [CONCEPTUAL] Implement the Maxwell model and play with it. Think of other models and implement them.

8. Configurations in color space [FOR MATHEMATICIANS] Consider a triad of distinct colors. By promoting them to primary color status, they become $\{\{1,0,0\},\{0,1,0\},\{0,0,1\}\}$; that is to say, you cannot distinguish such triples on the basis of the color coordinates. Consider tuples of $3 + n$ colors. How many different configurations are there? (Infinitely many of course, I'm obviously asking for dimensionality.) How would you characterize a tuple in a way that is invariant with respect to the choice of primaries? (It may help to remember the affine invariance of volume ratios.)

9. "Best" monochromatic primaries [HARD] Is there a "best" choice of wavelengths for monochromatic primaries? Intuitively, it would be stupid to pick three very similar wavelengths; you obviously have to "spread" them. The choice should be invariant against linear transformations though. All you can do is use the fact that linear transformations conserve volume ratios. Is this a defined problem? What if you constrain the radiant power of the primaries? Is the result (if any) unique? If so, how does it compare with conventional choices?

10. Maxwell's "color top" [KINDERGARTEN STUFF] Try to be as smart as Maxwell, that is to say, figure out how to cut cardboard disks for the color top in such a way as to make it easy to adjust the coefficients. Then build one!

11. Intrinsic geometry [FOR MATHEMATICIANS] The remarks on intrinsic geometry (section 4.8) are just the top of the iceberg: there is much more you can do. For instance, show how you can set up a duality between lines and points in the chromaticity diagram, how you can construct a unique third point to any pair of distinct points, and so forth. Try to interpret the colorimetric meaning of such constructions.

Chapter Notes

1. From a numerical perspective, some choices are better than others. But this is mere engineering.
2. G. Wyszecki, "Valenzmetrische Undersuchung des Zusammenhanges zwischen normaler und anomaler Trichromasie," *Farbe* 2, pp. 39–52 (1953).
3. On the physical dimension of the tristimulus values, see M. H. Brill, "Do Tristimulus Values have Units?," *Color Research and Application* 21, pp. 310–313 (1996).
4. I will denote "black" as "K," the "black space" as $\mathbb{K}$. The letter "B" is already in use for "blue." The "K" stands for "key," which is the black plate in CMYK color printing.
5. At least not at a single glance. If you have a set of spectacles, say "one for each wavelength," and are permitted to look through each of these in sequence, you're in business. This is simply spectroscopy of course.

   The idea is less far fetched than you might think, in scientific applications one frequently takes photographs through a set of assorted filters. If the scene is static there is no limit on their number.
6. Notice that the notation "$\mathbb{C}$" is not ambiguous because you will never work with complex numbers in colorimetry.
7. An affine space has no origin; that is to say, configurations that differ by a translation are considered to be the same. In a linear space, the origin is special though. In the literature of colorimetry, the concepts of affine space and linear space are commonly confused. Better watch it.
8. This makes much of the literature in colorimetry a mess. You will need an effort to interpret the various matrixes and vectors. It is usually not clear to what type of operator a matrix belongs and in which basis, whether a vector is indeed a vector or actually a linear functional, and so forth.
9. Look for the CVRL Color & Vision database at http://cvision.ucsd.edu/. You find the color matching matrix under CMFs (color matching functions).
10. The precise nature of the color matching functions near the spectrum limits is somewhat intricate, rendering the estimation of the limiting tangent nontrivial. These complications are of no relevance to this discussion, but you will encounter them when you try to find the limiting tangent directions. For the moment you may simply use a large wavelength increment for this estimate, or discard the tails of the color matching functions.
11. W. Blaschke, *Vorlesungen über Differentialgeometrie II, Affine Differentialgeometrie*, (Berlin: Springer, 1923).
12. In defense of Grassmann it has to be said that Helmholtz's initial report was not too sharp either. He

thought that there was only a single pair of complementaries. A short time later, Helmholtz had improved his experimental equipment significantly and measured the essentially modern branches of complementary pairs, clearly revealing the gap in the "green" part of the spectrum.
13. The CIE can be found at: http://www.cie.co.at/cie/.
14. Common mistakes are to use Newton's center of gravity (or barycentric) construction in the chromaticity diagram as if this were an affine space. It is certainly possible to do so, but you have to be very careful. It is much easier to go from the chromaticity diagram to color space, do the construction, and return to the chromaticity diagram. If you are unable to move to color space you may as well give up, because you will not be able to do the task in the chromaticity diagram either.
15. Everywhere in this book, I understand projective geometry as *oriented* projective geometry.
16. The convexity is conserved if you use oriented projective geometry. Oriented projective geometry has become an important topic in computer graphics, and excellent texts are now available. The book I like best is J. Stolfi, *Oriented Projective Geometry*, (Boston: Academic Press, 1991). Because the differentiation between inside and outside of the spectrum cone are of vital importance to colorimetry, *oriented* projective geometry is exactly what you need.
17. An "extreme point" is such that it does not occur as an interpolated point between other points that belong to the convex hull. Thus an extreme point does not occur as an interpolation between any pair of distinct points of the set. Intuitively, extreme points are necessary; indeed, the set of extreme points alone suffices to generate the convex hull. For example, the extreme points of a cube are its vertices. The convex hull of these (eight) vertices is the full cube. This is the Krein-Milman theorem; for a compact convex set in a linear topological space the convex span of the set of extreme points is the convex set itself.
18. While you're at it, you might as well apply a scaling and make the wavelength range 400–800 nm. But then, where's the end of it? You could tweak the widths of the curves and their relative heights. Not satisfied, you could change their shape. Eventually you'll end up with a model of the visual system that could be presented in a physiology textbook. Needless to say this would defeat the very purpose of the Maxwell model.
19. A good place to start is the Color & Vision Database (http://cvision.ucsd.edu/) of the Color & Vision Research Laboratories at the Institute of Ophthalmology, UCL.
   Actually, there are many complications of a physiological nature that bedevil colorimetry. Since this book is not about physiology, I have glossed over these. If you

are interested you may want to explore the following topics:

— suppose a retinal cone would contain multiple photopigments instead of just one. This would jeopardize Grassmann's laws;

— since the receptors are very small, they have to be regarded as dielectric waveguides with directional properties. So much is indeed evident from the so-called Stiles-Crawford effect. This leads to complications. Moreover, the waveguides are closely packed, thus electromagnetically coupled, leading to more complications;

— apart from the cones, the retina contains many "rod" receptors that contain a fourth photopigment. Look for "rod intrusion." You would expect tetrachromacy to cut in;

— there is a yellow filter in front of the fovea, leading to so-called macular screening;

— at very high radiation levels, appreciable amounts of photopigment are bleached. Photopigment depletion leads to a breakdown of Grassmann's laws;

— multiphoton effects might conceivably lead to nonlinear effects.

20. Of course you need the absolute sensitivities in order to be able to find the ratios, and such are not immediately available from the physiology. I simply did something reasonable here. This is a tricky issue that is hard (maybe impossible) to resolve. If you are somehow interested you are in for a major investigation.

21. No respectable color scientist will say such a thing as "red cone," but instead "long wavelength sensitive cone." The idea is that such cones *absorb* long wavelengths, thus will not *look* RED at all. If you say "red cone," people are certain that you don't understand the first thing about color. They will refuse to have a beer with you. But I'm sure you caught my meaning, didn't you? Perhaps we should allow people some leeway and not infer the worst on the least cue. But then, this is a way for experts to recognize each other. One silently nods.

22. Notice that this calculation again involves the absolute sensitivities (see previous note).

23. But watch out: Many of these mixing methods are only very approximately partitive. Most modern printing methods are not partitive at all.

Chapter 5

Colorimetry with an Achromatic Beam

5.1 The Notion of an Achromatic Beam

Although it is not at all necessary to consider achromatic beams in colorimetry (see chapter 4), this is rarely actually avoided. The early authors invariably add an achromatic beam (figure 5.1) without even discussing the issue. It is such a natural thing to do! Usually the achromatic beam will be natural daylight, of course. The classical authors believed that to be sufficient. This seems more complicated today, since you may load spectra of many flavors of "daylight" from the Internet. Yet most people probably (somewhat) understand what natural daylight means. The topic of the achromatic beam is a hairy one, so I will discuss it at length. This is the first remove from mere austere colorimetry. It is a very rewarding one.

In many (suitably simplified) settings you become aware of the properties of both a *source* and of *objects*. You become aware of the properties of the radiation from the source from its effect on white objects, for instance. A "white" object is a Lambertian[1] surface of unit albedo,[2] that is a surface that scatters all radiation impinging on it in a thoroughly random manner (notice that a mirror scatters all radiation too, but can hardly be called "white"). A piece of dull white paper is a good physical approximation.[3] "Colored" objects appear *colored* because they interact with the radiation in a spectrally selective manner (see chapter 17). They typically scatter *less* than a white object at any wavelength region of interest. In such (very much simplified) settings you may say that objects scatter "part of the light", that is to say, at any wavelength they scatter at most as much (though usually less) radiation to your eye as a white object would. In that sense the spectra of beams

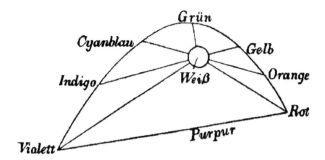

Figure 5.1 Helmholtz drew this spectrum locus in 1855. It is the first clear example of the generic horse shoe spectrum locus so characteristic for human vision. White (achromatic) is added without further thought. Notice that the hues are not ordered in complementary pairs.

scattered toward your eye are all "parts" of the beam[4] that would be scattered by a white object. That is why it is often of interest to define a fiducial "achromatic beam" and refer all other beams to that.

Notice that the choice of a Lambertian surface (ideally a matte surface) is essential here. More generally the radiance of the beam scattered to the eye depends on both the viewing and on the irradiation directions. People often define the bidirectional reflectance distribution function (BRDF) as a convenient approximation (it assumes no diffusion of photons in the bulk matter, but pure surface scattering). (See figure 11.6 and chapter 17.) The BRDF is defined as the radiance of the scattered beam over the irradiance caused by the incident beam, where both the scattered and the irradiating beam are considered "collimated" (that is, directional). For the Lambertian surface the BRDF is constant, thus the radiance of the scattered beam depends only on the magnitude of the irradiance, not on the direction of the incident beam, nor on the viewing direction. There are only few physical constraints on the BRDF, which is a function of two directions, that is, four angles. If a surface is not Lambertian, it may scatter much more radiant power in a certain direction than a Lambertian surface would (of course, this means that *less* would be scattered into other directions). Such can often been observed in the case of glossy surfaces or mirrors (figure 5.2). The analysis discussed in this chapter thus fails to apply to such surfaces. In face of the fact that the Lambertian surfaces are essentially nonexistent (although many matte surfaces come close) the discussion here might be termed "ideal" rather than fully realistic. However, I would be hard put to think of a better way to obtain a handle on colors, even in realistic scenes. You simply need to apply some good physical horse sense in order to augment the "ideal" case and render it useful in practice.

Figure 5.2 These sweet peppers have a rather non-Lambertian BRDF. The radiance from the specularities is actually higher than you would obtain from a piece of white paper at the same location and spatial attitude. Such—very common—effects are beyond the scope of the treatment in this chapter.

5.1.1 The relation of (spectral) domination

In order to deal with such cases, I will introduce a little formalism first. Suppose I have two beams **p** and **q** that are such that throughout the spectrum

$$p(\lambda)\,\mathrm{d}\lambda \le q(\lambda)\,\mathrm{d}\lambda.$$

Then I will say that beam **p** is "dominated by" beam **q**, or $\mathbf{p} \subseteq \mathbf{q}$. Clearly, you can have the generic cases

$$\mathbf{a} \ \subseteq \ \mathbf{b},$$
$$\mathbf{a} \ \supseteq \ \mathbf{b},$$
$$\mathbf{a} \ \nparallel \ \mathbf{b}, \text{ that is neither } \mathbf{a} \subseteq \mathbf{b} \text{ nor } \mathbf{a} \supseteq \mathbf{b}$$

for any pair of beams **a** and **b**. In case $\mathbf{a} \nparallel \mathbf{b}$ neither beam dominates the other, the beams are incompatible. If *both* $\mathbf{a} \subseteq \mathbf{b}$ *and* $\mathbf{b} \subseteq \mathbf{a}$, the beams **a** and **b** are identical copies.

Suppose I have a collection of beams $\{\mathbf{g_i}\}$, I will call such a collection a "gamut." Now suppose I have some beam **a** (say) such that $|a_\lambda\rangle > 0$ for any wavelength and

$$\mathbf{a} \supseteq \mathbf{g_i},$$

for any beam in the gamut. Then all colors in the gamut can be considered as parts of the beam **a**. Of course, there exist infinitely many such beams for any given gamut. I simply pick one and call it the achromatic beam. Perhaps the simplest of such beams has a spectrum that is the upper envelope of the spectra of all the beams in the gamut. (Though perhaps you may want to smooth it a bit if it turns out to be very rough.)

5.1.2   The concept of achromatic beam

Notice what a mean trick I played here. Achromatic suggests (literally) "colorless," but this relates to color *experiences*, and colorimetry has nothing to do with that. If you believe that each generator of the spectrum cone somehow relates to a unique hue experience, that hue experiences are invariant along lines through the origin of color space and that hue experiences vary smoothly with their location in color space, then you really believe a lot.[5] Moreover, you entertain beliefs on which colorimetry has nothing to say. But in case you do (as I do), then a topological argument (Brouwer's celebrated fixed point theorem[6]) ensures that some line through the origin in the interior of the (solid) spectrum cone, must be "hueless".   (See figure 5.3.) Thus it is perhaps not too strange to believe in hueless lines through the origin of color space. Of course, there is no way to figure out (colorimetrically that is) which line it is, so the very question is not defined. But you do not violate any part of colorimetry if you simply single out a line through the origin (*any* old line will do) and designate it achromatic hue, pick a point on it and designate it achromatic color, find any beam that produces that color and designate it achromatic beam. (What you do here isn't even wrong! But of course there is no notion of what it would mean for it to be "right"

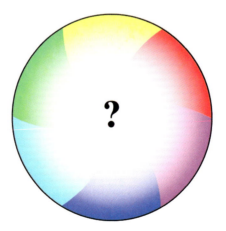

Figure 5.3 Brouwer's fixed point theorem suggests the existence of an achromatic color. The argument doesn't hang on qualia, it is sufficient to assume that all colors on the boundary of the chromaticity diagram look qualitatively different, the quality varying continuously.

either.) If you pick a beam with the properties of beam **a** discussed before, it also makes physical sense. All the beams in your gamut *could have* been produced by simple physical processes acting on the achromatic beam.

Obvious choices for the achromatic beam would be a uniform ("flat") spectrum, average daylight, a black body spectrum of a reasonable temperature, and so forth (figures 5.4 and 5.5). One thing I will insist on is that the spectral radiant power density of the achromatic beam is everywhere positive and different from zero. The achromatic beam is not permitted to have any "spectral gaps".[7] This requirement is necessary because such a spectral gap would introduce a blind spot in the spectral domain.

There will often be some "ecological reason" to pick an achromatic beam[8] (see chapter 17). Daylight and artificial (e.g., incandescent) light come to mind (figures 5.4 and 5.5). These are the examples I will use over and over in this book. For formal reasons (elegance) you might be inclined to pick a "uniform" (or "flat") spectrum. Such spectra actually occur, for instance, in high pressure Xenon arc sources. However, you need to be careful; this is not a mathematical issue, but a matter of physics. For instance, do you mean "uniform radiant power density on wavelength basis" or "uniform photon numerosity on photon energy basis"? The daylight spectrum is an obvious choice from a biological perspective.

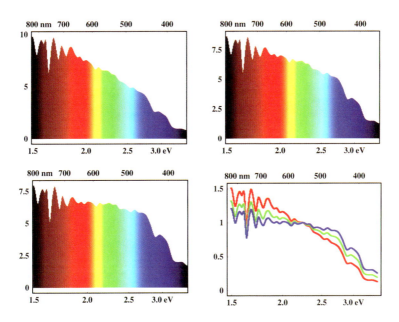

Figure 5.4 Varieties of "daylight." Depending upon time of day, cloud coverage, and so forth, the average daylight spectrum may vary a bit. These are CIE Standard Illuminant D55, D65, and D75.

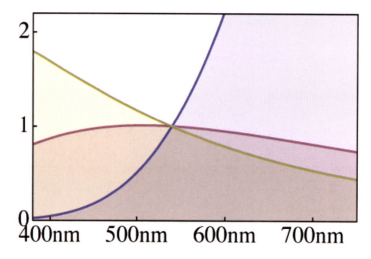

Figure 5.5 Varieties of black body spectra. The color temperatures are 2000°K (candle), 5700°K (nearly daylight), and 10,000°K (very bluish, some electronic flash units). The spectra have been normalized at 540nm for easy comparison.

Evolution has tuned your system to it. A Planckean spectrum of about 5700°K is roughly equivalent to "daylight" and has the advantage of being smooth and analytically defined. I will use it often in this book.

5.2   Colorimetry in the Presence of an Achromatic Beam

Once you have picked your achromatic beam, you can commence with colorimetry. It will turn out to be the case that the existence of an achromatic beam induces so much very useful additional structure that you would be hard put to ever again do without it.

Geometrically the achromatic beam is a *point* in color space, and thus also defines a *line* in color space (just connect the achromatic point to the black point) and is a point in the chromaticity diagram. The achromatic direction lies fully within the solid spectrum cone, and lies fully within the region of real beams in the chromaticity diagram.

Given a line through the origin in color space, you may immediately construct two unique planes, namely the planes spanned by the achromatic line and each of the two limiting tangents of the spectrum locus. These two planes again define two unique spectral generators of the spectrum cone (thus two unique wavelengths) and they divide the boundary of the spectrum cone into four simply connected regions (figure 5.6). All these constructions are fully invariant with respect to changes of the primaries.

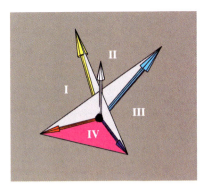

Figure 5.6 The planes through the achromatic axis and the spectrum limits divide the interior of the spectrum cone (and thus the totality of all real colors) into four partitions, roughly the "blues" (I), "greens" (II), "reds" (III) and "purples" (IV). Of course these terms should not be misunderstood; they stand for "short wavelengths region" and so forth. But it is much more pleasant to use colorful language, and no harm is done if you understand that colorimetry is not about color experiences.

5.2.1 Dominant and complementary wavelengths

Any generator of the spectrum cone defines a unique plane with the achromatic direction, which plane crosses the boundary of the spectrum cone again, at another generator. Thus the achromatic direction serves to associate the generators pairwise (figure 5.7).

What does it mean for two generators of the spectrum cone to be coplanar with the achromatic direction? It means that if I have two colors, one on each generator, then it is possible to mix such colors with positive coefficients so as to produce some scalar multiple of the achromatic beam. Conventionally, people call such generators *complementary*, and if both are spectral people call the corresponding wavelengths "complementary wavelengths."

If you have some color that is not achromatic but lies in the interior of the spectrum cone, then you may construct a unique half-plane with the achromatic direction as boundary. This half-plane meets the boundary of the spectrum cone in a unique generator. Thus any color can be decomposed[9] into a color from the boundary of the spectrum cone (either a monochromatic beam or a purple) and the achromatic beam. In case the color on the boundary is spectral, its wavelength is conventionally known as "the dominant wavelength" of the color (figure 5.7). This is an unfortunate term, because, as you can easily reason out yourself, the beam does not even need to have spectral radiant power at its "dominant wavelength."[10]

All these constructions depend upon the choice of the achromatic beam. A beam doesn't have anything like a dominant wavelength, except *with respect to some achromatic beam.*

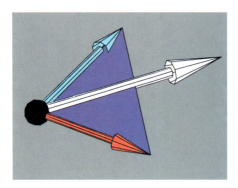 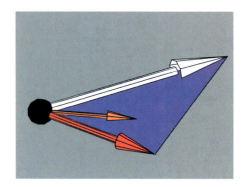

Figure 5.7 Left: Two monochromatic beams that happen to be coplanar with the achromatic beam are said to be mutual "complementaries." Right: For any given color you can find a spectrum cone generator (though it might turn out to be a purple) that is coplanar with it and with the achromatic color. The generator, if it is a spectral color, is said to define the "dominant wavelength" (really: the associated monochromatic beam) of the color.

The "attributes" of colors

A COLOR has various "attributes" that most visual observers find it easy to recognize or (approximately) estimate. Examples include the BRIGHTNESS, LIGHTNESS, or TONAL VALUE of a COLOR, the degree of COLORFULNESS or SATURATION of a COLOR, and the HUE of a COLOR. Do these notions have any correlates in the color coordinates? In austere colorimetry hardly, because there is very little to relate any given color to. Things are much improved once you introduce an achromatic beam. This means that color space gains a number of "landmarks" that can be used to relate any given color to. A color is a point in color space, a chromaticity an oriented half-line through the origin. You may relate these to the landmark elements, using any projective and affine invariants you may form. To form such invariants is indeed necessary, because the "attributes" should of course be invariant against changes of the primaries. After all, the primaries have been arbitrarily selected.

The "standard" constructs of colorimetric praxis typically rely on the introduction of additional structure, the "luminance functional" (see appendix E.7.1). This is not very elegant and has has a number of drawbacks. It is of some interest to explore invariants that do not rely on such additional structure and thus are of more general significance. This is explored in the appendix E.9 for "hue" and E.8 for "saturation." I will not use these constructs in the remainder of the text though, because (unfortunately) they are not in common use. Various additional constructions are possible. You may enjoy exploring these yourself.

Actually, quite a few of these constructions depend only on the *color* of the achromatic beam, and not on its *spectral composition*. It is important to be aware of this, but I will not pursue this issue here. The reason is that it is hardly worth the trouble. In most cases it is actually the achromatic *beam* rather than its *color* that is given in the first place. Many of the properties of object colors that I will consider later depend on the spectral composition of the achromatic beam. (See chapter 11.) Thus, I consider it most reasonable to pursue the case of a given achromatic beam throughout and perhaps mention what happens when you vary the spectral composition for constant color if the occasion calls for it.

5.2.2   Change of achromatic beam

In order to illustrate these concepts, I need to select achromatic beams. I will select two examples, average daylight and incandescent light (figures 5.8 and 5.9). These are two generic types of illuminant in the daily environment of most people. They are very different. The spectra of these sources are shown in figure 5.8. I have scaled these spectra such that the largest coordinate of their colors equals one. This is conceptually irrelevant; it simply makes it easier to compare the results. The colors of these sources are (I have used an arbitrary normalization in order to ease the comparison):

average daylight spectrum	$\{1.0000, 0.5053, 0.2938\}$
incandescent source spectrum	$\{1.0000, 0.2762, 0.0595\}$

Since the primaries are monochromatic beams of 645 nm, 526 nm, and 444 nm, you see that the radiation from a light bulb apparently contains much less blue than average daylight.

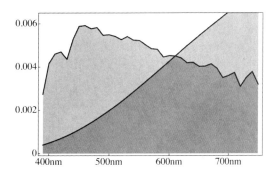

Figure 5.8 The spectrum of average daylight (the wriggly curve) and the spectrum of incandescent light (the smooth curve). Both are in common use as achromatic beams.

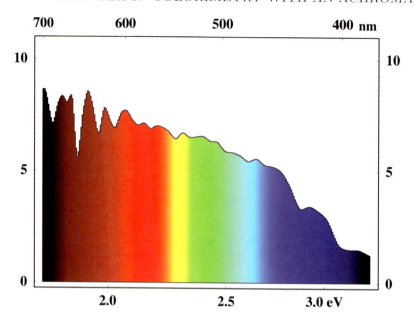

Figure 5.9 In terms of photon number density on a photon energy scale the daylight spectrum looks quite different to most people. Since the radiant power as a function of wavelength representation is most familiar, I also use it in this book. The photon representation is actually much to be preferred, though. In any case, you have to be aware of the differences.

This is even clearer if you plot the spectral contributions to the color for the two sources. To get this, you simply multiply the color matching matrix with the source spectra[11] (see figure 5.10).

Notice that the spectrum cone itself will not change when you change the illuminant. But the achromatic direction obviously does. As seen in figure 5.11, the directions are dramatically different as compared with the (average) semi-top-angle of the cone. As a result the complementary wavelength pairs are quite different for these illuminants (see figure 5.12).

5.2.3   The complementary pairs of monochromatic beams

Notice that the locus of complementary pairs falls apart into two branches. Of course, the branches contain the same pairs, only in reversed order. The asymptotes are interesting, they correspond to the complementaries of the spectrum limits and separate the wavelengths regions for which complementaries exist from those wavelength regions where

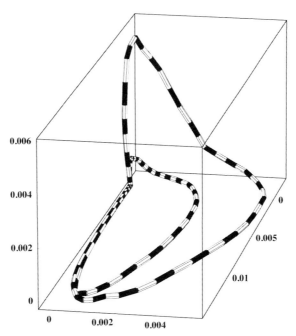

Figure 5.10 The spectral contributions to the color for the average daylight and incandescent illuminants. Notice that a light bulb emits so little in the short wavelength range that the curve almost degenerates into a planar one.

no complementaries exist. Here is a small table of the relevant wavelengths (here $\overline{\mu}$ stands for the complimentary of μ):

$$\text{Daylight}: \quad \overline{\lambda}_{\text{UV}} = 560.9 \text{ nm} \quad \overline{\lambda}_{\text{IR}} = 485.9 \text{ nm}$$
$$\text{Lightbulb}: \quad \overline{\lambda}_{\text{UV}} = 575.2 \text{ nm} \quad \overline{\lambda}_{\text{IR}} = 496.2 \text{ nm}$$

The visual wavelength region falls apart into three distinct ranges:

$\lambda_{\text{UV}} < \lambda < \overline{\lambda}_{\text{IR}}$, that is, the short wavelength region. (I'll often say "blue region.") For monochromatic beams in this region there exists a unique complementary in the long wavelength region;

$\overline{\lambda}_{\text{IR}} < \lambda < \overline{\lambda}_{\text{UV}}$, that is, the medium wavelength region. (I'll often say "green region.") For monochromatic beams in this region, there do not exist monochromatic complementary beams. This is again the fact that the Newtonian spectrum fails to be complete, that is to say, there exist nonspectral hues that look unlike any spectral hue;

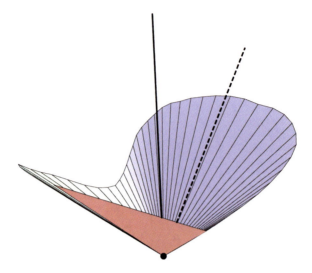

Figure 5.11 The achromatic direction for average daylight (drawn line) and a light bulb (broken line) inside the spectrum cone. The difference is appreciable.

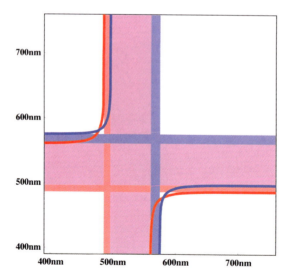

Figure 5.12 The complementary wavelength pairs for daylight (red) and incandescent light (blue). There exist two branches, with a gap in between. The gap contains the greens, which are opposite the line of purples. The reds are complementary to the blues and vice versa. Notice that the complementary pairs are different for the two illuminants.

$\overline{\lambda}_{\mathrm{UV}} < \lambda < \lambda_{IR}$, that is, the long wavelength region. (I'll often say "red region.") For monochromatic beams in this region, there exists a unique complementary in the short wavelength region.

For different illuminants these ranges have somewhat different boundaries, but they always exist. Notice that the spectrum limits λ_{UV} and λ_{IR} are ill defined. λ_{UV} is just "very short" (about 380 nm is usually small enough, the scale "starts from 0") and λ_{IR} "very long" (here about 740 nm is usually long enough, the scale "ends at ∞"). The complementaries of the spectrum limits are quite sharply defined, though.

5.3 The Spectrum Regions Defined by the Achromatic Beam

The monochromatic beams in the short wavelength region look "BLUE," in the medium wavelength region "GREEN," and in the long wavelength region "RED" to most people (under the right conditions, etc., etc.). I will use these often as *descriptive* terms, there can hardly be any misunderstanding if you constantly remember what colorimetry is all about.

The various regions can be clearly seen in the chromaticity diagram. First I calculate the extreme green point for these illuminants. It is 535 nm for average daylight and 551.5 nm for incandescent light. I plot the spectrum limits, their complementaries, the green point, and the achromatic point in the chromaticity diagrams. Since the diagrams are not all that clear I make use of the fact that the chromaticity diagram is defined only up to arbitrary projective transformations. Since projective transformations are determined by the location of four points, I may force the green point and the spectrum limits to lie on the vertices of an equilateral triangle, with the achromatic point at its barycenter. With this normalization, the difference between average daylight and incandescent light becomes readily apparent (figures 5.13 and 5.14).

In the part on "bare bones colorimetry" (colorimetry in the absence of an achromatic beam I mean, chapter 4) I introduced a method of parameterizing chromaticities via pairs of wavelengths $\{\lambda_1, \lambda_2\}$ with $\lambda_1 < \lambda_2$. It is clear that in this parameterization the achromatic chromaticity would correspond to the point $\{\overline{\lambda}_{IR}, \overline{\lambda}_{UV}\}$. You have the relations

$$(\lambda_{UV} < \lambda_1 < \overline{\lambda}_{IR}) \wedge (\overline{\lambda}_{IR} < \lambda_2 < \overline{\lambda}_{UV}) \quad \text{"blues"},$$
$$(\overline{\lambda}_{IR} < \lambda_1 > \overline{\lambda}_{UV}) \wedge (\overline{\lambda}_{IR} < \lambda_2 > \overline{\lambda}_{UV}) \quad \text{"greens"},$$
$$(\overline{\lambda}_{IR} < \lambda_1 > \overline{\lambda}_{UV}) \wedge (\overline{\lambda}_{UV} < \lambda_1 > \lambda_{IR}) \quad \text{"reds"},$$
$$(\lambda_{UV} < \lambda_1 > \overline{\lambda}_{IR}) \wedge (\overline{\lambda}_{UV} < \lambda_2 < \lambda_{IR}) \quad \text{"purples"},$$

thus the parameterization fits in very well with the structure induced by the achromatic beam.

5.3.1 Projective structure

There exist four cardinal points, namely the spectrum limits and their spectral complementaries. (The achromatic point is the intersection of the diagonals, thus it is not an independent fifth point.) Since four points in general position form a "projective basis," you can use this structure to introduce *projective coordinates* that are invariantly connected

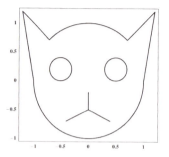 

Figure 5.13 Example of a "projective transformation." Although the transformation conserves rectilinearity, this is by no means a *linear* transformation. Projective transformations may even split a convex figure, for instance, map an ellipse on a hyperbola.

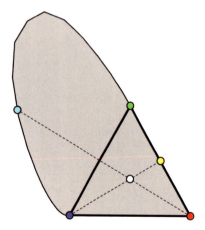 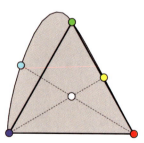

Figure 5.14 The chromaticity diagrams with cardinal points for average daylight (left) and incandescent light (right) after normalization with a suitable projective transformation. Here I place the spectrum limits and the extreme green point on the vertices of an equilateral triangle, with the achromatic point at its barycenter. Notice the blue boost of the daylight as compared with the light bulb.

with the structure, that is to say, invariant with respect to changes of the primaries or the choice of the "chromaticity plane." (Figure 5.15.)

The easiest way to appreciate the projective coordinates is to transform the configuration of the four cardinal points into the unit square. (See appendix E.1 for the method how to do this.) Then the projective coordinates appear visually as simply the usual Cartesian coordinates in the diagram. Notice that you have no need for the *wavelength* labels in this scheme, for each monochromatic beam is assigned unique projective coordinates.[12] This appears to be one of the most rational ways to introduce chromaticity coordinates. However, I have never seen it done. I will not use this attractive option in this book because it is perhaps too remote from established practice.

There are many projective invariants to be constructed that are by construction (being independent of all arbitrary choices apart from the choice of achromatic color) *meaningful*. Examples are the projective differential invariants of the spectrum locus and various multipoint invariants involving one or more fiducial points in combination with the cardinal

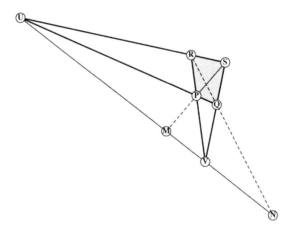

Figure 5.15 The quadruplet of points P, Q, R, S has the "vanishing points" V (as the intersection of RP with SQ) and U (the intersection of SR and QP) when you think of S as the "origin," R as the "unit X" and Q as the "unit Y" point, and P as the "unit point," that is, the point one unit from the origin in both the X and Y directions. The two vanishing points define the "line at infinity" UV. The two diagonals SP and RQ meet the line at infinity in M and N respectively, these are the vanishing points of the diagonals. Now you are in a position to draw lines parallel to the coordinate axes or the diagonal through *any* point, merely connect them with the corresponding vanishing points. This allows you to construct graph paper and plot points of arbitrary coordinates in the diagram. The four points thus indeed subtend a complete (projective) basis for the plane. The application to the chromaticity plane is immediate.

points. Again, conventional colorimetry has chosen to ignore such possibilities, and I will not pursue them here.

5.4   Arthur Schopenhauer's Parts of Daylight

A very interesting idea, due to Schopenhauer[13] (yes, the philosopher, author of *Die Welt als Wille und Vorstellung*), and inspired by Goethe (yes, the poet, author of *Die Leiden des jungen Werther*) is to think of colors as "parts of daylight." In the first analysis you identify "red," "green," and "blue" parts. The parts are defined through mutually nonoverlapping and abutting wavelength ranges. These primary parts may pairwise combine (through union of the spectral regions) to secondary parts, "cyan" (blue and green), "magenta" (red and blue), and "yellow" (red and green), whereas all wavelength regions taken together are just daylight again, thus "white." (See figure 5.16.) "Black" simply fits in as the "empty part." Since you have already identified well-defined short, medium, and long wavelength regions, you may identify these with the Schopenhauer parts of daylight (or incandescent light, and so forth, as the case may be). The colors of the parts are easily calculated and plotted in color space. You obtain a "RGB crate" that lies fully within the spectrum cone. For daylight, the dominant wavelengths of the "parts" are:

yellow 566 nm,

green 527 nm,

cyan 482 nm,

blue 460 nm,

magenta $\overline{527}$ nm,

red 592 nm,

rather close to the spectrum locations often quoted for the luminous experiences generally described by these color names. Of course, colorimetrically this is nothing but colorful language (pun intended).

   Notice that two vertices of the crate are achromatic, they are the origin ("black point" K) and the achromatic point (color of the source, or "white point" W). The remaining vertices lie on a closed polygonal arc, thus you obtain a periodic sequence of colors "Red" (R, a primary color), "Yellow" (Y, a secondary color), "Green" (G, a primary color), "Cyan" (C, a secondary color), "Blue" (B, a primary color), and "Magenta" (M, a secondary color). From magenta you arrive at red again, thus you have constructed a (at least topologically)

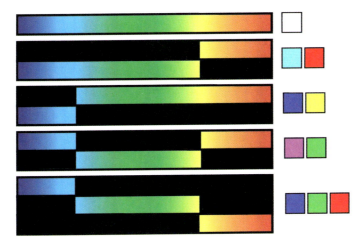

Figure 5.16 The Schopenhauer parts of daylight as spectral regions. The spectrum is divided into one, two, or three parts by a pair of cut loci.

color circle. Different from the spectrum locus, which lacks a magenta (or purple) range, the color circle contains *all* hues[14] and is naturally closed.

Suppose you take the red, green, and blue parts as your primaries. Then the color coordinates will immediately gain some intuitive meaning, for they become much like the RGB-pixel values of your digital camera. This is probably as close as you can ever get to an intuitive understanding of the colorimetric "colors." Notice that not every beam will give rise to a color that fits neatly into the RGB crate of the illuminant. However, the vast majority of beams encountered in natural scenes will fit into this format. The ones that don't (that are "out of gamut," as the jargon has it) will have some color coordinate that are (only slightly) out of bounds (zero to one). (See figure 5.17.) It is easy to check this by calculating the colors of spectral reflectance factors of a variety of surfaces found in some database (there are several such databases available on the Internet; I'll discuss some in detail later). Typically less than one out of a hundred items fails to be "in gamut" and if so the color is never significantly different from an RGB color.

Schopenhauer thought of colors as of "incomplete daylight." To Goethe the colors were shadowlike entities, that is to say, only *parts* of daylight. For instance, "yellow" is daylight with the blue part missing, and so forth. Although such notions are very coarse, they must be counted a very significant improvement over Newton's "colors of homogeneous lights." The Schopenhauer/Goethe notions are much closer to the way colors relate to beams in natural circumstances. For some reason this was not appreciated in the nineteenth century,

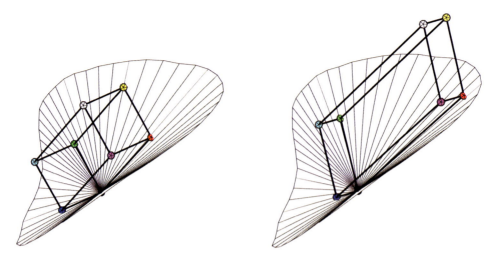

Figure 5.17 The RGB crates for average daylight (left) and a light bulb (right) in the spectrum cone. The volumes have been equated for easy comparison.

though. Even today, many people (perhaps especially physicists and philosophers) hold rather confused notions on this issue.

Schopenhauer had the idea that there exist "natural" boundaries of the parts. As I have shown, this is indeed the case. In order to appreciate the nature of these boundaries a little better, consider the following reasoning. A "yellow" beam might be due to a Newtonian "homogeneous light" of about 580 nm. Suppose this beam is due to daylight scattered from a paint layer. Then this beam is necessarily very dark, for if it reflects only a narrow part of the spectrum (ideally a part of zero width), then very little radiant power is scattered. As a result, the paint will not look YELLOW, but BLACK.[15] In order to improve the paint, more radiation needs to be scattered toward the eye, and the only way to do this is to widen the range of scattered wavelengths. In order not to change the hue, you may keep the dominant wavelength constant. Thus, if you admit additional long wavelengths in the scattered beam, you also will have to admit additional shorter wavelengths in order to compensate for the hue shift. As you make the width of the scattered region wider and wider, you run into a limit, though. From a certain width on there are no additional long wavelengths available because the region of scattered radiation has reached the IR-limit of the spectrum. Then you have *the brightest yellow paint* possible. Empirically, you find that all the best yellow paints are of this nature. They scatter all wavelengths except the short ones. Schopenhauer would say, "Yellow is daylight minus its blue part." The paint manufacturer has to agree or go broke. (See figure 5.18.)

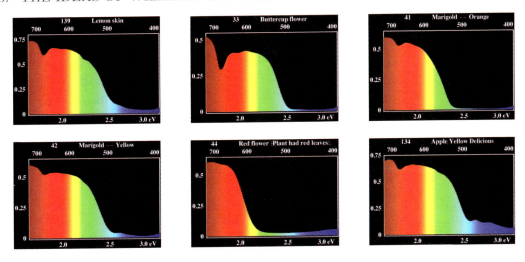

Figure 5.18 Some real-life "yellow" (pale yellow to red) objects.

5.5 The Ideas of Wilhelm Ostwald

Take an ideal yellow paint. What happens when you slightly shift its limiting wavelength? Well, if you shift it toward the long wavelengths you will darken the paint,[16] and if you shift it towards the short wavelengths you will lighten it. In neither case does the hue of the paint (that is its dominant wavelength) change at all. But in either case the paint becomes less vividly colored for you either add a little black or you add a little white to the paint. Thus the original (ideal) yellow paint is indeed "ideal" in the sense of being as vivid as it can be. Such notions can be made precise by formally expressing the color of the paint as a linear superposition of the achromatic beam and a monochromatic beam of the dominant wavelength of the paint.[17] This is always possible. For the ideal paint, the amount of monochromatic color apparently reaches the maximum possible value. This means that the Schopenhauer "part" is indeed an *optimal* choice.

In the early twentieth century Wilhelm Ostwald generalized these notions significantly with his concept of semichromes. Suppose I cut a part out of the daylight spectrum—for the moment you may think of a connected region of wavelengths between two limits. Such a part defines a color.

When is this color "most colorful"? According to Ostwald, obviously *when the limiting wavelengths are complementary.* For then widening the regions "adds white" whereas narrowing the regions "adds black," in either case making the color less colorful (white and black are not colorful at all, adding them "desaturates" a color). Such "full colors" (in German, *Vollfarben*) were called "semichromes" by Ostwald, who conceived of hues in terms of a "color circle" in which complementary hues are on diametrically opposite sides of the circle. (Figures 5.19 through 5.22.)

Ostwald's notion is very interesting indeed, because it solves a problem that even today continuous to puzzle people. The spectrum is evidently "incomplete" in that it lacks the magentas.[18] What then is the relation between the spectrum and the color circle? Ostwald's notion of semichromes, understood in the sense of Schopenhauer's parts of daylight, solves this problem in an elegant, intuitive manner. Although the spectrum is topologically a *finite segment*, the semichromes form a topological *circle*.

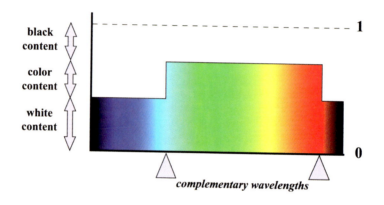

Figure 5.19 The reflectance spectrum of an Ostwald "semichrome." For a "pure" semichrome, both the white and black contents would vanish, and the color content would be one hundred percent.

Figure 5.20 If you "screen off half of the spectrum" and collect all remaining colors from the achromatic beam you obtain an Ostwald semichrome. Ostwald starts from the color circle (at left) and removes the nonspectral part (at right). In the color circle complementary hues are antipodes. If you screen of half of the remaining part (figure 5.21) you automatically get a semichrome.

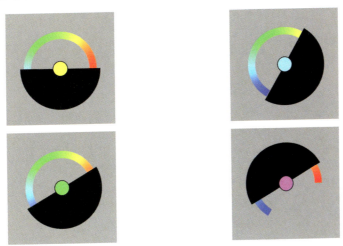

Figure 5.21 Yellow and cyan are low and high pass colors, green is a passband color and magenta a stop band color. These are all Ostwald semichromes. Think of the black sector as uniformly rotating; then this generates a periodic sequence of semichromes. Thus you generate the color circle from the spectrum! It may be Ostwald's most important insight.

Here is yet another interesting observation made by Ostwald, it concerns the mensuration of the color circle. Ostwald starts with the observation that the cardinal colors are like "beads on a string" and that they occur in mutually complementary pairs (figure 5.23). Thus the problem is how to space half of the cardinal colors, the others will fall into place automatically.

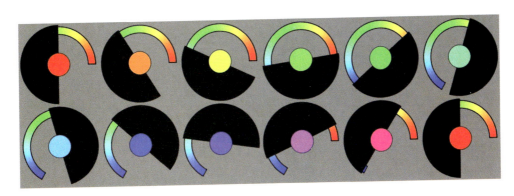

Figure 5.22 Generation of a full color circle.

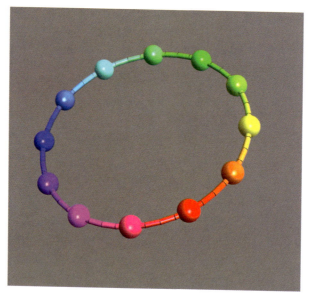

Figure 5.23 The cardinal colors are like beads on a string. Moreover, they occur in mutually complementary (antipodal) pairs.

Consider the periodic sequence YGCBMR. Notice that the equal mixture of two neighbors of any element has the same dominant wavelength as that element itself. For instance, R has the same dominant wavelength as the equal mixture of M, and Y, and so forth. You might say that each element "bisects the arc between its two neighbors." You may understand this as meaning that the arc length (of whatever curve it may concern) RY equals the arc length MR, and so forth. In other words, the Y, G, C, B, M and R are equally spaced on the arc YGCBMR. Ostwald calls this his "Principle of Internal Symmetry", of which more later. Whatever its meaning might be, intriguing it is (figure 5.24), because it introduces a *metrical* notion in a purely colorimetric (metricless!) way.

5.6   Colorimetry with Achromatic Beam

By now you will probably agree that the introduction of an achromatic beam (in a *deus ex machina* fashion, as you will remember) is well worth the initial embarrassment. It serves to sew some flesh on the bare bones (figure 5.25). Of course, all constructions that depend on it have the same drawback, namely that they indeed *depend on it*. Many entities in common use (for good reasons) exist only relative to some fiducial achromatic beam. It is

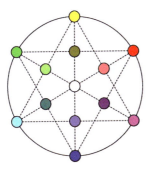 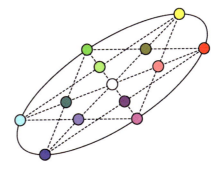

Figure 5.24 This illustrates the concept of Ostwald's principle of internal symmetry. Ostwald departed from the color circle and defined bisection of chords through equal mixture. This he uses to arrive at an arc-length parameterization. The method is invariant against affinities, hence independent of the choice of primaries. Although it is true, it is perhaps less clear where the color circle (as an affine rather than mere topological entity) derives from.

crucially important to remember this, and it is all too often forgotten. For instance, people often talk of "complementaries" or "dominant wavelengths" at the drop of a hat, as if these possessed some absolute existence. *Such is not the case.* Dominant wavelengths are much appreciated by many philosophers, probably because they remind one of Newton's ideas and the (mistaken) notion that color vision is seeing color by wavelength. Yet a beam may have

Figure 5.25 This is the basic chromaticity structure you need to remember. Here I arbitrarily drew a circular spectrum locus. In reality, this can assume a variety of (convex) shapes. The division into "green" and "magenta" (mutually complementary) and "red" and "blue" (mutually complementary) regions, where the magenta (or purple) region is not "spectral," is fundamental.

any dominant wavelength you fancy if you simply pick the appropriate achromatic beam. Only true monochromatic beams ("homogeneous lights") have a dominant wavelength in an absolute sense.

From the treatment of the Grassmann model (appendix E.4) it becomes evident that the "Schopenhauer parts" leave something to be desired. The reason is that the red and blue parts are indeed optimal in a true sense, but that the green part is *less than optimal*. (In the case of the Grassmann model, it even vanishes. Admittedly, this is a very singular occurrence.) The green part should really be an Ostwald semichrome. Unfortunately, it is not possible to optimize the various parts individually. Is there an optimum overall choice, then? *Yes*. If the achromatic beam is fixed, then you may try to build an RGB crate that captures as many of the colors possible under that illuminant, that is to say, a (hopefully unique) RGB crate of the maximum volume possible (see chapter 9). This is indeed a well defined problem. You may solve it through a brute force search algorithm, for instance. Simply pick ordered pairs of wavelengths $\{\lambda_1, \lambda_2\}$ with $\lambda_1 < \lambda_2$ and define the blue part as the range λ_{UV}–λ_1, the green part as the range λ_1–λ_2, and the red part as the range λ_2–λ_{IR}. Compute the volume of this crate. Now vary λ_1 and λ_2 over their full ranges (though constraining $\lambda_1 < \lambda_2$) and select the crate with the largest volume. Since volume ratios are affinely invariant, this is guaranteed to yield a result that is invariant with respect to changes of the primaries. Later I will show you a more elegant method to find such an optimum RGB crate. The concept is an important one.

For daylight, I find that the transition wavelengths for the best RGB crate are 480 nm and 565 nm, rather than 486 nm and 561 nm. The optimum crate has a volume that is only 1.12 percent more than that of the Schopenhauer crate defined earlier; thus, the improvement is very minor indeed. The optimum crate proves to be unique, much as expected. In most of the remainder of this book I will simply refer to the Schopenhauer crate as the maximum volume crate.

How good are such crates, that is to say, how much of the full gamut of possible colors lie within the crate? No doubt this question makes sense (there has to be an answer), but at this point in the exposition you don't have the tools to tackle it. I will return to it later. (See chapter 9.)

5.7   The Conical Order

Color space is a linear space, but it has actually more structure than that. The spectrum cone serves to define a "conical order" which makes color space a "partially ordered linear space" (figure 5.26). This is evidently more powerful than mere linearity. It induces a kind of "up-down" notion.

brighter

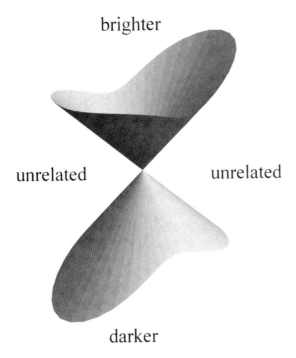

unrelated   unrelated

darker

Figure 5.26 The conical order.

Consider the spectrum cone with generators extended indefinitely toward both sides. The cone sits with its apex at the black point. Now consider translations of the cone parallel to itself. Then the apex may be positioned at any point, **c**, say. The cone divides color space into three regions, the colors in the "upper" cone, those in the "lower" cone, and those outside of the cone.

The colors in the upper region can be said to "dominate **c**", those in the lower region to "be dominated by **c**" and those outside the cone to "be unrelated to **c**". Thus you obtain a partial order of "domination", the colors form a "poset".

If a color **a** dominates color **c**, then the difference **a** − **c** lies inside the spectrum cone at the black point, thus it is a color (possibly) due to a real beam. Since adding a real color to a color can only make that color "brighter," it is reasonable to say that "**a** is brighter than **c**." Thus the conical order goes some way towards the definition of a heterochromatic brightness function. The relation remains undefined for mutually unrelated colors, of course.

The notion of conical order will gain a special importance in the theory of object colors (chapter 11).

The models

A good way to gain experience with formal colorimetry is to study simple models that allow you to handle things analytically, algebraically, or geometrically, rather than numerically. Moreover such models can be designed for low dimensions, and so forth, enabling you to exploit your intuition to the full. I treat a small number of models in some detail, but you will find it easy to design numerous interesting models yourself once you get the hang of it. The models treated throughout the book are:

The Maxwell model was already introduced on page 122. This is one of the simplest models in which everything can be done algebraically.

The RGB model is introduced on page 563. The RGB model is in many respects very similar to the Maxwell model. However, it can also be framed purely algebraically, omitting any notion of the "spectrum", and so forth. Then it becomes more than a model, it becomes a vital tool in computer graphics.

The Grassmann model is introduced on page 167. The Grassmann model is one on the most elegant analytical models imaginable in that is has the maximum possible symmetry: it is the same at any wavelength. You will gain much from studying it in depth.

The Helmholtz model is introduced on page 174. The Helmholtz model is in many respects similar to the Grassmann model, though with slight complications. It is best to consider it only after you have gained an understanding of the Grassmann model.

The Local model is introduced on page 178. This is also an analytical model. It has much in common with the Grassmann model, but it yields additional insight in the meaning of the color matching matrix. In the local model you "observe" certain spectral properties rather directly.

The Discrete model is introduced on page 163. The discrete model is purely algebraic and has been designed for the lowest interesting dimensions imaginable. This is the simplest possible nontrivial model of what "color vision" is about. You should thoroughly understand this model. Once you get it everything will fall in place.

Additionally, you may explore versions of the models for various dimensionalities. Thus you may study "monochromatic vision" (complete colorblindness), "dichromatic vision" (a condition of a sizeable fraction of the male population), "standard" trichromatic vision, or (more exotically) "tetrachromatic vision" (a small fraction of the female population might fall in this category, some animals certainly do), or general "polychromatic vision" ("God's Eye" perhaps?). This is done in appendix E.10.

5.1 How to find a projective transformation

In the context of colorimetry, you will often meet with the problem of how to construct a projective transformation such as to let four given points $\mathbf{a}$, $\mathbf{b}$, $\mathbf{c}$, $\mathbf{d}$ coincide with another set of four given points $\mathbf{p}$, $\mathbf{q}$, $\mathbf{r}$, $\mathbf{s}$ after the transformation. Thus you need to find a "homogeneous matrix" $\mathbf{M}$ such that

$$\begin{pmatrix} a_1 \\ a_2 \\ 1 \end{pmatrix} = \varrho_1 \begin{pmatrix} m_{11} & m_{12} & m_{13} \\ m_{21} & m_{22} & m_{23} \\ m_{31} & m_{32} & m_{33} \end{pmatrix} \begin{pmatrix} p_1 \\ p_2 \\ 1 \end{pmatrix},$$

and similarly for the other three pairs. Here "homogeneous matrix" means that the common factors $\varrho_{1\ldots4}$ of the matrix coefficients are arbitrary.

Notice that a relation as the above amounts to three linear equations for the coefficients. Since you have four such relations, it might seem that you have $4 \times 3 = 12$ linear equations for $3^2 = 9$ unknowns. But of course you still have to account for the fact that $\varrho_{1\ldots4}$ will remain indeterminate.

You may remove one unknown by (arbitrarily) setting $m_{33} = 1$. In most cases this is a good choice (for instance, in all cases where the transformation is not too far from the identity), if it is not, you have to figure out which coefficient you want to honor this treatment. Now you have 8 unknowns and 12 equations. The simplest way to proceed is to eliminate the common factors ϱ_i. For instance, consider the set of equations for the first point pair ($\mathbf{a}$ and $\mathbf{p}$) is

$$\begin{aligned} a_1 &= \varrho_1(m_{11}p_1 + m_{12}p_2 + m_{13}), \\ a_1 &= \varrho_1(m_{21}p_1 + m_{22}p_2 + m_{23}), \\ 1 &= \varrho_1(m_{31}p_1 + m_{32}p_2 + m_{33}). \end{aligned}$$

After elimination of ϱ_1 you have the two equations

$$\begin{aligned} a_1(m_{31}p_1 + m_{32}p_2 + m_{33}) &= \\ (m_{11}p_1 + m_{12}p_2 + m_{13}), \\ a_1(m_{31}p_1 + m_{32}p_2 + m_{33}) &= \\ (m_{21}p_1 + m_{22}p_2 + m_{23}), \end{aligned}$$

which are linear equations in the unknowns

$$\{m_{11}, m_{12}, m_{13}, m_{21}, m_{22}, m_{23}, m_{31}, m_{32}\}.$$

Taking the other three point pairs into account, you see that you end up with 8 linear equations in 8 unknowns; thus, you have (generically at least) a well determined problem with a unique solution.

Just for the record, such a problem is best solved using the pseudoinverse of the coefficient matrix for the problem. This automatically takes care of possible degeneracies.

5.2 The Maxwell model

The "Maxwell model was introduced earlier on page 122.

As was to be expected, the chromaticity diagram shows the spectrum locus as a "hook" (see figure 5.27). It looks very much like Maxwell's first attempt of the mid nineteenth century. It evidently captures the major structure of the human chromaticity diagram remarkably well. In this figure I have applied a projective transform to put the spectrum limits and the extreme green point at the vertices of an equilateral triangle. In order to do that, I needed the achromatic beam. I assumed a *uniform spectrum* for the achromatic beam to keep in line with the general simplifications of the model. The achromatic point is placed at the barycenter of the triangle. Notice that the "extreme green point" is indeed the vertex opposite to the purple gap.

The complementaries of the spectrum limits divide the spectrum locus into parts (figure 5.27). These define Schopenhauer's "parts of daylight."

This simple structure is well worth keeping in mind, since it sums up most of the relevant structure of the human color space. In remembering *structures* you remember *concepts*, and you do that by remembering *models* that capture the essence. As you shift your focus of interest, you jump from one model to the next, picking the one that suits the occasion best because it reveals the relevant structure of that moment in the clearest (and thus simplest) possible way. The Maxwell model is great for that, so you want to keep it in mind.

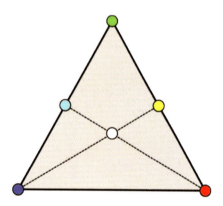

Figure 5.27 The chromaticity diagram for the Maxwell model. Here I assumed an achromatic beam with a uniform spectrum.

5.3 The discrete model

The "discrete model" is my attempt to construct the lowest dimensional model that is still of some interest (e.g., zero-dimensional models would be simpler, but hardly worth a second look). In the lowest dimensional case color space indeed *has* to be two-dimensional (for a one-dimensional system would be fully "color blind"), and the space of beams should be of low, but higher dimension (otherwise there would be no metamerism and colorimetry would be trivial). Thus the choice of a three-dimensional space of beams and a two-dimensional color space makes for the simplest interesting model.

As the "spectral basis" of the space of beams, I take $|m_1\rangle = \{1,0,0\}$, $|m_2\rangle = \{0,1,0\}$ and $|m_3\rangle = \{0,0,1\}$. In order to a obtain colorful description I will refer to these as the "short", "medium," and "long" wavelengths.

I assume that the visual system has two wavelength selective mechanisms, one with spectral sensitivity $S_1 = \{1, \frac{1}{2}, \frac{1}{4}\}$, the other one with spectral sensitivity $S_2 = \{\frac{1}{4}, \frac{1}{2}, 1\}$. I will refer to these as the "short" and "long" wavelength selective mechanisms. This is an arbitrary choice, of course. By playing with different choices, you may study variations on the model.

The responses of these two mechanisms to the monochromatic beams are $\{1, \frac{1}{4}\}$, $\{\frac{1}{2}, \frac{1}{2}\}$, and $\{\frac{1}{4}, 1\}$ respectively.

In colorimetric measurements, I will use the short and the long wavelength monochromatic beams of unit radiant power as my primaries. Thus the primaries are

$$|p_1\rangle = \begin{pmatrix} 1 \\ 0 \\ 0 \end{pmatrix}, \text{ and } |p_2\rangle = \begin{pmatrix} 0 \\ 0 \\ 1 \end{pmatrix}.$$

The response of the wavelength selective mechanisms to the primaries are $\{1, \frac{1}{4}\}$ for the short wavelength selective mechanism and $\{\frac{1}{4}, 1\}$ for the long wavelength sensitive one.

This choice of primaries is another arbitrary one (you need to assume an independent set of primaries though). You will find it instructive to play with a variety of choices.

Now you are ready to find the "colors" of the monochromatic beams, that is to say, to gauge the spectrum. The colors are $|c_1\rangle = \{1, 0\}$, $|c_2\rangle = \{\frac{2}{5}, \frac{2}{5}\}$, and $|c_3\rangle = \{0, 1\}$. This allows you to draw the spectrum locus in the (two-dimensional) color space. The spectrum cone is apparently the first quadrant of color space.

The color matching matrix for the discrete model is

$$\begin{pmatrix} 1 & 0 \\ \frac{2}{5} & \frac{2}{5} \\ 0 & 1 \end{pmatrix}.$$

As an achromatic beam I will take the spectrum $|A\rangle = \{1, 1, 1\}$. This is an okay choice, because nonnegative throughout. The achromatic color is $|a\rangle = \{\frac{7}{5}, \frac{7}{5}\}$. Again, you may want to explore alternative choices.

A reasonable choice for a "chromaticity" is to take the first coordinate of a color divided by the sum of the coordinates. The chromaticity of the achromatic beam is thus $\frac{1}{2}$, of the primaries 1 and 0, and of the monochromatic beams 1, $\frac{1}{2}$ and 0 respectively. Different choices will produce projective deformations.

The chromaticity diagram is one dimensional, it spans the segment $[0, 1]$, with the achromatic point at the middle $\frac{1}{2}$. (The "middle" is irrelevant, since bisection is not a projective property.) The short and long wavelength monochromatic beams are the spectrum limits, and the medium wavelength monochromatic beam turns out to be an achromatic color. (Figure 5.28.)

Notice that $|m_1\rangle$ and $|m_3\rangle$ are each other's complementaries, whereas $|m_2\rangle$ is (formally at

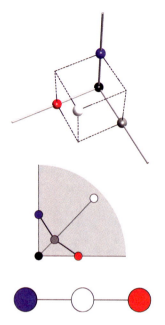

Figure 5.28 Top: The monochromatic and the achromatic colors in the space of beams; Center: The colors of the monochromatic and achromatic beams in color space. The thick line is the spectrum locus, the shaded area the spectrum cone; Bottom: The chromaticity diagram.

least) its own complementary. The achromatic point cuts the chromaticities into two sets that are mutually complementary. If you call $|m_1\rangle$ "blue" and $|m_3\rangle$ "red", and ($|m_2\rangle$ hueless, or "white," then all colors appear as either bluish or reddish, of various intensities and saturations. The "in between" chromaticies should not be considered "purple" (a habit you may have acquired through your lifelong experience with the human trichromatic system), there is no such a hue as "purple" in this model. For instance, the equal mixture of spectral blue with spectral red is hueless rather than purple. Thus, the simple discrete model turns out to be unexpect-

edly interesting. It is a good exercise to try to "grok"[19] what the color experiences of a "discrete" observer are like. This is to be considered a fully normal system, in no way degenerated or singular, but only low dimensional. Such experiences suggest perhaps that it might be rewarding to study different dimensionalities, such as tetrachromatic or pentachromatic systems. Such is indeed the case. There are plenty of surprises in store. What this boils down to is the realization that the human trichromatic system can be truly understood only if you know the effect of small dimensionality changes. A good many "properties" of color vision turn out to be specific to trichromacy whereas others hold true for any dimensionality. Such (extremely important) "comparative studies" are unfortunately not recorded in standard texts on human color vision.

This is a very simple system indeed and therefore a rich source of very useful exercises. One good exercise is to pick another set of primaries and repeat the computations, then do a transformation between the color spaces for the different primaries, and so on. Just play around with the model, and you will soon find all kinds of potentially interesting variations yourself.

For this system, it is not a problem to find the black space explicitly. The virtual beam

$$|k\rangle = \{2, -5, 2\}$$

has color $\{0,0\}$ and is thus "black." (Its chromaticity comes out as $\frac{0}{0}$ and thus is undefined, as it should be). If I add an arbitrary amount of black to any beam I do not change its color. The black space is simply the kernel of the color matching matrix displayed earlier.

With some thinking, you will see that "color space" looks exactly like the m_1m_3-plane of the space of beams if you project every point of the space beams on that plane using the black vector as the direction of projection. Thus color space appears graphically as a "shadow cast" of the space of beams. It is "visually" an oblique projection, though this is obviously a mere illusion, because I have not specified a scalar product in the space of beams. (Figure 5.29.)

This is a very nice illustration, enabled by the low dimensionality of the discrete model. Colors are "shadows" of beams, "thrown" by the direction(s) of the black space. Thus the human observer is revealed (colorwise that is) as being in the position of one of the prisoners in Plato's famous allegory of the prisoners in the cave who see reality only as the shadows on a wall.[20] Figure 5.29 yields a literal interpretation of this.

Another thing that is nice about the discrete model is that you can easily illustrate what the discrete observer sees. Notice that you can also illustrate the full spectrum image because the "spectrum" is exhausted by the RGB compo-

nents, which trichromatic human observers can indeed fully appreciate. In order to illustrate the discrete colors you need some convention of course. You easily show that the transformation

$$\begin{pmatrix} R_d \\ G_d \\ B_d \end{pmatrix} = \frac{1}{14} \begin{pmatrix} 10 & 4 & 0 \\ 5 & 4 & 5 \\ 0 & 4 & 10 \end{pmatrix} \begin{pmatrix} R \\ G \\ B \end{pmatrix},$$

does the trick. It changes the RGB values in such a way that white remains white ($\{1,1,1\}$) whereas the black color of the discrete model is mapped to $\{0,0,0\}$.

The "spectral image" shown in figure 5.30 is in fact just a normal RGB image. The interpretation of the "discrete colors" shown in figure 5.31 has been computed through the transformation given earlier. Notice that this rendering lacks a green-purple opponent channel and is indeed only dichromatic. The red balls come out reddish, the blue ball bluish, the white ball whitish, whereas the green balls and (light) purplish balls become neutral grays.

The discrete model is evidently trivial. However, it is not *essentially* different from the (in-

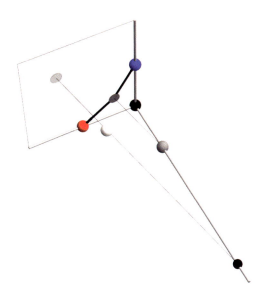

Figure 5.29 Color space as a "shadow" of the space of beams. Projection is by way of the black space (an oblique direction) on the plane spanned by the primaries.

Figure 5.30 A "spectral image" for the discrete model.

Figure 5.31 Rendering of the "discrete colors" for the spectral image.

finitely dimensional) case of human color vision. Understand the discrete model and you have understood the crux of colorimetry. This is as simple as I can possibly make it.[21]

5.4 The Grassmann model

5.4.1 History

In the mid-nineteenth century Newton's writings were gospel. When young Hermann Helmholtz published a paper on novel experiments on complementary colors in which he let the world share his observation that the "greens" (monochromatic beams with wavelengths in the midregion of the visible spectrum) did *not admit of (monochromatic) complementaries*, this was like cursing in church. The high school teacher Hermann Grassmann was so infuriated by this that he wrote a paper in response, stating in his introduction his aim of "proving mathematically" that Helmholtz was wrong and Newton (of course) right. This was perhaps not the best idea, for Helmholtz was a very careful experimenter, using very sophisticated instruments by his own design that made Newton's experimental design look like it came from the stone age. I don't think Helmholtz ever in his career got his *facts* wrong. But it is not that Grassmann was a nobody either. He invented most of linear algebra (and more) on his own steam and stands (from our present perspective) as one of the mathematical giants of all times. Few mathematicians understood him at the time, though, and eventually Grassmann, obviously disappointed, quit mathematics to become a famous philologist whose work on Sanskrit still stands today. Curiously, Grassmann's paper written against Helmholtz is also the first serious paper on theoretical colorimetry; in fact, there wasn't much that Grassmann missed. Only in the 1920s Grassmann's theoretical work was seriously extended, by Erwin Schrödinger of quantum mechanical fame.

Grassmann apparently fell for an elegant concept made up by Newton, Newton's concept of the *color circle*.

On the basis of his own experiments, Newton should have known better, but his concept of color space was pretty enough. To bad it didn't fit reality. Newton came to believe (as I say, in the face of his own observations) that the spectrum contains all imaginable hues (though he couldn't see all of these himself, he mentions a research assistant with "superior eyesight" who did), and that all imaginable hues are naturally arranged in a periodic sequence, the color circle.[22]

As artists have been aware for ages, the latter concept is indeed true; all imaginable hues *can* be arranged in a periodic sequence. But the former idea is wrong; not all imaginable hues appear in the spectrum.

Purples do not.

Newton populated the interior of the color circle with colors of lower vividness, placing white at the center. Apparently Newton already intuited Brouwer's fixed point theorem. Then Newton (for no obvious empirical reasons) dreamed up some laws for color mixture that amounted to linear combination. This is very awkward, since the interior of the color circle (why "circle" if only periodicity is a fact of experience?) has no obvious geometrical structure and can hardly be expected to be some variety of linear space. With these ideas in place Newton concluded that every "homogeneous light" (read "monochromatic beam") should have a complimentary mate, such that their even mixture should yield white. Indeed, if you think of monochromatic beams as beads of unit mass on the boundary of a circular disk (as Newton apparently did) then the center of gravity of two beads at diametrically opposite positions of the circle must lie at the center and thus be "white." Grassmann apparently found this Newtonian image irresistible.

5.4.2 The Grassmann model

The Grassmann model implements the Newtonian ideal. I have named it in the honor of Hermann Grassmann because Grassmann even defended the model in face of the facts. Indeed, the Grassmann model deserves to be true because of its elegance. Too bad the human race isn't perfect.

In figure 5.32 I have plotted a set of color matching functions that exactly implements Grassmann's Newtonian ideal. The functions are[23]

$$f_1(x) = \frac{1}{\sqrt{2\pi}},$$
$$f_2(x) = \frac{\sin x}{\sqrt{\pi}},$$
$$f_3(x) = \frac{\cos x,}{\sqrt{\pi}}$$

on $-\pi \leq x < +\pi$. The various factors derive from a convention that is irrelevant here (and indeed, they don't matter at all). Notice that the color-matching functions are the same at $x = \pm\pi$, thus the spectrum limits are not distinguishable, and this model will not give rise to a "gap of purples." This is indeed an implementation of the Newtonian notion that the spectrum is complete.

The color-matching functions do not gradually approach zero at the spectrum limits, which is of course an oddity of the model. Thus the spectrum locus does not issue forth from the origin, nor does it in any way approach it again. The spectrum locus simply "hovers in mid air" as seen in figure 5.33. The spectrum locus is a perfect circle.

Of course, this implies that the spectrum cone is a perfect right circular cone, without any need for a sector of purples. (See figure 5.34.) This indeed probably looks to be more of a "spectrum *cone*" to most people than the actual spectrum cone (for real human vision I mean) does. By now you should be convinced of the elegance of the Grassmann model.

Especially when you plot the chromaticity diagram (figure 5.35) it should become obvious

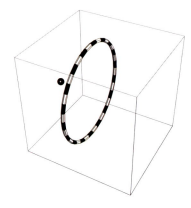

Figure 5.33 The spectrum locus of the Grassmann model is a perfect circle. Thus it is a closed curve, eliminating the need for a sector of purples. Notice that the locus is not connected to the origin (black sphere), which may be regarded as an oddity of the model.

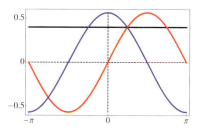

Figure 5.32 The color matching functions that define the Grassmann model are simply the truncated "fourier basis," that is, a constant, a sine, and a cosine function with wavelengths that fit the (finite) interval, here taken as $(-\pi, +\pi)$.

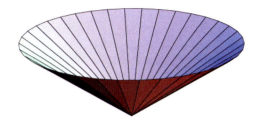

Figure 5.34 The spectrum cone of the Grassmann model is a perfect right circular cone, typically called just "cone" by non-mathematicians.

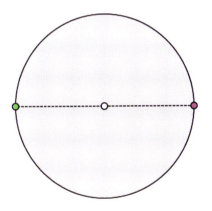

Figure 5.35 The chromaticity diagram of the Grassmann model. Here I defined chromaticity as the final two coordinates of a color, divided by the first one. The achromatic beam is assumed to have a flat spectrum. The spectrum locus in the chromaticity diagram is a closed circle, without a line of purples. As a result the usual configuration of cardinal points is highly degenerated.

to you that the Grassmann model—although indisputably nice and elegant—is going to throw you some problems. The nice "Schopenhauer parts" structure is completely degenerate in this model because

- the spectrum limits coincide

- the complementaries of the spectrum limits coincide

- the far green point is undefined,

and thus the whole scheme fails. Daylight simply falls apart into two parts, the "reds" and the "blues," say. This is unfortunate. Apparently the "awkward" gap of purples actually buys you something. This feature of the model suggests that you should reconsider the notion of the Schopenhauer parts of daylight, and I will do so soon. In this case an obvious solution is to build a Schopenhauer crate by dividing the spectrum into three equal parts.

In figure 5.36 I show the chromaticity diagram with fancy coloring. Don't get carried away with this, for the coloring is (of course) an arbitrary addition that has nothing to do with the Grassmann model per se. I have used a periodic sequence of hues.

A feature of the Grassmann model that is very convenient indeed is that all colors can be treated on exactly the same footing. This is a perfect setting in which to study Ostwald's notion of the "semichromes," because you need only study a single color and you have understood them all. Here I pick the dominant wavelength $x = 0$.

When you produce colors with the dominant wavelength $x = 0$ due to contiguous wavelength regions (simply varying the "slit width" of your spectroscope), then symmetry forces you to consider wavelength regions $-a/2 < x < +a/2$ for $0 < a < 2\pi$. I will consider an achromatic beam of uniform spectral density one.

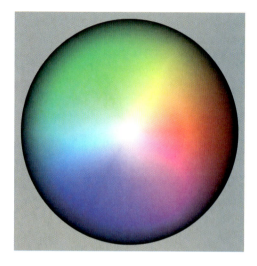

Figure 5.36 The Grassmann model chromaticity diagram with fancy coloring. Near the center (the achromatic point) the colors tend to gray; near the circumference (colors of very narrow passbands) they tend to black. The colors are most vivid at the Ostwald semichromes locus, which is a circle that lies fully in the interior of the chromaticity diagram.

Then the colors are:

$$\int_{-a/2}^{+a/2} \{f_1(x), f_2(x), f_3(x)\}\, \mathrm{d}x =$$

$$\frac{a}{\sqrt{\pi}}\{\frac{1}{\sqrt{2}}, 0, \mathrm{sinc}(\frac{a}{2})\},$$

where the "sinc function" is defined as

$$\mathrm{sinc}\, x = \frac{\sin x}{x}, \quad \mathrm{sinc}\, 0 = 1,$$

a function that assumes a maximum of one at the origin and has zeroes at $x = \pm\pi$. For $a = 2\pi$ you obtain the white color $\{\sqrt{2\pi}, 0, 0\}$. When you use a monochromatic beam with spectrum $\delta(x)$ (the Dirac delta function of unit weight) you

obtain a color $\{1/\sqrt{2\pi}, 0, 1/\sqrt{\pi}\}$. Thus you can write

$$\mathbf{a} = \frac{a}{2\pi}\left(1 - \mathrm{sinc}\frac{a}{2}\right)\mathbf{w} + 2\sin\frac{a}{2}\,\mathbf{m},$$

where $\mathbf{a}$ denotes the beam obtained for parameter a, $\mathbf{w}$ denotes the white beam, and $\mathbf{m}$ denotes the monochromatic beam of unit radiant power at dominant wavelength $x = 0$. This decomposition of a color in terms of a monochromatic and an achromatic component was proposed by Helmholtz. I will therefore refer to it throughout the text as the "Helmholtz decomposition."

As you plot the amount of monochromatic color and of the white color against the parameter a, you obtain the curves shown in figure 5.37. Notice that Ostwald's (intuitive) ideas are fully corroborated. You also find that the maximum amount of monochromatic color occurs for $a = \pi$, that is to say, exactly for the semichrome. For $a \to 0$ (the boundary of the chromaticity diagram) the colors tend to black, whereas for $a \to 2\pi$ (the center of the chromaticity diagram) they tend to white.

This result is "suggested" in figure 5.36. Indeed *suggested*, for colorimetry has nothing to say on qualia. In one sense, this figure looks like what must have been in Newton's and Grassmann's minds. The colors tend to white toward the center of the disk. But notice that the colors tend to black near the boundary of the disk. This is something neither Newton nor Grassmann envisaged. The reason must be obvious to anyone who ever looked in a conventional spectroscope. As you close the slit in order to make the spectrum purer, you lose sight of the colors because the field of view becomes darker and darker as you progressively close it.

Although an undeniable fact of experience, this fact is rarely acknowledged. The reason is no doubt Newton's uncanny power over people's minds.

Newton suggested that the best (in fact, the only real, uncontaminated) colors are the homo-

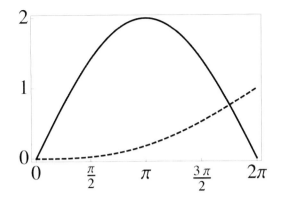

Figure 5.37 The monochromatic (drawn curve) and achromatic (dashed curve) content of a passband color, as a function of the width of the passband. The maximum of monochromatic contribution is reached for the Ostwald semichrome (full passband width of π).

geneous colors that populate the boundary of the color circle. Because Newton and almost anyone after him confused the interior of the "color circle" with what you now know as the "chromaticity diagram," most textbooks on color *even today* show suggestive pictures of the chromaticity diagram with the most vivid colors on the boundary, that is the spectrum locus.

This is an oddity that should be strongly discouraged. In any natural setting, the most vivid colors are the Ostwald semichromes, and their locus is a closed curve that encircles the achromatic point and is quite distinct from the spectrum locus (which is not even closed, except in case of the Grassmann model).

As you have seen, the Grassmann model is a great one to keep in mind. In a sense, it is an idealization" of human vision. The difference with the real system is so great that indiscriminate use of the Grassmann model easily gives rise to misunderstandings (as happened in the

case of Grassmann himself). On the other hand, the model is extremely nice in that it treats all wavelengths on an equal footing. Thus many researches on basic colorimetric structures profit from the use of the Grassmann model.

It is perhaps the preferred model if you want to obtain an understanding of the principles behind Wilhelm Ostwald's seminal (but often misunderstood) notion of "semichromes," since the explicit calculation is very simple in this model. I will return to this later.

5.4.3 "Partial models": Dichromacy in the Grassmann model

"Color blindness" is a favorite cocktail party topic, everybody's interested in it. Indeed, it is an obvious source of fascination that the world might appear differently to some of your fellow men. If you're a "normal trichromat" (and chances are that you are), then it is of interest to notice how some people (mainly males) confuse certain obviously different hues (mainly reds and greens). If you're a fairly typical dichromat yourself (if you're a male that is not extraordinary rare), you will guess forever at the rich world of "normal trichromats" (most people). The topic is of minor importance to this book though. I will simply show some examples of what dichromacy might be. I will not enter into the topic of actual dichromacies and various anomalies of generic trichromatic color vision in humans.

Suppose the three color matching functions of the Grassmann model were individually implemented as specific physiological "mechanisms" (obviously unlikely in view of the fact that they are not nonnegative throughout), and suppose it were possible (I won't go into the genetics) that there existed individuals that would lack one of these mechanisms. Then likely dichromatic subsystems could be imagined that would simply lack one of the three color matching functions. Of course, this would apply at most to the second and third mechanism, since the first one roughly implements the monochromatic "back-bone structure" of vision.[24]

If the second mechanism were missing, that is to say,

$$f_2(x) = \frac{\sin x}{\sqrt{\pi}},$$

you would be left with two even functions of wavelength. Thus, such an observer would be quite unable to detect odd functions of wavelength, for instance the spectral slope. On the other hand, if the third mechanism, that is to say,

$$f_3(x) = \frac{\cos x,}{\sqrt{\pi}},$$

were missing, the observer would be quite unable to detect even functions of wavelength, for instance the spectral curvature.

Of course the generic (trichromatic) observer is also be unable to detect many spectral features. The two dichromatic observers are more accurately said to miss the projection upon the second and third color matching function.

It is perhaps most informative to plot the loci of "confusion" for the dichromatic observers in the chromaticity diagram of the trichromatic observer (figure 5.38). It is a priori evident that:
— all chromaticities distinguished by a dichromat will be distinguished by the trichromat;
— some chromaticities that are distinguished by the trichromat will be confused by one of the dichromats;
— the two dichromats will rarely agree on the discriminability of beams.
Since everything is nicely linear, the dichromat sees the space of beams as a projection upon a basis that is a subbasis of the trichromatic basis. Thus the confusion loci in the chromaticity diagram of the generic trichromatic must be families of concurrent straight lines. These lines must pass through the chromaticity of the missing basis vector. Thus, in our "standard" chromaticity

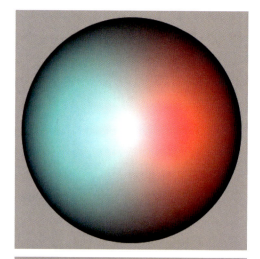

Figure 5.38 The two types of dichromacies in the chromaticity diagram of the generic trichromat.

diagram, the confusion loci are simply families of parallel straight lines, parallel to one axis for the first and parallel to the other axis for the other dichromatic observer.

Instead of drawing confusion loci in the chromaticity diagram of the trichromat, you may also plot dichromatic chromaticity diagrams. The latter would be one instead of two dimensional. Such chromaticity diagrams are typically finite line segments with the achromatic point somewhere in the interior (at the center say). Each end of the segment would be a certain hue for the dichromat and all chromaticities are desaturated versions of these hues. A transition from the one hue to the other necessarily passes through the achromatic chromaticity. In "colorful terms," the dichromatic observer who cannot see spectral slope will see greens and magentas that pass over into each other via white, and the dichromatic observer that cannot see spectral curvature will see blues, and yellows that pass over into each other via white. The trichromatic observer has greens, magentas, blues and yellows and need not necessarily pass from one to the other via white. Of course you should take such a description with a grain of salt since colorimetry has nothing to say about color appearances. The question whether the blue for the dichromatic observer is like the blue for the trichromatic observer has no answer in colorimetry.

In real life there exist (rare) observers with one trichromatic and one dichromatic eye.[25] You can simply ask these observers to compare notes on the appearances as experienced from either eye. It is not easy to make much sense of these reports, though, and of course even this "obvious" experiment is not at all conceptually trivial, when you come to think of it.

5.5 The Helmholtz model

Helmholtz was the first to discover the gap of purples, and thus a model designed to study the gap of purples (and its consequences) deserves to be named after him. The Helmholtz model is designed to be as simple as possible but (unlike the Maxwell model, which also sports a gap of purples) also be amenable to *analytical* treatment. Thus the color-matching functions are taken as low-order polynomials. (I assume a standard basis of Legendre polynomials.[26]) These color matching functions are (figure 5.39):

$$h_0(x) = \frac{1}{\sqrt{2}},$$

$$h_1(x) = \frac{3}{\sqrt{2}}x,$$

$$h_2(x) = \frac{5}{\sqrt{2}}\left(\frac{3}{2}x^2 - \frac{1}{2}\right).$$

Thus the color-matching functions are a constant, a linear slope, and a parabola. Notice that the colors of the endpoints of the interval are different; thus there is indeed a gap of purples. Notice also that the Helmholtz model has a nice symmetry with respect to the center wavelength $x = 0$.

Because the color matching functions do not approach zero near the end points of the interval, the spectrum locus again hovers in the air. In figure 5.40 the spectrum locus plotted. It is a parabolic arc. Notice that—unlike the Grassmann model—this spectrum locus is an open instead of a closed arc.

In contrast to the Grassmann model, the spectrum cone (figure 5.41) of the Helmholtz model requires a sizeable sector of purples in order to close it. This is indeed a simple and geometrically nice model on which to study the effect of the gap of purples.

The chromaticity diagram is shown in figure 5.42. It is made up of a parabolic arc, the spectrum locus, and a straight segment, the line

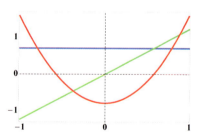

Figure 5.39 The color matching functions that define the "Helmholtz model" are confined to the interval $(-1, +1)$. They are composed of the constant function, a linear slope, and an on-axis parabola.

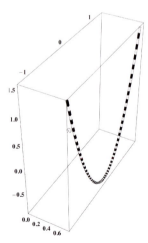

Figure 5.40 The spectrum locus of the Helmholtz model is a parabolic arc. It is not connected to the black point (also plotted in this figure).

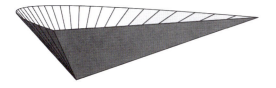

Figure 5.41 The spectrum cone for the Helmholtz model sports an appreciable sector of purples.

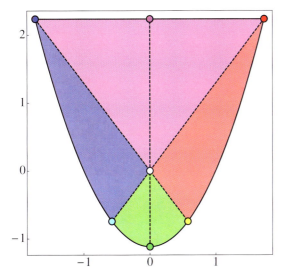

Figure 5.42 The chromaticity diagram of the Helmholtz model. In this case I defined the chromaticity as the second and third coordinates of the color divided by the first one, this yields a pleasant layout. The configuration of cardinal points is as usual, in this respect the Helmholtz model is very similar to the Maxwell model. Here I assumed that the achromatic beam has a flat spectrum.

of purples. You obtain the familiar geometrical configuration for the extreme green point, the complementaries of the spectrum limits, and

the spectrum limits themselves. Thus you obtain the Schopenhauer parts of daylight, much as you found it for human vision and quite unlike the case of the Grassmann model.

For the Helmholtz model everything is so nice that you can explicitly derive the relation for the complementary wavelengths explicitly, you have

$$x\overline{x} + \frac{1}{3} = 0,$$

where x and $\overline{x}$ are mutually complementary. The relation is plotted in figure 5.43. Because you need $|x| \leq 1$ as well as $\overline{x} \leq 1$, you obtain only solutions if $|x| > 1/3$. The hyperbolic relation $x\overline{x} + 1/3 = 0$ is formally identical to the one used by Helmholtz to describe his empirical findings for human vision, a formula still found in many textbooks.[27] Of course, Helmholtz's formula is

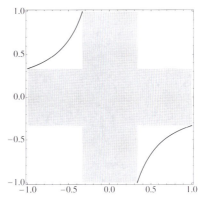

Figure 5.43 A plot of the complementary wavelength pairs for the Helmholtz model. Again, the achromatic beam is assumed to have a flat spectrum. Wavelengths in the range $-1 < x < -1/3$ have complementary mates in the range $1/3 < x < 1$ and vice versa. Wavelengths in the range $-1/3 < x < 1/3$ ("greens") fail to have complementaries, hence the "gaps" plotted in gray tone.

only an approximation; moreover, the relation depends upon the achromatic beam used as reference. However, the roughly hyperbolic relation is generic, and the Helmholtz model shows the reason for this in the most lucid manner.

Because the Helmholtz model is so simple, you can derive the Ostwald semichromes analytically, although the expressions are not particularly enlightening. In figures 5.44 and 5.45 I show plots of the semichrome locus in color space and in the chromaticity diagram. As you see, the curve is a closed one, in contradistinction to the spectrum locus, which is an open, convex arc. The semichrome locus is a curvilinear, polygonal arc made up of four smooth arcs. The curve

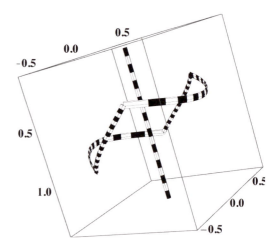

Figure 5.45 The semichrome locus for the Helmholtz model in color space encircling the achromatic axis. Notice that the locus is a closed, twisted space curve. It is made up of four smooth stretches.

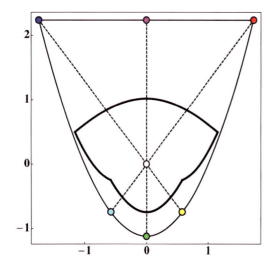

Figure 5.44 The semichrome locus for the Helmholtz model in the chromaticity diagram. Notice that the locus is made up of four smooth, curvilinear arcs that meet at an angle. Two of the arcs are parts of the edge color loci. The locus in color space is a twisted (nonplanar) space curve, that is centrally symmetric about the median gray point (figure 5.45).

is (by construction) centrally symmetric about the median gray point, that is the equal mixture of the achromatic color with black. This is a very important locus, especially in the theory of object colors, and I will return to its discussion repeatedly. For the moment, you should simply get familiar with its shape.

Of the four smooth pieces of which the locus is made up, two are pieces of edge color loci (one cool, one warm), one is made up of the passband (or "Newtonian," or "spectral") colors, and one is of the stopband (or "non-Newtonian," or "non-spectral," perhaps "inverted-spectral") colors.

For the case of actual human vision, you will find that the semichrome locus is very similar indeed to the one here found for the Helmholtz model. Thus the Helmholtz model is very convenient in studying the semichrome structure in one of the simplest possible (analytical) settings. The Helmholtz model is more interesting (because less symmetric) than the Grassmann model, though.

It is also very instructive to derive Schopenhauer's parts of daylight for the the Helmholtz model. The cut loci suggested by the basic structure due to the spectrum limits and their complementaries are $\{-\frac{1}{3}, +\frac{1}{3}\}$. However, the maximum volume inscribed RGB crate is obtained for the cut loci $\{-\frac{1}{\sqrt{5}}, +\frac{1}{\sqrt{5}}\}$. The ratio of volumes is $\frac{100\sqrt{5}}{243} \approx 0.920\ldots$, thus the maximum volume crate has roughly ten percent more volume than the one obtained from the spectrum limits. This is more extreme than the case of human vision, thus the Helmholtz model is perfect to investigate the issue.

5.6 The Local Model

Consider the following engineering problem: Radiant power density spectra in the range 360–840 nm have to be encoded for transmission. Bandwidth limitations force you to limit the code to three numbers (say in the range 0–255). Little is known concerning the type of spectra to be transmitted, but general physical considerations suggest that the spectra will be fairly smooth distributions. Such a problem might turn up in the design of an autonomous robot designed to scout the Martian surface for instance.

How would you encode the spectra? Notice that this problem is not unlike the design problem for color vision of terrestrial animals, at least, when the "only three numbers" and the wavelength window constraints have already been decided upon.

In this section I will use the range $(-\infty, +\infty)$, with the center of the "visual range" at the origin. Remember that it is easy enough to transform this to a realistic human visual range if you feel the urge to do so.

An obvious idea to attack the problem would be to simply divide the wavelength range into three parts and to measure the total radiant spectral power in each. This is essentially nothing but the RGB-model and could be easily implemented with three dichroitic filters in front of photocells. Another idea would be to measure in three strategically placed narrow bands. This could also easily be implemented via interference filters in front of photocells.

A little thinking shows that the second method is in many ways inferior to the first, because large parts of the spectrum are ignored. You may remember the engineering course on discrete sampling and start worrying about the effects of aliasing. It is also terribly wasteful on photons. That is unfortunate because the noise

levels in the system are determined by the photon counts. The more photons you catch, the better.

Here is an example of the type of things you may do. In figure 5.46 I show a certain rough spectrum. How do you summarize such a spectrum?

Well, the spectrum is very rough indeed, but if you look through your eyelashes you perceive a roughly parabolic profile (figure 5.47). So why not fit the spectrum with a linear mixture of polynomials of low order? This is a very general method, clearly *any* spectrum can be fitted with an average value, a linear trend and a curvature, though with variable degrees of success. (see figures 5.48 and 5.49).

You obtain a smooth but intuitively reasonable approximation to the rough spectrum. Clearly such a method will often turn out to work well. The "official" method would be to simply do a *Taylor expansion* (but of a sufficient smoothed spectrum of course) near the center of the visual range. This is one particular form of summarizing.

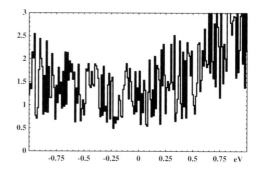

Figure 5.46 A very rough spectrum. Notice that I use the full real axis for photon energies. This is physically unrealistic, but it is only a model.

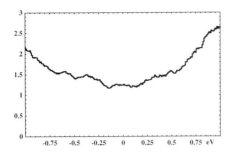

Figure 5.47 A smoothed (or low passed) version of the rough spectrum. After sufficient smoothing any spectrum will start to look like a parabolic arc.

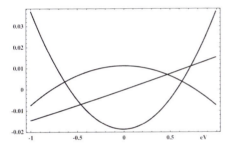

Figure 5.48 An average value (constant), a trend (linear slope), and a curvature can be combined to approximate *any* curve.

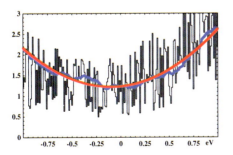

Figure 5.49 A second-order approximation to the rough spectrum. Notice that the fit is not bad.

What all problems boil down to is that

any method will introduce a selective blindness of some type

You should use all prior knowledge available to you in order to minimize the harmful effects. In this case, the only bit of prior knowledge available is that the spectra are expected to be fairly *smooth*. This suggests that you might perform a local Taylor approximation and encode the differential structure, say average radiance, spectral trend and spectral "curvature." However, Taylor series involve *differentiation*, and you were told in engineering school to avoid differentiation of real signals because it is well known to be unstable.

In order to make some progress, it is necessary to consider the sampling problem from a general and abstract perspective. Suppose you have some one-dimensional distribution of stuff (in our application radiant power density). How do you measure the radiant power density at a specific location (in our example: at a given wavelength)? This is the problem of quantitative spectroscopy in a nutshell.

The first observation to make is that the task is *impossible*. It is physically impossible to measure the density of stuff with infinite precision. Moreover, no theory in physics has anything to say concerning measurements performed with infinite precision. In all cases this would extrapolate from the domain of empirical knowledge upon which the theory was built. In the case of radiant power spectra, the best theories predict that the variance of the spectral power will grow without bounds when the precision (inversely proportional to the slit width of the monochromator) is increased indefinitely.[28] About the only option is to select the best tool for the job and accept its finite slit width.

Let the slit be centered at a certain wavelength and let the slit width be set at a certain value. You measure the total radiant power that passes the slit and summarize this observation as "The radiant power density at the given wavelength is the total radiant power that passed the slit divided by the slit width". Usually the radiant power density is specified as a *single number*. There is a problem here, for when you use a different apparatus (different slit width) you obtain a different number. The more cautious and correct way to specify the result of the observation is "The radiant power density at the given wavelength is so and so much *for the specified slit width*". Although the difference might appear slight, it is really enormous. You don't speak any more of *the* spectral radiant power density, but you specify a *one parameter family* of such densities, the slit width being the parameter. When you measure the spectrum (vary the fiducial wavelength) you obtain different spectra for different slit widths.

Most people are not satisfied with such a (perfectly correct) account. They want the *real* spectral radiant density, not some one-parameter family (whatever that may be). The point to grasp here is that the "real" spectral radiant density is a *nonentity*. The reason is that the "scale" (in this case: slit width) is essentially *arbitrary*. What is the "correct" slit width: 10^{-10} nm, 1 nm, or 10^{10} nm? Who knows? It all depends on the context. There is *no such a thing* as the "real spectral density" and that is why you *have* to specify a one parameter family. At least when the context is undecided. If the resolution is decided upon a priory there is no problem of course.

The one-parameter family radiant power density as a function of slit position and width is by now generally known as the "scale space" of (in this case) radiant power density spectra.[29] Given the scale space you may ask for the *relations* between spectra at different scales. How does the spectrum change when you change the slit width? (See figures 5.50 and 5.51.)

Intuitively, at least part of the answer is obvious enough. You should certainly expect that:

Decreasing resolution should at most destroy detail, but certainly not generate it.

This is perhaps the very idea of "resolution." In order to make this more precise, you should specify what you mean by detail. In this case you may use a very simple and immediately obvious criterion (the situation is slightly more complicated in higher dimensions):

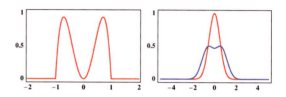

Figure 5.50 When you take a weird "slit" like the doubly peaked one on the left, nobody is surprised to find that the observation of a peak (red curve on the right) reveals a spurious *doubly peaked* shape. This shows that the shape of the slit should be an important consideration.

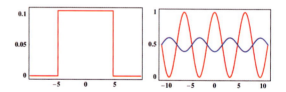

Figure 5.51 Even with a "normal" slit like the boxcar shape on the left, things may go terribly wrong. In the example on the right a series of peaks (red curve) is being observed. The observation (blue curve) has multiple peaks (as expected), but peaks are "observed" that really are valleys and vice versa. Thus the boxcar slit yield "spurious resolution" too. You can prove that the Gaussian slits are the only ones that behave well.

Decreasing the resolution should only lead to the destruction of local extrema, but in no case to their creation.

This is a simple constraint that lends itself readily to formal analysis when you combine it with some obvious constraints that derive from our assumption of a lack of prior knowledge.

The method of sampling should not depend on spatial position, nor on scale, nor on the actual density. This essentially constrains the sampling method to the application of a translation invariant and self-similar kernel. The slit of the monochromator is one instance. The kernel is of the simple "boxcar" type.

It turns out that the constraints I mentioned above completely determine the admissible kernel. It *has* to be a Gaussian (bell-shaped) curve; thus you need a slightly "blurry" slit. Only then the destruction of extrema but not their creation under progressive blurring is guaranteed (figure 5.52). Thus even the simple, pure "boxcar" shape turns out to be a bad choice.

The Gaussian kernels are bell-shaped functions specified via their half-width and total unit weight (figure 5.53).

When you start from the "actual" spectral power density (here the "actual" is symbolic for "zero slit width," clearly a physical impossibility), the observed density will be the "convolution" with the Gaussian weight. Here the "convolution operator" is a "blurring," or "weighted running average". At every wavelength I arrive at the "observed spectral power density" by integration over the "actual" spectral power density with a weight equal to a Gaussian with unit weight centered on that wavelength. The width of the Gaussian is the analog of the slit width and specifies the "resolution" of the observation.

This picturesque description should be taken with a grain of salt though. The reason is that the description refers to the *actual* spectrum,

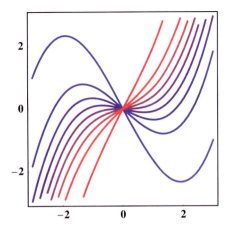

Figure 5.52 A typical example of Gaussian blurring. The minimum/maximum pair (blue) become a mere uniform slope on blurring. Blurring removes extrema in pairs.

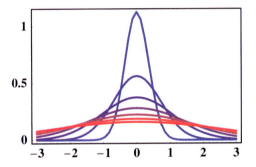

Figure 5.53 Gaussian kernels of various widths.

which is not observable, and thus has a shaky status at best:

the "actual" spectral power density is quite unknown to you. It might even be said not to exist in an operational sense. All you can ever know are samples at—necessarily finite—resolution.

The picture of "blurring the actual power density" should be understood as a symbolic way of indicating a certain physical operation, namely the observation of the spectral radiant power density at a certain wavelength and at a certain resolution. Notice that *any real observation* is necessarily at some finite resolution (when the slit is closed no observation is possible).

The blurring is a *linear operation* (essentially a weighted integration), just like differentiation. This is important because linear operations have many properties that enable you to make a large number of general statements concerning their effect. In this case I am especially interested in the result of the concatenation of differentiation and blurring. You expect the *sequence* of operations to be irrelevant to the result of the combined operation. More specifically, you expect the results of

— blurring and consequently differentiating;

— differentiation and consequent blurring;

— and convolving (blurring) with the derivative of the blurring kernel

to be the same. In some (very) loose formal sense this may be true. In a stricter formal sense it is not, and in practice these sequences of operations are widely different. It is very important to understand the differences. I belabor this point because even so much of the technical literature seems to miss the point completely. Apparently all three sequences of operations mean to indicate the derivative of the spectral radiant power density at some level of resolution. But

the first operation yields the derivative of the observed density. The problem here is that the observed density is only empirically known. It probably only exists as a list of observed values. The warnings from engineering classes apply here. Don't even try to differentiate such entities. Thus the

first "method" actually has no reasonable operational meaning;

the second operation doesn't even make any sense to start with. You don't know the "real" density let alone that you could differentiate it. The issue of differentiability doesn't even arise. Thus the second "method" cannot be considered to be a useful operational meaning either. However,

the third operation has a very simple operational meaning. You use a certain kernel (or weighting function) to *observe* the derivative. The actual observation is implemented via an *integration* (blurring). This type of computation is quite stable, well defined and poses no particular problems. Notice that differentiation is applied to the Gaussian blurring kernel, which is an *ideal*, not an empirically *observed* entity. Unlike real things you may differentiate ideal things without thinking twice. Differentiability is part of the very *definition* of this ideal entity.

This is good news. It is possible to *observe* (not *compute*) the derivatives of the various orders (figure 5.54) in an inherently stable manner. The only *proviso* is that the observations are at a particular level of resolution. But this is necessary *anyway*. In this respect the derivatives are not different from the density itself. *The* density is a nonentity. The only *real* densities are *observed* densities, and this implies that they are densities at some finite level of resolution.

In summary, it is certainly feasible to *observe* the initial part of a Taylor series in an inherently robust manner. Such an observation yields a local description of the shape of the spectrum. The precision of the shape description depends on the order at which the Taylor series has been truncated, and the region of validity of the de-

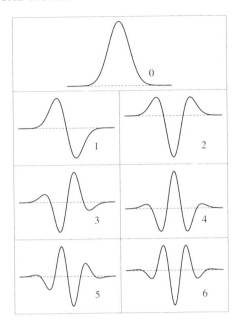

Figure 5.54 The derivatives of order zero to six. These operators yield the exact derivatives at the scale defined by the width of the zero order operator.

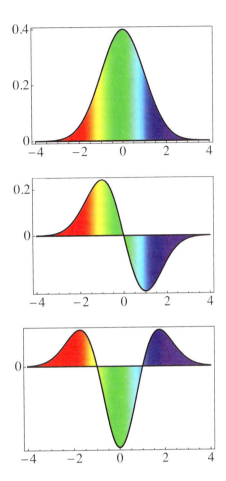

Figure 5.55 The "fundamental response curves" of the local model. The first fundamental is simply a Gaussian, the second fundamental is the first derivative of a Gaussian, it is much like a yellow-blue opponent mechanism, the third fundamental is the second derivative of a Gaussian, it looks much like a green-magenta opponent mechanism.

scription is set by the level of resolution at which the observations took place. In the specific case of trichromacy, the order has to be truncated at two (this yields three terms, the constant level, the trend and the "curvature"), whereas the required level of resolution is set by the width of the range: 380–830 nm. The resolution ("blurry slit width") has to be in the 200–400 nm ballpark.

In figure 5.55 I show the fundamental response curves (G_1, G_2, G_3) for the local model. Notice that I haven't worried about the issue of nonnegativity. It is easy enough to amend that. The first fundamental function is a Gaussian with (roughly) the visual region as support, the second fundamental is the first derivative and the third fundamental is the second derivative of this Gaussian. The only weird thing here is that I let the "photon energies" range over the full real axis (minus to plus infinity), with the center of the visual region at the origin. I simply didn't think it worthwhile to implement transformations to try to fit the real world. The thing to do would be to fit the Gaussians on the logarithm of the photon energies normalized by the energy of the center of the visual region. Although this would look more "realistic," I think it not worth the trouble because the local model is only a *model* anyway. In the present form the model is simply more transparent and easy to use.

This (very general and abstract) reasoning immediately leads to a very practical implementation. When you apply these kernels at the center of the visual region (say 550 nm), with a resolution $\Delta\lambda = 100$ nm, you obtain an interesting model set of sensitivity functions, very similar to certain linear transformations of the human fundamental response curves. I will denote this set of three functions (Gaussian bell-shaped curve, first derivative of the Gaussian, second derivative of the Gaussian) as the *sensitivity matrix* of the local model.

You may complain that the response curves cannot be physically implemented because they fail to be nonnegative. There simply are no filters with such characteristics. The solution is to form three linear combinations that are nonnegative throughout the range. This is not hard (you can do it in infinitely many ways). You may obtain three bell-shaped (unimodal) curves, mutually staggered along the wavelength axis, curves that are quite reasonable.

Notice that you will not encounter the term local model in any standard text on colorimetry. I just made it up for this occasion. It is an ideal model, that is, one aimed at conceptual understanding, not at physiological or psychophysical modeling at all. This is an *analytic* model, and thus you can do all kinds of analytic calculations and prove nice theorems. Moreover, this ideal model is remarkably close to the actual

human color vision system, though—of course—it is different in *detail*. That the human system is very much like the ideal local model means that the real color system comes very close to specifying the initial three terms of a Taylor series of the spectral radiant power density, centered at about 540 nm and with a resolution of roughly 100–200 nm. This is a valid and intuitively fruitful way to think of the human color system, though various other perspectives yield additional insights.

In figures 5.56 and 5.57 I show the spectrum locus and two views of the spectrum cone of the local model. It turns out that the cone is closed, so there is no need for a plane of purples. Because the model is so simple, it is possible to construct (very simple) analytic expressions for all these entities.

It is not hard to work out the spectrum cone and the chromaticity diagram (figure 5.58) for the local model. Many convenient definitions of the chromaticity will work well, but it is possible to pick one in which the spectrum locus is a full circle. The chromaticity of the uniform spectrum doesn't lie in the center of this circle, though. (This may be a disappointment if you

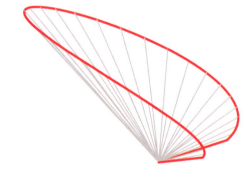

Figure 5.56 The spectrum locus for the local model. The limiting spectrum cone generators coincide.

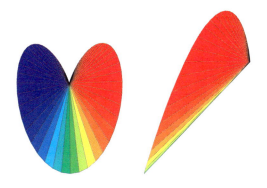

complementary pairs are related via a simple hyperbolic curve of two branches (figure 5.59).

The local model provides you with yet another and intuitively very attractive perspective on what color vision achieves:

> *Human color vision roughly determines the value, slope, and curvature of the spectrum at some photon energy with some (rather low) resolution.*

Figure 5.57 Two views of the spectrum cone for the local model. The limiting tangents turn out to coincide. Thus the cone is closed, and you don't need a sector of purples.

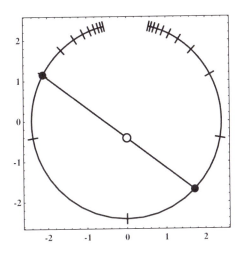

Figure 5.58 Chromaticity diagram for the local model. It is possible to pick a representation such that the spectrum locus becomes a perfect circle. The circle is actually closed. I have only drawn a (large) part of it.

are of an aesthetically oriented mind.) Thus the local model does not need a "line of purples" at all, and *every hue has a complementary hue*. The

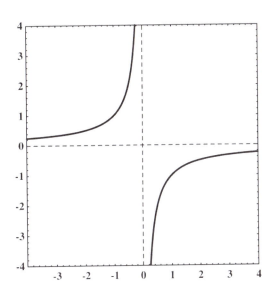

Figure 5.59 The complementary pairs for the local model. For any photon energy I can find a complementary mate. No problem with greens or purples here. Newton and Grassmann would be happy.

5.7 Luminance

I will start the discussion of luminance with a topic that doesn't address luminance directly, but is somewhat related to it. It involves the "far green point."

The concept of the "extreme green point" can be generalized. Intuitively, the extreme green point is the monochromatic beam that is at the largest distance from the plane of purples. Is is possible to define such a distance for *any* color? Evidently you would like to have a *linear functional* whose contraction upon a color would yield a measure of the distance of that color from the plane of purples. Let's call this desired linear functional $\langle\beta|$. What can you say about it?

Since points in the plane of purples are evidently at zero distance from it, you have that $\langle\beta|c_{UV}\rangle = 0$ and likewise $\langle\beta|c_{IR}\rangle = 0$, where c_{UV} and c_{IR} are colors of the spectrum limits. In order to determine $|\beta\rangle$, you need one additional condition, a "calibration."

The calibration is somewhat arbitrary, you have to stipulate for some color that it is at unit distance from the plane of purples, then the linear functional is fully determined. You may define $\langle\beta|c_G\rangle = 1$ where $|c_G\rangle$ is the color of the monochromatic beam of unit radiant power at the largest distance from the plane of purples (the extreme green point).

Thus you have the three linear equations

$$\langle\beta|c_{IR}\rangle = 0,$$
$$\langle\beta|c_{UV}\rangle = 0,$$
$$\langle\beta|c_G\rangle = 1,$$

and consequently the linear functional $\langle\beta|$ is fully determined.

I have defined the linear functional in color space, but it can easily be defined on the space of beams too. Define the linear functional $\langle\beta^\star|$ on the space of beams as follows

$$\langle\beta^\star| = \langle\beta|\left(\sum_{i=1}^{3}|p_i\rangle\langle\mu_i|\right),$$

where the $|p_i\rangle$ are the primaries and $\langle\mu_i|$ the color matching functions. Evidently $\langle\beta^\star|k\rangle = 0$ for any black beam $\mathbf{k} \in \mathbb{K}$. The actual location of the green point is not relevant since it only affects the scaling of the linear functional.

In figure 5.60 I show the effect of this functional on the monochromatic beams of unit radiant power. In figure 5.61 I attempt to sketch the functional in the chromaticity diagram.

In a sense the linear functional measures how luminous a beam is; thus you may well call it "luminosity." (There exists also a technical meaning of luminance, via a CIE convention, but I will consider that later. At this moment there is no connection.) Since the colors in the plane of purples are mixtures of (just about) IR and (just about) UV, it is perhaps reasonable that

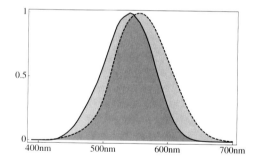

Figure 5.60 The "greenness" function compared with the luminance function (dashed curve). This function is based on a flat radiant power spectrum, it will be different for daylight, and so forth.

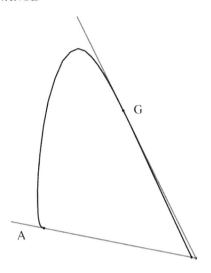

Figure 5.61 The linear functional in the chromaticity diagram, drawn by way of the alychne and the tangent plane at the far green point spectrum cone generator. A is the alychne, G the far green point.

they should be assigned zero luminosity. After all both the UV and IR regions are supposed to be invisible.[30]

The meaning of this luminosity is geometrically obvious. Given a color, I construct a plane that contains the color and is parallel to the plane of purples. This plane intersects the achromatic direction in a point, an achromatic color. This color can be directly compared to the color of the achromatic beam, since they are multiples of each other. This is a simple way to compare colors that are not simple multiples; you transfer them all to the achromatic direction via planes parallel to the plane of purples.

It is somewhat more intricate to locate luminosity in the chromaticity diagram. The only entity you can use is the plane of purples, because it is mapped on the line of purples in the chromaticity diagram. All chromaticities on this line

have zero luminosity, thus you have located the locus of vanishing luminosity in the chromaticity diagram. It might be referred to as the alychne or (literally) lightless line, but conventionally the alychne is associated with the luminance functional. Different definitions of a "luminosity function" are obtained by redefining the alychne.

Notice that there is very little freedom in the construction of the luminosity functional. Your only choice is in the alychne and in the calibration; otherwise there is no freedom. But in the choice of alychne you have much of a choice, for the only hard constraint is that it should not intersect the spectrum locus in the chromaticity diagram.

The extension of the luminosity functional to the space of beams is fully constrained by the black space. Notice that any deviation (apart from the calibration) of the definitions given here must lead to inconsistencies in the sense that there would exist black (invisible) beams with finite (positive or negative as the case may be) luminosity. But the notion that a black beam might have *some* (as different from none) luminosity assigned to it is evidently preposterous. You simply cannot admit such inconsistencies.

This means that the luminosity functional is largely determined by the color matching functions, it has to be some nonnegative linear combination of these. Anything else is automatically inconsistent.

Any linear functional on color space can be extended to the space of beams in a consistent manner. You just have to make sure that any black beam will yield zero, thus the functional must be a linear combination of the color matching functions. Any linear functional whose zero level plane does not intersect the spectrum cone and yields positive values for all monochromatic beams is a possible candidate for a luminance functional. Thus the simple constraint is that the alychne (in the chromaticity diagram) does not intersect the spectrum locus.

Unfortunately, "the" luminance has been standardized early (by the CIE in 1924) and became one of the defining parts of the standard photometric unit "candela." This is indeed unfortunate, because you can hardly redefine the candela as that would imply that many calibrated instruments in daily use all over the globe would become useless, but on the other hand, the conventional luminance function no longer is a linear combination of the most dependable color matching functions. Thus photometry and colorimetry have become incompatible.[31] In order to get around this, you need to redefine the "luminance functional" as a linear combination of the color matching functions that is (in some sense) "closest" to the empirically determined functional.[32]

Empirically, you find that the alychne based upon the luminance functional is quite different from the plane of purples, thus the "luminosity" as defined above is quite different from "luminance." The former luminosity functional remains of interest, though, because it is a linear functional that is invariantly implied by the color matching functions. It may perhaps be called "greenness."

5.7.1 The "official" luminance functional

In the old days photometric units were defined via standard candles, that is to say, candles (or various other flames or glowing bodies) prepared according to a certain fixed protocol, using standard ingredients. Many physicists became famous by having a candle named after them.

This unsatisfactory state of affairs changed for the better in 1909 when the national laboratories of the United States of America, France, and Great Britain decided to adopt the international candle represented by carbon filament lamps. Interestingly, Germany decided to stay with the Heffner candle, defined by a flame standard. The Hefner candle amounted to about nine-tenths of the international candle. It is hard

to say who was right; after all carbon filament lamps are just as arbitrary as candles, hard to produce and run reproducibly and probably even less stable over time.

The ideal solution was well known in theory, namely a black body at some standard temperature, but could not be produced. In 1933 the principle was adopted that new photometric units would be based on the luminous emission of a blackbody at the freezing temperature of platinum (2045°K). This eventually led to the adoption of the *candela* in 1948 (amended in 1967).

In 1979, because of the experimental difficulties in realizing a Planck radiator at high temperatures and the new possibilities offered by radiometry, a new definition of the candela arose:

> *The candela is the luminous intensity, in a given direction, of a source that emits monochromatic radiation of frequency 540×10^{12} hertz and that has a radiant intensity in that direction of 1/683 watt per steradian.*

Notice that this definition leaves open the wavelength dependence.

The "need" for a "luminance" in photometry does in no way imply that "heterochromatic photometry" is a viable enterprise to begin with. But for reasons altogether mysterious to me the scientific community has come to adopt the beliefs that

- heterochromatic photometry is possible;
- there exists a luminance functional that is of global applicability;
- this luminance functional is linear.

In my view there is not a shred of evidence to sustain such cheerful beliefs. All that exists (at best) are local patches of "equibrightness surfaces" that do not necessarily globally mesh. But there you are. The photometry community firmly subscribes to these myths.

In order to deal with the wavelength dependence, you really need the visual efficiency function, that is, the luminance functional. In 1924, the CIE adopted a standard photopic luminous efficiency function for 2° angular subtense photopic viewing conditions, CIE 1924 V_λ, which is still used today to define luminance (figures 5.62 and 5.63). In retrospect the definition was most unfortunate, because the "official" curve is known to fit modern psychophysical data badly.

For instance, it is not a linear combination of the best, generally adopted color matching functions, thus is inconsistent. Even the data on which the definition was based is not all too clear. Several laboratory results (obtained via

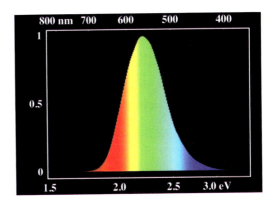

Figure 5.63 The official CIE "V_λ-curve" as a function of photon energy. A figure like this yields some insight into the extent of the visual range. It is about the range of chemical binding energies. Higher photon energies destroy vital biological tissue, while lower energies do little of interest with vitally important materials.

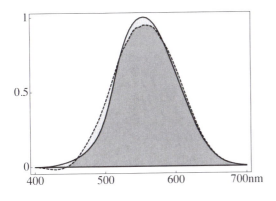

Figure 5.62 The official CIE "V_λ-curve" (drawn) and an approximation that is a possible luminosity functional (dashed curve). This "best approximation" shown here is slightly odd (or useless if you want) in that (small) negative values occur in the blue (below 455nm). This is odd, because anything you *see* should conceptually possess positive luminosity. Thus you need to apply the nonnegativity constraint. Of course "best approximation" is ill defined anyway, since you have no metric.

widely different methods) were somehow "combined" in order to arrive at the "best" estimate, a dangerous procedure as there was (or is) absolutely no guarantee that the various methods measured the same entity.

If you are not forced (for instance, legally) to use the visual efficiency function it is probably best to leave it alone. In a great many important applications you won't even miss it.

5.7.2 Hering's notion that pure colors have no brightness

Ewald Hering entertained the idea that "pure colors" have no brightness; only achromatic colors have brightness. I will substitute "luminance" for "brightness" here. It is usually thought that this is a belief concerning intrinsic properties of colors. But is it?

The luminance functional foliates color space into planes of constant luminance. The alychne is such a plane, the one in which all colors have vanishing luminance. Apparently only those colors would satisfy Hering's constraint. All but the black color are necessarily virtual, of course.

Consider a basis such that two of the basis vectors span the alychne. As the third basis vector I take the achromatic point. It is clearly transverse to the alychne, thus independent of the former pair of basis vectors. Being virtual, these two basis vectors are the differences of pairs of real colors, thus they are Hering's "opponent colors," which arise naturally in this context.

In this basis, any color is decomposed in an achromatic component and a "pure color" component that lies fully in the alychne. Only the achromatic component has nonzero luminance, whereas the "pure color" component is not luminous at all. Thus I have constructed a description in which Hering's notion holds true without exception.

Notice that this renders Hering's notion somewhat trivial. It is seen to be not so much an interesting intrinsic property of colors as a mere result of the existence of a luminance functional with positive luminance for all spectral colors. (But, of course, the latter is itself problematic.)

5.8 Saturation

If the achromatic beam is "hueless" and monochromatic beams the most vivid hues ever seen, then you may wonder whether the "amount of vividness" of arbitrary colors (for instance the Ostwald semichromes) can be quantified? Such a measure is conventionally known as the "saturation" of a color.

The conventional definition is based upon the luminance functional. You can always decompose[33] a color as the sum of an achromatic component and a monochromatic component of the correct dominant wavelength (for the moment I will disregard the case of the purples). This is fine as it goes, but in order to define a saturation measure you need to be able to compare these components. The luminance functional solves this problem, for you simply compare the luminances of the components. For instance, the luminance of the monochromatic component divided by the sum of the luminances of the monochromatic and the achromatic components yields a convenient measure that runs from zero (achromatic) to one (monochromatic).

This conventional definition is fine as it goes, but you perhaps wonder whether it would be possible to define a notion of saturation in the absence of a luminance functional. After all, it is not that there are no conceptual problems with the luminance functional, since it runs counter to the standard type of response reduction on which colorimetry is founded. It turns out to be the case that such a definition is indeed possible.

In considering the concept of "saturation," you deal with a planar geometry, namely the section of color space with the plane spanned by the fiducial color and the achromatic beam. This plane meets the boundary of the color cone in two generators, for instance (disregarding the case of purples and greens for the moment) a certain monochromatic color and its complementary mate. Thus you deal with a planar configuration of four concurrent lines (figure 5.64). For such a configuration there exist a well-known projective (and thus affine) invariant, namely the *cross-ratio*[34] of the four lines. You can set it up in such a way that the cross-ratio equals zero if the fiducial color coincides with the complementary monochromatic color, with unity if the fiducial color coincides with the achromatic direction, and with infinity if the fiducial color coincides with the monochromatic color. (Figure 5.65.) Then the logarithm of the cross-ratio

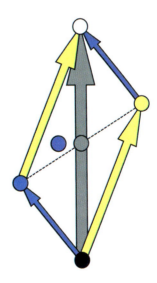

Figure 5.64 The desaturated blue color is located with respect to the blue (same dominant wavelength) and yellow (complementary dominant wavelength) spectral generators. Notice the coplanar configuration of these colors with the achromatic color. The invariant cross-ratio of this quartuple can be used to define "saturation" in a natural way.

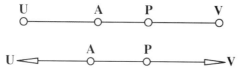

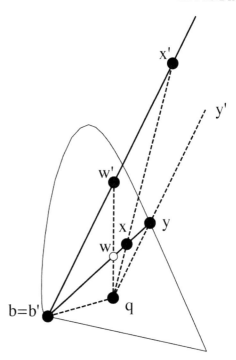

Figure 5.65 Let A be the achromatic color and V the monochromatic color at the dominant wavelength of the fiducial color P. Let U be the complementary of V. The quadruple UAPV is coplanar. Here I have drawn it (at top) in the chromaticity diagram. It is a finite line segment with four points, the monochromatic colors at the extremes, the achromatic color somewhere in between, and the fiducial color between the achromatic color and the spectral color V. Using the projectively invariant cross-ratio, you assign numbers to points on the line UV, such that U corresponds to 0, A to 1, P to something, and V to infinity. Taking the logarithm throws U to minus infinity with V still at plus infinity, whereas A becomes the origin of the full line UV (at bottom). Now the location of P varies between minus and plus infinity. Either infinity implies maximum saturation whereas the origin implies zero (minimum) saturation. You have obtained a very useful saturation scale for the fiducial color P.

Figure 5.66 Construction of the saturation of a color $\mathbf{x}$ in the chromaticity diagram. The point $\mathbf{q}$ is arbitrary, though not on the line $\mathbf{wx}$.

will equal zero for the achromatic direction, infinity for the monochromatic, and minus infinity for the complementary monochromatic direction. (If you wish, it is easy enough to get rid of the infinities,[35] but they don't bother me.) What I especially like here is that this definition of saturation depends on both the generator at the dominant wavelength of the fiducial beam as well as on the generator at the complementary of the dominant wavelength. This introduces a tighter structure than the conventional definition does.

A simple construction in the chromaticity diagram is shown in figure 5.66. Here I take simple

case of a desaturated yellow $\mathbf{x}$, the spectral color of which is $\mathbf{y}$, the complementary of $\mathbf{b}$. Thus the quadruple $\{\mathbf{b}, \mathbf{w}, \mathbf{x}, \mathbf{y}\}$ is colinear. I pick an arbitrary chromaticity $\mathbf{q}$ (say) and draw the line through $\mathbf{b}$ parallel to the line $\mathbf{qy}$. I find the intersections $\mathbf{x}'$ and $\mathbf{w}'$. Consider the colinear points $\{\mathbf{b}', \mathbf{w}', \mathbf{x}'\}$ (with $\mathbf{b} = \mathbf{b}'$) and the "infinite point" (not drawn $\mathbf{y}'$. The ratio of Euclidian distances $\mathbf{x}'\mathbf{b}'$ and $\mathbf{w}'\mathbf{b}'$ is the cross-ratio you need; its logarithm is the saturation of $\mathbf{x}$. This is an example of a projective construction in the chromaticity diagram, rarely (if ever) attempted in the standard accounts, yet elegant and useful.

Notice that this sophisticated definition will work just as well for the greens and purples. It really makes no difference; you simply select generators from the boundary of the spectrum cone, whether they happen to be "spectral" or not. In figure 5.67 I show a simple example, the saturations of Schopenhauer's parts of daylight. Notice that the two definitions yield results that are in a roughly monotonic relation. There exists no *simple* relation between these measures of course.

Although I prefer this definition over the conventional one, I will not use it much in this text. The reason is simply the sad fact that it seems at too large a remove from conventional praxis. Too bad.

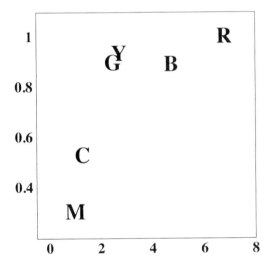

Figure 5.67 The "saturation" of the Schopenhauerian parts of daylight. Here I compare the intrinsic definition of saturation (ordinate) with the one based on the luminance functional (abscissa). In either definition the saturation of cyan and magenta is much lower than that of green, yellow, blue, and red, red being the overall winner.

5.9 Hue

Hue is generally understood as the unique, special quale that is evoked by the generators of the spectrum cone. The "REDNESS" (or "what is feels like to experience RED") is the "hue" associated with very long wavelength monochromatic beams. Of course colorimetry has nothing to do with this. Objective measures for the location of the generators are the dominant wavelengths. For the purples you have to set up a different scheme; for instance, you could indicate them by the wavelength associated with the complementary generator (there is guaranteed to be such a generator). This is somewhat awkward though, because "wavelengths" are descriptive labels of monochromatic beams, but have no obvious colorimetric relevance or explanation. Colorimetry could easily have been developed in absence of the notion of "wavelength." For instance, you may conceive of a civilization that labeled monochromatic beams "monoenergetic" and used photon energy as label, the idea of "wavelength" being foreign to their thinking. (For instance, they might consider photon energy the biologically relevant parameter and wavelength essentially irrelevant with respect to visual processes. In fact, I would sympathize with them.) Thus, it would be far more elegant to do away with this wavelength labeling business. Is this possible?

It is indeed possible once you have decided on an achromatic beam. For then you have a triple of well defined invariant planes at your disposition, namely:

- the plane spanned by the achromatic direction and the infrared limiting tangent to the spectrum locus;

- the plane spanned by the achromatic direction and the ultraviolet limiting tangent to the spectrum locus;

- the plane spanned by the achromatic direction and the generator of the extreme green point of the spectrum locus.

Notice that these three planes are concurrent, for they hold the achromatic direction in common.

Now an arbitrary generator of the spectrum cone also spans a plane with the achromatic direction. Thus you obtain a fourth plane, concurrent with the other three. (Figure 5.68.) Remember that the cross-ratio of four concurrent planes is a projective (thus certainly an affine) invariant. Thus the cross-ratio of the above four planes is invariant against changes of the primaries. Since three of the four planes are fixed, the value of the cross-ratio depends only on the

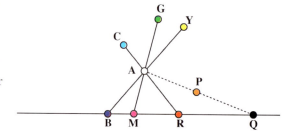

Figure 5.68 The idea of a projective hue label: B and R are the spectrum limits, Y and C the complementaries in the chromaticity diagram. G is the far green point and M the complementary purple. Notice that R, B, and M are co-linear; they lie on the "line of purples". Take any spectral color P; then you find its trace Q through projection from the achromatic point on the line of purples. Since BMRQ are a co-linear quadruple, you may compute a cross-ratio that measures the position of Q (and thus P). This cross-ratio can be used as a label instead of wavelength.

spatial attitude of the fourth one. *Thus you can use the invariant cross-ratio to measure the position of any generator with respect to the spectrum limits and the achromatic direction.*

A simple way to see how it might work is to consider the configuration in the chromaticity diagram. Use the achromatic point to map any point of the spectrum locus or the line of purples on the line of purples itself, you do this by construction of the intersection of the line defined by the achromatic point and the fiducial point with the line of purples. This is evidently a one-to-one map if you use *oriented projective geometry*.[36] On the line of purples you have a triplet of fixed points, namely the spectrum limits and the complementary of the extreme green point. The image of the fiducial point is variable (you may pick the fiducial point at your convenience) and the cross-ratio of there four collinear points (which is the same as the cross-ratio of the planes considered above) is a function of the fiducial point. Thus you obtain a nice parameterization of the "hues" that is quite independent of any notion of "wavelength."

The arctangent of the cross-ratio yields an angle, and if you are careful (use oriented projective geometry) you can assign angles to the boundary of the locus of real chromaticities in a consistent manner. In figure 5.69 I show the hue angles as a function of wavelength for human vision. Notice that though the relation is (by construction) neatly monotonic, it is (of course) far from linear.

Notice that the slope of the hue angle function is very shallow near the spectrum limits. It is of some interest to plot the slope as a function of wavelength (see figure 5.70). There apparently exist two spectrum regions where the angle varies very fast with wavelength. This is reminiscent of the wavelength discrimination curve for human observers, which has often been determined psychophysically.[37] If I plot the reciprocal of the slope (figure 5.71), I indeed obtain a relation that

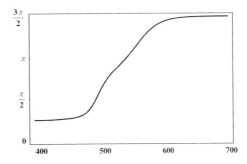

Figure 5.69 The hue angle (computed from the cross-ratio) for human vision.

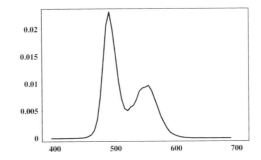

Figure 5.70 The slope of the hue angle as a function of wavelength.

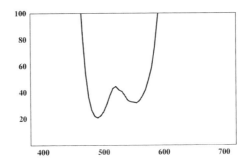

Figure 5.71 The reciprocal of the rate of change of the hue angle with wavelength ("slope") looks much like the familiar human wavelength discrimination threshold function.

is similar to the wavelength discrimination curve. There exist differences in detail, though. What this shows is that the intrinsic manner of denoting locations along the spectrum locus is very different from a parameterization by wavelength.

Nice and elegant as the construction may be, conventional colorimetry has decided to ignore it (I have never even seen it mentioned in the literature). Thus I will not pursue this notion in the text.

5.10 Polychromacy

5.10.1 Monochromatic vision

"Monochromatic color vision" perhaps sounds like wooden iron. It is not, though. It is indeed the simplest possible color system, implemented over and over again in various animal species. In essence, monochromacy has most of the complexities (perhaps I should say perplexities) of the higher-dimensional cases. It would have made sense to start this book with the discussion of monochromatic systems; however, that would perhaps have been considerably less motivating.

For monochromatic vision, the chromaticity space shrinks to a single point; there is only a single hue. To call this hue "achromatic" makes little sense, though: achromatic with respect to what? Color space is just a line, so colors differ along a "brightness" dimension. All monochromatic colors may be made to look the same, simply through adjustment of their intensities. When two colors are discriminated one is necessarily brighter than the other.

Two distinct monochromatic systems differ in their fundamental response curves. This often makes a huge difference. In order to be able to appreciate that you may consider photographs made with "orthochromatic" or "panchromatic" emulsions (such films can still be bought, even in this era of digital cameras). (See figure 5.72.) In ancient photographs (orthochromatic emulsion) the sky is very light and landscapes look misty. In later photographs (using panchromatic emulsion with a yellow or orange filter) the blue sky is dark, clouds stand out in high contrast, and you look much farther into the landscape. The differences are striking. There are indeed infinitely many different "monochromatic" visual systems possible. Throughout the animal kingdom you encounter many distinct implementa-

Figure 5.72 A scene (top) and an orthochromatic (center) and panchromatic (bottom) monochrome rendering.

tions. Most are roughly tuned to the solar spectrum, of course.

Human monochromatic vision is a such rare phenomenon as to be of little interest here. (Moreover, that interest would be largely of a physiological nature.) However, monochromatic images in printing are still current enough, and they offer excellent opportunities to study the varieties of monochromacy.

5.10.2 Dichromatic vision

Dichromacy is not particularly rare among the male population. Some forms of dichromacy are much more rarely encountered than others, though. The reasons for this are understood, but they do not particularly concern me here.

Cases of dichromacy have been important in physiological optics. Since colorimetry reveals only arbitrary linear combinations of the fundamental response curves, dichromacy has been used as an important handle on the actual curves.[38] König simply considered dichromacies to be subsystems of the more generic trichromacy, an assumption that indeed led to useful results. In this view, the trichromat has three types of "elementary sensations" (*Grundempfindungen* in German) and consequently you expect three types of dichromats, each characterized through the lack of one of these elementary sensations. Hence the terms "protanope" (lacking the first sensation), "deuteranope" (lacking the second sensation), and "tritanope" (lacking the third sensation) for dichromats. Although the idea remains useful, modern research has added numerous important complications and additions.[39] Since this book is not about physiology, I skip these details.

In order to obtain a feeling for what it might be like to be in the condition of a dichromat[40] (or monochromat), you may take an RGB image and tweak it such as to become dichromatic. This can be done in numerous ways, but—taking your cue from König—you may try to

substitute GGB for RGB (that is, "protanopia"), or RRB for RGB (that is, "deuteranopia"), or RGG for RGB (that is "tritanopia") for instance (figure 5.73). It is instructive to play with such substitutions.

You will notice numerous differences between the dichromatic renderings and the original, or the dichromatic renderings compared with each other. All these dichromatic renderings still contain much chromatic information though. Dichromats must confuse many chromaticities that are effortlessly discriminated by trichromats, but their worlds are by no means dull. But then, even monochromacy is not too bad, you could easily survive on monochromatic vision. This is obvious from the fact that we all effortlessly deal with black and white photographs, for instance. You don't even miss the color, except in special cases.

Notice how similar the protan and deutan renderings appear. For most people the (distinct) dichromacies are simply "red-green blindness." Indeed, the short wavelength mechanism stands far apart from the medium and long wavelength mechanisms, which overlap heavily.

A very simple model for dichromatic vision is obtained if you omit either the second or third dimension from the Helmholtz model. In these "stripped Helmholtz models," color space is only two-dimensional, and the chromaticity space one-dimensional. If you plot the "confusion loci" in the chromaticity diagram of the trichromatic Helmholtz model (figure 5.74) you find that they form concurrent line bundles (because of the projective structure, these may also appear as parallel line bundles).

The dichromatic chromaticity spaces are one-dimensional, and the dichromats enjoy only two different (and complementary) hues at various saturations. For dichromats there exists a wavelength such that the monochromatic beam of that wavelength appears achromatic, a "hueless" monochromatic beam.

Figure 5.73 Protan, deutan, and tritan dichromatic images of the scene shown in figure 5.72 top.

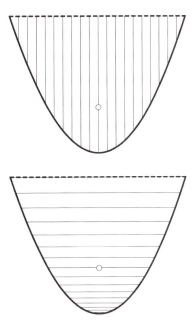

Figure 5.74 Two types of dichromacy of the dichromatic Helmholtz model. Here loci of confusion are plotted in the chromaticity space of the trichromat. Thus the dichromats have only one-dimensional chromaticity spaces.

Most monochromatic beams have many "complementaries" in these dichromatic systems. There exists a monochromatic beam that appears achromatic; monochromatic beams at either side in the spectrum have the same hue and differ only by saturation. Thus, there exist only two hues (apart from achromatic) for the dichromats, a very much reduced "color circle" (pair of points)!

Such simple models indeed suffice to illustrate the essence of dichromacy. There exists an extensive literature, but that is concerned with the physiology and psychophysics, not colorimetry.

5.10.3 Tetrachromatic vision

The understanding of any object is much illuminated by the study of its substructures (or its specializations) as well as its superstructures (or generalizations). It is only through perspective and focus that you have any hope to understand *anything* at all.

The understanding of human color vision is deepened when you consider various substructures such as dichromats, monochromats and even blind people. The various models used in this book are likewise examples of such specializations.

Far less has been done in the direction of generalization. For instance, human color vision might be studied in the context of visual systems throughout the animal kingdom, technical camera systems, and so forth. The simplest generalization in a purely colorimetric context is probably to consider *tetrachromacy*. More specifically, it is simple (and rewarding!) to consider the tetrachromatic models of which the Grassmann and Helmholtz models are "specializations." Indeed, I hold the view that it is not possible to understand human color vision fully without having appreciated the tetrachromatic (or polychromatic of any order) embedding. I consider it to be somewhat of a scandal that the literature has so little to offer here. Color science is embarrassingly anthropocentric.

The simplest model is probably an extension of the RGB model (chapter 14). Suppose you have a "fourth channel," say Q (for "quag"), then you may consider the "RGBQ color (hyper)cube." The RGBQ hypercube has sixteen vertices. The significance of two of these is immediately obvious: $R^{00}G^{00}B^{00}Q^{00}$ is "black," or the "origin," whereas $R^{99}G^{99}B^{99}Q^{99}$ is the white point. The primary colors are

Red $R^{99}G^{00}B^{00}Q^{00}$,

Green $R^{00}G^{99}B^{00}Q^{00}$,

Blue $R^{00}G^{00}B^{99}Q^{00}$,

Quag $R^{00}G^{00}B^{00}Q^{99}$.

Apart from these primary colors, there are six binary mixtures. I will stick to the familar "yellow" for "red-green", "cyan" for "green-blue," and "magenta" for "red-blue," but (since names are not available) I will use "red-quag", "green-quag," and "blue-quag" for the others. Apart from the binary mixtures there exist four ternary mixtures that are truly "colored" (not monochromatic).

A three-dimensional projection by way of the achromatic axis (see figures 5.75 and 5.76) shows a convex polygon with fifteen vertices (of course, white and black coincide in this view) and twelve faces. Such a figure clearly reveals the vertices that are as far from the achromatic axis as possible, which are the *full colors* (see figure 5.77). Thus the polyhedron may be considered the equivalent of the *color circle*. Apparently you have to reckon with a *color sphere* in the tetrachromatic case. The full color locus has the topology of $\mathbb{S}^2$, and thus the possible relations between hues are much richer than in the trichromatic case. This makes tetrachromacy an attractive case to study.

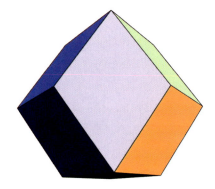

Figure 5.75 The RGBQ cube projected into three dimensions along the achromatic axis.

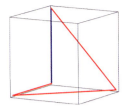

Figure 5.76 The RGBQ spectrum locus in the chromaticity diagram. It is like Maxwell's hook, only one dimension up. The spectrum locus is the red line, and the blue line is like a line of purples, though in this case it fails to complete the boundary of the convex hull of the spectrum locus. The convex hull is a tetrahedron, the spectrum locus only accounting for a measure zero part of its boundary.

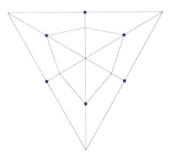

Figure 5.77 The RGBQ full colors in the chromaticity diagram. This is an especially pleasant, symmetric view.

The "spectral sequence" is evidently the sequence

$$R^{00}G^{00}B^{00}Q^{00}$$
$$R^{99}G^{00}B^{00}Q^{00}$$
$$R^{00}G^{99}B^{00}Q^{00}$$
$$R^{00}G^{00}B^{99}Q^{00}$$
$$R^{00}G^{00}B^{00}Q^{99}$$
$$R^{00}G^{00}B^{00}Q^{00}.$$

The spectral sequence visits five of the sixteen

vertices of the RGBQ cube, starting and ending in the black point (see figure 5.76).

In order to obtain a more intuitive picture, it makes sense to consider the chromaticity diagram. This is a three-dimensional space, so you can at least form "visual pictures" for your mind's eye. However, all my warnings against the indiscriminate use of chromaticity diagrams fully apply here. An obvious way to define chromaticity of the color $\{r, g, b, q\}$ is as

$$\frac{\{r + g - b - q, r - g - b + q, r - g + b - q\}}{r + g + b + q}.$$

Chromaticity space is a tetrahedron (the obvious generalization of the triangle) with the primary colors at the vertices. The binary mixtures are on the midpoints of the edges and the ternary mixture (such as red-green-quag") at the barycentra of the faces. The barycenter of the tetrahedron itself is of course the achromatic point. This structure is just as intuitive as the familiar color triangle in the trichromatic case. It is nothing but the obvious generalization.

Only three of the six edges of the RGBQ-tetrahedron are on the spectrum locus and only two of the four faces are made up of ternary mixtures of spectrum colors. Thus three edges and two faces are like the line of purples in the case of the RGB triangle. All colors in the interior are tetrachromatic mixtures of spectrum colors. Thus only a small part of the boundary of the RGBQ-tetrahedron is spectral. This is indeed typical of tetrachromatic systems in general.

Even the tetrachromatic Grassmann model (figures 5.78 through 5.80) has a chromaticity space such that the set of real colors has a boundary on which the spectral colors have measure zero. Spectral hues are very rare indeed, except in the (from the polychromatic perspective singular) trichromatic case.

Connected with this finding is the fact that the notion of "complementary wavelengths" becomes next to useless. In the tetrachromatic

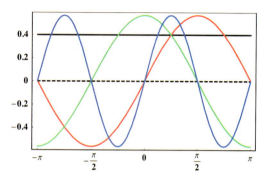

Figure 5.78 The color-matching functions of the tetrachromatic Grassmann model.

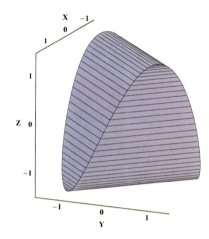

Figure 5.80 The region of chromaticities of real beams in chromaticity space of the tetrachromatic Grassmann model. The crease is the spectrum locus. It exhausts merely an area of measure zero of the boundary. This is a major difference with the trichromatic case. Notice that points on the boundary are typically binary mixtures of monochromatic colors.

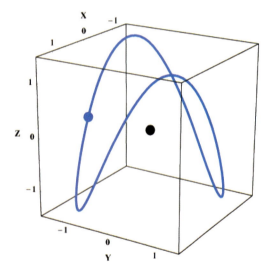

Figure 5.79 The spectrum locus in chromaticity space of the tetrachromatic Grassmann model. It is a closed space curve. The black point denotes the origin (the achromatic point), the blue point (on the curve) the (coinciding) spectrum limits.

case, you may find triples of monochromatic beams from which you may obtain an achromatic

ternary mixture, but an arbitrary monochromatic beam is not necessarily a member of such a triplet.

Such relations are conveniently studied in the tetrachromatic extensions of the Grassmann (shown in figures 5.78 through 5.80) and Helmholtz (shown in figures 5.81 through 5.84) models. It is a trivial matter to extend these models, you simply add the next fourier (in the Grassmann case) or Legendre polynomial (in the Helmholtz case) component to the set of color-matching functions.

It is not particularly surprising that the notion of "complementary wavelength" becomes less useful. The notion of "supplementary spectrum" is much more generic, a "supplementary spectrum" corresponding to a given spectrum

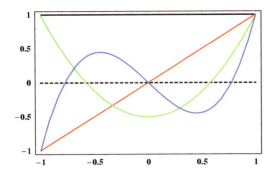

Figure 5.81 The color-matching functions of the tetrachromatic Helmholtz model.

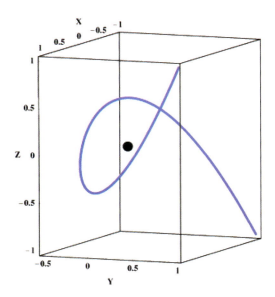

Figure 5.82 The spectrum locus in chromaticity space of the tetrachromatic Helmholtz model. It is a general ("doubly curved" or "twisted") space curve whose convex hull is a finite volume.

being the spectrum that yields the achromatic beam when added to the fiducial spectrum. The "complementary beams" of trichromatic col-

Figure 5.83 The region of real beams in chromaticity space of the tetrachromatic Helmholtz model. The crease is the spectrum locus; it exhausts only an area of measure zero of the boundary. Generic boundary points are binary mixtures of monochromatic beams.

orimetry are simply defined because the spectral resolution of a trichromatic system is so low. In polychromatic systems you need more degrees of freedom to describe supplementary spectra. For very high resolution, such summary descriptions are almost identical to that of the supplementary spectrum itself. This is an important observation. Even in the case of trichromacy, the notion of "supplementary beam" is much more useful and significant than that of "complementary beam." In retrospect, you should probably forget about the old-fashioned "complementarity" altogether and reserve the term for "supplementarity."

5.10.4 Polychromatic vision

If "tetrachromacy" is interesting, then how about "pentachromacy," "hexachromacy," and so forth?[41]

In the limit you would have spectroscopic vision, of course. As a "spectroscopist," you would never confuse two different spectra. Of course

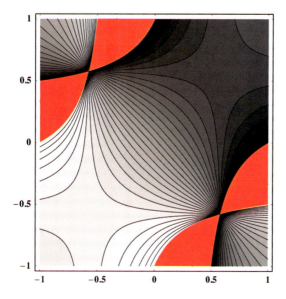

Figure 5.84 The complementary triples of the tetrachromatic Helmholtz model. Each point in the diagram specifies two wavelengths; the gray tone specifies a third wavelength such that the triple is "complementary." In the red regions no such a triple exists.

this is mere science fiction, for in real life there is necessarily a finite resolution.

Wouldn't polychromatic vision with nanometer resolution be great? The answer depends on the ecological relevancy. Which resolution would be so useful as to be worth the additional cost? Answers are not forthcoming, this is a very intricate question indeed. Most spectra you encounter in daily life are not very highly structured. Half a dozen (or perhaps a little more) spectral bands amply suffice to do effective spectroscopy of natural materials and phenomena. My guess is that humans would barely benefit from tetrachromatic, pentachromatic, let alone hexachromatic vision. The relatively minor improvement would almost cer-

tainly not pay for the loss of spatial resolution, the increased complexity of the neural circuitry, and so forth. We should probably be grateful for our trichromatic fate.

Christine Ladd-Franklin[42] pioneered a theory of evolution of the human visual system (figure 5.85). As vision developed, it was merely monochromatic and served primarily as a distance sense sensitive to optically revealed spatiotemporal events in the space in front of you. Then a short wavelength-sensitive channel was added, making the system sensitive to spectral slope. This is very useful to distinguish effects of distance and atmosphere in the landscape[43], discriminating between various (broad)

Figure 5.85 The progression in human vision according to Christine Ladd-Franklin: Monochromatic, dichromatic (short wavelength sensitive channel added), and trichromatic (medium wavelength channel split in a long and medium wavelength subchannel) renderings of a scene. It is not possible to extrapolate into the evolutionary future here, and I wouldn't know how to render tetrachromacy, and so forth.

classes of potential foodstuffs, and so forth. Only rather recently the long wavelength (originally monochromatic) channel split into two heavily overlapping channels, which might now be referred to as long and medium wavelength-sensitive channels. This is often believed[44] to be especially useful to fructivorous primates, since it allows you to differentiate between various colors of vegetation. Of course, plenty of room for speculation and experimentation is left here.

Exercises

1. Dominant and complementary wavelengths [ELABORATE] Write an algorithm that lets you find the dominant wavelength of a color (don't forget to handle the purples and the achromatic point correctly). It is a small extension to write an algorithm that will find the complementary wavelength of any given wavelength (again, handle all cases correctly). Study how these entities change when you vary the achromatic beam.

2. The basic structure [REQUIRED] Construct the spectrum limits and their complementaries. Plot them in the chromaticity diagram; include the achromatic point and the far green point. Repeat this for a number of achromatic beams. Plot the chromaticity diagram in a canonical way in order to be able to study the changes.

3. Parts of daylight [ELABORATE] Find Schopenhauer's parts of daylight for a number of achromatic beams (starting with daylight of course). Plot the RGB crates in the spectrum cone. Given any color, express it in terms of the Schopenhauer parts. In most cases the coordinates will be in the zero to one range (the color being located inside of the crate); thus you have a method to render the points of color space in the interior of the crate onto RGB colors. Render the colors of a database of object spectral reflectances (to be found on the Internet) as illuminated by daylight in this manner. Check whether bananas are properly yellow, and so forth.

4. Semichromes [PARTLY ELABORATE] Find the colors of all semichromes and plot them in color space and in the chromaticity diagram. Repeat for a number of illuminants. (This is easy if you construct an algorithm that selects all spectrum cone generators on one side of any plane through the achromatic axis.) To calibrate this locus via Ostwald's principle of internal symmetry is a more challenging exercise.

5. The discrete model [REQUIRED] Play with variations of the discrete model. You can change the fundamental spectral sensitivities, the primaries, the achromatic beam and/or the definition chromaticity.

6. Analytic models [CONCEPTUAL] Implement the Grassmann, Helmholtz, and the local model in an environment that lets you do symbolic algebra and calculus (e.g., Mathematica or Maple; MatLab is less suitable here, though it is great for numerical studies). If you don't have such an environment, you probably like to do algebra and calculus by hand, and the models are sufficiently simple for that. Compute whatever you can and visualize the results. Compare the models with each other and with the actual visual system. A more challenging exercise is to extend the models to the case of tetrachromatic vision.

7. Put color on the models [ENTERTAINING] The models are *colorimetric* entities and have nothing to do with colors as you experience them. You may enjoy "painting" the various geometric configurations of the models though. This is especially interesting if you try to capture various properties you observe in real life. I have "painted" some of the model results in this book, though—by design—not in the same manner in all instances. A good point to start is to think about the spectral colors in the Grassmann model. What should happen at the spectrum limits?

8. Luminance, hue, and saturation [CONCEPTUAL] Study the CIE definitions. Try a number of obvious alternatives and compare them. Study their transformation properties under changes of primaries, definition of chromaticity or change of illuminant. How useful are the conventional definitions? Study their detailed dependence on the luminance functional.

9. The physical correlate of "color" [FOR PHILOSOPHERS] Show that the space of "spectral shapes" as defined by spectral slope and curvature has the structure of color space, especially that it implies the "color circle." Compare the view that color is "seeing spectral shape" to that of "seeing by wavelength." For instance, study the problem that no wavelength corresponds to magenta hues. Are spectral shapes "in the world"? What is their ontological status as compared to "wavelength"?

10. Polychromacy [CONCEPTUAL] Guess the general dependence on dimensionality from the examples given in the appendix. Now try to prove your conjectures. Is trichromacy in any sense "special"? What qualitative changes have to be made in the standard account of (trichromatic of course) human color vision in order to render it "generic"?

11. Construct complementary triples   [REQUIRED] Explicitly construct the complementary triples for the Grassmann and the Helmholtz tetrachromatic models.

12. Explore multipoint projective invariants   [FOR MATHEMATICIANS] How many points do you need in order to obtain a unique projective configuration (i.e., one that is projectively different from other configurations with that number of points)? How would you define the "projective shape" of a multipoint configuration?

13. The best slit   [EASY] Prove to your own satisfaction that the unique slit shape that avoids spurious resolution is the Gaussian window function.

14. The response curves of the models   [EASY] Explicitly rebuild the models such that the fundamental response curves are nonnegative throughout in each case.

15. Colorimetry without luminance   [CONCEPTUAL] Is a "complete" colorimetry possible without the notion of *luminance*? Investigate what is lost and try to patch the holes. Discuss the coherence and consistency of the CIE structure with these facts in mind.

16. The colors of the rainbow   [FOR PHYSICISTS] Plot the "colors of the rainbow" in the chromaticity diagram. Assume "average daylight," or simply use a black body radiator at the temperature of the sun's photosphere. Look up the theory of the "Airy rainbow integral" if you have forgotten or never heard of it.

Chapter Notes

1. A Lambertian surface looks equally bright from any viewing angle, irrespective of the particular direction of illumination. This is very special and doesn't apply to most physical surfaces.
2. The albedo is the fraction of the incident radiation that is scattered. A surface of "unit albedo" thus absorbs no radiation but remits all.
3. Glossy paper is a mixture of a diffuse scatterer and a (somewhat dull) mirror. For glossy surfaces the specularities can be much brighter than expected, think of the twinkles you see as the sun is reflected from a rippled water surface.
4. This is important in relation to Arthur Schopenhauer's theory of colors as "parts of daylight."
5. This is what Newton believed on the basis of his experiments.
6. Eric W. Weisstein, "Brouwer fixed point theorem," MathWorld—A Wolfram Web Resource. http://mathworld.wolfram.com/BrouwerFixedPointTheorem.html.
7. It would indeed be a bad thing to have any spectral gaps. There are a number of technical reasons for this, but the main point is that the achromatic beam will often be the *illuminant*, as I will discuss in the chapter on object colors. If the illuminant has zero radiant power (in some spectral region), it cannot reveal object colors due to reflectance variations in that region. It would be like looking at things in a pitch-black room (in that spectrum region).
8. Whereas there is no such thing as a "white" singled out by basic optics, there is certainly such a thing as a "preferred white." Kruithof published a graph (in 1942) that shows "preferred" color temperature as a function of light level. In a space that is uniformly illuminated at 20 footcandles by daylight, a color temperature of $6000°$K appears gloomy and overcast, whereas the same space, at the same light level, but illuminated with light bulbs at $3000°$K appears comfortable and quite pleasant. At higher light levels the "pleasing" color temperature rises. See: A. A. Kruithof, "Tubular luminescence lamps for general illumination," *Philips Technical Review* 6, pp. 65–96 (1941).
9. This particular decomposition is due to Helmholtz. Throughout the book I will refer to it as "the Helmholtz decomposition."
10. It is easy enough to cook up examples. Simply take the addition of two distinct monochromatic beams for instance. Then the dominant wavelength will be different from the wavelengths of the two monochromatic beams; thus, it is not in the spectrum of their composition.
11. When you try such calculations yourself, you will notice (no doubt to your chagrin) that the various tables you load from the Internet are rarely in sync, that is to say, the tables cover different wavelength ranges and the wavelengths increments are different, too. You have to restrict the range to the intersection of wavelength regions and to do (linear) interpolations to the smallest wavelength increment where required. A package such as Mathematica makes such transformations relatively painless. You simply turn *tables* into interpolation *functions*, and you restrict the ranges or use padding by zeroes to enlarge them. Either method may be useful to arrive at a common range on which all entities are defined.
12. Of course, the projective coordinates assigned to a monochromatic beam of a given wavelength will vary with the color of the fiducial achromatic beam though.
13. Goethe publicly acclaimed Schopenhauer's *Über die vierfache Wurzel des Satzes vom zureichenden Grunde*, and when Schopenhauer was at Weimar they collaborated on color theory. However, Schopenhauer's manuscript on color did not meet with Goethe's approval. Schopenhauer actually had to ask Goethe to return the manuscript so he could publish it (*Über das Sehn und die Farben*, 1816).
14. Remember that Newton wrapped the spectrum around a circle (thus omitting the purples) and tried to convince the world that this "color circle" was "complete." The important point is that the purples look every bit as a "true" color as the spectrum colors do, yet lack in the spectrum. For Newton (any so many after him) this was unpalatable. Too bad for the facts. Apparently Newton saw no other solution, although it is quite evident that he felt not too good on the issue.
15. This is another one of these trivial facts that so often go unacknowledged, even by scientifically educated people. It is absolutely vital that you should appreciate this point.
16. Here I consider only *infinitesimal* shifts.
17. This is, of course, nothing but the Helmholtz decomposition. At this point it becomes clear how useful this decomposition can be.
18. Remember that Newton (erroneously) denied this.
19. The dictionary on my Apple notebook has:

> "**grok**, verb (**grokked, grokking**) [trans.] informal: understand (something) intuitively or by empathy: *because of all the commercials, children grok things immediately.* ●[intrans.] empathize or communicate sympathically; establish a rapport. ORIGIN mid 20^{th}c.: a word coined by Robert Heinlein (1907–88), American science fiction writer, in *Stranger in a Strange Land.*

20. Plato, *The Republic*, beginning of book 7 (514a–520a).

21. Sad experience has taught me that many students take a long time to catch on, even though I usually teach students of physics with some background in linear algebra, and the like.

22. A good case can be made for the possibility that Newton was inspired to this through the graphical representation of the musical octave by Descartes.

23. This simple fourier model was proposed by Yilmaz, but his analysis is completely different from what I present here. H. Yilmaz, "Color Vision and a New Approach to General Perception," in *Biological prototypes and synthetic systems*, edited by E. E. Bernard, and M. R. Kare (New York: Plenum Press, 1962), pp. 126–141.

24. Of course it is easy enough to build a "third type" of dichromat by introducing a blindness to some linear combination of the color matching functions. You may want to explore this for yourself. One way to proceed is to transform the color matching functions to a triple of thoughout non-negative functions, and to define the dichromats as each missing on of these "fundamentals." The model leaves plenty of room for experimentation if you are the adventurous type!

25. An example of such a unilaterally dichromatic observer is descibed in C. H. Graham and Yun Hsia, "Studies of color blindness: A unilaterally dichromatic subject," *Proc.Natl.Acad.Sci U.S.* 45, pp. 96–99 (1959).

26. The Legendre polynomials are a complete set of orthogonal polynomials, conventionally denoted $P_n(x)$ that satisfy the differential equation

$$(1 - x^2)y'' - 2xy' + n(n + 1)y = 0.$$

(See M. Abramowitz, and I. A. Stegun, "[Legendre Functions]", and "[Orthogonal Polynomials]," in *Handbook of Mathematical Functions with Formulas, Graphs, and Mathematical Tables*, edited by M. Abramowitz, and I. A. Stegun (New York: Dover, 1972), Chs. 8 and 22)

An explicit expression is

$$P_n(x) = \frac{1}{2^n n!} \frac{\mathrm{d}^n}{\mathrm{d}x^n} (x^2 - 1)^n.$$

They are not normalized, but rather

$$\int_{-1}^{+1} P_n(x) P_m(x)\, \mathrm{d}x = \frac{2\delta_{nm}}{2n + 1}.$$

27. A formula often quoted is $(\lambda_1 - 559)(498 - \lambda_2) = 424$, where $\lambda_{1,2}$ are the wavelengths in nanometers of a pair of mutually complementary wavelengths. The formula is due to V. Grünberg, "Farbengleichungen und Komplementärfarben im Young-Helmholtz-System," *Wien. Ber.* IIa, Bd. 113, pp. 627–636 (1909).

It makes no sense to commit this formula (convenient though it is) to memory as the complementary pairs depend on the choice of achromatic beam.

28. You will find a good discussion in M. Born, and E. Wolf, *Principles of Optics*, (New York: Pergamon, 1975).

29. Some good texts on the topic of "scale space" are available now. L. Florack, *Image Structure*, (Dordrecht: Kluwer, 1997); T. Lindeberg, *Scale-space Theory in Computer Vision*, (Dordrecht: Kluwer, 1993); B. ter Haar, *Front-End Vision and Multi-Scale Image Analysis*, (Dordrecht: Kluwer, 2003).

30. This is a weak argument of course, because the chromaticity diagram doesn't carry any intensity information! The only constraint that you really *have* have to impose on the alychne in the chromaticity diagram is that it does not intersect the spectrum locus. But the point is of minor importance here. I consider the alychne is more detail later.

31. CIE, *Commission internationale de l'Éclairage proceedings*, (Cambridge: Cambridge U.P., 1924); Committee on Colorimetry, "The psychophysics of color," *J.Opt.Soc.Am* 34, pp. 245–266 (1944); P. Lennie, J. Pokorny, and V. C. Smith, "Luminance," *J.Opt.Soc.Am. A* 10, pp. 1283–1293 (1993); L. T. Sharpe, A. Stockman, W. Jagla, and H. Jägle, "A luminous efficiency function, $V^\star(\lambda)$, for daylight adaptation," *J. of Vision* 5, pp. 948–968 (2005).

32. Of course you need a metric for that.

33. This is just the "Helmholtz decomposition" again.

34. The "cross-ratio" is the ratio of ratios of distances. Given four collinear points $\mathbf{p}_{1,...,4}$ in the projective plane, denote the Euclidean distance between two points $\mathbf{p}_{1,2}$ as Δ_{12}. Then a definition of the cross-ratio is

$$[\mathbf{p}_1, \mathbf{p}_2; \mathbf{p}_3, \mathbf{p}_4] = \frac{\Delta_{13}\Delta_{24}}{\Delta_{14}\Delta_{23}}.$$

The number you obtain is a projective invariant. Notice that the cross-ratio depends upon the *order* in which the points are taken.

35. For instance, the "compressive isomorphism" defined as $\varphi(x) = \tanh x$ maps the full line $(-\infty, +\infty)$ on the interval $(-1, +1)$.

36. J. Stolfi, *Oriented Projective Geometry: A Framework for Geometric Computations*, (New York: Academic Press, 1991).

37. On human wavelength discrimination see F .H. Pitt, "The Nature of Normal Trichromatic and Dichromatic Vision," *Proc. R. Soc. B* 132, pp. 101–117 (1944); W. D. Wright, *Researches on Normal and Defective Color Vision*, (London: Henry Kimpton, 1946).

38. A. König, and C. Dieterici, "Die Grundempfindungen und ihre Intensitäts-Vertheilung im Spectrum," *Sitz.Ber.d.Akad.Wiss.Berlin* 2, pp. 805–829 (1866).

39. For instance, these papers lead beyond the discussion here: K. A. Jameson, S. M. Highnote, and L. M. Wasserman, "Richer color experience in observers with multiple photopigment opsin genes," *Psychonomic Bull. and Rev.* 8, pp. 244–261 (2001); H. Hofer, B. Singer, and D. R. Williams, "Different sensations from cones with the same photopigment," *J. of Vision* 5, pp. 444—454 (2005).

40. This is a moot subject, of course. You (if you are a trichromat) cannot see like a dichromat and compare. Some rare cases of people with one trichromatic and one dichromatic eye occur: C. H. Graham, and Y. Hsia, "Studies of color blindness: A unilaterally dichromatic subject," *P.N.A.S.* 45, pp. 96–99 (1959).

41. The current champion is apparently a mantis shrimp with at least ten channels. See T. W. Cronin, and N. J. Marshall, "A retina with at least ten spectral types of photoreceptors in a mantis shrimp," *Nature* 339, pp. 137–140 (1989).

42. C. Ladd-Franklin, *Colour and Colour Theories*, (London: Kegan Paul, Trench, Trubner, 1932). Christine Ladd-Franklin was a remarkable woman, you may find it of interest to look up her biography on the Internet.

43. M. Guillot, "Vision des couleurs et peinture." *J.Psychol.norm.path.* 64, pp. 385–402 (1967).

44. J. D. Mollon, "Cherries among the leaves: The evolutionary origins of colour vision," in *Colour Perception: Philosophical, Psychological, Artistic, and Computational Perspectives*, edited by S. Davis (Oxford: Oxford U.P., 2000), pp. 10–30.

Chapter 6

The Goethe Edge Colors

It is generally believed that there is only *one* reasonable choice for a basis of the space of beams $\mathbb{S}$, which (obviously) is the spectral basis due to Newton (the "spectrum"). Sophisticated people will grant you that (in principle at least) an infinity of bases is possible, and even that all such bases are (at least formally) equivalent, but they will often be hard put if you ask them to give an example of an alternative basis. The reason is that the spectral basis makes so much sense in a *physical* setting, and the reason for that is simply that Maxwell's equations are linear and that the spectral basis is a basis of eigenfunctions of the time shift operator. Since the absolute epoch should not matter in physical problems, this is an important property. The spectral basis makes very good sense if you are dealing with problems of physics. But notice that *the invariance under time shifts is irrelevant to colorimetric problems*. Thus there exists no obvious reason why the spectral basis is to be preferred in colorimetry. The only reason for its almost universal acceptance is *convention*.

6.1 History

In 1808, Johann Wolfgang von Goethe proposed what is essentially an alternative basis for the space of beams $\mathbb{S}$, albeit not in a mathematical framework. This basis has a number of properties that render it quite attractive for many colorimetric problems. It is hardly used in the literature, though, probably because Goethe has gained (under scientists at least) the reputation of an author who can safely be skipped, except, perhaps, to be enjoyed as belles lettres. I will not enter into the historical side of the matter here, although that is of much interest. I will instead offer a modern view on the matter that may render Goethe's insights of value to the sciences.

6.1.1   Newton Reconsidered

Consider Newton's early experiments on the spectrum. He constructed what is essentially a *camera obscura*, a dark room with a little hole in the blinds through which sunlight entered, such that the sun's image was projected on a whitewashed wall. Then he held a glass prism in the beam and noticed that the sun's image became an oblong (no longer circular) image, and that it appeared in different hues, ordered along the axes of elongation of this oblong. (Figures 6.1 and 6.2.) From his drawing, you see that the length to width ratio of the oblong was about one to five; from this you may estimate the effective resolution of this "spectroscope" to be about 50–100 nm. This is a lousy spectroscope by modern standards,[1] and Newton himself improved on the set up in later experiments. The point I want to make here is that *the colors seen in this low resolution spectrum were probably very vivid.* Contrary to what might be expected, you don't come to see *more*, nor more *vivid* hues if you increase the spectroscopic resolution. Very strange. This should perhaps bother you if you still believe that color is seeing by wavelength.

If you happen to work in a physics laboratory, as I do, you may be able to scare up an old-fashioned "hand spectroscope" dumped in some dusty cabinet. Otherwise you may be able to borrow one or find one in a science museum shop. Such instruments usually let you control the "slit width," which is a way to change the spectral resolution. Actually, the slit width[2] controls both the amount of radiation that pass through the instrument to your eye and the spectral resolution. This is what happens when you adjust the slit width from very wide to very narrow:

Figure 6.1 Newton selects a "homogeneous color" and applies a prism to it again: the selected "monochromatic" beam does not split up into a spectrum. Notice that there are no lenses in the drawing, it is all done through apertures (holes cut into a shutter blind, hole cut in a paper screen) in a dark room. (Contrary to modern usage, in this drawing the rays enter from the right!)

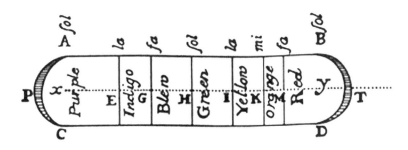

Figure 6.2 The spectrum drawn as by Newton. Notice the musical notation at top and the "seven colors of the rainbow."

If the slit is very wide, the spectrum looks washed out. The hues are not very vivid. You may see a "white" center part with colored fringes on both sides.

If the slit has some "optimal" size, you see the "spectrum" as advertised in the text-books.

If the slit is very narrow, the field of view becomes very dark. Sure, you *know* that at this superior resolution the spectrum should be nearly perfect, but in real life it looks lousy because the colors tend to black.

If you try to set the slit to a width such that the spectral colors appear to best advantage, you will find that the slit has to be set to a surprisingly wide aperture—much wider, for instance, than you would set it to make out spectral details such as the Fraunhofer lines in the sun's spectrum. This is to be expected in terms of of Ostwald's "semichromes." Roughly speaking, a semichrome is a beam with spectral support of about half the visual spectrum.[3] Ostwald made this notion precise in the early twentieth century. The colors in Newton's primitive spectrum were actually not too far from being such semichromes, rather than true "monochromatic" beams. As Ostwald showed, the semichromes are the "best" (most "colorful") object colors (section 7.3.1).

Goethe took a lead from this observation. If the slit was very wide, he noticed an achromatic center with colored fringes at both sides.[4] (See figure 6.3.) This suggested to him that the fringes might be due to the left and right *edges* of the slit separately, rather than the slit as a whole. This again induced him to look at *single edges* through a prism. What he saw were the edge colors (German: *Kantenfarben*). There were two families of such edge colors, you see them when you look at a light-dark or a dark-light transition ("edge"), relative to the base of the prism. He then concluded that Newton's "spectral colors" were actually "mixtures" of edge colors and not at all "homogeneous" (or atomic in the sense of "smallest part") colors. (Figure 6.4.)

 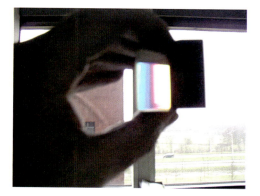

Figure 6.3 You see the families of Goethe edge colors when you look at light-dark transitions (as here, the edge of a window in my office) through a prism holding the "refracting edge" of the prism parallel to the contour. You can hold the prism in any of two orientations. One yields the family of "warm," and the other the family of "cool" edge colors (the same orientation at the other edge of the window does the same thing).

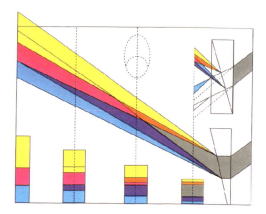

Figure 6.4 A figure by Goethe showing "mixture" of the two families of edge colors, such as to give rise to the *inverted* Newtonian spectrum.

The physicists ridiculed Goethe's explanation because Newton's *experimentum crucis* was generally held to be absolutely decisive and indeed paradigmatic of the very crux of the "scientific method." It showed without the shadow of a doubt that the homogeneous colors were *atomic* in the sense that they could not be split (as sun light is split into the homogeneous colors) with an additional prism.[5]

Yet Newton's reasoning, based on his *experimentum crucis* can easily be demonstrated to be defective. A good way to do this is to invoke Babinet's principle, which is a very basic principle of both geometrical (ray) and physical (wave) optics[6] (appendix F.1). According to Babinet's principle,

> *complementary apertures give rise to complementary images.*

The principle depends only on linearity; thus it is very fundamental and "certain." Now

> *Newton's spectrum is nothing but the image of the slit (or, in the first experiment, the image of the hole in the window shutter).*

If you use a "complementary slit" (that is, an obstruction) you should find a complementary image, and (of course) you do. This is the so-called inverted spectrum.[7] The colors of the inverted spectrum are different from the spectral colors, to wit the complementaries of these. Where Newton's spectrum fails the purples, the inverted spectrum fails the greens. If you vary the (complementary) slit width, this is what you see:

If the complementary slit is very wide, the spectrum looks dimmed out. The hues are not very vivid. You may see a "black" center part with colored fringes on both sides.

If the complementary slit has some "optimal" size, you see the "inverted spectrum" as described by Goethe.

If the complementary slit is very narrow, the field of view becomes very light. Sure, you *know* that at this superior resolution the inverted spectrum should be nearly perfect, but in real life it looks lousy because the colors tend to white and become washed out.

If you try to set the complementary slit to a width such that the inverted spectrum colors appear to best advantage, you find that the slit has to be set to a rather wide aperture. The colors in a vivid inverted spectrum are actually not too far from being Ostwald semichromes.

The *experimentum crucis* works just as well with the inverted spectrum as it does with the "real" spectrum. Babinet's principle indeed forces this result on you, so there is no reason for surprise. Following Newton's logic, the beams of the inverted spectrum are to be regarded as "homogeneous" or "atomic." But this cannot be, since you can easily show that the spectrum of an inverted spectrum component is not "monochromatic" (of course the complementary statement is also true), hence Newton's original conclusion is fallacious, and the *experimentum crucis* fails to be "crucial" in Newton's sense.

All these facts are easily demonstrated to an audience, as I have often done in the classroom.[8] There is nothing mysterious about them; in fact, they make for a rather nice demonstration of Babinet's principle. Yet they are only to be found in the fringes of the literature, instead of being mentioned in any introductory text book on optics. The inverted spectrum is far from popular. It is because (erroneously) thought to be "unscientific" because it is supposedly at odds with very basic and well-established physical principles.

6.2   The Edge Colors

A simple way to see what happens is to start to construct the edge colors[9] from the spectral basis (figure 6.5). The spectrum of daylight $\mathbf{d}$ is

$$\mathbf{d} = \left(\sum_0^\infty |\lambda\rangle\langle\lambda| \right) |d\rangle = \sum_0^\infty \langle\lambda|\mathbf{d}\rangle|\lambda\rangle = \sum_0^\infty |d_\lambda\rangle.$$

This is the achromatic color. The "summation" is of course to be replaced with an integral because the basis is continuous. Now we can define the Goethe edge colors in terms of their spectra. The "low pass edge colors" are

$$\underline{\mathbf{e}}_{\lambda_0} = |\ulcorner_{\lambda_0}\rangle = \sum_0^{\lambda_0} |d_\lambda\rangle,$$

where λ_0 denotes the wavelength of the transition. The "high pass edge colors" are

$$\overline{\mathbf{e}}_{\lambda_0} = |\lambda_0\urcorner\rangle = \sum_{\lambda_0}^\infty |d_\lambda\rangle,$$

and you clearly have that $\underline{\mathbf{e}}_{\lambda_0} + \overline{\mathbf{e}}_{\lambda_0} = \mathbf{d}$. The "passband colors" are defined as ($\lambda_1 < \lambda_2$):

$$\overline{\mathbf{s}}_{\lambda_1,\lambda_2} = |\lambda_1\sqcap\lambda_2\rangle = \sum_{\lambda_1}^{\lambda_2} |d_\lambda\rangle,$$

and the "stop band colors" as ($\lambda_1 < \lambda_2$):

$$\underline{\mathbf{s}}_{\lambda_1,\lambda_2} = |^{\lambda_1}\sqcup^{\lambda_2}\rangle = \sum_0^{\lambda_1} |d_\lambda\rangle + \sum_{\lambda_2}^\infty |d_\lambda\rangle.$$

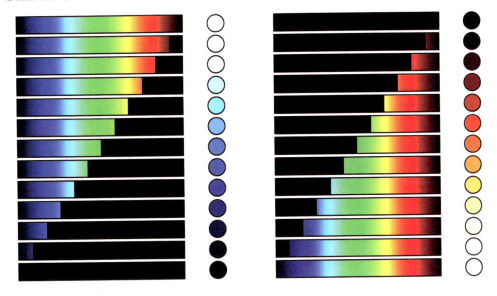

Figure 6.5 The generation of the "cool" (right) and "warm" (left) edge color sequences.

Of course the low pass and high pass colors with the same transition wavelength are mutually complementary:

$$\bar{\mathbf{s}}_{\lambda_1,\lambda_2} + \underline{\mathbf{s}}_{\lambda_1,\lambda_2} = \mathbf{d}.$$

(See figure 6.6.)

The passband colors are what you generate with a spectroscope. The edges of the slit define the two transition wavelengths. Thus these colors may also be referred to as "spectral colors" or "Newtonian colors." The stop band colors are what you prepare via a complementary slit. Thus they may also be referred to as "inverted spectrum colors" or "non-Newtonian colors." Notice that, because of the complementarity,

> *there exist exactly as many non-Newtonian colors as there are Newtonian ones.*

It is perhaps interesting to notice that Newton apparently managed to *ignore exactly half of the colors* in his investigation.

A spectral color can evidently be obtained as the *difference* of two (low-pass or high-pass) edge colors, whereas an inverted spectrum color can be obtained as the sum of two edge colors, a low-pass and a high-pass one. Thus Goethe's notion that the spectrum colors are mixtures of edge colors is at least not preposterous or evidently wrong, as it is often

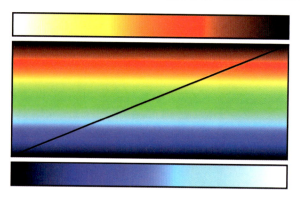

Figure 6.6 The "cool" (lower) and "warm" (upper) edge color sequences. The edge colors come in complementary pairs that are effectively two parts of daylight. The center part of the figure illustrates the split.

construed. Yet it is not a particularly good way to state the fact either, as unfortunate indeed as Newton's statement that daylight is a "confused mixture" because it is *made up* of "homogeneous lights." As I mentioned before, there is a crucial difference between "being composed of" and "being analyzable into." A sausage is not composed of slices before you slice it. Likewise, spectral colors are not mixtures of edge colors, nor is daylight composed of monochromatic beams.

In case of a linear space any basis is equivalent to the next one (page 87). The monochromatic beams can be expressed in terms of Goethe's edge colors (here the old-fashioned "color" of course should be read as "beam") and vice versa. Thus either set is a basis, and the two descriptions are equivalent.

Although the descriptions are indeed formally equivalent, there do exist differences of a physical nature. For instance, the edge colors can easily be produced in the laboratory, whereas the monochromatic beams cannot. If you close the slit of a spectroscope you get indeed closer to a "monochromatic" beam, but it is not possible to ever arrive there.[10] If you close the slit (true monochromacy), you obtain the empty beam. Indeed, "monochromatic paints" do not exist, for they would be black. In contradistinction, as I will show later, the best paints are of the edge color, or pass or stop band semichrome variety (chapter 7). Thus Goethe's alternative basis has much to recommend it. Moreover, the Goethe colors have unique optimality properties that make them figure prominently in the theory of object colors (chapter 11).

Consider what the series of edge colors will look like to the generic human observer (figure 6.6). If you remember the Schopenhauer's "parts of daylight" (section 5.4), then you intuit that the low pass edge colors will start out as BLACK (when you vary the transition from the short to the long wavelengths, starting at the low end), then (gradually) grow in intensity and become BLUE. Next the green part is added, thus you obtain a gradual progression from BLUE to CYAN. The next part to be added is red; thus you get a gradual progression from CYAN to WHITE. The low pass edge colors thus pass through the series K-B-C-W. A similar reasoning applies to the high pass edge color series. It must start out as WHITE; then, when it loses the blue, it gradually becomes YELLOW. Next, when the green is omitted it turns gradually into RED. Then, finally, the RED becomes dimmer and turns into BLACK. The high pass edge colors then compose the series W-Y-R-K. Notice that the colors of the low pass and high pass series (for the same transition wavelength) are mutually complementary colors. If you know one series, you know the other. The edge colors will prove to be very important in later chapters. Usually I will pick one of the two series for the formal development.

Artists often divide colors into two major families, the "COOL" and the "WARM" colors (figure 6.7). This division apparently coincides with the edge color series. All of the low pass edge colors are in the "COOL" family, whereas all of the high pass edge colors are in the "WARM" family. Thus you may without ambiguity talk of the "warm" or "cool" edge

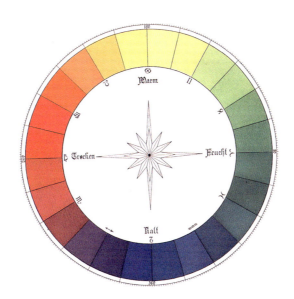

Figure 6.7 A figure from Hans Adolf Bühler, *Das innere Gesetz der Farbe, eine künstlerische Farbenlehre* (Berlin-Grunewald: Horen-Verlag, 1930). The "warm" and "cool" families are indicated in the color circle.

colors.[11] In fact, the edge color families are likely candidates for the objective correlates of the artist's subjective notions. I have not seen this mentioned in the literature, but it certainly seems worth serious consideration.

I see only two immediate alternatives. One—which I will discard offhand—is to argue that the ideas of the artists are bogus anyway. The other is to identify the artistic warm/cool dichotomy with the sign of *spectral slope* in the coarsest possible sense. Perhaps I should say, "long or short wavelength spectral excess," but "spectral slope" is a convenient term that I will use frequently throughout the text. The outcome of this is very similar to the edge color idea. A difference is that the GREENS and PURPLES do not occur among the edge color series and thus would obtain no "warm/cool" label. In the "spectral slope" picture they obviously do fit in, although the labels may turn out very "weak" if the spectrum is almost balanced. However, the artistic praxis is sufficiently fuzzy and varied that this difference does not allow me to make a choice.

The warm edge color series has often been used as a "natural color scale" (often called "temperature scale") that can be used in scientific visualization instead of a gray scale (see figure 6.8). It supposedly offers a larger number of discriminable levels than a gray scale does, which might explain the popularity of this scale. Although I have doubts regarding these supposed advantages, it must be admitted that the scale is at least fairly "natural,"

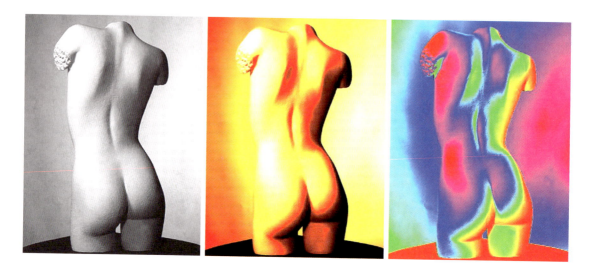

Figure 6.8 A gray scale image (left), visualized with the "temperature scale" (center). The temperature scale yields an easily appreciated, "natural" rendering. At right the gray scale image rendered in "psychedelic colors" ("rainbow scale"). The rainbow scale renders the image hard to parse.

much more so than the (unfortunately also quite popular) "rainbow scale" for instance. The rainbow scale makes it very hard to "read" even a straight photograph, thus is all but "natural". It was popular at the time when Jimi Hendrix played (or rather, took apart) "The Star Spangled Banner" at Woodstock, for in that period photographs of pretty girls in long white dresses with flowers in their hair were often colored this way and enjoyed as "psychedelic." (Figure 6.8.) It is an amazing fact that this scale is very popular in scientific visualization applications today. It must be due to the indiscriminate use of crummy software and perhaps a regrettable lack of judgment on the side of the scientists.

It is important to have a grasp of the geometry of the edge color loci in color space. Clearly, both the warm and the cool edge color sequences must connect the black point to the achromatic point. Moreover, the curves must be related via a central symmetry about the mid point of the "achromatic axis," that is the straight line segment that connects the black point to the achromatic point. Since the curves move in all three directions of color space, they must be "doubly curved," or "twisted" space curves. Finally, in the vicinity of the black point the curves must have tangents that coincide with the UV and IR limiting tangents. From the central symmetry you may conclude that near the achromatic point the limiting tangents must be reversed. For instance, the cool edge colors start out from the black point in the direction of the UV limiting tangent and arrive at the achromatic point from the opposite direction of the IR limiting tangent. From a differential geometric perspective, you appreciate that the tangents of the edge color loci are just the daylight spectrum locus generators. Thus you can guess quite a bit of the geometry a priori.

A calculation reveals that the edge color loci are general *helixes* of opposite (due to the central symmetry) chirality. Thus, looking from the direction of the achromatic axis, you see a figure eight. Each series of edge colors describes one loop of the figure eight, and at the intersection the directions are smoothly joined. (See figures 6.9–6.11.))

The warm edge color locus departs from the black point K in the direction of the spectrum cone generator for the IR spectrum limit and it arrives at the achromatic point A from the direction of the inverted spectrum cone generator for the UV spectrum limit. This is obvious from the fact that the warm edge color series is nothing but the cumulative spectrum of the achromatic beam. As you start to "accumulate spectral power," you have merely power near the IR spectrum limit, and as you are nearly done with the accumulation you have almost the achromatic color but are lacking some power near the UV spectrum limit. Since the cool edge color series is nothing but the inverted warm edge color series, it has to be the case that the cool edge color locus departs from the black point K in the direction of the spectrum cone generator for the UV spectrum limit, and it arrives at the achromatic point A from the opposite direction of the spectrum cone generator for the IR spectrum limit. Thus,

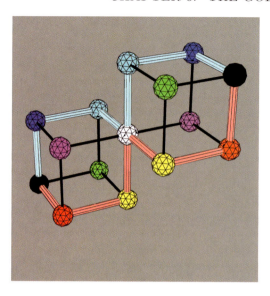

Figure 6.9 Putting two mutually rotated RGB cubes (chapter 14) in series, the warm edge color locus naturally completes the path of the cool one and vice versa.

> *the direction into which the warm edge color locus departs from the black point (the spectrum generator of the IR-spectrum limit) is opposite (thus parallel) to the direction from which the cool edge color locus arrives at the achromatic point (the inverted spectrum generator of the IR-spectrum limit).*

This allows you to do an elegant geometrical construction. You repeat the achromatic axis periodically, arriving at the "achromatic line" on which the (infinitely many!) black and white points alternate as ...KAKAKA.... The edge color loci are also periodically repeated, and because of the aforementioned property they carry on seamlessly without any discontinuity in their direction. Thus you obtain an infinite smooth helix that intersects the achromatic line at the K and A points. The idea is shown in figure 6.9 for the simple case of the RGB cube, while two cycles of the human Goethe edge colors for daylight are shown in figure 6.10. Remarkably, the two edge color loci now appear as a single, continuous curve. Over the interval KA it appears as the warm edge color locus and at the intervals AK as the cool edge color locus. If you look along the direction of the achromatic line this helix appears as a characteristic figure-eight shape as shown in figure 6.11. This neat geometrical construction is no mere plaything either. Many important colorimetric calculations involving the color solid can be performed graphically on these geometrical entities (either the figure-eight or the helix). Thus these figures served as convenient "nomography"[12] in the

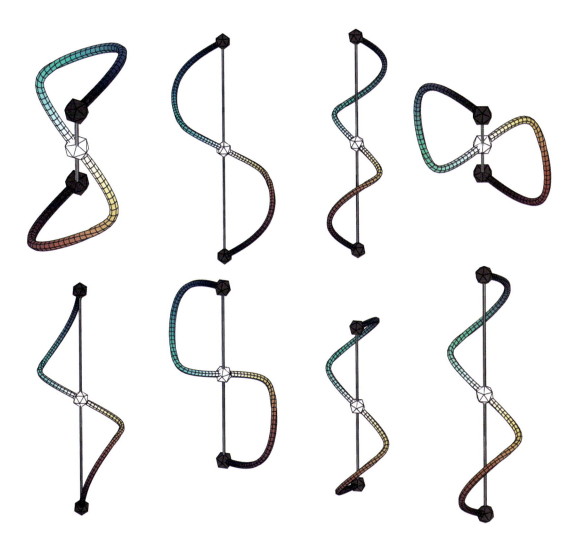

Figure 6.10 Various views of the edge color loci in color space. When you put several copies of the achromatic axis in series the edge color loci carry on smoothly. This is often convenient in differential geometric work. In this figure the achromatic axis has been repeated, thus you see K-A-K. This allows you see that the edge color locus smoothly carries on. Indeed, for the infinitely repeated axis ...K-A-K-A-... you obtain a smooth helix with infinitely many turns, of which I show only two here.

Figure 6.11 As seen from the direction of the achromatic axis, the edge color locus has a characteristic "figure eight" appearance. In this view, the white and black points coincide.

era when people depended on their heads, tables, and slide rules for numerical calculations. In our era such calculations are easier done by computer, with a gain of accuracy,[13] but perhaps a loss of intuitive understanding. You will find that it pays to study these structures in detail. Not only will you enjoy to discover numerous pretty relations, but you will also strengthen your intuitive grasp of the structure of color space.

For the models you can find the edge color loci analytically. The Grassmann model is especially nice, because the edge color loci become perfect circular helixes. In all cases you find the same qualitative structure.

Because the edge colors include the B, C, Y, and R parts of daylight, the edge color loci run through these vertices of the RGB crate defined earlier. (See figure 6.9.)

The edge colors in the models

If you want to study the edge colors in a convenient setting, it is best to use one of the models introduced in a previous chapter. In the appendixes to this chapter I study the edge colors for the various models in some detail.

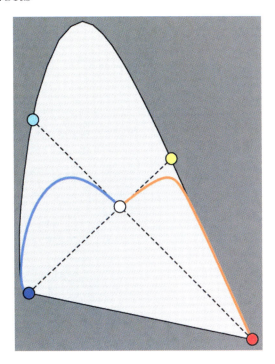

Figure 6.12 In the chromaticity diagram, the edge color loci issue forth from the spectrum limits, starting tangentially to the spectrum locus. They end in the achromatic point and enter it from the direction of the complementaries of the spectrum limits. This is the case in any chromaticity diagram. Here you have to remember that *any* point in the chromaticity diagram can be due to a color that is arbitrarily near black, Thus the "achromatic point" may double for the black point, and so may the spectrum limits.

In the chromaticity diagram the edge color loci leave the spectrum limits in the direction of the spectrum locus, then turn toward the achromatic point, which they reach from the direction of the complementaries of the spectrum limits (figure 6.12). From this configuration it is visually evident that neither edge color sequence contains GREENS or PURPLES. As I will show later, GREENS occur only among the passband colors, and PURPLES among the stop band colors.

6.1 Babinet's principle

Babinet's principle is a very general principle in optics that depends only on linearity of superposition of (in cases of interest to colorimetry) incoherent beams. It applies equally well to the ray or the wave picture of the propagation of radiant power. The main idea is trivial. Suppose some imaging process (in the most general sense) can be described as $\Phi(A) = B$, where A is the input image and B the output image, whereas Φ describes some operator of which you require linearity, that is,

$$\Phi(\alpha A + \beta B) = \alpha \Phi(A) + \beta \Phi(B),$$

then you conclude that if $A + B = I$, where $\Phi(I) = I$, then $\Phi(A) + \Phi(B) = \Phi(I) = I$, thus

$$B = I - A \quad \rightarrow \quad \Phi(B) = I - \Phi(A).$$

Think of I as a featureless white "image" and A as any image except that at any point x of the image $A(\mathbf{x}) \leq I(\mathbf{x}) = I$, then B is the "negative image" or "complementary image" of A. Babinet's Principle then says that *the image of the negative of an object is the negative of the image of that object.* (See figure 6.13.)

Although Babinet's principle is conventionally framed in terms of complementary apertures, that is, white or black instead of gray-tone images, this is in no way necessary. Due to its generality it works with *any* image, also with color images. Simply apply Babinet's principle per color component (R, B and B) or per wavelength interval (see figure 6.14).

Only linearity is required; thus you need not think of some "imaging apparatus" as a camera, projector, or photographic enlarger. Anything will do. Even the bottom of an empty beer glass may be used as an imaging device. In particular, a spectroscope may well be understood as

Figure 6.13 An image, its negative image, and their images.

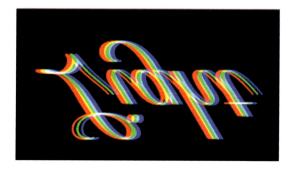

Figure 6.14 An image, its negative image, and their images, this time with chromatic aberration present.

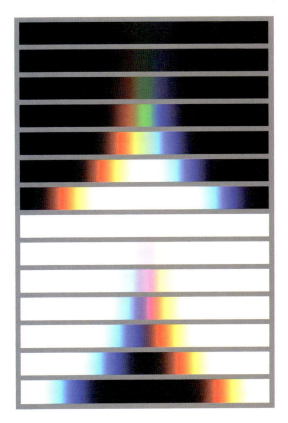

Figure 6.15 Complementary slits as used by Goethe. In fact, Goethe never used "slits" in the strict sense. He simply drew these figures in black ink on white paper. You can do the same thing and look at them through a prism. This will allow you to verify all observations easily. As Goethe noticed, Newton's darkened room was totally unnecessary.

an imaging device, where the input image is a narrow slit and the output image the spectrum.

This is where Goethe comes in. He considers a "complementary spectroscopic slit" (figure 6.15). Such an idea would have been totally alien to Newton's mindset. But nobody keeps you from invoking Babinet's principle (Goethe simply *looked*, of course).[14]

If you look at the regular spectroscopic slit you see the Newtonian spectrum (figure 6.16). Goethe goes to great length to show that this Newtonian trick works only if the slit is not too narrow (for then you see only darkness) or too broad (for then you see only white light but no spectrum). But at reasonable slit widths you certainly see Newton's spectrum as advertised.

Babinet's principle now enforces the complementary observation (figure 6.16). If the "complementary slit width" is at all reasonable, you see the "inverted spectrum" in which every "spectral component" is the complementary color of

Figure 6.16 At the top the Newtonian spectrum, at the bottom the inverted spectrum at various slit widths. The third to fifth samples are "about right" for a nice spectrum to be seen. For narrow slits you get darkness (Newtonian case) or white (inverted case), for wide slits you see the Goethe edge colors at either side. Slit width was doubled between samples.

the corresponding component in the Newtonian spectrum. It cannot be otherwise, and it is hardly worth the effort to actually do the experiment. Yet the (perhaps amazing) fact is that the notion of the inverted spectrum remains somehow repulsive to many physicists.

Perhaps surprisingly, it has often been necessary for me to demonstrate these simple facts experimentally for an audience of disbelieving physicists. Explaining things via Babinet's principle often helps. It is a shock to many fellow physicists that this shows that Newton's famous *experimentum crucis* (that veritable jewel of the scientific method) is thus shown to be inconclusive. When I go on and explain that it is incorrect to hold that sunlight "is composed of monochromatic components" but that you can hold only that "sunlight can be decomposed into monochromatic components," I usually lose the audience members again. Either they don't get the difference or they think I'm splitting hairs. I'm not: the difference is important. Come to think of it, the erroneous belief is actually nonsense from the perspective of basic physics, too (though not particularly relevant here). "Sunlight" is due to numerous events, each of short duration, happening at random intervals. Their superposition is a stochastic process of which the "original parts" are pulse-like phenomena rather than monochrome waves of infinite duration.

6.2 The edge colors in the models

The various models yield an opportunity to study the edge color loci analytically. Especially the Grassmann model (page 5.7) yields a very clear and beautiful picture of the essential geometry. In the case of the models I will study the *Euclidian* differential geometry of the curves since this is most helpful to the intuition. It remains to study the *affine* differential geometry. Although an attractive topic in itself I will skip it in this book. The important qualitative features carry over from the Euclidian case, so there is little immediate need. I will discuss the nature of metrical colorimetry in later chapters.

6.2.1 The Grassmann model

The Graßman model (page 167) yields the prettiest representation of the edge color loci. The differential geometry of the curves is particularly simple.

I start with the warm edge color series. Cumulation of the achromatic spectrum can be done immediately, since the integrals are elementary. You obtain the spatial representation of the warm edge color locus, parameterized by the transition locus a (say),

$$\mathbf{w}(a) = \left(\frac{(a+\pi)}{\sqrt{2\pi}}, -\frac{1+\cos a}{\sqrt{\pi}}, \frac{\sin a}{\sqrt{\pi}} \cdot \right).$$

I can write this as

$$
\begin{aligned}
\mathbf{w}(a) \;=\; & \left(\sqrt{\frac{\pi}{2}}, -\frac{1}{\sqrt{\pi}}, 0 \right) + \\
& + \frac{a}{\sqrt{2\pi}} \, (1,0,0) + \\
& + \frac{1}{\sqrt{\pi}} \, (0, -\cos a, \sin a),
\end{aligned}
$$

that is, an offset, a uniform movement in the achromatic direction, and a uniform circular

movement in a plane orthogonal to the achromatic axis. As the linear movement goes from the black to the white point (a from $-\pi$ to $+\pi$, the orbit is traversed exactly once. Thus the curve is a circular helix that has one turn as it connects the black point to the white point. The radius of curvature is $1/\sqrt{\pi}$ and the pitch (length of the achromatic axis) is $\sqrt{2\pi}$. Thus both the curvature and the torsion are constant. Since both the translation in the direction of the helix axis and the circular movement are uniform, the edge color locus is essentially (up to a constant factor) "rectified", that is to say, parameterized by arc-length.

The cool edge color locus can be deduced through symmetry considerations or derived via the analogous calculation. The curvature is the same as that of the warm edge color locus whereas the torsion has opposite sign. Thus the two loci are circular helixes of opposite chirality.

In figures 6.17–6.19 I show pictures of the two edge color loci from various viewpoints. As

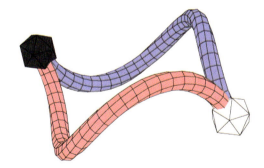

Figure 6.17 The edge color loci for the Grassmann model are circular helixes. They spiral between the black and the achromatic points. Here the warm edge color locus is tinted reddish, the cool edge color locus bluish.

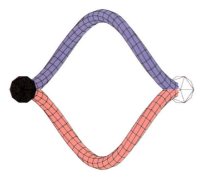

Figure 6.18 Another view of the edge color loci for the Grassmann model.

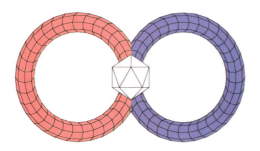

Figure 6.19 Yet another view of the edge color loci for the Grassmann model.

seen from the achromatic axis the loci appear as a pair of touching perfect circles. Of course, this tangency is non-generic.

In figure 6.20 I show a picture of the edge color helix. Although only two cycles are shown, this illustrates perfectly how the two edge color loci can really be regarded as a single continuous curve, in this case a circular helix.

Because the edge color loci are circular helixes, they lie on the surfaces of right circular cylinders. There are two such cylinders, one for the warm and one for the cool edge color series, their generators parallel to the achromatic axis. The two cylinders touch along the achromatic

Figure 6.20 The edge color curve for the Grassmann model. Here two cycles of the infinite helix are shown.

axis. This is a useful structure to keep in mind, as it provides a "scaffold" for the edge color loci that is of considerable heuristic value.

In figure 6.21 I show the edge color loci in the chromaticity diagram. Together they divide the chromaticities into two areas. These areas are *congruent*, something you would perhaps not immediately conclude at a glance. Such things

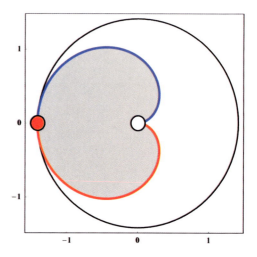

Figure 6.21 The edge color curves for the Grassmann model in the chromaticity diagram. Use your "projective eye" to "see" that the two adjacent areas are actually congruent.

are one reason why I try to discourage the (often unnecessary) use of chromaticity diagrams in this book.

The Graßman model yields no doubt the prettiest picture of the edge color loci. It is a good model, for the loci for the case of generic human vision are qualitatively very similar. Most of the important geometrical properties of the edge color loci can be conveniently studied in the context of the Grassmann model.

6.2.2 The Helmholtz model

The Helmholtz model is simple in principle (low order polynomial color matching functions), but in practice the formalism soon begins to be intractable. The edge color loci provide an example. It is easy enough to cumulate (an elementary integration) the achromatic spectrum and thus to find the edge color loci. The warm edge color locus is

$$\mathbf{w}(a) =$$

$$\frac{1}{\sqrt{2}}\left(1 + a, \frac{\sqrt{3}}{2}(a^2 - 1), \frac{\sqrt{5}}{2}a(a^2 - 1)\right),$$

and the cool edge color locus is just as simple. I have not been able to rectify the curve (its total length can be obtained in terms of elliptic functions) which makes it a chore to do the differential geometry.

It is easy enough to plot the edge color loci of either series (see figures 6.22–6.24) explicitly. Notice that the "figure-eight figure" shows a generic intersection,[15] in contradistinction with the Grassmann model where you had a tangency. The transverse crossing is indeed the generic case and is also encountered in the case of human vision.

The infinite helix curve (see figure 6.25) shows the familiar shape. The Helmholtz model is in almost all respects representative of the generic human visual system.

In figure 6.26 I show the edge color loci in the chromaticity diagram. The loci depart from

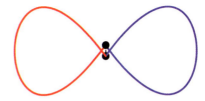

Figure 6.22 A view of the edge color curves for the Helmholtz model. The warm edge color series is tinted red, the cool series blue.

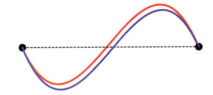

Figure 6.23 Another view of the edge color curves for the Helmholtz model.

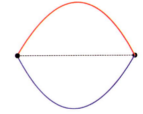

Figure 6.24 Yet another view of the edge color curves for the Helmholtz model.

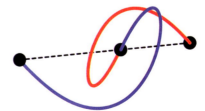

Figure 6.25 A view of the edge color curve for the Helmholtz model. Two cycles are displayed.

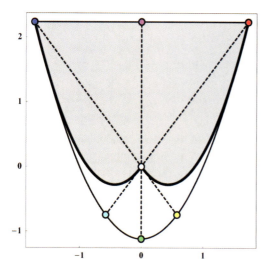

Figure 6.26 The edge color loci of the Helmholtz model in the chromaticity diagram. Notice the intimate relation to the spectrum limits, complementary spectrum limits, the spectrum locus, and the achromatic point.

the spectrum limits in the direction of the spectrum locus and arrive at the achromatic point in the direction toward the complementary spectrum limits. This is entirely typical of the human case.

Notice that the edge colors contain neither purple nor green hues. In Goethe's view, the edge colors are basic and the purples and greens mixtures of edge colors. This is one possible (indeed valid) view.

6.2.3 The local model

In a local model with color matching functions

$$s_0(x) = \frac{e^{-x^2/2}}{\sqrt{2\pi}},$$

$$s_1(x) = \frac{ds_0(x)}{dx},$$

$$s_2(x) = \frac{ds_0(x)^2}{d^2 x},$$

and uniform achromatic spectrum, the warm edge color locus is[16]

$$\mathbf{w}(a) = \left(\frac{1}{2}(1 + \operatorname{erf}(\frac{a}{\sqrt{2}})), \frac{e^{-a^2/2}}{\sqrt{2\pi}}, -a \frac{e^{-a^2/2}}{\sqrt{2\pi}} \right),$$

and the cool edge color locus very similar. This model is rather difficult to trace analytically, I merely show the shape of the loci.

In figures 6.27–6.29 I show the edge color loci. Notice how similar this looks to the case of the Helmholtz model, despite the fact that the wavelength range is the full real line in the case of the local model. Evidently the parameterization must be quite different from what it is in the Helmholtz model, although the shapes are very similar indeed.

In figure 6.30 I show the loci in a chromaticity diagram in which the spectrum locus is a parabola. This reveals the difference with the Helmholtz model, because the spectrum limit structure for the local model happens to be nongeneric. The local model is actually similar to the Grassmann model in many respects, and a suitable projective transformation may turn the

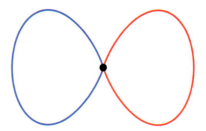

Figure 6.27 A view of the edge color loci for the local model. The warm edge color series is plotted in red, the cool one in blue.

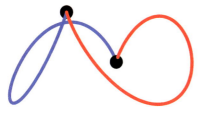

Figure 6.28 Another view of the edge color loci for the local model.

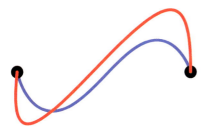

Figure 6.29 Yet another view of the edge color loci for the local model.

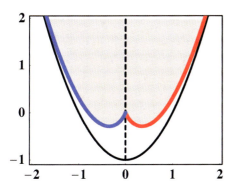

Figure 6.30 The edge color loci of the local model in the chromaticity diagram. Notice the intimate relation to the spectrum limits, complementary spectrum limits, the spectrum locus, and the achromatic point. Here the parabola extends to infinity. The two regions are—again—congruent.

parabolic spectrum locus into a circle. You may want to try this for yourself.

6.2.4 The discrete model

In the discrete model, the spectrum of the achromatic beam is

$$|a\rangle = |m_1\rangle + |m_2\rangle + |m_3\rangle,$$

where $|m_1\rangle = \{1,0,0\}$, $|m_2\rangle = \{0,1,0\}$ and $|m_3\rangle = \{0,0,1\}$. The warm edge color locus is the polygonal arc

$$\{|m_3\rangle, |m_2\rangle + |m_3\rangle, |m_1\rangle + |m_2\rangle + |m_3\rangle\},$$

in the space of beams. Projected into color space you obtain the polygonal arc

$$\{\{0,0\}, \{0,1\}, \{\tfrac{2}{5}, \tfrac{7}{5}\}, \{\tfrac{7}{5}, \tfrac{7}{5}\}\}.$$

The achromatic color is $\{\tfrac{7}{5}, \tfrac{7}{5}\}$ and the cool color edge locus is simply the inverted warm color edge locus. Thus you obtain the configuration shown in figure 6.31.

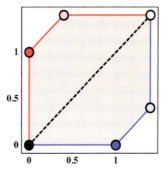

Figure 6.31 The edge color loci for the discrete model are polygonal arcs. Because color space is only two dimensional they are planar curves instead of helixes in this case. Here the edge color loci form the boundary of the set of all colors dominated by the achromatic beam.

In this case, the edge color loci are planar polygonal arcs. There is no way they could be helixes because the dimensionality of the discrete color space leaves them no room and prevents the curves from twisting and permits curvature only. For the discrete model, the edge color loci form the boundary of the set of all colors whose spectra are dominated by the spectrum of the achromatic beam.

Exercises

1. The edge colors   [EASY] Construct both series of the Goethe edge colors. Repeat for various illuminants. Plot the loci in color space, and view them from various directions. Plot the loci in the chromaticity diagram too. Why do the loci meet at both ends in color space but not in the chromaticity diagram?

2. The edge colors in the models   [ELABORATE] Construct the edge color loci for the models. Study the differential geometry of the curves. If you cannot derive the curvature and torsion analytically, then plot them numerically. Compare the Grassmann and the Helmholtz model and discuss the results.

3. The semichromes and the edge colors   [EASY] Construct the semichromes from the edge colors. This is easy, and it vividly illustrates Goethe's notion that all colors (the semichromes include nonspectral colors, the purples) are mixtures of edge colors.

4. The edge colors in polychromatic vision   [ELABORATE] Discuss general properties of the edge colors in general, polychromatic vision. Why are monochromatic, dichromatic and trichromatic systems special? What happens in higher dimensions (the tetrachromatic system is an easy initial target)?

5. The human condition   [TAKES NUMERICAL EFFORTS] Find the curvature and torsion of the edge color locus for the case of human vision (numerically, of course). If you object that there is no metric, so this makes no sense you are right. I will come to the problem of metrics later. For the sake of this computation, simply assume the Euclidian metric.

6. The *experimentum crucis*   [FOR PHILOSOPHERS] Carefully analyze the *experimentum crucis*. What could Newton have concluded? Can you think of potentially more conclusive experiments?

7. The "warm" and "cool" families   [CONCEPTUAL] Look up descriptions of the "WARM" and "COOL" colors in various books[17] on how to paint.

Do you consider my identification of "warm" and "cool" with the edge color series reasonable? Would it be possible to make this stick? Is this situation any different from that of the status of the color circle?

8. The beauty and the beast   [CONCEPTUAL] The edge color curves are beautiful spirals in all models. This may cause you to believe that it could not be otherwise. By way of an antidote you may try to produce "monsters" that fail to be nice. It is easy enough to do so. For instance, depart from the Grassmann or Helmholtz model and change the color matching functions. By playing around for a while you will hit on a variety of monstrosities. The nice models are indeed nice by design, not by necessity. That the human visual system is "nice" is an accident or due to evolutionary pressure, take your pick.

Chapter Notes

1. "Lousy" is of course a relative measure. Here it relates to the width of the visual region. Newton got no more than about a half-dozen independent samples. This is a "lousy" description if you consider the numerous spectra humans effortlessly discriminate.

2. The nomenclature is not standard. One often uses "spectroscope," "spectrograph," and "monochromator" interchangeably. More correctly, a spectroscope is an apparatus you look into to see a spectrum, a spectrograph a camera that records a spectrum, and a monochromator an apparatus that selects a certain monochromatic beam. A "monochromator" consists of a collimator that collects the radiation incident on the *entrance slit* into an (approximately) parallel beam. This beam is made to pass a prism (with the refracting edge parallel to the entrance slit). The beam that leaves the prism is made to enter a collecting objective. In the focal plane of this lens you find the images of the entrance slit, one image for each wavelength ("homogeneous light"). Thus the continuous set of images forms a broad band of "monochromatic colors," that is the *spectrum*. In a "spectroscope" you use an ocular to view the spectrum, in a "spectrograph" you put a detector such as a photographic plate or a (linear) CCD array at the focal plane, and in a monochromator, you place an *exit slit* at the focal plane in order to select the radiation of a certain narrow wavelength band. Thus, you have *two slits* in a generic monochromator. It makes sense to set the exit slit to the width of the image of the entrance slit (taking account of a possible magnification factor that differs from unity); thus typically the slits are adjusted in tandem.

3. Although Ostwald's theory properly applies to object colors (I will treat these later), you are in much the same situation here because the spectroscope can only *remove* spectral radiant power, not add it.

4. Goethe had borrowed a prism, and when obliged to return it he decided to perform a quick experiment. He put the prism before his eye and looked through it toward a white wall. Because the wall showed no spectrum (apparently he had expected that, for some reason), he "knew" right away and with full conviction that Newton had it wrong. More detailed observation revealed edge colors that appeared at light/dark transitions, hence the name.

5. Newton's spectroscope implemented the projector on the spectral basis. Since projections are idempotent the second projection did not change anything (this explains the *experimentum crucis*). But a projection on *another* basis (for instance the Goethe edge colors) *would* have changed things! All bases are equivalent, which means that no "atoms" exist in linear spaces.

Newton apparently failed to appreciate this (or rather *the*) nature of linear spaces.

6. This is due to Jacques Babinet (1794–1872), a French physicist. The principle is usually stated in terms of wave optics.

7. The philosophers have their own notion of "inverted spectrum" that is quite different from the use in this chapter, so beware. They use the term inverted spectrum for the hypothetical case of an observer whose color experiences are the inverses of those of a generic observer. Such an observer, when confronted with a yellow beam, "sees" a blue one, but because he or she is raised to speak in an understandable manner, will declare this "blue" experience to "look yellow." Such hypothetical cases open up a can of worms that I gladly keep closed in this book.

8. Here is a simple but effective setup if you want to try it. Use a linearly modulated interference wedge filter. (These are expensive, but you may be able to scare up one from a forgotten drawer, since nobody uses them anymore.) You can put it on an overhead projector and project the spectrum on a screen. It works best if you screen off all but the wedge filter, using black paper, or some similar material. Then prepare a filter to be put over the lens of the overhead projector. This filter is made up of a number of thin, parallel glass rods. (It is easy enough to assemble yourself if you get some piece of glass rod from the chemistry glass workshop.) Such a filter can take the heat shock (typical overhead projectors concentrate a lot of IR heat at the spot of the lens) and draws out the image in a direction orthogonal to the axes of the glass rods. If you draw out the spectrum along its length, you effectively mix all monochromatic components and obtain an elongated achromatic band on the screen. Now you are all set to start. Using black paper, you can block some monochromatic beams and let others pass. If you remove the glass rod filter, the audience can see exactly what selection you take; if you replace it, they see the mixture. By using the top and bottom halves of the height of the spectrum you can effectively demonstrate complementary spectra. I found this to be a great demo that works for a large audience, though only if you can darken the room completely. (I usually arrive early in order to investigate the lecture room (be sure to dark adapt!). Then I tape black paper over the illuminated exit signs, and so forth.)

9. Here I start using the conventional but potentially misleading terminology where "colors" and "beams" are both denoted "colors." If I do not do this, much of the literature would appear alien to you, so I don't have much of a choice. It isn't necessarily a problem either; you merely have to exercise your good judgment. The "edge colors" (in the conventional parlance) can denote either "colors" or "beams" (in the technical—correct—

sense). In this section I often mean "beam" or "spectrum" when I use "color." Better watch it!

10. In order to obtain an approximation to a "monochromatic beam" by selection, using a spectroscope, you have to increase the radiant spectral power density in inverse proportion to the slit width. This requires extremely bright sources if you require high spectral resolution, and, of course, most of the radiant power of the source is being wasted. This is how it worked in the physics laboratory before the advent of lasers. I remember working in atomic physics, using dangerous high power Xenon arc sources (they had to be water cooled, and I experienced occasional explosions) in order to obtain a meagre "monochromatic" beam. Today one uses laser sources, of course. Because the narrow spectrum is an automatic result of the generation of the beam, selection is rarely necessary. Even laser sources have a finite spectral width, true "monochromatic" beams do not exist. I do not consider laser sources in this book, as they play no role in daily life vision.

11. Notice that GREENS and PURPLES are not to be found among the edge colors, thus "WARM" and "COOL." does not apply to them. Indeed, in the arts the treatment of these colors in terms of "WARM" and "COOL" is rather varied.

12. The dictionary on my Apple computer has:

> "**nom.o.gram** (also **nom.o.graph**) noun—a diagram representing the relations between three or more variable quantities by means of a number of scales, so arranged that the value of one variable can be found by a simple geometric construction, for example, by drawing a straight line intersecting the other scales at the appropriate values."

Nomograms were commonly used in the days before computers. They went the way of slide rules: Some of my students have seen one in a museum, but most never heard of them. Time flies. I was reared on them. Most people my age have a strong conviction that the use of slide rules should be forced upon beginning students of physics, as it promotes good horse sense. I fully agree.

13. The accuracy obtained by nomographical methods and slide rules is of the order of a fraction of a percent, generally sufficient for most colorimetric calculations.

14. Goethe made quite an issue out of this and ridiculed Newton, who in order to study *light* went into a *dark* room:

> "Freunde, fliet die dunkle Kammer,
> Wo man euch das Licht verzwickt

> Und mit kümmerliche Jammer
> Sich verschobnen Bildern bückt"

which may be translated (not without fancy) as

> Friends, leave behind that darkened room
> Where light of day is much abused,
> And, bent low by crooked thought and gloom,
> Our sight is anguished and confused.

(Translation by T. Roszak, *Where the Wasteland Ends*, (New York: Doubleday, 1972).)

15. The "figure eight" configuration is composed of pieces of a "Newton's cubic parabola," also known as "Tschirnhausen's cubic."

16. The Cartesian form for the "figure eight" configuration is

$$x^2 \log 2\pi x^2 + y^2 = 0.$$

It doesn't seem to be one of the "classic" lemniscates, but I'm not sure about that.

17. Or simply use Google on the Internet. Here is my first hit, it is fairly typical of what you may expect to find:

Warm colors are vivid in nature. They are bold and energetic. Warm colors are those that tend to advance in space; therefore, caution needs to be taken so you do not overwhelm your content with eye catching hues. If an element in your design needs to pop out, consider using warm colors to do that.

Cool colors are soothing in nature. They give an impression of calm and rarely overpower the main content or message of a design. Cool colors tend to recede; therefore, if some element of your design needs to be in the background, give it cool tones.

Chapter 7

Schrödinger Optimal Colors

The Schrödinger optimal colors are usually considered in the context of object colors (not the colors of beams[1]). Indeed, I will return to the topic in the chapter on object colors. It is quite natural to consider the topic in the context of colorimetry in the presence of an achromatic beam, though (see chapter 5).

7.1 The Hypercrate in the Space of Beams

Remember that I consider only beams whose spectra are dominated by the spectrum of the achromatic beam. Then the gamut of possible colors must be a *finite volume*, which, again, means that for any given chromaticity there must exist a unique color such that no color of that chromaticity is further away from the origin. Such colors are denoted "(Schrödinger) optimal colors." The optimal colors must be on the boundary of a finite volume that contains all colors admitted by the fiducial achromatic beam.

An intuitive way to think of these relations starts with the space of beams $\mathbb{S}$. Since I consider only beams such as $\mathbf{s}$, with (figure 7.1)

$$s(\lambda) \leq a(\lambda),$$

(where $\mathbf{a}$ denotes the achromatic beam), all such beams lie in an infinitely dimensional *hypercrate* (or hypercuboid) with edge length $a(\lambda)$ along the dimension $|\lambda\rangle$. Since color space is just a projection of the space of beams parallel to the black space, the volume of colors that are admitted by the achromatic beam is simply the three-dimensional projection of this infinitely dimensional hypercrate. Since projections are linear maps and thus preserve convexity, whereas the hypercrate is evidently a convex volume, the projection also has to be a convex volume. Since the central symmetry of the hypercrate is also preserved under

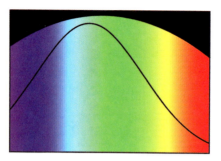

Figure 7.1 "Spectral domination" involves an all-positive achromatic beam. Spectra such as that indicated by the black line are "spectrally dominated" by that beam. The ratio of spectral radiant power in the beam to that of the achromatic beam is the "spectral reflectance factor" and can take values on $(0, 1)$. The dominated beams are a (very simplified) model of *object colors*. Think of the achromatic beam as the beam scattered to the eye by a piece of white paper. Then the dominated beams represent all possible colored papers. They are contained in a hypercrate with edge lengths equal to the spectral radiant power of the achromatic beam.

linear (affine) transformations (affine transformations leave the bisection of linear segments invariant), you obtain a centrally symmetric, convex "color solid". (The projections of general hypercubes are explored in appendix G.1.)

The black point and the achromatic point evidently lie on the color solid. For very dim beams (in the infinitesimal environment of the empty beam), the situation is really not different from generic colorimetry (without achromatic beam I mean), because the achromatic beam poses no limit. Thus, near the black point, the color solid must look exactly like the spectrum cone. Because of the central symmetry, you may conclude that near the achromatic point the color solid must appear like the *inverted* spectrum cone. (See figure 7.2.)

The intersection of the spectrum cone and the inverted spectrum cone defines a finite convex body. The color solid must lie in the interior of this body, such that it is tangent to it (has conical singularities) at the black and the achromatic point. Thus, you gain the view of the color solid as a *fusiform* (spindle, zeppelin, or cigar-shaped) body, which will turn out to be quite apt. Without any calculation you have already intuited most of the relevant structure of the color solid.

The geometrical insight that the color solid is a low-dimensional projection of a hypercrate in the space of beams is crucial. It allows you to intuit many important properties of the optimal colors (and—as I will show later—of the object colors) without any formal derivation.

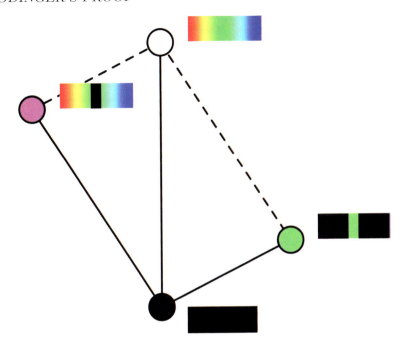

Figure 7.2 The black point corresponds to the empty beam, the white point to the achromatic beam, here symbolized by the full (flat) spectrum. The (green) "almost monochromatic" color is both complementary and supplementary (the spectra add to that of the achromatic beam) to that of the magenta color. Thus you obtain a planar parallelogram of colors. Likewise, the spectrum cone at the black point is mirrored by the inverted spectrum cone at the white point.

7.2   Schrödinger's Proof

Since the color solid is a convex body and the black point lies on it, a "chromaticity" (section 4.7), that is, a line through the black point, either is a generator (lies on the surface of) the color solid, or it has one intersection with the color solid that differs from the black point. That intersection is an *optimal color*. If it is unique (and it will turn out to be the case that the optimal colors for generic human vision are essentially unique[2]), then Schrödinger has shown that their spectrum is of a very specific type,[3] namely:

- the spectral radiant power density is either zero, or is equal to that of the achromatic beam;

- there exist no more than two transitions throughout the spectrum.

This means that the optimal colors are either Goethe edge colors (one transition) or passband ("spectral") or stop band ("inverted spectral") colors. Notice that Ostwald's "semichromes" and the Schopenhauer "parts" are also of these types. Apparently both the edge color loci and the semichromes locus lie on the boundary of the color solid.

The proof of Schrödinger's theorem can be done most elegantly via the theory of "vector measures" and the "bang-bang principle",[4] however, in my view, Schrödinger's original proof has more charm and actually yields some useful insights.[5]

Schrödinger proves the theorem[6] by showing that if a color does not satisfy one of his conditions, then there must exist (he shows explicitly how to *construct* one) a color that is farther away from the origin; thus, colors on the boundary necessarily satisfy both conditions.

Suppose that, for some fiducial beam, there exists a region in which the radiant spectral density of beam is in between zero and the radiant spectral density of the achromatic beam (figure 7.3). Then I can select three non-overlapping subregions such that (in each subregion) I can lower or raise the radiant spectral density *ad libitum*. Each subregion has a certain color, because the subregions don't overlap these colors are independent, and I can take them as primaries. If I express the fiducial beam in terms of these primaries, I see that I can always find a perturbation of the spectrum such that I obtain a *positive increment* of the fiducial color. This perturbation will increase the intensity of the fiducial beam. Therefore, the fiducial beam cannot be an optimal color. Notice the importance of the first condition. It prohibits arbitrary perturbations, for if the radiant spectral density is zero you cannot lower it, and if it equals the radiant spectral power density of the achromatic beam you cannot raise it.

Figure 7.3 On the left, a spectrum that fails to be either zero or equal to the illuminant at every wavelength (for simplicity assume a flat spectrum for the illuminant). At right, the spectrum has been perturbed at three different locations. Notice that I can either "add" or "subtract" spectral radiant power at each of these locations. By linear combination, I can let the perturbation be a color of any chromaticity I may fancy.

The second condition is handled in essentially the same way. If there exist more than two transitions I select three of them (figure 7.4). By slightly perturbing the locations of these three transitions, I can once again increase the intensity and thus show that such a beam cannot be optimal. Notice the importance of the second condition. It prohibits arbitrary perturbations, for if there are only two (or even one) transitions I will not be able to synthesize the fiducial color and thus increase its intensity.

Schrödinger's proof is exemplary in its simplicity and immediate intuitive content.

7.3 The Geometry of the Color Solid

The color solid[7] is a convex body with two conical points connected by two creases (figures 7.5–7.10). The passband and the stop band colors depend upon two parameters; thus, they will make up the two dimensional regions of the boundary of the color solid. The edge colors depend upon only a single parameter; thus, they define a curve (you have seen that already) on the boundary of the color solid. From the symmetry you see that the boundary falls apart into two congruent parts (figures 7.7 and 7.9), mutually bounded by the edge color loci. One part contains the passband colors, this is the "spectral," or "Newtonian" part, whereas the other part contains the stop band colors, thus the "inverted spectral" or "non-Newtonian" colors. The structure is in many respects reminiscent of the RGB color cube (figures 7.10 and 7.11, see chapter 14). This is a key observation that can be greatly expanded upon. I will do so later.

The color solid can be cut into two mutually congruent volumes one containing the "spectral," the other the "non-spectral" colors (figure 7.8). The cut surface is a conical

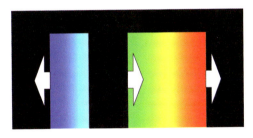

Figure 7.4 Here is a spectrum that fails to jump at most two times between zero and the spectral radiant density of the illuminant (for simplicity assume a flat spectrum for the illuminant). The spectrum can be perturbed at three different transition locations. Notice that I can either "add" or "subtract" spectral radiant power at each of these locations: By linear combination I can let the perturbation be a color of any chromaticity I may fancy.

surface with the central gray point (center of the color solid) as apex and that contains the edge color loci.

Colors near the black and the white point are clearly almost black or almost white as the case may be, thus hardly strongly "colored." Indeed, as you find the color content of the passband colors for a given dominant wavelength, there turns out to be a unique optimum of fairly broad spectral width. These "most colorful" colors are Ostwald's semichromes. (Figures 7.12 and 7.13.)

The semichromes depend upon a single parameter, thus specify a curve on the surface of the color solid. If you slightly perturb a semichrome by shifting its transition locations such as to keep the hue constant, you either add or subtract an achromatic component. Geometrically, this implies that:

> *The circumscribed cylinder with generators parallel to the achromatic direction touches the color solid at the semichrome locus.*

The semichromes are indeed very special, for they make up the "equator," the point at greatest remove from the "polar (achromatic) axis," of the color solid.

Thus you obtain a very nice geometrical insight in the relations between the various remarkable beams and colors that I have considered separately thus far.

At this point I am in a position to clear up a problem left when I discussed the "Schopenhauer crates": How much of the full gamut of colors admitted by the achromatic beam falls inside the interior of the crate? The answer is simple, you merely have to find the ratio of volumes of the crate and the full color solid. It also suggests a more systematic way of finding a maximum volume crate than brute search (ughh). Notice that the vertices R, G, and B already specify a crate. Intuitive geometry reveals that the crate will be of extremal volume in case R, G, and B lie on the boundary of the color solid and moreover

- the tangent plane at the color solid at R is parallel to the plane spanned by G and B;

- the tangent plane at the color solid at G is parallel to the plane spanned by B and R;

- the tangent plane at the color solid at B is parallel to the plane spanned by R and G,

the reason being that the volume of the crate is invariant to small perturbations of the crate if these constraints are fulfilled. In chapter 9 I will find another formulation of these constructions.

The shape of the surface of the color solid depends on the color matching functions, weighted by the spectrum of the achromatic beam. Thus the Goethe edge colors (chapter 6) are very prominent features on the boundary of the color solid (figures 7.14 through 7.16).

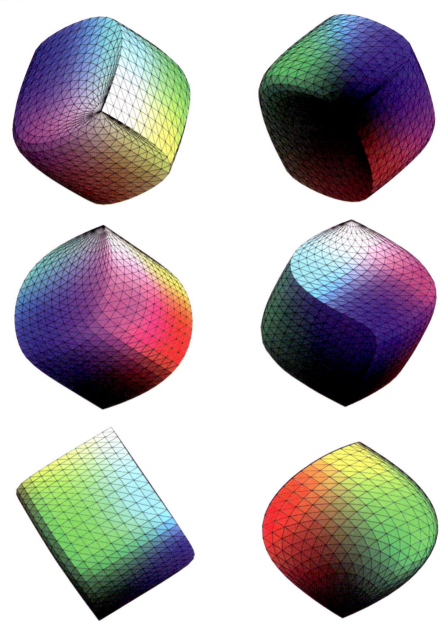

Figure 7.5 Some views of the color solid for average daylight. The particular parameterization and representation will be explained later. In the conventional CIE representation, the solid looks much less regular.

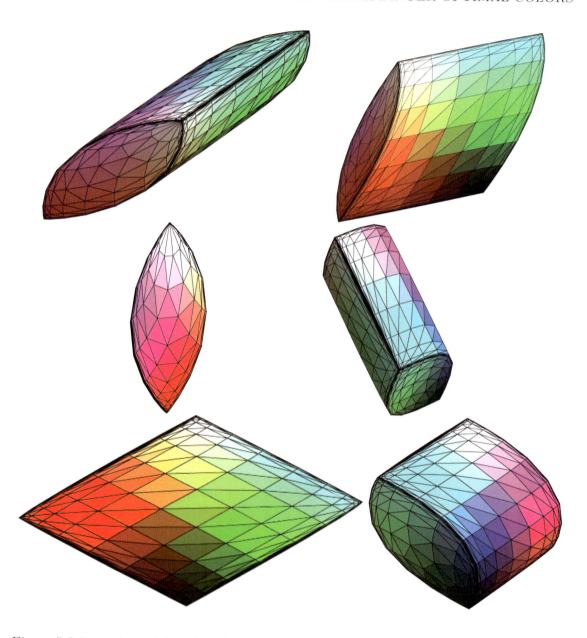

Figure 7.6 Some views of the color solid for a achromatic beam that has the same color as average daylight, but a (very) different spectrum. The representation is the same as with the previous figure; thus, the figures are directly comparable.

A common problem

Several people asked me how it can be that the optimal colors separate cleanly into two *disjunct groups* (figure 7.7; "spectral" and "non-spectral" colors), whereas the hues of the spectrum and the inverse spectrum *partly overlap*. The spectrum contains no purples, the inverse spectrum no greens, but they share the reds and blues. The reason is that the locus of best colors ("semichromes") that represent the color circle on the color solid (figure 7.15, its "equator"), runs along the edge color curves in the blue-cyan and the red-yellow regions. It is only in the greens and purples that the semichrome locus breaks away from the edge color curves and runs through the interior of the spectral and non-spectral patches.

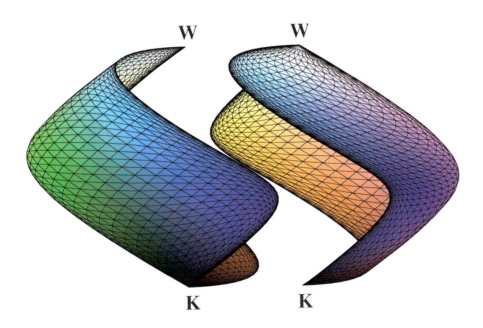

Figure 7.7 The Schrödinger color solid is a centrally symmetric, convex body. It is smooth throughout, except at the black and white points, where the solid is tangent to the spectrum cone and the inverted spectrum cone respectively and at the edge color loci which subtend creases. The boundary falls apart into two congruent parts, one containing the "spectral" or "Newtonian", the other the non-spectral" or "non-Newtonian" optimal colors. Both parts are smooth surface patches.

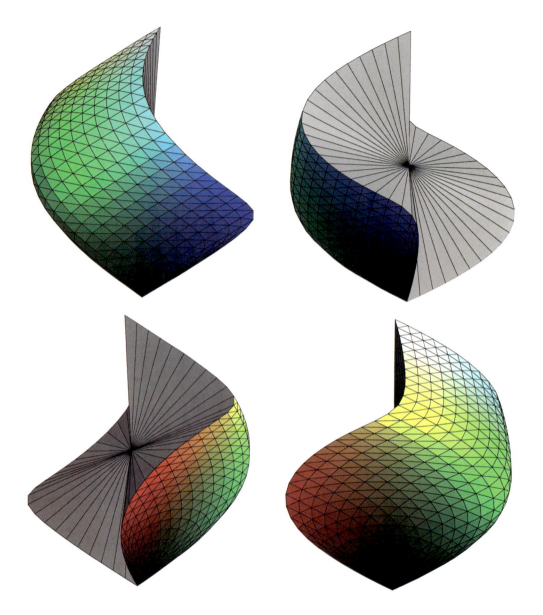

Figure 7.8 The color solid can be divided into two congruent volumes by cutting it with the surface obtained by connecting mutually complementary edge colors. This surface is a "wobbly" (non-convex) cone.

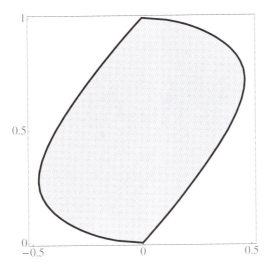

Figure 7.9 A planar cross-section of the color solid at 540nm, clearly revealing the structure.

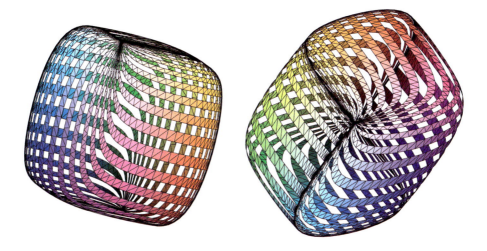

Figure 7.10 This type of rendering shows the structure of the Schrödinger color solid to better advantage. Each ribbon is for a certain location of the pass or stop band, the running parameter along each ribbon is the width of the band. Constant location clearly does not imply constant dominant wavelength. In order to keep dominant wavelength constant, you have to fiddle both bandwidth and band location simultaneously.

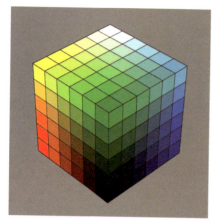 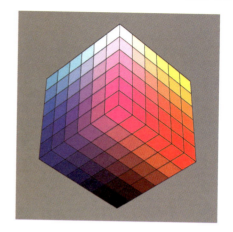

Figure 7.11 The familiar "RGB cube" also has two congruent sides, one containing the Newtonian, the other the non-Newtonian hues, exactly like the Schrödinger color solid. In fact, the simple RGB cube is an excellent model for the color solid.

A generic point on the Newtonian half of the surface can be parameterized as

$$\mathbf{s}(u,v) = \overline{\mathbf{e}}_{\mathbf{u}} - \overline{\mathbf{e}}_{\mathbf{v}};$$

thus, the tangents in the parameter directions are

$$\mathbf{t}_u = \frac{\partial \mathbf{s}(u,v)}{\partial u} \quad = \quad +\frac{\partial \overline{\mathbf{e}}_{\mathbf{u}}}{\partial u},$$
$$\mathbf{t}_v = \frac{\partial \mathbf{s}(u,v)}{\partial v} \quad = \quad -\frac{\partial \overline{\mathbf{e}}_{\mathbf{v}}}{\partial v}.$$

The tangent $\mathbf{t}_u$ does not depend on the parameter v; thus, along any parameter curve $u = \text{constant}$ the tangents in the u-direction are parallel. The same story applies as you interchange the parameters. Differential geometers say that the surface of the color solid is a "surface of translation" (in two ways) and thus very special. Surfaces of translation[8] have many special properties, even in affine geometry, like in colorimetry. Not only are the tangents parallel, but they also have the same magnitude (notice that this makes affine sense, since they are parallel). Thus an infinitesimal parameter quadrangle is a little parallelogram, this makes the net a "Chebycheff net," which is again quite special.[9]

Now consider what the tangent vector

$$\mathbf{t}_u = +\frac{\partial \overline{\mathbf{e}}_{\mathbf{u}}}{\partial u},$$

Figure 7.12 An "almost monochromatic" color must be close to the black point and thus appear almost black, hardly colored at all. Likewise, a spectrum with an "almost monochromatic" part missing must be very near the achromatic point and thus appear almost achromatic, again hardly colored at all.

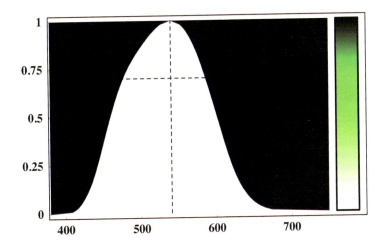

Figure 7.13 Suppose you keep the dominant wavelength fixed and—starting from a monochromatic beam—widen the passband. The color has to go from black (monochromatic beam) to achromatic (illuminant). Evidently, there will be one or more intermediate widths for which the color looks best. Indeed, the color content turns out to assume a single maximum. Here the case is illustrated for the far green. Notice how surprisingly wide the passband of the optimal color is. The semichrome passband is half of the spectrum.

Figure 7.14 The gray axis is the "polar axis" of the Schrödinger color solid.

 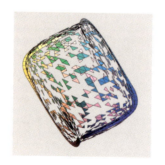 

Figure 7.15 The edge color loci on the boundary of the Schrödinger color solid.

really is. You have that $\overline{\mathbf{e}}_{\mathbf{u}}$ is the integral of spectral power density from $-\infty$ to u, thus the partial derivative is simply the integrand in u. But that means that $\mathbf{t}_u$ is some multiple of the monochromatic color at wavelength u. A similar discussion applies to the other tangent. Thus the tangent plane is spanned by the two spectrum cone generators at the transition loci of the passband.

The two vectors $\mathbf{t}_u$ and $\mathbf{t}_v$ span the *tangent plane* at $\{u, v\}$. The tangents are also tangents of the edge color locus; thus, the edge color locus by itself alone describes the full geometry of the surface of the color solid in a simple, geometrically transparent manner. Notice that the parameterization is nowhere singular because the edge color locus has no points with parallel tangents. It is easy to see this, because the tangent image of the edge color locus is simply the spectrum locus for the achromatic beam. The condition is thus simply that the spectrum locus in the chromaticity diagram should be a "simple, convex curve," that is, should not inflect or self-intersect and change direction by less than a full (2π) angle. Since it does not, this parameterization of the boundary of the color solid is indeed nowhere singular.

This is not to say that color vision systems with singular color solids are not possible. Indeed, it is easy enough to find models that show all kinds of weird (that is as compared to generic human vision) behavior. You should be thankful that the human visual system is as nice as it is.

The differential geometry of the color solid is further explored in appendix G.2.

7.3.1 The semichromes

The semichromes form the "equator" of the color solid and are thus a prominent feature (see figures 7.16 and 7.17).

The generation of the semichromes is best understood in terms of Ostwald's original ideas (figures 7.18–7.20). This construction is most remarkable, because it formally derives the *color circle* from the *spectrum* (which is a mere linear segment). Surprisingly, this fundamental construction is most often omitted from modern textbooks. Ostwald fully understood (and pushed) the importance of the concept.[10]

The semichromes are of various types, that is to say, either warm or cold edge colors, or pass or stopband colors. Moving around the color circle you have warm edge colors in the R-Y range, next passband colors in the G range, then cool edge colors in the C-B range, followed by stopband colors in the M range. The relation between the dominant wavelength of a semichrome and the location of the transition(s) of its characteristic wavelength band is not simple (figure 7.20). A calculation reveals a complicated pattern that becomes somewhat understandable only after considerable study.

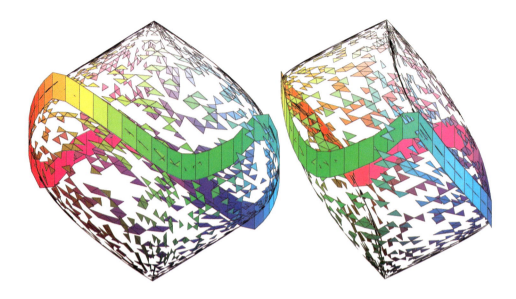

Figure 7.16 The semichrome locus is the "equator" of the Schrödinger color solid.

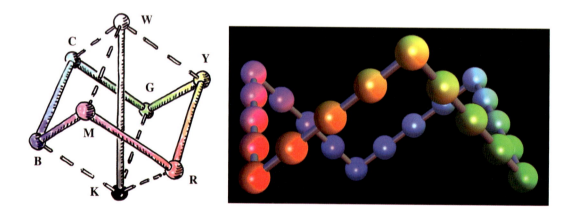

Figure 7.17 Left: The full color locus in the RGB color cube has exactly the same structure as the full color locus for human vision. It is the perfect model to have in mind if you try to visualize the full color locus. Right: Some views of the full color locus in the RGB color cube (chapter 14) should make its geometrical shape fully transparent to you.

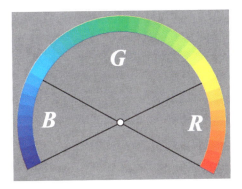

Figure 7.18 The "Ostwald horseshoe" is a partial color circle such that complementaries appear as antipodes. Since the greens have no complementaries this circle has a gap. It is a very natural idea, though the geometrical shape of the circle is pulled out of thin air.

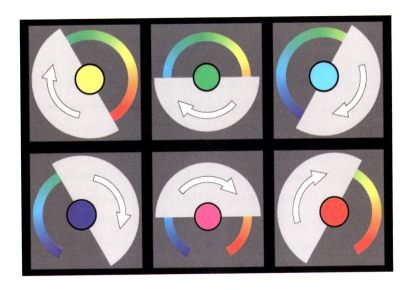

Figure 7.19 The semichromes are produced by integrating over a semicircle (shown by occluding the other half). Now it is visually obvious that the totality of semichromes forms a topological circle. Thus Ostwald showed—for the first time in history—how the "color circle" as a topological circle $\mathbb{S}^1$ is generated by the spectrum, which is a topological segment $\mathbb{I}^1$.

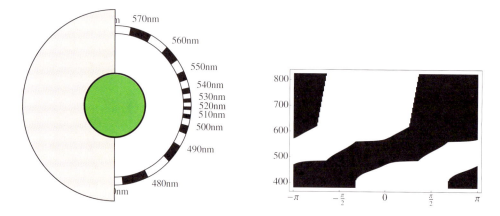

Figure 7.20 Left: How "Green" is generated as a semichrome. You simply admit one half of the color circle (I'll explain the peculiar wavelength scale later); Right: The passbands of the semichromes. On the abscissa the hue angle, on the ordinate wavelength.

It is much simpler to form a vivid image of the shape of the semi-chrome locus in color space (figure 7.17). It is a closed space curve that encircles the achromatic axis. Its characteristic up-down movement is best understood by having the RGB cube in mind (chapter 14). The full color locus of the RGB cube (a closed polygonal arc made up of cube edges that avoids the black and white vertices) is indeed the perfect model for the real thing.

Since the color solid is a convex body that touches the origin at the spectrum cone, it must be the case that the chromaticities of the optimal colors neatly fill the interior of the spectrum locus in the chromaticity diagram in a one-to-one fashion. (See figure 7.21.) For each chromaticity you have a unique optimal color.

It is evidently of interest to consider the loci of Goethe edge colors and Ostwald semichromes in the chromaticity diagram. These loci partly coincide because some optimal colors are also Goethe edge colors. The locus of semichromes encircles the achromatic point and lies within the interior, at many places quite apart from the spectrum locus (figure 7.22).

You have to conceive of the spectrum locus as representing *blacks*, the achromatic point as representing the achromatic color itself (white), and the semichrome locus as the locus of most colorful colors. This is quite a different picture from what is usually suggested in standard textbooks on basic colorimetry.

Figure 7.21 A regular array of optimal colors in transition location squared space (in arc length parameterization according to the Schopenhauer parts of daylight metric) mapped upon the chromaticity diagram. Near the center the colors tend to white, near the spectrum locus to black. The most vivid colors lie near the full color locus (figure 7.22). Notice how the dot density varies over the chromaticity diagram: most of the surface of the color solid is mapped within the full color locus. The RGB colors plotted here were computed directly from the Schopenhauer parts of daylight representation. This is very different from conventional representations were the coloring is due to someone's artistic eye.

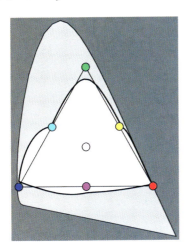
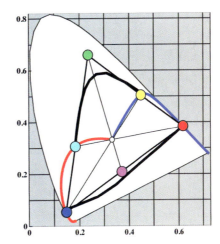

Figure 7.22 The full color locus in the chromaticity diagram. The edge color loci (blue: cool edge colors; red: warm edge colors) partly coincide with the semichrome lous. Notice their relation to the spectrum limits and complementaries thereof. On the right I show the same configuration in the "official" CIE xy diagram, it is obviously "distorted" as is evident from the RGB triangle.

This concept (figure 7.23 left) is very well suited as a "thought model" that serves to quickly intuit the COLOR (yes, I mean the *quale* here!) for a given spectrum or the spectrum for a given COLOR. Of course, this can be done only in a very coarse fashion, and you must not think that colorimetry actually predicts such things. Far from it; colorimetry has nothing to say about *qualia*. Thus these models are in the domain of psychology. Yet they are obviously useful. The semichrome model is perhaps most useful in the setting of "object colors" and the "spectra" should be understood as essentially spectral reflectances.

The other model shown in figure 7.23 right is more suitable when you think of the COLORS of beams (or "aperture colors" as the psychologist would say). It is essentially the same idea though. You look for spectral excess, this time via the spectral slope and curvature. If you understand this in terms of the "scale-space model" you understand that the slope and curvature describe the "second order jet" of the spectrum,[11] that is the vector composed of the first and second derivatives (at the midpoint, say). The jet lives in a two-dimensional space of which the origin stands for the flat spectrum (the GRAYS). All spectral variations correspond to directions from the origin, thus the hue domain is periodic, just as that of the semichromes. As a simple analytical model you may use ($\mu > 1$)

$$S(\lambda) = S_0 e^{\mu(\lambda \cos \varphi + \frac{1}{2}\lambda^2 \sin \varphi)}.$$

As you vary the parameter φ, the spectrum will change as indicated in the figure.

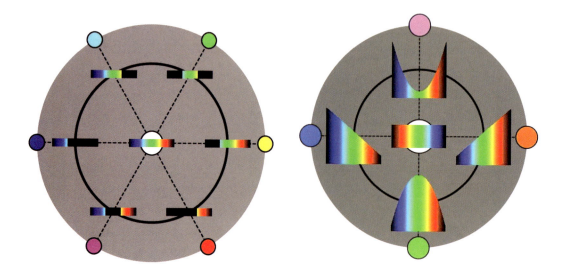

Figure 7.23 Two intuitive "models" for perceived color. The left one is Ostwald's semichrome picture. The right one considers colors as manifestations of spectral slope and curvature.

Notice how different these models are from the standard notion. The YELLOW of a yellow flower is not due to the presence of monochromatic radiation in the yellow region of the spectrum. Not at all. It is due to spectral excess in the medium and long wavelength parts of the spectrum (or lack of the short wavelengths, which amounts to the same). The "seeing by wavelength" notion is void, for *there need not even be* radiant power at the dominant wavelength of the yellow flower color and if the petals reflected only a narrow band at that wavelength the flower would appear black (for lack of luminance)!

Of course, these models have to be understood sympathetically. Here are some necessary grains of salt to take with them:

— you have to "see" the spectrum in a very blurry way, in terms of excess power at one side, the other side, the middle, or a lack in the middle. All finer modulations have no causal effect on the COLOR. Your "visual acuity for spectral variations" is very limited indeed;

— in real life you never see a beam or an object in isolation (see chapter 11). You are "adapted" to your general environment and anything is embedded in a surround. The general environment tends to appear achromatic, even if its spectrum is not at all flattish. The COLOR of a minor element is due to the spectral differences with the larger environment. Thus "spectral excess" has to be understood *relative* to the spectrum of the environment.

Although these models are very crude and cannot be given a sound foundation in colorimetry, you will find them very useful and much more effective than the (spurious) "seeing by wavelength" concept.

7.3.2 Colors and beams

The color solid surface is remarkable in that the *colors* contained in it correspond in a one-to-one fashion to *beams* (see chapter 4). Remember that a (colorimetric) color stands for an *equivalence class* of beams of infinite cardinality. A color gives you little of a handle on the beam that created it. But it is not even that simple to find *any* real beam that might have created a given color. The obvious method is to use a linear combination of the primaries, of course. But in many cases this will not get you a real beam; there will be stretches of negative radiant spectral power density in the spectrum. But, using the color solid, you may describe any color as a suitably attenuated optimal color. Since an optimal color corresponds uniquely to some real beam, you find a unique real beam for any color in the color solid. If the color is not optimal, then the beam you get will not be the beam that caused the color (at least, that probability is zero). But it will be a real

Goethe's notion of colors as shadowlike entities reconsidered

Goethe's notion that colors are shadowlike entities that somehow live in a realm between white and black is often ridiculed. However, in view of Ostwald's identification of certain special spectra connected with colors the idea can actually be given some body. This is especially clear if you think of Ostwald's method to physically generate colors, which is nothing but an inverted spectroscope. If you shine daylight on the input aperture (usually the *output*, that is the spectrum side, of the instrument!) and look through the output aperture (in normal use the entrance slit!), you see "WHITE," that is, the color of daylight. If you block the input aperture you see "BLACK" (no radiation enters the instrument). Interesting phenomena are seen when you block *part* of the entrance aperture. You are able to produce all possible COLORS this way. This is indeed nothing special, for you can produce arbitrarily attentuated instances of arbitrary Schrödinger optimal colors, that is to say, *any* color within the color solid, that is the total gamut of colors enabled by daylight. These colors apparently are generated by blocking certain parts of daylight, and in that respect they are truly "shadows of daylight". All these "shadows" are contained within the daylight color solid, between WHITE and BLACK. Thus Goethe is literally right in describing colors the way he does. Moreover, it is a description that makes much sense, because the colors of matte surfaces, illuminated by daylight, are exactly generated in this way (by selectively removing certain parts of daylight that is).

beam nevertheless. This is the principle behind Ostwald's "color science". It is entirely correct and consistent, though Ostwald only partially understood this and the physicists from his time misunderstood it entirely, leading to rather unfortunate and (in retrospect) embarrassing polemics.[12]

7.4 Ostwald's Principle of Internal Symmetry

The color solid induces yet more structure of considerable interest. Remember how Wilhelm Ostwald devised a clever way to mensurate the "color circle," actually the locus of semichromes, via bisection, his so-called principle of internal symmetry (figures 7.24 and 7.25). There are several problems with this method, both conceptually and computationally.[13] The color solid lets you do the job in a conceptually simpler and computationally much more convenient and stable way. Since the ratio of volumes is an affine invariant, you

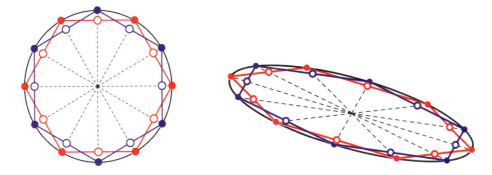

Figure 7.24 This may have been Ostwald's intuition when he framed his principle of Internal symmetry. The "affine arc length" of the circle is invariant under affinities.

may cut the color solid into equal volume sectors, a section being defined by two cutting planes that contain the achromatic axis. Thus you divide the color solid much as you would cut honest pie slices. It is easy to show that this is equivalent to Ostwald's principle of internal symmetry if the slices are infinitesimally thin,[14] that is to say, in the continuous limit. In this way you can put a continuous, periodic scale on the semichromes, and it yields a "hue" measure for any color in the color solid, because any color can be written as a unique linear combination of the achromatic color and some semichrome (figures 7.26–7.28).

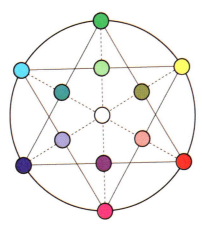

Figure 7.25 The vertices of the Schopenhauer RGB crate yield cardinal colors that automatically satisfy the Principle of Internal Symmetry.

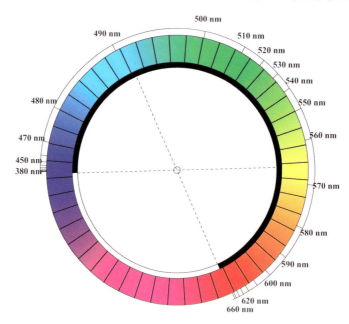

Figure 7.26 The "color circle" mensurated via Ostwald's principle of internal symmetry.

The Schopenhauer parts of daylight R, G, and B may be combined so as to generate the secondary colors $Y = R + G$, $C = G + B$ and $M = B + R$. Notice that the periodic sequence $\dots YGCBMR \dots$ is automatically mensurated by Ostwald's principle of internal symmetry.

For instance, Y has the same dominant wavelength as $(R + G)/2 = Y/2$, R has the same dominant wavelength as $(M + Y)/2 = ((B + R) + (R + G))/2 = A/2 + R/2$ and so forth. Thus the Schopenhauer parts of daylight automatically generate a well tempered color circle.[15]

What goes for the Schopenhauer parts also holds for the RGB colors. The crate structures with one vertex in the black, one in the achromatic point all automatically generate well tempered color circles. Most people profit from this on a daily basis as they kill time in front of their RGB computer monitors or TV sets.

This method of hue mensuration can be used to define a finite number of "cardinal" colors. This is useful in tasks like color *naming*. The number of colors should be even, because the cardinal colors should occur in complementary pairs. It should also be a multiple of three, reflecting the trichromatic nature of human vision. Thus the minimum number of cardinal colors is six. In order to define them you have to decide on the "phase" on

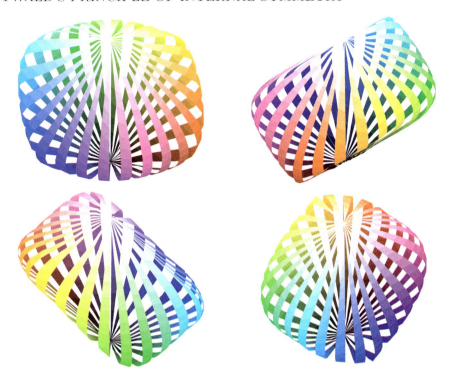

Figure 7.27 Here the color solid has been foliated with constant dominant wavelength sections that are equally spaced according to Ostwald's principle of internal symmetry.

the color circle. An obvious choice is to define the "yellow" cardinal color as "daylight minus blue", that is, the edge color that contains all wavelengths longer than the complementary of the IR limit. In numbering the cardinal colors you must also decide on a sense, for instance from yellow to green. This proposal corresponds with the choice Ostwald made in constructing his color atlas in the early twentieth century. Then the six-step cardinal color sequence is

        0: Yellow,
        1: Green,
        2: Cyan or turquoise,
        3: Blue,
        4: Magenta or purple, and
        5: Red

(of course, the next cardinal color is 6 = 0, that is, Yellow again).

Figure 7.28 "Constant hue planes" (left) are best generated via the "honest pie slice principle" (right).

When you do the calculations for average daylight you find results that are interesting in that the dominant wavelengths of the cardinal colors correspond very well to the spectrum locations of the beams corresponding to the luminous experiences generally indicated by these color names ("yellow", "green", and so forth). (Figure 7.29.) Thus you have a purely colorimetric method that lets you generate color names that prove to be quite acceptable to generic human observers.

Since colorimetry has *nothing* to do with color experiences, it is remarkable that the space of hues (the color circle) can be colorimetrically mensurated in such a way that the cardinal colors of the scale agree very well with pure "eye measure." Indeed, when you compare the various eye measure scales that have been colorimetrically calibrated among each other, you find variations of one to two steps on a twelve point scale (most such scales contain 12, 24, or 48 "cardinal" colors). The mensuration via the (amended) principle of internal symmetry agrees about equally good with the overall average, indicating that it is just as useful as any of these scales. At first blush this appears little less than miraculous. However, few people find it at all remarkable that your "eye measure" for lengths corresponds reasonably well with the results you obtain through purely physical methods (yardsticks and such). (Figure 7.30.) In my view the latter fact *is* remarkable, and I would suggest that there is conceptually not much difference with the case for hues.

If colors in nature were "uniformly distributed" (I will discuss the hairy topic of the statistics of spectral reflectance factors and spectral radiant power densities later), then the "honest pie-slices" would contain equal numbers of them. The uniform prior thus is what you need in a Bayesian treatment. This may actually be learned over a long time span. In

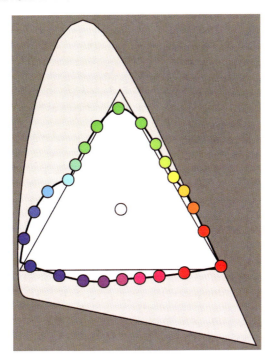

Figure 7.29 The semichrome locus in the chromaticity diagram with "24 cardinal colors" defined by way of Ostwald's "Principle of Internal Symmetry" for CIE D65 daylight.

this view the "honest pie-slices" metric is an optimal adjustment to the environment.[16] The notion perhaps succeeds at least partially to remove the "magick".

Figure 7.30 Top: A uniform wavelength scale, divided by 10nm. This shows the region 395–825nm, the midpoint is 527.5...nm. It is evident that the wavelength parameterization is perhaps not a particularly fortunate one; Bottom: A spatial extent mensurated by progressive bisection. Notice that it looks uniformly divided by "eye measure." Why is it that nobody is surprised by such an example, whereas the results of Ostwald's principle of internal symmetry meet with general contempt?

For color naming, the six cardinal colors suffice in daily life, though a twelve-point scale is required if you need to discuss more subtle variations (e.g., recognize "ORANGE" between YELLOW and RED). A twelve point or twenty-four point scale is all you will ever need, and a forty-eight point scale is clearly overkill.[17] From a colorimetric perspective you have a continuous scale of course. It is only for color naming that you will use a set of cardinal colors at all.

7.5   Variation of the Achromatic Beam

Ostwald did not understand the fact that the semichromes depend upon the illuminant and that they are by no means "full colors" in general. Thus he defined the set of all colors as the volume claimed by the convex hull of the union of the semichrome locus with the achromatic axis (figure 7.31). Although perhaps roughly in the ball park, it is by no means correct. Even for a fixed illuminant the color solid contains colors that fail to be contained in Ostwald's double cone.[18] In order to deal with the problem, you need some additional geometry.

In considering the structure of the color solid, it is very convenient to think of it as contained within the intersection of the spectrum cone with the inverted spectrum cone (defined via the central symmetry of the color solid). (Figures 7.32 and 7.33.) Remember that the color solid is tangent to these cones at both the black and the achromatic point; thus the color solid fits very snugly in this "double cone".

The double cone is to be thought of as the envelope of all color solids obtained when you vary the spectrum of the achromatic beam, keeping its color (the achromatic point) fixed. The envelope is never reached for achromatic beams that have nowhere zero radiant spectral power density throughout the spectrum. The only way to let the color solid touch the generators of the double cone over their full extent is to consider *very singular* achromatic beams, namely the mixture of two suitable (complementary) monochromatic beams. When you do this, the color solid collapses to a planar parallelogram that indeed touches the double cone along the corresponding generators. Such a "color solid," would not really be a "solid" since it would have zero volume.

From such considerations you may appreciate that a measure of the "quality" of the achromatic beam would be the ratio of the volume of the color solid that it generates to the volume of the spectrum double cone (evidently an affine invariant). This "quality" is something like the total number of colors enabled by the achromatic beam. You have already seen that some achromatic beams are better than others in this respect. Thus the problem arises as to what would be the "best" achromatic beam for a given achromatic color? I don't know the answer to that (a good science project), but the best beam will probably not be too different from average daylight. It is not hard to find the best beams for some of the models. For instance, for the Grassmann model the best beam for the achromatic color $\{1, 0, 0\}$ is (as you would have guessed) the beam that has a flat spectrum. (Try to prove this!)

It may be useful to consider some numbers at this point. For average daylight the color solid fills 64 percent of the volume of the spectrum double cone. It seems likely to me that the best illuminant (I am not sure what it is, not having calculated it, but it is certainly not

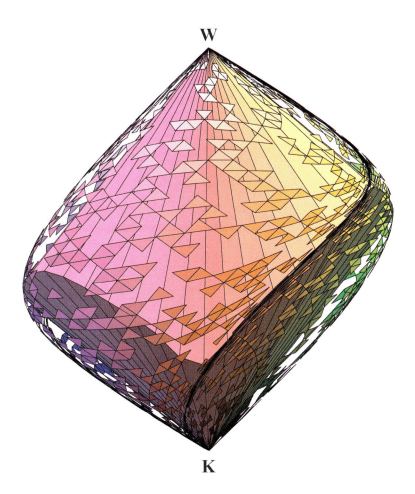

Figure 7.31 The "Ostwald double cone" is generated by the semichrome locus and the gray axis. Ostwald erroneously identified it with the color solid. It is indeed inscribed in the color solid, but it fails to exhaust its volume.

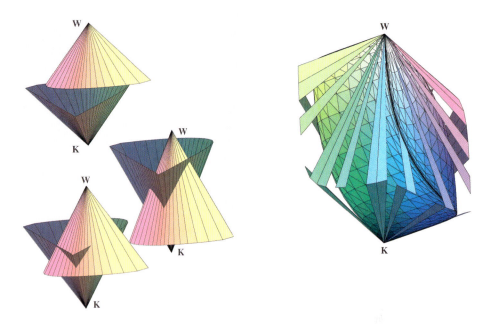

Figure 7.32 The intersection of the spectrum cone with the inverted spectrum cone forms the "spectrum double cone", which is the envelope of all color solids for a given white color (left). On the right the color solid inside the spectrum double cone for the case of average daylight.

too different from a uniform spectrum) will not do *much* better. Ostwald's full color double cone exhausts 77 percent of the volume of the color solid; thus Ostwald missed 23 percent of all possible colors in his atlas. Fortunately, these missing colors are very close to optimal

Figure 7.33 The intersection of the spectrum double cone with the inverted spectrum double cone.

colors and you are very unlikely to meet with them in real life. Possibly they can't even be printed with current technology. Schopenhauer's RGB crate (see chapter 9) based on the optimum tripartition of daylight exhausts 65 percent of the color solid. Again, the missing 35 percent of colors are close to optimal colors and unlikely to occur in natural scenes. For almost all practical purposes, the Schopenhauer parts of daylight subtend an ample gamut. In order to put these numbers in perspective you would have to apply a weight to the colors, but a prior probability distribution is not available, neither from theory, nor from empirical data. If you study the collections of colors sampled from nature or daily life available from the internet you find that they comfortably fit the Schopenhauer RGB crate (chapter 9). Apparently Schopenhauer's (derived from Goethe's) notions are not without practical interest.

The "crease" that is the equator of the spectrum double cone defines a closed space curve that contains colors that are as remote from the achromatic axis as possible, *no matter what the spectrum of the achromatic beam might be*. Thus they are—in a sense—a generalization of the full colors. (The full colors are as remote from the achromatic as as possible *for the given spectrum of the achromatic beam*.) I will call them—for want of a conventional term—"ultimate colors." They are distinct from the full colors (figure 7.34).

If Ostwald had thought of it he would probably have put the ultimate colors in place of the full colors. It would have saved him quite a bit of hassle. From a formal point of view it is indeed the rational step to take, because the ultimate colors are invariant against spectral changes of the illuminant—always keeping the achromatic *color* fixed of course. Any variation under such "invisible changes" is a major headache. Thus the ultimate colors

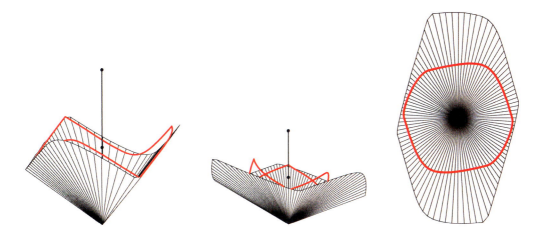

Figure 7.34 The spectrum double cone with the full color locus.

are the perfect landmarks to use (figure 7.35), you can even use them to indicate the change of full colors under variations of the illuminant spectrum.

An especially interesting application of the ultimate colors is the analysis of the full colors in terms of (ultimate) color, white and black content. This is something Ostwald was unable to do, since he mistook the full colors to be "ultimate". Such an analysis is of interest because many observers believe to "see" white and/or black in certain a full colors, reports that irritated Ostwald no end. If you do the analysis, you see that some full colors indeed contain quite substantial amounts of white and/or black. Such an analysis turns out to corroborate the "eye witness reports" at least in a qualitative sense. (Figures 7.36–7.39.)

In figure 7.36 I show another analysis of the full colors, namely the distance from the achromatic axis and the height of the orthogonal projection on the achromatic axis as a function of the hue angle and achromatic content. Notice that the primary colors (red, green, and blue) have the lowest, the secondary colors (cyan, magenta, and yellow) the highest achromatic content. Blue has especially little, yellow especially much achromatic content. The distance from the achromatic axis is another measure of the "amount of color" and is seen to vary quite a bit (low for yellow and blue, high for magenta and green). This analysis is different from that shown in figure 7.37, which depends on the relation of the full colors to the ultimate colors.

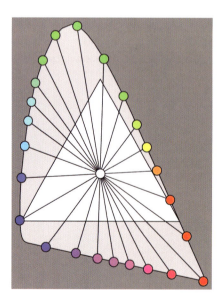

Figure 7.35 The twenty-four cardinal colors defined in terms of the ultimate colors. These have the advantage that they are fixed for a given white color.

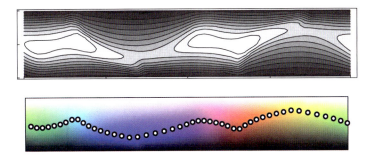

Figure 7.36 An analysis of the full colors. In the upper figure the distance from the achromatic axis is shown in gray tone, in the lower figure the height of the orthogonal projection on the achromatic axis have been plotted as a function of the hue angle (horizontally) and achromatic content (vertical axis).

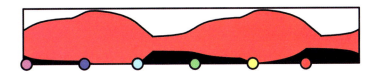

Figure 7.37 The full colors analyzed in terms of their white, black, and (ultimate) color content.

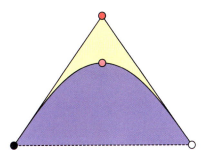

Figure 7.38 A section at a certain dominant wavelength for the Grassmann model. Notice how the Schrödinger color solid fits snugly into the spectrum double cone. All possible colors are contained in the convex hull of white, black, and the ultimate color, whereas they are not in the convex hull of white, black, and the full color, as Ostwald erroneously assumed.

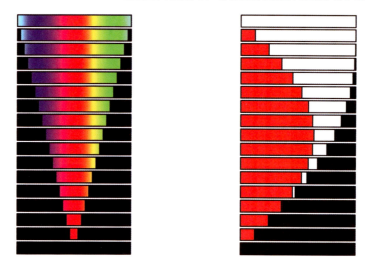

Figure 7.39 The optimal colors for a fixed dominant wavelength analyzed in terms of their white, black, and (ultimate) color content, here for the Grassmann model. Left the passbands of the optimal colors, right their composition in terms of color, white, and black content. Of course the "colors" for the Grassmann model are mere fancy.

There do not exist illuminants (except extremely singular ones) for which all ultimate colors are real, though what exactly the possibilities are is unclear to me. Since this has to do with the "color rendering properties" of light sources, the topic is by no means irrelevant. It seems likely that one has to live with virtual ultimate colors, though.

A perusal of figures 7.36 and 7.37 suggests that not all full colors are equal. Some are evidently "special" because of some extremal property. This is of interest because of the fact that *wavelength* and the *spectrum* are irrelevant to visual experience. You simply do not "see by wavelength", and purples are every bit as true hues as the so called "spectral" hues are.

The color circle (or the locus of full colors) does have neither beginning nor end, there seems to be no special phase singled out. Yet throughout the centuries people have singled out certain hues as special, certainly RED, YELLOW, and BLUE, for instance. Here you have an instance of purely colorimetric properties that are also located at certain specific points on the full color locus. Of course, colorimetry does not predict COLOR, it was never designed to do that. But nothing keeps you from promoting colorimetry to the status of a "model" for certain optical *qualia*. Thus it becomes a theory in the psychological realm. This is very speculative, of course, but no one who is a hard-nosed colorimetrist has to buy in to this.

In terms of this "theory" you are able to "predict" the special colors on the color circle. They seem to fit experience rather well. Of course, the pattern found here is ultimately due to the shape of the cone action spectra, thus, if that makes you feel better, you can go one step further and turn the colorimetric structure into a physiological account, then end up with a physiology based psychological theory.

7.5.1   Example of a "bad" illuminant

It is easy enough to find examples of "bad" illuminants. By way of an example, I construct a bona fide illuminant, that is to say, with overall positive radiant spectral power, that is really bad. I take two mutually complementary monocromatic beams and "blur" the spectrum so as to be very uneven (two dominant peaks) yet overall smooth. See figure 7.40.

The edge color loci for this illuminant are *very* different from that of average daylight, even though the illuminants have the same color (you can't tell them apart by eye!). See figure 7.41 left.

The Schrödinger color solids for average daylight and the bad illuminant are strikingly different (see figure 7.41 right). The volume of the "bad" solid is much less than that of the "nice" one, suggesting that the "color rendering properties" of the bad illuminant are rather less than those for proper daylight.

7.6   The Approximate Bilateral Symmetry

The full color locus has many symmetrical properties in the Grassmann and Helmholtz models, due to the symmetries of the color matching functions. As is to be expected, the

Figure 7.40 Example of a really bad illuminant that has the same color ("white") as average daylight.

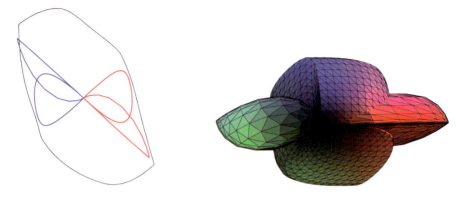

Figure 7.41 Left: The crease of the spectrum double cone (black curve) and the edge color curves for average daylight (the "nice" figure eight loop) and the bad illuminant (the strongly peaked loop). (This is a perspective view from a finite distance; the edge color loci of course lie fully within the spectrum double cone!) Right: The color solids for average daylight (the "slightly inflated cube") and the bad illuminant (the flattish thing that sticks out) plotted as they interpenetrate each other. This vividly illustrates how very different the two solids are, even though they share the black and white points.

actual semichrome locus for human vision is rather less regular. However, it has a rather marked approximate bilateral symmetry (figure 7.42).

The plane of bilateral symmetry is the yellow-blue plane; thus the symmetry leaves yellow and blue invariant, but interchanges red with green and cyan with magenta. One reason

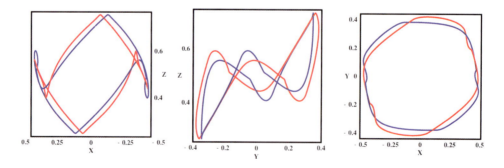

Figure 7.42 The approximate bilateral symmetry of the semichrome locus (here for a flat spectrum) is rather marked. Here I show the result from a brute-force search for bilateral symmetry in canonical coordinates, using Euclidian distance measure.

> ### The set of real colors for beams dominated by a given achromatic beam
>
> One way to study the gamut of all colors obtained from beams that are dominated by any given achromatic beam is to generate random instances and see where they end up in color space. This turns out to be a tricky topic because the various spectra are constrained in non-trivial ways. I explore this in some detail in appendix G3.

that this works as well as it does (figure 7.42) is no doubt that the relative brightnesses are roughly invariant under the symmetry. A bilateral symmetry that would interchange yellow and blue would look unnatural for that reason.

In principle, you might expect most of the symmetries of the cube to apply to the color solid—albeit approximately of course. The great difference in luminosity between yellow and blue changes this markedly, though.

7.7   Definitions of Saturation and Color Content Compared

The most convenient way to indicate any color $\mathbf{x}$ (say) is to refer it to the frame defined by the two complementary generators of the spectrum cone $\mathbf{u}$ and $\overline{\mathbf{u}}$ that are coplanar with the color $\mathbf{x}$ and the achromatic axis.

In terms of an Ostwald-style description, you would write,

$$\mathbf{x} = c\mathbf{u} + w\mathbf{a} + k\mathbf{k} = c\mathbf{u} + w\mathbf{a},$$

where $\{c, w, k\}$ are the color, white and black content (with $c + w + k = 1$), $\mathbf{a}$ the achromatic (or white) point, and $\mathbf{k}$ the black point.

Given any color $\mathbf{x}$ that is not achromatic, there is a unique plane that contains the color and the achromatic axis. This plane cuts the double cone in a parallelogram, spanned by the monochromatic beams $\mathbf{u}$ and $\overline{\mathbf{u}}$. The vertices of the paralellogram are the black beam $\mathbf{k}$, the monochromatic beam and its complementary $\mathbf{u}$ and $\overline{\mathbf{u}}$, and a white beam $\mathbf{a}$. This "white beam" is rather special, though, for although it has indeed the same *color* as your regular white beam, it is simply the superposition of the two monochromatic beams $\mathbf{a} = \mathbf{u} + \overline{\mathbf{u}}$.

Suppose I write,

$$\mathbf{x} = \mu\mathbf{u} + \overline{\mu}\,\overline{\mathbf{u}} = \frac{\mu + \overline{\mu}}{2}(\mathbf{u} + \overline{\mathbf{u}}) + \frac{\mu - \overline{\mu}}{2}(\mathbf{u} - \overline{\mathbf{u}}),$$

The optimal colors in the models

By far the most convenient setting to study the "optimal colors" is in the context of analytical or algebraic models. This is done in appendixes G.4–6.

In appendix G.7 I consider "pathological models," this clearly shows that the models studied in this book (including the human visual system!) are quite special.

Finally, in appendix G.8 I consider a number of perturbations to the generic Grassmann model. These studies are very useful in gaining an understanding of things that might happen in the real world. Even many experts are vague on some of the issues.

---

which again can be rewritten,

$$\mathbf{x} = I\left(\mathbf{a} + s(\mathbf{u} - \overline{\mathbf{u}})\right),$$

with the "intensity" $I = (\mu + \overline{\mu})/2$ and the "saturation" $s = (\mu - \overline{\mu})/(\mu + \overline{\mu})$. This is geometrically an evidently reasonable definition of something like "saturation". In this description the "intensity" I is fully accounted for by the achromatic component, which is simply Hering's intuition.

Unfortunately, current practice is not likely to change its definitions, awkward as they may be.

Notice how both the ultimate color *and its complement* appear in these constructions. This is indeed how you should conceive of the hue dimension: A "color" is really composed of a *pair*, a hue and its complementary hue, suitably tempered by white and black. This is reminiscent of Hering's *Gegenfarben*, (opponent colors)[19] but it is different in that Hering only uses two pairs of opponent colors (apart from black-white), whereas here I use a continuum. In a sense, Hering's system is simply a reparametrization, as discussed by Schrödinger[20], but I propose that you think of *all* hues as being members of an opponent pair. Thinking of colors means thinking in complementary color pairs. This is clearly brought out by the formalism, but is also most conducive to your intuition.[21] (You should try.) The major pairs to keep in mind are red-cyan, yellow-blue, and green-magenta.

7.1 Low-dimensional projections of high-dimensional hypercubes

The concept that Schrödinger's color body is nothing but the projection of an infinite dimensional hypercube[22] from the infinite dimensional space of beams into the three-dimensional fundamental space or—if you want—color space, is a very powerful one. Of course, three is rather less than infinity; thus you have necessarily only a very impoverished glimpse of reality. Nevertheless, many of the striking geometrical properties of the color solid derive rather directly from those of the infinitely dimensional hypercube.

The simplest way to obtain some notion of the shape of projections of high-dimensional hypercubes is to consider graphical representations of them. It is rather easy to draw these. One of the simplest methods is to project all vertices of a high dimensional hypercube and find the convex hull of the projection. This yields the silhouette of the projection.

The main drawback is that high dimensional hypercubes have very many vertices. The number of vertices is 2^n, where n denotes the dimension.[23] You may generate random orthogonal projections by generating vectors with random normally distributed, coordinates and orthonormalizing them[24]. Algorithms that let you find convex hulls are readily available. In this way you can easily find projections of hypercubes up to dimension 20 or so in reasonable time on a laptop computer.

Such outlines tend to look similar to the Schrödinger color solid already (figure 7.43), exactly as expected. It is instructive to study various directions of projection.

However, the projections you obtain with such an algorithm turn out to be degenerate (despite being randomly generated) in a special way. Consider the "primitive vertices" of the hyper-

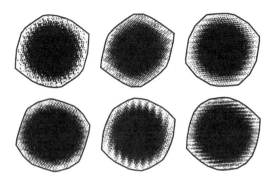

Figure 7.43 These are six random views of the 16-dimensional hypercube. Notice the low density of vertices near the boundary.

cube, by which I mean the vertices with coordinates

$$\{0, 0, \ldots, 1, \ldots, 0, 0\},$$

that is to say, only one coordinate unequal to zero. I call these vertices "primitive" because they form an orthonormal basis in which you can conveniently (and indeed conventionally) express all other vertices. "Translated" to colorimetry, the primitive vertices are just the "monochromatic beams" of course. The projection of the primitive vertices is a highly *over-complete* basis of the low-dimensional space; thus there is strong linear dependency between them. However, in colorimetry you are most interested in *convex*, rather than *linear* combination. You would like the projections of the primitive vertices to be "independent" in *that* way, that is to say, you want them to be vertices on the boundary of the convex hull of the projection. For then none of the directions of the projections of the primitive vertices is in the convex hull of any triple of directions of projections of other primitive vertices.

Translated toward colorimetry, all primitive projections are *discriminable*, and thus have different "hues". The spectrum is a series of hues in which there are no repeats.

Whereas it is indeed a requirement that there will be no repeats, this does by no means guarantee that the spectrum locus will be a nice, smooth curve. For suppose it were; then I could simply do a random permutation on the basis of primitive vertices and destroy this pleasant geometry. Arbitrary permutations of basis vectors do nothing special, at least not in terms of basic linear algebra. Thus you need something in addition to the mere linear algebra, something that does not derive from mathematics, but from physics. You need a "well ordered" basis to begin with, and mathematics gives you no clue. From the perspective of physics, an example of a "well ordered" basis would be the monochromatic beams of unit radiant power ordered by wavelength or photon energy. I will assume that the start point is such a physically well ordered basis of primitive vertices. Then the obvious requirement is that there should be no repeats.

This is indeed desirable. However, the hard reality of the matter is that such a pleasant state of affairs is very rare. Almost all of the randomly generated projections violate the requirement formulated above (figures 7.44 and 7.45). They represent spectra with one or more "repeated hues". I will call a projection that happens to be "nice" a projection that "fully displays" all dimensions (read: shows the hues without repeats). Such *revealing* projections satisfy the requirement:

A revealing projection *is such that any direction of a projection of a primitive vertex is not in the convex hull of any triple of directions of the projections of primitive vertices.*[25]

Such revealing projections are models for "good" visual systems. Thus good visual systems are

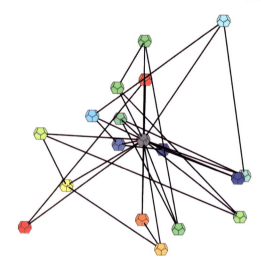

Figure 7.44 The projection of a primitive basis by a random projection. Dimension 17.

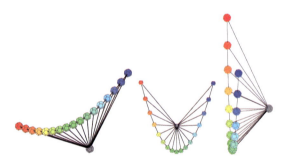

Figure 7.45 The projection of a primitive basis by a "revealing" projection (here the Helmholtz model). Dimension 17.

very rare indeed, if you draw a random example it will be "pathological" with probability one. On the other hand, you have already seen that revealing projections *exist*. Simple examples include the Grassmann and the Helmholtz models, and—of course—the human visual system (figures 7.46 and 7.47).

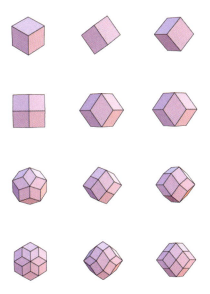

Figure 7.46 Grassmann-type (page 167) projections of hypercubes of dimensions three to six.

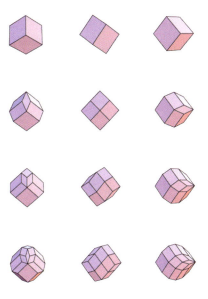

Figure 7.47 Helmholtz-type (page 174) projections of hypercubes of dimensions three to six.

In order to obtain such a revealing projection, the "color space" has to be at least *three dimensional*. In dichromacy it is impossible to "display" the spectrum, that is, to have each monochromatic component have its own hue. Thus the human (trichromatic) system is the lowest dimensional implementation of such a "good" system. Then the projection is revealing if and only if all 3×3 subdeterminants of its matrix have the same sign. In the continuous case, this amounts to the condition that the three tangents to the edge color locus at different points span a tetrahedron of the same orientation for any triple (taken in their natural order along the curve). This again translates to the condition that the edge color locus is a curve with nonzero curvature throughout (remember that the curvature of space curves is an *unsigned* quantity) and that the torsion has the same sign along the curve. Thus the edge color locus has to be a segment (at most half of a full turn) of a generalized helix.[26]

Here is an example. Consider projections of the four-dimensional hypercube. The primitive vertices are

$$
\begin{aligned}
\mathbf{v_1} &= \{1,0,0,0\}, \\
\mathbf{v_2} &= \{0,1,0,0\}, \\
\mathbf{v_3} &= \{0,0,1,0\}, \\
\mathbf{v_4} &= \{0,0,0,1\}.
\end{aligned}
$$

Consider the projection Π with matrix

$$
P = \begin{pmatrix}
1 & 0 & 0 & 0 \\
0 & \frac{1}{\sqrt{3}} & \frac{1}{\sqrt{3}} & \frac{1}{\sqrt{3}} \\
0 & 0 & -\frac{1}{\sqrt{2}} & \frac{1}{\sqrt{2}}
\end{pmatrix},
$$

it is clearly an orthogonal projection. The projections of the vertices are

$$
\begin{aligned}
\Pi\mathbf{v_1} &= \{1,0,0\}, \\
\Pi\mathbf{v_2} &= \{0, \frac{1}{\sqrt{3}}, 0\}
\end{aligned}
$$

$$\Pi \mathbf{v_3} = \{0, \frac{1}{\sqrt{3}}, -\frac{1}{\sqrt{2}}\},$$

$$\Pi \mathbf{v_4} = \{0, \frac{1}{\sqrt{3}}, \frac{1}{\sqrt{2}}\}.$$

Now you have that

$$\Pi \mathbf{v_2} = \frac{1}{2} \left(\Pi \mathbf{v_3} + \Pi \mathbf{v_4} \right),$$

thus the second vertex is actually superfluous (not "revealed") because the convex hull of the projection is already spanned by the projected vertices $\{\Pi \mathbf{v_1}, \Pi \mathbf{v_3}, \Pi \mathbf{v_4}\}$ alone. You easily check that this projection fails to make the grade, since the subdeterminants are not all positive.

On the other hand, the projection Λ with matrix

$$L = \frac{1}{2} \begin{pmatrix} 1 & 1 & 1 & 1 \\ -1 & -1 & 1 & 1 \\ -1 & 1 & 1 & -1 \end{pmatrix},$$

it is clearly also an orthogonal projection and it projects the vertices as

$$\Pi \mathbf{v_1} = \frac{1}{2}\{1, -1, -1\},$$

$$\Pi \mathbf{v_2} = \frac{1}{2}\{1, -1, 1\},$$

$$\Pi \mathbf{v_3} = \frac{1}{2}\{1, 1, 1\},$$

$$\Pi \mathbf{v_4} = \frac{1}{2}\{1, 1, -1\},$$

which are all revealed. Indeed, the determinants of all the 3×3 submatrixes equal $+4$, showing that no projected vertex lies in the convex hull of any triple of others. (See figure 7.48.)

For the case of revealing projections it is a simple matter to draw the projection of the hypercube. First you find the projections of the primitive basis. Next you construct the "edge color curve" (figure 7.49), you do this by constructing the cumulative sums of the projections.[27] Then the visible faces can be constructed

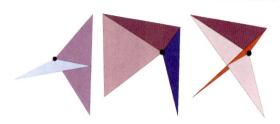

Figure 7.48 At left the spectrum cone generated by the matrix P, at center that generated by the matrix L. At right I have swapped the second and third primitive vertices.

from this curve since the exposed projected vertices are the passband and stop band colors.

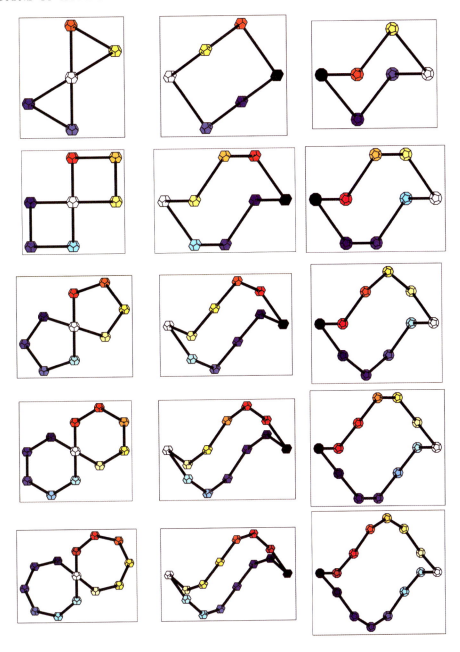

Figure 7.49 The cumulative projected primitive vertices for the Grassmann-type projection of hypercubes of dimension three to seven.

7.2 Differential geometry

The color solid is an overall smooth body
except for two creases (the edge color loci) and
two conical points (the white and black points).
The differential geometrical structure of the edge
color loci has already been covered in some detail. The conical points are also familiar enough,
locally they are exactly like the spectrum cone
(the inverted spectrum cone having the same geometry as the spectrum cone of course). What
remains to be done is to study the structure of
the smooth areas in relation to the singularities.

I will discuss only the affine differential structure here. Given a metric, there is a much richer
differential geometry to explore. I will cover that
later in chapter 8.

7.2.1 Geometry induced by the edge color spirals

The boundary surface of the color solid is
made up of two smooth patches, sewn together
at the creases (the edge color loci). The two
patches are mutual images through the central
inversion symmetry. Since these patches are congruent it suffices to study the structure of just
one of them. One patch contains only passband
colors, the other stopband colors. I study the
structure of the passband colors patch here, that
are the "spectral" colors. The geometry of the
patch of nonspectral or non-Newtonian colors is
fully analogous.

The structure of the surface of passband colors is closely connected to that of the cool edge
color locus. Let $\underline{\mathbf{e}}_\lambda$ denote the cool edge color of
transition wavelength λ. That is to say,

$$\underline{\mathbf{e}}_\mu = \int_\mu^\infty A(\lambda)\mathbf{cmm}(\lambda)\,\mathrm{d}\lambda,$$

where $A(\lambda)$ is the radiant spectral power density of the achromatic beam and **cmm** the color

matching functions. Then a passband color such
as $\bar{\mathbf{s}}_{\lambda_1,\lambda_2}$ with transition wavelengths $\lambda_1 < \lambda_2$ is
(as Goethe would have it) is simply an additive
mixture of edge colors

$$\bar{\mathbf{s}}_{\lambda_1,\lambda_2} = \underline{\mathbf{e}}_{\lambda_2} - \underline{\mathbf{e}}_{\lambda_1}.$$

Here is the geometrical meaning

*The passband colors are the
chords (linear segments between the
points at the corresponding transition wavelengths) of the edge color
curve.*

This is easily generalized to the band-stop colors
and warm edge color series. You can combine the
two and it is here that the significance of the periodically extended edge color helix becomes evident. Many of the nomographical methods proposed in the early twentieth century ultimately
derive from this simple geometry.

The special form of $\bar{\mathbf{s}}_{\lambda_1,\lambda_2}$ implies that (in
terms of the classical theory of surfaces) the sur-
face of the color solid is a *surface of translation*
in two ways, that is to say, with respect to both
parameters. "Surfaces of translation" are easily
generated and have been important in ship hulk
construction and so forth. Consequently, there
exist a huge literature on their special properties.

Consider a point $\bar{\mathbf{s}}_{\lambda_1,\lambda_2}$ of the spectral patch.
The curves

$$\bar{\mathbf{s}}_{\lambda,\lambda_2} = \bar{\mathbf{s}}_{\lambda_1,\lambda_2} + \bar{\mathbf{e}}_{\lambda_2} - \bar{\mathbf{e}}_\lambda,$$

and

$$\bar{\mathbf{s}}_{\lambda_1,\lambda} = \bar{\mathbf{s}}_{\lambda_1,\lambda_2} + \underline{\mathbf{e}}_\lambda - \underline{\mathbf{e}}_{\lambda_1},$$

are the $\lambda_{1,2}$-parameter curves on the patch that
pass through the fiducial point $\bar{\mathbf{s}}_{\lambda_1,\lambda_2}$. These parameter curves are simply the edge color curves

again, translated by $-\overline{\mathbf{e}}_{\lambda_1}$ and $-\underline{\mathbf{e}}_{\lambda_1}$ respectively. Thus the patch is covered with a parameter mesh made up of shifted edge color curves. The patch is evidently smooth, since the parameter curves are. (See figure 7.50.)

The edges of the meshes of the parameter net on the surface are apparently mutually parallel, thus the meshes have to be *planar*. If you look along the direction of one parameter curve, the contour has to be the other parameter curve, and thus the parameter curves are in mutually con-

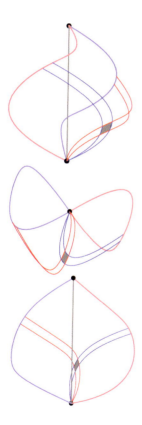

Figure 7.50 A single element of the parameter mesh with four parameter curves, two of each family. The parameter curves are simply translated copies of the edge color helix.

jugate directions. Such parameter nets are like the weave and weft of tight fitting sexy suits and are known as "Chebyshev nets".[28] Thus

> *The optimal color patch is a surface of translation (in two ways). The transition wavelength parameter curves form a Chebyshev net.*

What is of primary interest here is that you can construct the boundary surface of the color solid by suitable shifts of segments of the edge color helix. *All of the relevant geometry is already contained in this helix.* It is in that sense Goethe's intuition was absolutely right, the edge color series are of fundamental importance to color science. (Because of the central symmetry a *single* edge color curve already suffices, of course.)

Given that the pass-band colors are just differences of edge colors it is a simple matter to find the directions of the tangents of the surface in the directions of the λ_1 and λ_2 parameter curves

$$\frac{\partial \overline{\mathbf{s}}_{\lambda_1,\lambda_2}}{\partial \lambda_1} = +\frac{\partial \mathbf{e}_\lambda}{\partial \lambda}\big|_{\lambda_1} = +\mathbf{g}(\lambda_1),$$

$$\frac{\partial \overline{\mathbf{s}}_{\lambda_1,\lambda_2}}{\partial \lambda_2} = +\frac{\partial \mathbf{e}_\lambda}{\partial \lambda}\big|_{\lambda_2} = -\mathbf{g}(\lambda_2).$$

The partial derivatives $\frac{\partial \overline{\mathbf{s}}_{u,v}}{\partial u,v}$ are *directional derivatives* or *tangent vectors* in the direction of the parameter curves u, v. The partial derivative $\frac{\partial \mathbf{e}_\lambda}{\partial \lambda}$ is nothing but the spectrum cone generator $\mathbf{g}(\lambda) = A(\lambda)\mathbf{cmm}(\lambda)$ of the spectrum locus defined by the color matching functions times the achromatic radiant power spectrum. The directional derivatives are also tangents of the edge color locus.

> *The tangents of the parameter curves are the spectrum cone generators.*

The spatial attitude of the tangent plane is found as the plane spanned by the two tangents, that is, the bivector

$$\mathbf{n}(\lambda_1, \lambda_2) \propto -\mathbf{g}(\lambda_1) \wedge \mathbf{g}(\lambda_2).$$

This works out fine, except at the black and the achromatic points and the edge color loci. Notice the geometrical meaning:

> *The tangent plane at a point of the boundary of the color solid is the plane spanned by the spectrum cone generators of the transition wavelengths of the corresponding passband color. (See figure 7.51)*

It is convenient to consider a parallel projection along the direction of the gray axis. The edge color helix appears as a "figure eight" loop. It is conveniently parameterized by the transition wavelength of the edge colors. This configuration allows a number of very useful constructions.

Consider a line through the achromatic point. It is the projection of a plane through the achromatic axis. It meets the cool edge color loop in a unique point, consider the tangent to

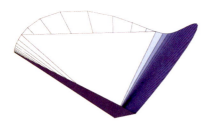

Figure 7.51 Two spectrum cone generators span a plane that is parallel to the tangent plane at the color solid at the location of the optimal color defined by the wavelengths corresponding to these of the generators (the generators have been arbitrarily scaled).

the curve at that point. The tangent is a limiting chord of vanishing length, thus represents a monochromatic beam, its wavelength has to be the parameter of the curve at the point.

Consider all optimal colors $\bar{\mathbf{s}}_{\lambda_1, \lambda_2}$ of some fixed dominant wavelength λ_0. They must all lie in the plane spanned by the achromatic axis and the corresponding spectrum cone generator $\mathbf{g}(\lambda_0)$. Thus you have the constraint

$$[\mathbf{g}(\lambda_0), \mathbf{w}, \bar{\mathbf{s}}_{\lambda_1, \lambda_2}] = 0.$$

The corresponding chords of the edge color helix thus form a family of parallel linear segments, since the monochromatic member is a limiting case, they must be parallel to the tangent defined earlier. Thus you may immediately read off the transition wavelengths of the optimal colors of some given dominant wavelength from the parameterization of the loop. (See figure 7.52.)

The optimal colors of fixed dominant wavelength form the *meridians* of the boundary of the color solid. The curves that are conjugate to the meridians are closed curves that act as "latitude circles". Thus you obtain a parameterization that is very similar (a slight generalization) to that of Runge's color sphere.

Consider chords of the loop that have parallel tangents at their end points (figure 7.52). These chords apparently correspond to *full colors*, since an infinitesimal perturbation (keeping the dominant wavelength fixed) either adds or subtracts an infinitesimal amount of white. For a given dominant wavelength such a chord may or may not (then the full color is an edge color) exist, if it exists it is unique. (If it does not exist the corresponding full color is an edge color.)

Consider the two optimal color patches at their common boundary, the edge color spiral. The tangent planes at either side are the planes spanned by the spectrum cone generator at the transition wavelength and the UV-limiting generator on one side, the IR-limiting generator on the other side.

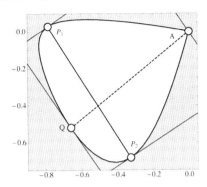

Figure 7.52 An edge color loop (edge color locus seen from the achromatic direction) with a tangent at Q and a chord $P_1 P_2$ parallel to the tangent. Wavelength (the edge color transition wavelength) is the parameter of the loop. The chords of the loop yields optimal colors with dominant wavelength equal to the transition wavelength of the tangent parallel to the chord, for parallel chords correspond to optimal colors of the same dominant wavelength. The chord $P_1 P_2$ is special because the tangents at its endpoints are parallel, thus the transition wavelengths of the optimal color $P_1 P_2$ are mutually complementary, and hence the chord corresponds to the full color of that dominant wavelength. All optimal colors corresponding to parallel chords lie on a single meridian between the black and white poles.

Thus,

> The edge color curves are creases on the boundary of the color solid. The dihedral angle is that subtended by the planes spanned by the spectrum cone generator at the transition wavelength and the UV-limiting generator on one side, the IR-limiting generator on the other side.

Next, consider the environment of the black point. The optimal colors of infinitesimal length are "monochromatic" and must have infinitesimally different transition wavelengths; thus they correspond to the tangents of the loop. This means that they are proportional to the spectrum cone generators. Hence,

> The black point is a conical singularity of the boundary of the color solid. The tangent cone is the spectrum cone. By the central symmetry the white points is also a conical singularity. Its tangent cone is the inverted spectrum cone.

These simple characterizations of the two conical singularities of the color solid serve to connect the geometry of color space (as given by the spectrum cone) to that of the color solid. To keep this in mind is often a great aid to the intuition.

7.3 Random beams

Suppose I consider a huge number of "random" beams, what positions do they occupy in color space?

It all depends upon what you mean by "random beam." Suppose the radiant power density varies randomly with wavelength or photon energy. Then the color that corresponds to such a "random" beam will be "achromatic" on the average. If the power density varies a lot over the visual region, then random beams will lead to colors that are invariably very close to gray. All such random beams look the same, that is to say, they hardly look random. In order to obtain colors that spread widely in chromaticity, you need spectra that have just a few degrees of freedom, (about three) over the visual range. If there are too many degrees of freedom you will in all probability end up with hardly any spread in color space at all.

Since spectra are nonnegative throughout, you cannot simply use a normally distributed random generator (say) to generate random spectra. It helps to remember that—in the absence of any prior knowledge other than nonnegativity—the "natural scale" is logarithmic. Any other prior would favor some particular level and there exists (by hypothesis) no rational for that. This is well known among statisticians[29] and it is frequently applied in Bayesian estimation.

Thus a good way to generate "random beams" is to use

$$s(\lambda) = A e^{a\,n(\lambda)},$$

where A and a are amplitudes and $n(\lambda)$ is a Gaussian random process with zero average and unit variance. The number of degrees of freedom is controlled via the width σ (say) of the autocorrelation function of the noise. The amplitude A sets the total radiant power, and the amplitude a the spectral roughness. In this way you generate only regular beams (positive throughout).

A useful extension of this method is to include an overall spectral slope and curvature (figure 7.53). The typical "illuminants" conventionally used in colorimetry can be well approximated by such simple spectra and it is extremely useful to have a parameterization of illuminants (see figure 7.54) instead of a mere discrete set. Such a parameterized set often used in colorimetric studies is that of the black body (Planckean) spectra. Unfortunately, the Planckean locus is only a one-parameter set and, moreover, it runs into a limiting chromaticity for very high temperatures. Since actual illuminant spectra tend to deviate from the "ideal" illuminants anyway, it makes more sense to simply use spectral slope

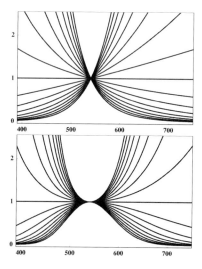

Figure 7.53 The upper curves show the influence of the "spectrum slope" parameter and the bottom curves the influence of the "spectrum curvature" parameter.

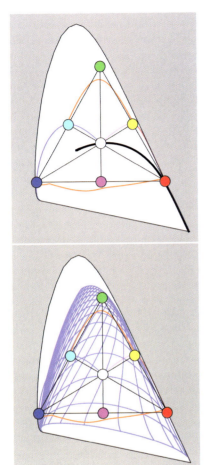

Figure 7.54 The control parameters slope and curvature nicely parameterize the chromaticities. The fat curve at top is the Planckean locus, from 1000°K to 20000°K. At bottom the nexus of parameter lines of spectral curvature and slope. The slope parameter has approximately the same influence as the color temperature.

and curvature. This is recognized by some manufacturers of digital cameras[30] who offer an additional degree of freedom to set the "white point"

instead of mere "color temperature". Then the random illuminant generator would become

$$s(\lambda;\lambda_0,a,b,c) = \mathrm{e}^{a+b\log\frac{\lambda}{\lambda_0}+\frac{1}{2}c\left(\log\frac{\lambda}{\lambda_0}\right)^2+n(\lambda;\alpha,\sigma)},$$

where

a sets the total radiant power (mostly irrelevant);

λ_0 sets the "middle" of the visual spectrum;

b sets the overall slope of the spectrum;

c sets the overall curvature of the spectrum;

α sets the spread of the normally distributed stochastic component;

σ sets the spectral width of the autocorrelation function of the stochastic component.

Notice that I have also introduced a logarithmic wavelength scale as "natural" and indeed convenient. For "reasonable" values of these parameters (that you may estimate from an analysis of the spectra of conventional illuminants for instance) the resulting samples look pretty much like spectra you might actually encounter in natural scenes. The curve around which the chromaticities of beams that vary mainly through their spectral slope cluster is very much like the "black body locus" that you find in the literature. Black body spectra (Planck's formula) indeed vary mainly through spectral slope over the visible region. Thus this curve is roughly representative for the illuminants you typically encounter in natural scenes.

The average chromaticity is set via the overall spectral slope and curvature. In figure 7.54 this parameterization is compared with the black body locus. Spectral slope largely corresponds to color temperature, but gives wider control since there is no limiting chromaticity. The curvature gives a control that is nicely "transverse" to that of the slope parameter. Notice that I say "transverse" instead of "orthogonal", since there is no such a notion in chromaticity space. This is a fact that makes that various conventional attempts to construct an "orthogonal control" to color temperature are necessarily of an ad hoc character[31].

If you plot some thousands of samples in the chromaticity diagram for various values of σ for a globally flat spectrum you observe the phenomenon mentioned above. Chromaticities of very articulated spectra cluster near some point at the center of the chromaticity diagram. Weakly articulated spectra, essentially differentiated only through spectral slope, cluster about a curve in the "yellow-blue" direction. Weakly articulated spectra, essentially only differentiated through spectral curvature, cluster about a curve in the "green-magenta" direction. Fairly smooth spectra with random slope and curvature are scattered all over the place, their position mainly determined by the overall slope and curvature. Any fast wriggles—irrespective of their amplitude—merely fuzzify locations a little bit.

If you add random perturbations to an otherwise smooth spectrum (figure 7.55) you obtain (precisely for small perturbations, approximately

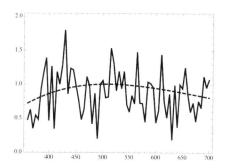

Figure 7.55 Although the random spectra look a mess, they end up in a narrow region of the chromaticity diagram. They have (almost) the same color as the daylight spectrum (the dashed curve).

for large ones) a cloud of colors in a triaxial ellipsoid centered upon the average color. Not much of a surprise there. It is not too hard to predict the covariance ellipsoid. The influence of large perturbations may seem surprisingly small, but remember that any spectrum that rapidly fluctuates about zero must be approximately black.

Such a simple state of affairs changes completely if you add additional *constraints*. For instance, you may constrain the spectra such as to be dominated by that of some fiducial spectrum. A good way to implement such a constraint[32] is to set

$$s'(\lambda) = s(\lambda)\,\frac{1}{2}(\mathrm{Erf}(n(\lambda)) + 1),$$

where $s(\lambda)$ is the fiducial spectrum, $s'(\lambda)$ the perturbed spectrum, and $n(\lambda)$ a zero mean, normally distributed random spectrum. The colors you obtain in this case must of course be contained in the corresponding Schrödinger color solid. If you increase the width of the autocorrelation and also increase the amplitude, you will obtain a greater fraction of spectra that are not unlike Schrödinger optimal colors. Thus you expect that the colors will accumulate against the boundary of the color solid. This is indeed the case. If the width of the autocorrelation function is quite large, many spectra will be similar to the Goethe edge colors, and indeed, from a simulation you see a cloud that extends along the edge color locus. (Figure 7.56.)

In figure 7.57 I show random Schopenhauer RGB colors for spectral reflectances of various statistics, illuminated with average daylight. If the spectral reflectances are very rough all colors cluster about the median gray point as expected, if the amplitude is very large and the spatial change more gradual they cluster about the Goethe edge color loci, again as expected. This once again vividly illustrates the fundamental importance of Goethe's edge colors.

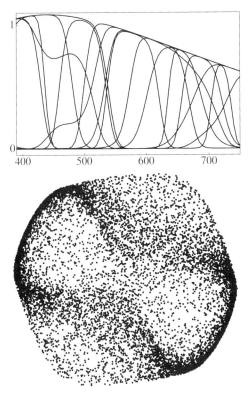

the wavelength axis. These preferences are due to the specific structure of the human system, mere trichromacy being no explanation (for instance, the trichromatic Grassmann model fails to yield such preferences). This observation goes at least some way toward an "explanation" of the *focal colors* as described by Berlin and Kay.[33]

Figure 7.56 When the width of the autocorrelation function is appreciable the spectra (top) start to tend towards the Schrödinger optimal colors. The colors thus concentrate near the boundary of the color solid. Notice the strong concentration near the Goethe edge color locus.

The hue of the edge colors changes only very little near the black and white points. Thus you expect to find a preponderance of red and blue shades and cyan and yellow tints. You also expect vivid colors in the range red to yellow (oranges) and blue to cyan, but a relative scarcity of greens and magentas. Notice that you must expect this even for spectra with statistics that are invariant with respect to translations along

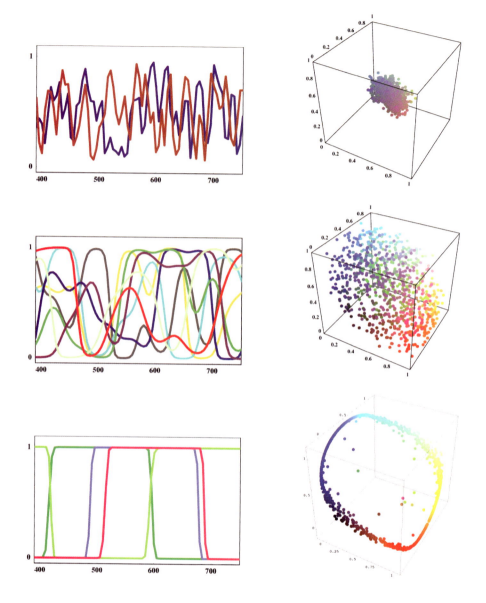

Figure 7.57 In the left column, some instances of random spectral reflectance functions drawn from three (very) different probability density functions (two parameters, the amplitude and the spatial variation were varied). In the right column, I plot a thousand colors generated from these statistics, in terms of the Schopenhauer RGB parts. In the top row the colors cluster about the median gray point, in the middle row they more or less fill the RGB crate uniformly, whereas in the bottom row they cluster about the edge color loci.

7.4 The discrete model

The discrete model has the color matching matrix

$$\begin{pmatrix} 1 & 0 \\ \frac{2}{5} & \frac{2}{5} \\ 0 & 1 \end{pmatrix}.$$

Suppose you consider spectra $\{a, b, c\}$ dominated by the beam $\{1, 1, 1\}$. The colors of these spectra are

$$\begin{pmatrix} 1 & 0 \\ \frac{2}{5} & \frac{2}{5} \\ 0 & 1 \end{pmatrix}^{T} \begin{pmatrix} a \\ b \\ c \end{pmatrix} = \begin{pmatrix} a + \frac{2}{5}b \\ \frac{2}{5}b + c \end{pmatrix}.$$

In order to find the optimal colors you simply project the hypercube of spectra into color space. You obtain the vertices

$$
\begin{aligned}
\{0,0,0\} &\rightarrow \{0,0\} \\
\{1,0,0\} &\rightarrow \{1,0\} \\
\{0,1,0\} &\rightarrow \{\frac{2}{5},\frac{2}{5}\} \\
\{1,1,0\} &\rightarrow \{\frac{7}{5},\frac{2}{5}\} \\
\{0,0,1\} &\rightarrow \{0,1\} \\
\{1,0,1\} &\rightarrow \{1,1\} \\
\{0,1,1\} &\rightarrow \{\frac{2}{5},\frac{7}{5}\} \\
\{1,1,1\} &\rightarrow \{\frac{7}{5},\frac{7}{5}\},
\end{aligned}
$$

that you may plot in color space (figure 7.58). The color solid is the convex hull of these colors.

The color solid turns out to be a centrally symmetric, irregular hexagon. Notice that it is simply the hypercube in the space of spectra as seen from the direction of the black beam. Thus the vertices that are not on the boundary (thus fail to be optimal colors) are $\{1, 0, 1\}$ and

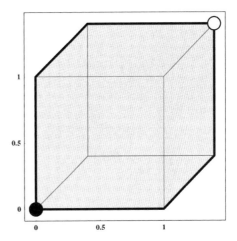

Figure 7.58 The color solid for the discrete model. The temptation is to see it as a picture of the 3D cube. Resist that temptation. This is the rendering of a flat hexagon. The hexagon is the projection of a 3D hypercube in the space of beams. Notice that only some edges of the hypercube make it into boundary elements of the hexagon. Also notice that not all primitive vertices are "revealed" in the hexagon, one being a convex combination of others.

$\{0, 1, 0\}$, indeed to be expected given that the black space is spanned by $\{2, -5, 2\}$.

In this case it is not surprising that you obtain a hexagon, because you can immediately see that the hexagon is the outline of the projection of a cube. In the actual human system it is perhaps much less obvious that the color solid is the outline of the projection of an infinitely dimensional hypercube, which of course it is. The great advantage of the low dimensional model is that you can easily overview all of the relevant dimensions. Once you are used to this, the real system will appear much less mysterious to you.

It is really worth your while to devote time to the simple model, even though it might (at first blush) seem hardly worth the trouble.

The discrete model is a good testing ground if you want to study the influence of a change of illuminant, much simpler than the Grassmann example I will discuss later. This makes a good exercise.

The discrete model is of such a great heuristic value because you can *see* everything in space with dimensions two or three. The good news is that what you see here works in exactly the same way in the realistic case of human vision, with the only exception that the dimensions are three and infinity. The Schrödinger color solid is nothing else but a low-dimension projection of an infinitely dimensions hypercube, or rather "hypercrate", because the edge lengths need not be equal. By plotting projections of high dimensional crates (a fun exercise) you indeed obtain shapes that look exactly like the typical color solid. Of course, the shape will depend somewhat on the projection. For instance, in the case of the discrete model a change of primaries yields slightly different shapes. However, these shapes are linearly related, so nothing dramatic happens.

7.5 The Grassmann Model

The color matching functions of the Grassmann model are

$$f_1(x) = \frac{1}{\sqrt{2\pi}},$$

$$f_2(x) = \frac{\sin x}{\sqrt{\pi}},$$

$$f_3(x) = \frac{\cos x,}{\sqrt{\pi}},$$

on $-\pi \le x < +\pi$.

Given the symmetry, I can without loss of generality concentrate on a single wavelength, and for convenience I assume $x = 0$. For all other wavelengths a simple shift suffices. Again, because of the symmetry, the dominant wavelength of any passbandpassband optimal color that passes wavelength in the interval $(-\Delta x/2, +\Delta x/2)$ will be equal to 0 to. This is extremely convenient, since in general you need to pick the transition wavelength very carefully in order to keep the dominant wavelength fixed.

You obtain the color of such a passband color through integration over the band:

$$\mathbf{p}(0, \Delta x) = \int_{-\Delta x/2}^{+\Delta x/2} \{f_1(x), f_2(x), f_3(x)\} \, \mathrm{d}x =$$

$$\{\frac{\Delta x}{\sqrt{2\pi}}, 0, \frac{2}{\sqrt{\pi}} \sin \frac{\Delta x}{2}\},$$

thus the dominant wavelength is indeed $x = 0$, all these colors lie in the plane $x = 0$ through the achromatic axis $\{1, 0, 0\}$.

The color of the achromatic beam is evidently $\mathbf{a} = \mathbf{p}(0, 2\pi) = \{\frac{1}{\sqrt{2\pi}}, 0, 0\}$, whereas the color of a monochromatic beam $\mathbf{m}$ of unit radiant power is

$$\mathbf{m}(0) = \lim_{\Delta x \to 0} \frac{\mathbf{p}(0, \Delta x)}{\Delta x} = \{\frac{1}{\sqrt{2\pi}}, 0, \frac{1}{\sqrt{\pi}}\}.$$

In an attempt to do a Helmholtz decomposition of the band pass color I write $\mathbf{p}(0, \Delta x) = \alpha \mathbf{a} + \beta \mathbf{m}(0)$ and obtain two linear equations for $\{\alpha, \beta\}$. Thus I write:

$$\mathbf{p}(0, \Delta x) = \frac{(\Delta x - 2 \sin \frac{\Delta x}{2})}{2\pi} \mathbf{a} + 2 \sin \frac{\Delta x}{2} \mathbf{m}(0).$$

This immediately reveals the shape of the color solid (figures 7.59 and 7.60). It is obtained by rotating half a sinusoid over 2π about is axis. Thus the color solid is a spindle-shaped body, smooth throughout except at the white and black points. For very narrow passbands you obtain

$$\mathbf{p}(0, \Delta x) = \Delta x \, \mathbf{m}(0) + \mathrm{O}(\Delta x^3),$$

that is exactly the *spectrum cone*. The symmetry proves that the tangent cone at the white point is the inverted spectrum cone.

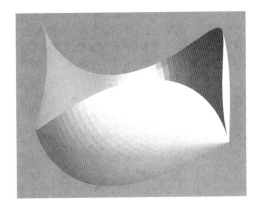

Figure 7.59 The spectral part of the Grassmann color solid. Notice the way the edge color curves join. Near the black point essentially all colors are spectral, near the white point they are all nonspectral. This is of course due to the lack of a plane of purples in the Grassmann model.

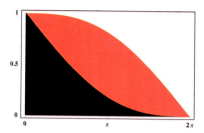

Figure 7.61 The Ostwald color, white, and black content of the optimal colors of a certain dominant wavelength as a function of the width of the passband. The maximum color content is arrived at for the semichrome of width π.

The full color locus is a circle with the median gray point as center and in a plane perpendicular to the achromatic axis (figure 7.62).

The white content increases monotonically as you increase the width of the passband. Thus the semichrome does indeed have a finite achromatic content as measured via the Helmholtz decomposition. This may be one (of the few) unfortunate judgments by Ostwald, who failed to detect any white in a full color.

Figure 7.60 The spectral (in green) and nonspectral (in purple) halves of the Grassmann color solid. The parts join without a crease. The color solid is perfectly smooth except for the poles, despite the fact that it is a projection (by the Grassmann color matching matrix) of a unit hypercube in the space of beams.

Notice that the passband color is at most a distance 2 from the achromatic axis, this occurs for the case $\Delta = \pi$ in which case the transitions are π apart, thus complementary. Thus the full color is indeed a semichrome. (Figure 7.61.)

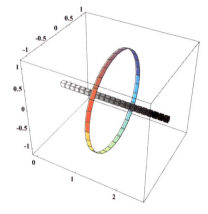

Figure 7.62 The full color locus is a circle.

You can also analyze the optimal colors in terms of their white, black, and ultimate color composition. (Figure 7.61.) The ultimate colors are found by intersecting the spectrum cone with the inverted spectrum cone (figure 7.63). Here I only need the ultimate color for $x = 0$ (all others are found by rotation about the achromatic axis). The ultimate color is

$$\mathbf{u}(0) = \{\sqrt{\frac{\pi}{2}}, 0, \sqrt{\pi}\}.$$

An analysis of the optimal colors in terms of the ultimate color, white and black requires you to write $\mathbf{p}(0, \Delta x)$ as a linear interpolation between $\mathbf{a}$, $\mathbf{u}(0)$ and $\mathbf{k}$. You obtain the color, white, and black amounts (figure 7.61)

$$c = \frac{2}{\pi} \sin \frac{\Delta x}{2},$$
$$w = \frac{1}{2\pi} \left(\Delta x - 2 \sin \frac{\Delta x}{2} \right),$$
$$k = \frac{1}{2\pi} \left(2\pi - \Delta x - 2 \sin \frac{\Delta x}{2} \right).$$

The color content of course reaches a maximum for the semichrome (the full color). For a very small passband width the color is almost completely black ("monochromatic paints" are black!) whereas very wide passbands are almost completely white (just the full achromatic

beam). The full color does contain quite a bit of black and white, its composition is 63.66 percent color content, 18.1 percent white content and 18.1 percent black content. Thus as much as 36.3 percentof the full color is median gray.

In the chromaticity diagram, the full color locus is a circle that is concentric with the spectrum locus, but that has a much smaller radius (figure 7.64). The edge color loci are spirals and they cut out conchoid-shaped areas of Newtonian and non-Newtonian colors (figures 7.65 and 7.66).

It is hard to pass up the opportunity to draw a fancy (because colored) chromaticity diagram (figure 7.67). I will not go through the list of objections here; it is very long. By now you should understand that such a picture is really nonsensical. What is good about this silly picture is that it is visually apparent that both the region near the achromatic point and the region near

Figure 7.63 The spectrum double cone.

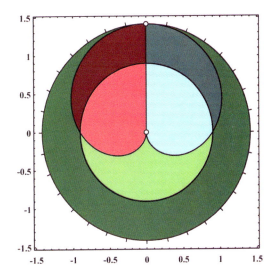

Figure 7.64 The structure defined by the full color and edge color loci in the chromaticity diagram of the Grassmann model.

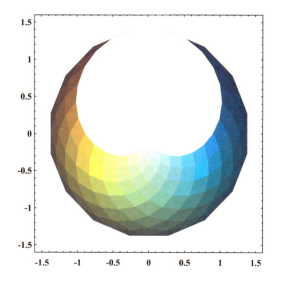

Figure 7.65 The spectral, or Newtonian colors in the chromaticity diagram of the Grassmann model.

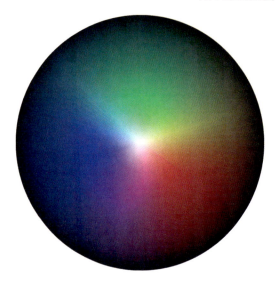

Figure 7.67 A painted chromaticity diagram, for all it is worth.

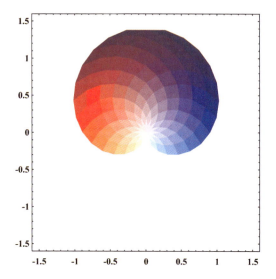

Figure 7.66 The nonspectral, or non-Newtonian colors in the chromaticity diagram of the Grassmann model.

the spectrum locus are hardly "colored" at all. One is whitish, the other blackish. The "best colors" are encountered in the neighborhood of the semichrome locus. This is essentially Ostwald's insight. I consider it very important in practical intuition, and I consider most renderings I find in the standard textbooks to be thoroughly misleading.

The illustrations I used here are all for the "nice" (uniform spectrum) illuminant. When you change the spectrum of the achromatic beam, even when you keep the achromatic color fixed, the Grassmann model becomes rather less nice and symmetric. The formal description becomes much more involved in such cases. It is of interest to pull yourself together and go through the unpleasant math though, for the Grassmann model is an almost unique laboratory to study the influence of the achromatic spectral composition with comparative (everything is relative) ease.

7.6 The Helmholtz Model

The Helmholtz color matching functions are:

$$
\begin{aligned}
h_0(x) &= \frac{1}{\sqrt{2}} \\
h_1(x) &= \frac{3}{\sqrt{2}} x \\
h_2(x) &= \frac{5}{\sqrt{2}} \left(\frac{3}{2} x^2 - \frac{1}{2} \right)
\end{aligned}
$$

on the interval $(-1, +1)$.

This model is more complicated than the Grassmann model because not every wavelength has a complementary one. You have

$$
x \, \bar{x} = -\frac{1}{3} \quad (|x|, |\bar{x}| > \frac{1}{3}).
$$

where x and $\bar{x}$ are mutually complementary wavelengths. In order to find the semichrome locus you now have to split the range and consider the lowpass, highpass, passband, and stopband branches separately. All the integrals are elementary, but the resulting expressions are surprisingly complicated.

The edge color loci are

$$
\int_{-1}^{x}
\begin{pmatrix}
h_0(u) \\
h_1(u) \\
h_2(u)
\end{pmatrix}
du =
\begin{pmatrix}
\frac{1+x}{\sqrt{2}} \\
\frac{1}{2}\sqrt{\frac{3}{2}}(x^2 - 1) \\
\frac{1}{2}\sqrt{\frac{5}{2}}x(x^2 - 1)
\end{pmatrix},
$$

and

$$
\int_{x}^{+1}
\begin{pmatrix}
h_0(u) \\
h_1(u) \\
h_2(u)
\end{pmatrix}
du =
\begin{pmatrix}
\frac{1-x}{\sqrt{2}} \\
-\frac{1}{2}\sqrt{\frac{3}{2}}(x^2 - 1) \\
-\frac{1}{2}\sqrt{\frac{5}{2}}(x^2 - 1)
\end{pmatrix},
$$

an expression that is more simply obtained by remembering that you simply need to take the complement, that is, the achromatic color minus the first branch.

The first semichrome branch is

$$
\int_{x}^{-\frac{1}{3x}}
\begin{pmatrix}
h_0(u) \\
h_1(u) \\
h_2(u)
\end{pmatrix}
du =
\begin{pmatrix}
-\frac{1+3x^2}{3\sqrt{2}x} \\
\frac{1-9x^4}{6\sqrt{6}x^2} \\
\sqrt{\frac{5}{2}}\left(\frac{-1+9x^2+27x^4-27x^6}{54x^3} \right)
\end{pmatrix},
$$

and the second is obtained by complementing the first, that is

$$
\begin{pmatrix}
\frac{1+3x(x+2)}{3\sqrt{2}x} \\
-\frac{1-9x^4}{6\sqrt{6}x^2} \\
\sqrt{\frac{5}{2}}\left(\frac{-1+9x^2+27x^4-27x^6}{54x^3} \right)
\end{pmatrix}.
$$

Thus the expressions are (perhaps somewhat disappointingly) complicated. This soon becomes irritating if you attempt deeper investigation of the optimal colors, for example, find their dominant wavelengths and so forth. Yet I doubt whether a much simpler model (that still has all the generic traits of human color vision) could be constructed.

In figure 7.68 I show the edge color loci in the chromaticity diagram. The structure that arises here is more complicated than that seen in the Grassmann model because not every wavelength has a complementary mate. This means that there must exist finite branches of edge colors, whereas in the Grassmann model each wavelength has a complementary mate, and therefore the full colors are proper semichromes without exception. The structure you see in the Helmholtz model is typical for the case of human color vision. The Helmholtz model offers a convenient because formally tractable (though the expressions become complicated the symbolic calculations are easily done with a package such as Mathematica) model of the real thing.

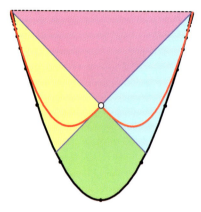

Figure 7.68 The edge color loci in the chromaticity diagram.

Using the first edge color branch, you find that the Newtonian optimal colors for transition wavelengths $\{u, v\}$ ($v > u$) are

$$n(u,v) = \begin{pmatrix} \frac{v-u}{\sqrt{2}} \\ \frac{1}{2}\sqrt{\frac{3}{2}}(v^2 - u^2) \\ \frac{1}{2}\sqrt{\frac{5}{2}}(u - v + v^3 - u^3) \end{pmatrix}.$$

Thus it is easy enough to plot the color solid or do differential geometry of it. By complementation you find the non–Newtonian optimal colors, they are

$$nn(u,v) = \begin{pmatrix} \frac{2+u-v}{\sqrt{2}} \\ -\frac{1}{2}\sqrt{\frac{3}{2}}(v^2 - u^2) \\ -\frac{1}{2}\sqrt{\frac{5}{2}}(u - v + v^3 - u^3) \end{pmatrix}.$$

In a generic view of the color solid (figure 7.69), you see clearly that it has a "crease." Both the Newtonian and the non–Newtonian parts of the surface are smooth, but they are joined at a sharp seam. This is also found in the Schrödinger color solid of human vision, and for much the same reason.

In figures 7.70 through 7.72 I show some singular views of the color solid that bring out its symmetry properties clearly. The color solid is a centrally symmetric body, the non–Newtonian half being the image under this symmetry of the Newtonian half. Note especially how the two halves lock together at the white and black points. The creases are nothing but the Goethe edge color loci.[34]

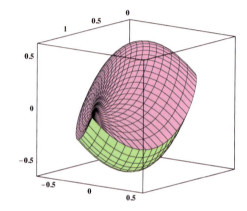

Figure 7.69 A generic view of the color solid. The newtonian part is colored greenish, the non–Newtonian part purplish.

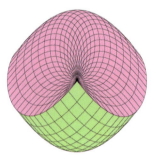

Figure 7.70 A rather singular view of the color solid that brings out the symmetry. Notice the sharp crease.

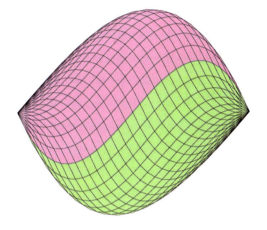

Figure 7.71 Another singular view of the color solid that brings out a different aspect of the symmetry.

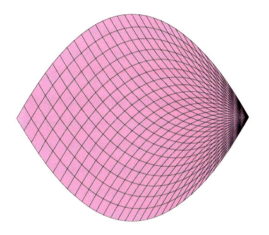

Figure 7.72 Another singular view of the color solid that brings out yet another view of the symmetry.

the cone with a sector of purples. Thus the spectrum cone and the inverted spectrum cone look rather different from a given viewpoint because the sector of purples occurs on opposite sides. Of course, this means that the spectrum double cone is a more complicated object. In figures 7.73 and 7.74, I show the color solid with tangent spectrum and inverted spectrum cones.

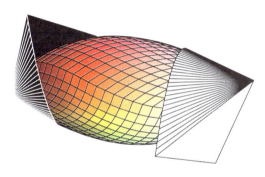

Figure 7.73 The color solid nestled snugly in the spectrum double cone.

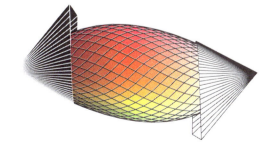

Figure 7.74 The color solid nestled in the spectrum double cone, yet another view.

In the case of the Helmholtz model, the spectrum cone is rather less symmetrical than it is in the case of the Grassmann model. This is because the Helmholtz model forces you to close

7.7 Pathological color solids

It is easy enough to construct "pathological" color vision systems. There are so many possibilities that I will not even attempt to discuss them exhaustively; I merely offer some examples. The study of such pathologies is an interesting and rewarding topic in itself. Unfortunately there hardly exists any literature.

For a start, consider the Grassmann model with two turns instead of a single one. In order to make the turns more easily visible in the chromaticity diagram it is convenient to add an amplitude modulation. Thus you obtain the color matching functions

$$\mathrm{e}^{\frac{x}{2}}\{1, \sin 4\pi x, \cos 4\pi x\},$$

on the wavelength region $0 \le x \le 1$. In figure 7.75 I show the chromaticity diagram. In this case you need a short line of purples, but these are hardly purples, because very close to only one of the spectrum limits.

The other spectrum limit, and indeed much of the spectrum locus, is in the interior of the total gamut. There are many monochromatic colors that are desaturated versions of other ("true") monochromatic colors. It is of some interest to study the type of pathological color vision enabled by this system in detail. Notice that the spectrum locus is a nice convex curve. The pathology occurs merely because it is too long.

In figure 7.76 I show the surface generated by the optimal colors. It is evidently not a convex volume, thus you will need some ingenuity to figure out the region of all colors of beams that are dominated by the achromatic beam.

It is also easy to let the spectrum locus in the chromaticity diagram self-intersect. An example of such a model is given through the color matching functions

$$\{1, x - 2x^3, 2x^2\},$$

on the wavelength interval $-1 < x < +1$. I show the chromaticity diagram in figure 7.77. Notice that the spectrum locus isn't so bad; it is convexly curved throughout. Its only fault is again that it happens to be too long.

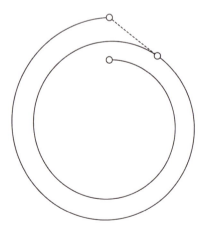

Figure 7.75 A spectrum locus of two turns.

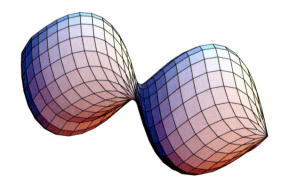

Figure 7.76 The surface of optimal colors for the (pathological) two turns model.

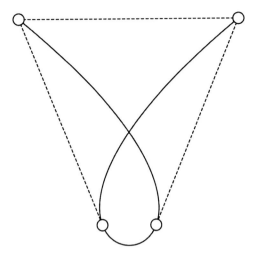

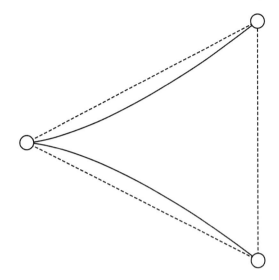

Figure 7.77 The chromaticity diagram of the "self–intersection model".

Figure 7.78 The chromaticity diagram of the nowhere convex model.

You need three straight line pieces to bound the gamut. One of these is much like the "line of purples", connecting the two spectrum limits. The other two connect one spectrum limit with some point on the spectrum locus. Notice that the spectrum locus runs largely within the gamut, only a small stretch being "exposed".

In the case of the self-intersection model, at least *part* of the spectrum locus is exposed, but it is easy enough to construct models where essentially *all* (except for a finite number of points) of the spectrum locus runs within the gamut. An example is generated through the color matching functions

$$\{1, x^2, \frac{1}{2}x^3\},$$

on the wavelength interval $-1 < x < +1$. I show the chromaticity diagram for this "nowhere convex model" in figure 7.78. Notice that the spectrum locus is nowhere convex and has a singularity (a "cusp") at $x = 0$. Only the three points $x = -1, 0, +1$ are exposed. The boundary of

the gamut is made up of three straight line segments, one a line of purples, the other two each connecting one spectrum limit to the midpoint of the spectrum ($x = 0$). All monochromatic colors are desaturated. Perhaps surprisingly, except for these oddities this is not a particularly pathological model, a color vision system based on it would serve its possessor quite decently.

I have only superficially indicated some of the possible pathologies here. A full study should include the nature of the edge colors, analogies to the full colors, and so forth. When I teach colorimetry, I often use such pathological cases for exam problems. You really have to understand the colorimetric structures in order to be able to perform detailed analyses. They make great exercises.

7.8 The perturbed Grassmann model

The Grassmann model is the perfect platform from which to explore the effects of various changes in parameters. The model is so convenient because it lets you study almost anything analytically, thus you obtain exact results and can draw valid conclusions. About the only mistakes you are likely to commit in this setting are *overgeneralizations* due to nongeneric symmetries of the model. Thus some care in the interpretation of results is (as always) necessary.

In this appendix I consider the effects of various changes of the illuminant on the structure of the color solid. These changes can be divided into two categories:

— specific changes of the spectral composition of the illuminant for an invariant illuminant color;

— more general changes that also affect the white point.

The former type is interesting because an observer is not visually aware of the change from a change in the white point. Yet the change in spectral composition will affect the color solid.

I will start with the base line, that is to say, the generic Grassmann model with the default illuminant, a uniform spectrum (figure 7.79).

In this generic case the color solid is rotationally symmetric. The full colors have 64 percent color content, 18 percent white content and 18 percent black content. The color solid exhausts 62 percent of the volume of the spectrum double cone. I illustrate a cross-section of the color solid (figure 7.80), some projections of the color solid itself (figure 7.81), and the chromaticity diagram (figure 7.82).

I will consider perturbations of the illuminant of the general type

$$I(x; n, m, a, \varepsilon, \varphi) = a\left(1 + \varepsilon \sin^m (nx + \varphi)\right),$$

with $n, m \in \mathbb{N}$ and $a, \varepsilon, \varphi \in \mathbb{R}$. Such perturbations are sufficiently general to study almost anything of immediate interest, yet they lead to analytically traceable problems.

I will use large perturbations by way of demonstration, setting the value of the perturbation parameter $\varepsilon = 1$. Such perturbations are dramatic, much more so than what you typically may expect in realistic circumstances.

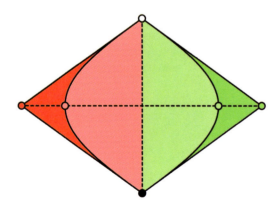

Figure 7.80 A planar section of the color solid and the spectrum double cone in the generic case. All planar sections are congruent.

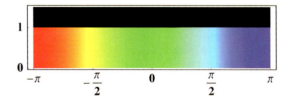

Figure 7.79 The default illuminant spectrum for the generic Grassmann model is uniform.

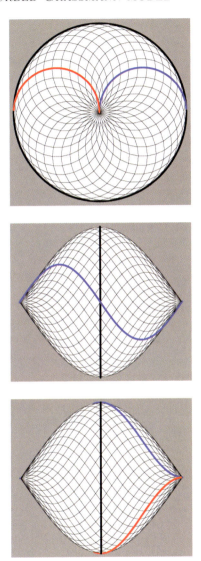

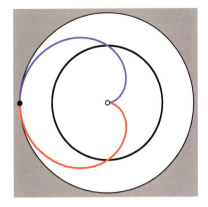

Figure 7.82 The chromaticity diagram in the generic case. The white disc contains the interior of the spectrum cone. The black circle is the full color locus, it is concentric with the spectrum locus and the achromatic point. The edge color loci are drawn in red (warm family) and blue (cool family). Again, the two regions are congruent.

The first system I will consider is the illuminant

$$I(x) = 1 + \cos 2x,$$

(figure 7.83). It is strongly modulated and even has "blind spots" in the spectrum. (Since these are mere *points*, they are harmless.)

Figure 7.81 Three views of the generic color solid. The views are from the directions of the first (achromatic), second and third dimension. The edge color loci are drawn as red (warm family) and blue (cool family) curves, the full color locus (in this case a planar circle) is drawn in black. The full color locus is a (planar) circle.

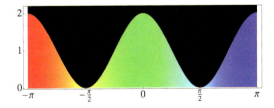

Figure 7.83 The illuminant for the first perturbed case.

This case is already *much* more complicated than the generic model. Although it is not difficult to do all calculations analytically, the resulting expressions are not very enlightening. The full colors are

$$\mathbf{f}(\xi) = (\sqrt{\frac{\pi}{2}},$$

$$\frac{4\sin^3 \xi}{3\sqrt{\pi}},$$

$$\frac{9\cos \xi - \cos 3\xi}{3\sqrt{\pi}}),$$

just to give an idea of the rather moderate complexity involved here.

The white point being fixed, the transition locations for the semichromes are exactly the same as in the generic case. That is to say, the reflectance spectra of the semichromes are not at all affected. However, for the same reflectance spectrum you get a *different dominant wavelength*. The dominant wavelengths for the case of the perturbed illuminant differ from those of the generic case by an amount

$$-\arcsin\left(\frac{4}{3\pi}\sin 2x\right),$$

that may run up (depending on the fiducial wavelength x) to 25.1135°.

The color solid and the chromaticity diagram look very squashed as compared with the generic case (figures 7.84 through 7.85).

The color solid is indeed squashed, and as a result the structure of the full colors is quite different from that of the generic model. The color, white and black contents of the full colors (relative to the ultimate colors of course) varies with hue (figure 7.86). For some hues the color content is higher, for others lower than for the unperturbed case.

For this perturbation the color solid exhausts only 52 percent of the volume of the spectrum

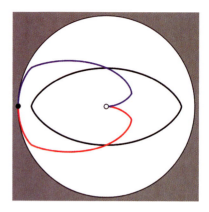

Figure 7.84 The chromaticity diagram for the first perturbed case. Notice the strong deformations of the full color and edge color loci.

double cone, thus the "color rendering properties" of this illuminant are significantly less than those of the uniform spectrum.

Notice that the semichromes are "full colors," for when you see the color solid from the direction of the achromatic axis, the semichrome locus coincides with the contour.

The second type of perturbation I consider is the illuminant

$$I(x) = 1 + \sin 3x,$$

(figure 7.87).

It happens to be the case—*mirabile dictu*—that the semichromes have the same dominant wavelength (for the same spectral reflectance) as those of the generic case. (When you think it over for a while, this may become obvious to you without any calculation.) The full colors are of the form

$$\left(\frac{3\pi + 2\cos 3(\xi - \pi/2)}{3\sqrt{2\pi}}, \frac{2\sin \xi}{\sqrt{\pi}}, \frac{2\cos \xi}{\sqrt{\pi}}\right).$$

Helmholtz decomposition reveals that the color content of the full colors is 64 percent, just as

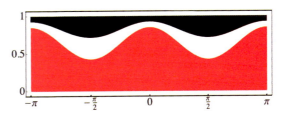

Figure 7.86 The color, white and black contents of the full colors for the first perturbed case. As compared to the generic case in some regions colors are better, in others worse.

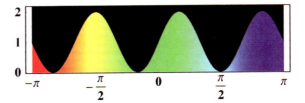

Figure 7.87 The spectrum of the illuminant for the second perturbed case.

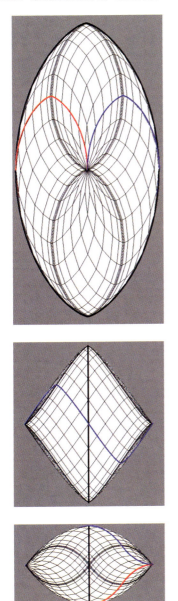

Figure 7.85 Three views of the color solid for the first perturbed case. Notice that the full color locus is a planar oval, that appears as the contour as seen from the achromatic direction.

it is in the case of the generic model. However, the white and black contents are not fixed, but depend upon the dominant wavelength of the full color. In figure 7.88 I show the composition of the full colors as a function of their dominant wavelength.

As you might expect, the color solid is very lopsided. This is evident from planar sections, renderings of the solid and the chromaticity diagram (figures 7.89–7.91).

The full color locus in the chromaticity diagram has turned from a circle (or at least a convex oval) into a nonconvex, closed curve with three "modes".

It is in itself not at all remarkable that the full color locus in the chromaticity diagram is not convex. The same is true for the case of the human observer in average daylight.

Since the interior of this locus represents the chromaticities of the interior of the Ostwald double cone erected on the full color locus, you draw the conclusion that Ostwald's double cone is not

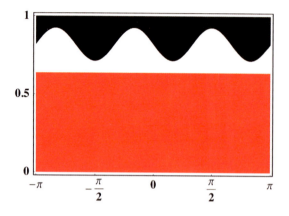

Figure 7.88 The color, white, and black contents of the full colors in the second perturbed case. Notice that though the color content is constant, the white and black contents vary with wavelength.

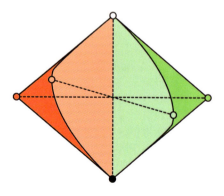

Figure 7.89 A planar section through the color solid of the second perturbed case. Not all sections are congruent. This is an especially lopsided section.

necessarily convex either. This is seen in figures 7.92 and 7.93, for instance.

This is not apparent from the usual representation, where the Ostwald double cone is erected upon the color circle. Whereas this latter representation indeed looks pretty, it has lost the affine structure; thus an equal mixture no longer maps on the midpoint of the connecting straight segment. The representation in color space is much to be preferred for this very reason, but then the nonconvexity of the double cone is perceived to be a real nuisance. An equal mixture of two colors inside of the Ostwald double cone is not guaranteed to again lie in the interior of that double cone. In practice the issue is not very pressing, since the nonconvexity is only slight, but in principle it is really disturbing. The obvious way out of this problem is to erect the double cone upon the ultimate colors rather than the full colors, that is to say, to identify the Ostwald double cone (in the novel sense) with the spectrum double cone. This yields a principled and coherent structure that has all the advantages of Ostwald's construction, and indeed many more. For instance, it allows you to deal with *all* colors without using the nonintuitive device of possible

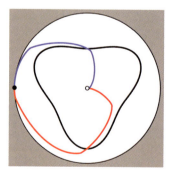

Figure 7.90 The chromaticity diagram for the second perturbed case. Notice that the full color locus is not convex.

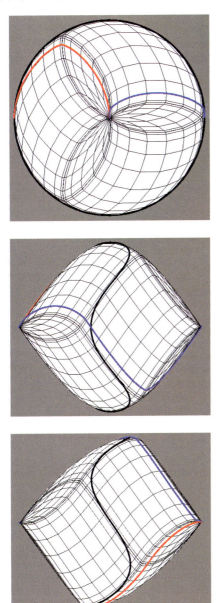

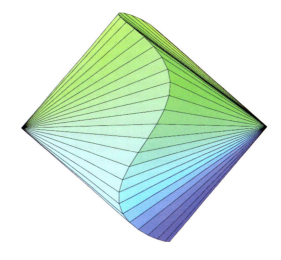

Figure 7.92 The Ostwald double cone for the second perturbed case. Notice that the boundary fails to be convex (look at top right).

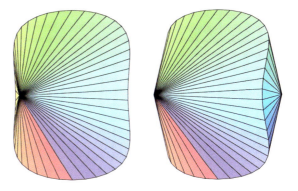

Figure 7.93 Two more views of the Ostwald double cone for the second perturbed case. Again, notice the failure of convexity.

Figure 7.91 Three views of the color solid for the second perturbed case. The full color locus is not a planar curve here, but is very significantly doubly curved.

negative white or black contents, it allows you to identify the white and black content of the full colors, and it allows you to deal with changes in the spectral composition of the illuminant at fixed achromatic color.

In this case, the color solid exhausts only 58 percent of the volume of the spectrum double cone; thus this particular illuminant also has color rendering properties that are markedly inferior to those that apply to the uniform spectrum.

Notice (again) that the semichromes are full colors, for when you see the color solid from the direction of the achromatic axis, the semichrome locus coincides perfectly with the contour. Small wonder, of course, but good to notice.

So you may draw a number of important conclusions (which are indeed general, except for the final one):

In case the achromatic point is fixed, but the spectral composition of the illuminant varied you have that:

— the shape of the color solid depends upon the spectral composition of the illuminant, even if the color of the illuminant does not change;

— the spectral reflectances of the semichromes are the same as in the generic case;

— the semichrome dominant wavelengths in general differ from those for the generic case for the *same* spectral reflectance;

— the color, white, and black content of the semichromes vary with the dominant wavelength and differ from those for the generic case;

— the color solid exhausts a fraction of the volume of the spectrum double cone that depends upon the spectral composition of the illuminant;

— the semichromes are full colors;

— the color solid of the generic case exhausts the maximum volume of the spectrum double cone.

Notice that these properties are by no means generally acknowledged in the literature. For instance, I recently got involved in heated discussion over the issues of whether the semichromes (as a family) are invariant (they are) and whether the semichromes are indeed full colors (they are) in the case the achromatic point is fixed. It is easy enough to prove these properties generally (Ostwald did it by intuitive reasoning; a formal proof could simply follow his lines).

Next, I will consider a case were the achromatic point is allowed to change. This will happen for perturbations of the type

$$I(x) = 1 + \varepsilon \cos x.$$

I will again consider large perturbations ($\varepsilon = 1$). (See figure 7.94).

In this case the achromatic point changes, it becomes $\{\sqrt{2\pi}, 0, \varepsilon\sqrt{\pi}\}$. For $\varepsilon = 0$ you have the generic case. What happens is that the white point shifts in the direction of the third dimension. This of course implies that the complementary pairs change, hence that the semichromes change. Consequently, spectral reflectances of the semichromes change too.

The transition wavelengths of the semichromes become

$$\left(x - \arccos \frac{\cos x}{2}, x + \arccos \frac{\cos x}{2}\right),$$

instead of $x \pm \pi/2$. The width of the passband may change by as much as 60°. In figure 7.95 I

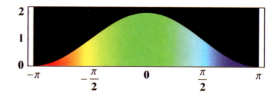

Figure 7.94 The spectrum of the first illuminant with variant achromatic point.

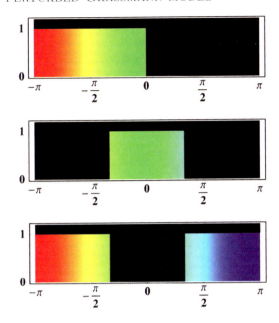

Figure 7.95 Three semichromes for the case of a variant achromatic point. These semichrome spectral reflectances are (in general) different from those of the generic case.

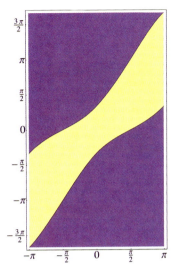

Figure 7.96 The semichrome passbands for the case of a variant achromatic point as a function of the semichrome dominant wavelength.

plot a few semichromes, in figure 7.96 I show the passbands as a function of dominant wavelength. They are *very* different from the semichromes of the generic case.

The dominant wavelengths of the semichromes (though perhaps of different width) centered on the same wavelength differ from these of the generic model.

The structure of the solid is dominated by a pronounced flatness about a certain plane through its center. It is only to be expected that the color solid structure is quite different from that of the generic case. I illustrate it in figures 7.97 and 7.98.

Again, if you look along the achromatic axis it is apparent that the semichromes are full colors (figure 7.99). This is a general observation and

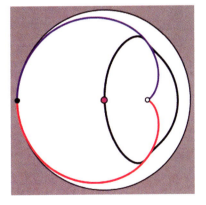

Figure 7.97 The chromaticity diagram for the case of the variant achromatic point. The generic achromatic point has been indicated as the purple dot. Notice that it is very different from the actual achromatic point.

it was indeed how Ostwald came on the notion of the semichromes. He was interested in
the full colors and tried to characterize these
in some formal way. Intuitive thinking then revealed that broadening the passband should have
the effect of adding white, while reducing the
width of the passband should have the effect of
adding black. Slightly changing the width of
the passband should not affect the color content.
This is evidently the definition of a full color;
its color content should be an extremum (maximum in this case). But the criterion implies that
the transition wavelengths should be mutually
complementary, for only then can their mixture
be achromatic. Hence the full colors should be
semichromes.

As was to be expected, the dominant wavelengths of the semichromes (irrespective of their
widths) centered at the same wavelength differ
from those of the generic case (figure 7.100).
Since the hue is periodic the hue shift has to
feature at least two fixed points, like is the case
in the figure. Notice that the hue shifts are quite
large, over 30°.

From this demonstration you gather some
observations that may indeed be shown to have
general applicability. They are useful, because
much harder to glean in the general (human)
case.

*In case the illuminant is varied in a general
way, not leaving the achromatic point invariant
you have that*:

— several of the observations that applied to
   the specific case of invariant achromatic
   point still apply;

— the semichromes are not conserved as a
   family. Thus a semichrome for the generic
   case is not necessarily a semichrome for
   the perturbed case;

— semichromes have various passbands (not
   just 180° as in the generic case);

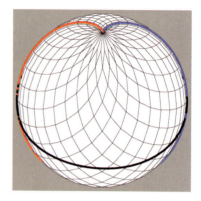

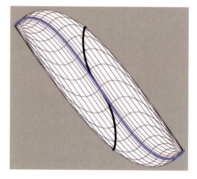

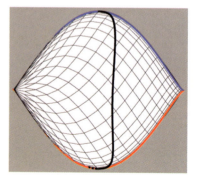

Figure 7.98 Three views of the color solid for
the example of a variant achromatic point. Notice that the full color locus is not on the contour
of the first view, which is along the first dimension, which differs from the actual achromatic
direction.

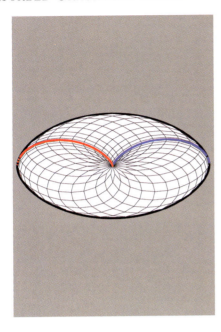

Figure 7.99 A view of the color solid from the actual achromatic direction for the case of a variant achromatic point. The full color locus coincides with the contour, thus demonstrating that the semichromes are full colors.

— the dominant wavelength of a semichrome does not (in general) coincide with the center of its passband;

— the semichromes are full colors.

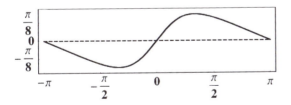

Figure 7.100 The hue shift of the semichromes as a function of the center of the passband.

The latter property is quite general. In fact, the distinction between "semichromes" and "full colors" is an artificial one, the two are to be considered the same. The distinction merely derived historically from early independent definitions.

Finally, I give an example of a "bad" illuminant for the Grassmann model, I take

$$I(x) = \frac{128}{35}\left(1 + \sin^4(2x - \frac{\pi}{2})\right).$$

Why the apparently odd factor 128/35? Well, in this way the illuminant is set up to yield the standard achromatic color. The illuminant is almost everywhere non-negative so it is a "good" illuminant, and it is nicely smooth so all calculations can be done analytically. So what is "bad" about it? It is because it concentrates most radiant power in two mutually complementary regions and has very little spectral radiant power over most of the visual region. Its "coverage" is bad. The illuminant is shown in figure 7.101. If you want you may make the illuminant even worse by increasing the power (here four) but at the cost of (much) more complicated analytical expressions for the interesting entities. In this case the expression for the full colors already contains polynomials in $\sin x$ and $\cos x$ of degree nine with awkward coefficients.

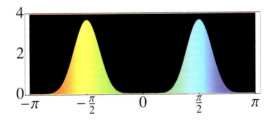

Figure 7.101 The spectrum of a bad illuminant with standard achromatic point. Notice that the spectral radiant power is low at most of the range, but very high at a pair of mutually complementary locations.

The chromaticity diagram (figure 7.102) shows a badly squashed full color locus and edge color loci that run along the boundary for much of their course.

This is indeed a bad illuminant as becomes evident when you calculate the color solid volume. It is only thirty-three percent of that of the ultimate color double cone, thus only about half of that of the volume for the generic case. The color solid is indeed squashed flattish, it looks a bit like a lozenge-shaped pancake (figure 7.103).

The investigations that I have shown here for the special case of the Grassmann model can be extended to the other models. For the case of the local model and the Helmholtz model, most calculations can be performed analytically if you make sure that the illuminants are picked such as to render the various integrations that are in-

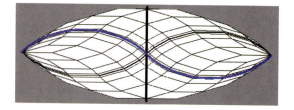

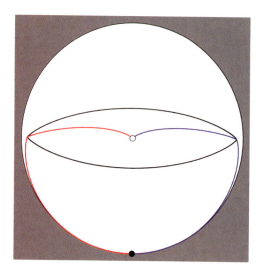

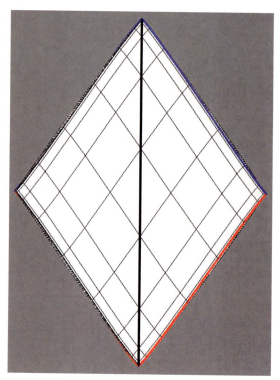

Figure 7.102 The chromaticity diagram for the case of the bad illuminant. Notice that the full color locus is severely quashed and that the edge color loci run along the boundary for a large part of their courses.

Figure 7.103 Three views of the color solid for the case of the bad illuminant. The full color locus is a nice planar curve. The edge color loci run largely (roughly) along the rim of the lozenge-shaped pancake.

volved elementary. This is a most worthwhile exercise. For the case of the discrete model it is also fairly easy to do exact calculations, in this case of a mere algebraic nature. In order to render the results interesting, you may have to move to higher (though still fairly low) dimensions though. This type of calculation is bound to be a little more painful than what you may expect from the "straight" models, so it perhaps takes some true grit to carry them through, although packages like Mathematica certainly help to ease the pain and invite experimentation. It is very much worth it, though, since the effects shown here for the Grassmann model are not documented in the literature.

If you are mainly interested in numerical results, most probably in the form of computer graphics, it makes much sense to set up the calculations in numerical schemes right from the start. This allows great flexibility in all choices and leads to fast implementation. It is the method of choice for initial free experimentation and most conducive to developing a keen intuition. Once you know what you are after, you may attempt analytical calculations for the simplest possible models and—eventually—general proofs.

Exercises

1. Schrödinger's proof [CONCEPTUAL] Consider cases of spectra that violate Schrödinger's criteria and explicitly construct perturbations that increase their intensity, leaving their chromaticity invariant. Identify spectrum regions where Schrödinger's proof doesn't work very well.

2. The color solid [COMPUTER GRAPHICS] Render the color solid, departing from the edge color series. Plot the edge color loci and semichrome locus on the surface of the color solid. Repeat for a number of illuminants.

3. Mensuration of the color solid [ELABORATE] Parameterize the optimal colors by dominant wavelength (this involves a lot of interpolation, easy in Mathematica). It is most convenient to define a "hue angle" and find dominant wavelength as a function of hue angle. Build interpolation functions either way. (In defining the hue angle it is convenient to put the branch cut in the purple region.) Now construct the semichromes. You should be able to specify the semichrome for a given dominant wavelength. Finally mensurate the semichrome locus. Construct the cardinal colors and plot a color circle. Repeat for different illuminants.

4. The spectrum double cone [EASY BUT WORTHWHILE] Construct the spectrum double cone. Calculate the structure of the full colors in terms of the ultimate colors, the achromatic color and black.

5. Volume ratios [ELABORATE] Find the volumes of the interior of the spectrum double cone, the color solid, Ostwald's double cone on the full colors and Schopenhauer's RGB crate. Repeat for various illuminants. Which illuminant has the best "color rendering" properties?

6. Projections of hypercubes [FOR MATHEMATICIANS]
Plot low-dimensional projections of high dimensional hypercubes. You may construct projection operators randomly or constrain their structure.

7. Nomographical methods [FOR MATHEMATICIANS] Devise nomographical methods to find the semichrome for a given dominant wavelength and so forth, using the "figure eight" curve. Think of other computations that might be done graphically in this way.

8. Random beams [CONCEPTUAL] Random beams are used in the standard handbook of Wyszecki and Stiles.[35] Look it up and comment on the choice of their random beam generator. How does the choice influence their conclusions?

9. Random beams and spectral reflectances [ELABORATE] Construct random generators for beams and spectral reflectances. They should yield physically realistic samples, that is to say, nonnegative radiant power spectra and reflectances in the range zero to one. Think of good parameterizations. For certain ranges of parameter values the samples should look like instances you might encounter in real life. Study color and chromaticity of large sets of samples.

10. The models [ELABORATE] Construct the color solids for the Grassmann, Helmholtz, and local models and study the differential geometry analytically. Explicitly construct the Ostwald representation and the Schopenhauer RGB crates for the models.

11. Pathologies [CONCEPTUAL] Construct examples of pathological systems. Can you think of a natural way to order the various pathologies? Can you think of ways to let an agent "live" with particular pathologies? How would a pathology manifest itself in behavior?

12. Explicit expression for the transition wavelengths [FOR MATHEMATICIANS] Show that an explicit expression for the spectral reflectance (thus the band pass limits) of the Newtonean optimal colors is

$$R(\lambda) = \frac{1}{2} \left[1 + \mathrm{sgn}(\mathbf{n} \cdot \psi(\lambda))\right],$$

where the "signum function" is defined as $\mathrm{sgn}(x) = +1$ for $x > 0$, $\mathrm{sgn}(x) = -1$ for $x < 0$ and undefined for $x = 0$, $\mathbf{n}$ is the normal of the boundary of the color solid at the fiducial optimal color, and $\psi(\lambda)$ is the illuminant spectrum times the color matching functions. (This expression is closely connected to the classical nomographic method. You may wish to show the connection explicitly.)

13. Implement the principle of internal symmetry [ELABORATE]
Implement Ostwald's principle of internal symmetry so as to achieve a discrete arc length parameterization of the semichrome locus. Aim at the positions of twelve or twenty-four cardinal colors. Remember that the locus is a *closed* curve! You will have to set up two iterative algorithms in series. It will take some computation, but it can be made to work. (You should really do it on a computer. Bouma did it half a century ago by hand calculation, and Ostwald did it a century ago by eye measure). Compare the result with that of the (much easier) "equal pie slice" algorithm.

14. Schopenhauer RGB crates [TAKES COMPUTATION] Find both the inscribed crate of maximum volume and the circumscribed crate of minimum volume to the color solid. What is their volume ratio? What is the fraction of real colors outside the small and the fraction of virtual colors inside the larger crate? How useful are these constructions? Repeat this for the fourier model (easy exercise).

15. Conditions for a "good" color system [CHALLENGE] Prove the statement that in order to obtain a "good" color vision system it is necessary and sufficient that the edge color locus be a generalized helix of at most a half turn.

16. A formally "cleaner" proof of the optimality of Schrödinger's colors [IF YOU ARE FAMILIAR WITH VECTOR MEASURES AND INTEGRATION] Improve on Schrödinger's proof by invoking Liapunov's bang-bang principle.

17. The "special hues" of the color circle [REQUIRES CONSIDERABLE (NUMERICAL) COMPUTATION] Find the dominant wavelengths of the full colors that are in some way "special" because of some extremal property. How do the results compare to the wavelengths conventionally quoted for "unique YELLOW", and so forth? How do they depend upon the spectrum of the achromatic beam?

18. Parameterization by slit width and location [ELABORATE] Parameterize all chromaticities in terms of—either real or complementary—

slits. Draw parameter curves of equal slit width and slit position in the chromaticity diagram.

19. Colorimetric calculations [FOR ZOMBIES] Buy, collect, or write algorithms that let you compute the answer to virtually any question you might be confronted with and that indeed has an answer in terms of colorimetry. As a customer hands you a problem, you simply push a few buttons and the answer pops out. This puts you in business as a colorimetric practitioner. Can you be said to "understand colorimetry"?

20. Visualization [MENTAL GYMNASTICS] Visualize all colorimetric entities you have met so far (spectrum cone, inverted spectrum cone, gray axis, plane of purples, edge colors, full colors, ultimate colors, cardinal colors, chromaticities, ...) in the space of beams, color space and the chromaticity diagram. Your visualizations should be invariant with respect to affinities of projective transformations as the case may be. When you are able to do this with ease, easily switching between representations, you may be said to be on your way understanding colorimetry.

21. The discrete model [SIMPLE BUT (VERY) INSTRUCTIVE] Study the effect of changes of the spectrum of the achromatic beam (keeping its color constant) as well as changes of the achromatic color on the shape of the color solid. Find the spectrum of the "most revealing" achromatic beam, that is the one that maximizes the volume of the color solid. (Here you need to constrain the set among which you look for the best spectrum, for example, you may define a "luminance" and constrain it.)

22. Color rendering properties [CHALLENGING] Some possible achromatic beams for a given achromatic color are clearly more natural than others. This is the problem of the color rendering properties of light sources. Look up the conventional (CIE of course!) way to grade the "quality" of achromatic beams. Can you think of characterizations that are better rooted in the fundamental colorimetric structure (I'm thinking of geometry, of course)? How might one implement these?

23. Mensuration of the color circle [ELABORATE AND CONCEPTUAL] Collect examples of color circles that were mensurated by eye measure. With some effort you may come up with half a dozen. Compare these among each other and with Ostwald's principle of internal symmetry for natural daylight. This calls for some statistics of course.

Do you agree with my verdict that Ostwald's principle yields results that are in the ballpark of the eye measure attempts? Discuss various meanings one may attach to this observation.

Chapter Notes

1. Good accounts are available in many texts. For instance see M. Richter, *Grundriss der Farbenlehre der Gegenwart*, (Dresden and Leipzig: Theodor Steinkopff, 1940).

2. Minor exceptions occur near the spectrum limits and in the yellow-red where the spectrum cone mantle is nearly planar. However, such minor effects are best ignored on an initial investigation.

3. Of special interest are papers by Rösch, who based colorimetry (both theoretically and empirically) completely on Schrödingers "optimal colors": S. Rösch, "Die Kennzeichnung der Farben," *Pysik.Z.* XXIX, pp. 83–91 (1928); —, "Notiz über Optimalfarben," *Naturwiss.* 19, pp. 615–617 (1931); —, "Darstellung der Farbenlehre für die Zwecke des Mineralogen," in *Fortschritte de Mineralogie, Kristallographie und Petrographie*, edited by W. Eitel (Berlin: Deutschen Mineralogischen Gesellschaft, 1929), Bnd. 13, pp. 753–900.

4. A "vector measure" is defined, analogously to the more familiar measures, as a map from a σ-algebra of subsets of a space to a linear vector space (instead of the nonnegative real numbers). It is called a "Liapunov measure" if the measure of any subset is convex and compact. In analogy with the conventional measure theory one defines (Liapunov) integrals. The integration map is again a Liapunov measure; thus its range are convex and compact sets. From this the "bang-bang principle" follows. Any point of the range of the integration map can be reached by a characteristic function. The extreme elements are indeed unique characteristic functions (Liapunov's bang-bang theorem). See I. Kluvánek, and G. Knowles, *Vector measures and control systems*, (Amsterdam: North-Holland, 1975).

5. The idea of "vector measures" is rather natural in colorimetry. Think of the human photopigments, which define nonnegative weight spectral functions. Each such weight defines a measure on the space of spectra, the three of them defining an ordered set of three distinct measures which may be regarded as a three-dimensional vector measure. It would be natural to do this here if there were any need for generalization. However, since there is no such a need it would merely add unnecessarily to the formalism.

6. E. Schrödinger, "Theorie der Pigmente von grösster Leuchtkraft," *Ann.Phys. IV*, 62, pp. 603–622 (1920).

7. Important sources are: R. Luther, "Aus dem Gebiet der Farbreizmetrik," *Z.f.techn.Physik* 8, pp. 406–419 (1927); N. Nyberg, "Zum Aufbau des Farbenkörpers im Raume aller Lichtempfindungen," *Z.Physik* 52, pp. 406–419 (1928); C. Dolland, "Über den Luther-Nybergschen Farbkörper," *Die Farbe* 5, pp. 113–136 (1956).

8. Surfaces that can be represented as

$$\{x(u,v), y(u,v), z(u,v)\} = \begin{pmatrix} f_1(u) + g_1(v) \\ f_2(u) + g_2(v) \\ f_3(u) + g_3(v) \end{pmatrix},$$

are known as surfaces of translation. They have many special property that can be found in the classical literature of differential geometry of surfaces in three-dimensional space.

9. A "Chebycheff net" is a net in which the meshes are equilateral paralellograms. P . L. Chebyshev, *Collected works*, (Moscow-Leningrad: Acad.Sci. U.S.S.R., 1951), pp. 165–170.

10. See the section "*Warum bilden die Farbtöne einen Kreis*" in W. Ostwald, "Die Attribute der Farben," *Sitz.Ber. d. Preuss. Akad.d.Wiss. Phys.-Math.Klasse* 30, pp. 417–436 (1939).

11. The "n^{th} order jet" of a function at a point is the set of 1st to n^{th} derivatives at the point. Together they specify the local behavior of the function.

12. For instance, see K. W. F. Kohlrausch, "Bemerkungen zur Ostwald'schen sogenannten Farbentheorie," *Phys.Z.* 22, pp. 402 (1921).

13. Ostwald never performed such computations. Attempts by Richter (M. Richter, "Zur Einteilung des Ostwaldschen Farbtonkreises (Aufstellung eines Normalkreises)," *Das Licht* 13, pp. 12–15 (1943), and M. Richter, and K. Witt, "The story of the DIN color system," *Color Res. Appl.* 11, pp. 138–145 (1986)) related to the construction of the DIN-system were not very successful and turned into a kind of travesty when computation was somehow merged with eye measure. The first one to use Ostwald's principle successfully was Bouma (P. J. Bouma, "Zur Einteilung des Ostwaldschen Farbtonkreises," *Experientia* II/3, pp. 99–103 (1946); —, *Physical Aspects of Colour*, (Eindhoven: Philips, 1947)). Unfortunately, lack of interest (or rather, misunderstanding) put an end to such researches. Bouma used the principle of internal symmetry in the (well, almost) infinitesimal domain in order to arrive at an arc length parameterization. Richter, who participated in the development of the DIN-system, applied the principle of internal symmetry to large segments of the color circle (quadrants) and—not surprisingly—arrived at different results. In M. Richter, "Die Beziehung zwischen den Farbmaßzahlen nach DIN 6164 und den Ostwald-Maßzahlen," *Die Farbe* 6, pp. 49–62 (1957) Richter remarks that the "the principle of internal symmetry does not yield a unique solution," apparently ignorant of the fact that the discrepancy with Bouma's work was only to be expected.

14. Let **k** and **a** denote the black and achromatic points, let $\mathbf{f_1}$, $\mathbf{f_2}$ be two full colors. The principle of internal symmetry indicates $(\mathbf{f_1} + \mathbf{f_2})/2$ as a color of the

same dominant wavelength as the full color $\mathbf{f_{12}}$ on the full color locus that bisects the arc $\mathbf{f_1 f_2}$. Consider the color $\mathbf{x} = \mathbf{f_1} + \mu(\mathbf{f_2} - \mathbf{f_1})$ on the linear segment $\mathbf{f_1 f_2}$. The volume of the tetrahedron $\{\mathbf{k}, \mathbf{a}, \mathbf{f_1}, \mathbf{x}\}$ is $[\mathbf{a}, \mathbf{f_1}, \mathbf{x}]/3! = \mu[\mathbf{a}, \mathbf{f_1}, \mathbf{f_2}]/3!$, whereas the volume of the tetrahedron $\{\mathbf{k}, \mathbf{a}, \mathbf{f_1}, \mathbf{f_2}\}$ is $[\mathbf{a}, \mathbf{f_1}, \mathbf{f_2}]/3!$. Thus the color $(\mathbf{f_1} + \mathbf{f_2})/2$ coincides with the color $\mathbf{x}$ for $\mu = 1/2$. Thus the "honest pie slice" principle yields the same result as Ostwald's principle of internal symmetry.

15. I use "well tempered" of course with the example of Johann Sebastian Bach's *Das wohltemperierte Clavier* in mind. In 1722 Bach published a book of preludes and fugues in all twenty-four major and minor keys, starting with a pair in C-major and ending with a B-minor fugue.

The arabesque that starts Bach's book is thought by some to indicate a tuning scheme.

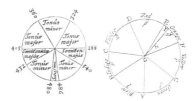

Left a circular representation of the tempered diatonic octave by Descartes (Compendium Musicæ, 1618), at right Newton's color circle (Opticks, 1702).

Bach probably wrote for a 12-note well-tempered tuning system. The precise tuning system used by Bach is still hotly debated. (See R. Kirkpatrick, *Interpreting Bach's Well-tempered Clavier: A Performer's Discourse on Method*, (New Haven: Yale University Press, 1987).) Analogies between the color circle and the musical octave have been proposed throughout the history of color science. Remember that Newton's spectrum is fitted out with musical notation! Newton is likely to have modeled spectrum and color circle on a circular drawing of the musical octave by Descartes. Although the analogy is spurious, you find it in serious books on physiology and psychophysics (for instance, Helmholtz's Handbook). It is perhaps of some interest that interpretations of Bach's "well tempered" division of the octave are reminiscent of Ostwald's principle of internal symmetry.

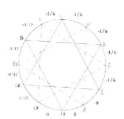

A speculative scheme of Bach's tuning system by Bradley Lehman (http://www.larips.com) that shows a remarkable likeness to Ostwald's principle of internal symmetry.

16. Such reasonings are popular nowadays. For instance see Z. Yang, and D. Purves, "Image/Source statistics of surfaces in natural scenes," *Comput.Neural Syst.* 14, pp. 371–390 (2003); F. Long, and D. Purves, "Natural scene statistics as the universal basis of color context effects," *P.N.A.S.* 100, pp. 15190–15193 (2003); A. Lewis, and L. Zhaoping, "Are cone sensitivities determined by natural color statistics?," *J. of Vision* 6, pp. 285–302 (2006).

17. Ostwald was a strong believer in the decimal system, which is why he started out with a 100-step scale, which is *really* awkward because it is not a multiple of three and is far too large. Later he settled on 96 (multiple of 6, yet close to 100), then 48 or 24 (much more sensible resolution).

18. This was already fully understood by Nyberg. N. Nyberg, "Zum Aufbau des Farbenkörpers im Raume aller Lichtempfindungen," *Z.Physik* 52, pp. 406–419 (1928).

19. E. Hering, *Zur Lehre vom Lichtsinn*, (Vienna: Gerolds, 1878).

20. E. Schrödinger, "Über das Verhältnis der Vierfarben zur Dreifarben Theorie," *S.B. Akad. Wiss., Wien* IIa, pp. 471–490 (1924).

21. S. Quiller, *Color Choices*, (New York: Watson-Guptill, 1989).

22. Actually a crate or cuboid, rather than a true hypercube, that is to say, the edge lengths are in general not all equal. This makes only a trivial difference though and—for the sake of clarity—I am going to ignore it in this appendix.

23. This is easy to see if you consider the coordinates of the vertices of the unit hypercube in standard position. The coordinates are $\{\xi_1, \xi_2, \ldots, \xi_n\}$, where the ξ_i are either zero or one and all combinations occur. Thus you can simply take the binary digits of numbers in the range $[0, 2^n]$ and you are done. This is a one-liner in an environment like Mathematica.

24. Use the Gram-Schmidt procedure. The Gram-Schmidt procedure turn a general basis into an orthogonal one. Simply take all vectors in turn, subtract the projections

on all preceding ones and normalize. Although this is indeed the principle, modern algorithms yield numerically much better results because they prevent the accumulation of errors.

25. This translates into a simple algorithm to check the matrix of the projection operator for its capability to "reveal" the primitive basis. You simply check the signs of the determinants of its 3×3-submatrixes. In geometrical terms, the spectrum locus in a reasonable chromaticity diagram is a simple convex arc.

26. See also G. West, and M. H. Brill, "Conditions under which the Schrödinger object colors are optimal," *J.Opt.Soc.Am.* 73, pp. 1223–1225 (1983).

27. The primitive vertices are naturally ordered, so the sequence of cumulative sums is well determined.

28. Introduced by Chebyshev in 1878 in a report *On the cutting of our clothes* in P. L. Chebyshev, *Collected works* 5, (Moscow-Leningrad: Fizmatgiz, 1951), pp. 165–170.

29. E. T. Jaynes, *Probability Theory: The Logic of Science*, (Cambridge: Cambridge U.P., 2003); H. Jeffreys, "Fisher and Inverse Probability," *International Statistical Review* 42, pp. 1–3 (1974).

30. For example, see the review of the Olympus E-500 at `www.wrotniak.net/photo/oly-e/e500-rev.html` on the topic of "WB (White Balance) adjustments".

31. The "official" (CIE) method is "correlated color temperature." (CIE/IEC 17.4:1987, International Lighting Vocabulary (ISBN 3900734070).)

32. If there is a good model of the physics, you should use that, of course. In many cases the Kubelka-Munk theory of turbid layers will do (it applies, to colored papers, paint layers, and so forth). In that case you should use the "spectral signature" instead of the error function, as I do here. The error function is merely a convenient substitute in case no model of the physics is available. Whatever you use, you should use the idea of "homomorphic filters" to take care of the nonlinearities that constrain the reflectance to the range $(0, 1)$. This is a major problem with the relevant literature. For instance, you will often encounter "Principal Component" analyses of spectral reflectances. That makes little sense because PCA is a linear method. Using the homomorphic filtering, you solve this problem. In any physical setting it can be made to make sense; in the general setting used here it at least prevents nonsense. (On homomorphic filters, see A. V. Oppenheim, R. W. Shafer, and T. G. Stockham Jr., "Nonlinear filtering of multiplied and convolved signals," *Proc.IEEE* 56, pp. 1264–1291 (1968).)

33. P. B. Berlin, and P. Kay, *Basic Color Terms: Their Universality and Evolution*, (Berkeley: University of California Press, 1969).

34. I conjecture that generic three-dimensional projections of infinitely dimensional hypercubes have two conical singular points and a pair of creases like you have in the Helmholtz model. However, I see no obvious manner to prove this.

35. G. Wyszecki, and W. S. Stiles, *Color Science: Concepts and Methods*, (New York: Wiley, 1982).

Part III

Metrical Color Space

Chapter 8

Metrical Colorimetry

So far I have treated colorimetry without any reference to a metric (chapters 3 through 7). Yet a metric, in the form of a scalar product (also commonly known "inner product" or "dot product") for example, is extremely desirable in many contexts. (I give formal background in appendix H.1.1.) Here are some[1] of the major opportunities that would become available if you had one:

- an inner product lets you relate any vector (ket) to a unique linear functional (bra) and *vice versa*;

- an inner product lets you move linear operators from the vector space to its dual space and vice versa;

- an inner product lets you find a unique point of an affine subspace that is *closest* to a given point;

- an inner product lets you find a unique point of a given convex set that is *closest* to a given point;

- an inner product lets you define *orthogonal* projections. This is the major advantage from the perspective of colorimetry (see figure 8.1);

- an inner product lets you construct orthonormal bases that factor a vector space in a most convenient way;

- an inner product lets you find an essentially unique orthonormal basis that diagonalizes a given (generic) symmetric matrix;

- an inner product lets you define congruences or isometries.

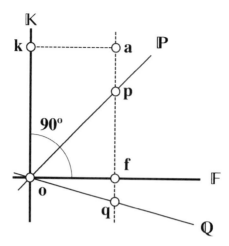

Figure 8.1 Given any beam **a** you obtain various colorimetric representations, for instance **p** in the space $\mathbb{P}$ spanned by primaries $\{\mathbf{p_i}\}$ or **q** in the space $\mathbb{Q}$ spanned by primaries $\{\mathbf{q_i}\}$. Although the projection is always parallel to the black space $\mathbb{K}$ these representations are not unique and there exists no unique "black component" of **a**. The "fundamental space" $\mathbb{F}$ is the orthogonal complement of the black space, thus the "fundamental component" **f** of **a** is well defined and so is its "black component" **k**. This is the advantage of having a scalar product. Without it there is no way to define a unique subspace on which to project.

These are a few of the desirable things that would become available. In the colorimetric context, this would mean that you could find a basis in which the color matching matrix (section 4.2) would become diagonal (with only three nonzero diagonal elements), or that to any color you might assign a unique (not necessarily real) fundamental beam, that you might strip a spectrum of its black content, and so forth. Plenty of good things.

Of course all of these nice operations make sense only *relative to the inner product*; thus metrical colorimetry is necessarily less general than nonmetrical colorimetry.[2]

This loss of generality need not be a problem. For instance, in many cases a problem might be solved either in a metrical or an affine setting, whereas the result would be guaranteed to be the same. In such cases, the metrical method might be the more convenient (through perhaps less "elegant" from a purist perspective) to use, and no harm will be done. It is different if you construct colorimetric entities whose very meaning requires the metrical framework. Such things would be meaningless in nonmetrical colorimetry, and you should consider very seriously whether you have something you actually want. For instance, did you use an inner product that made sense to your problem? Such worries are not void, for there exists freedom in the definition of an inner product.

8.1 The Choice of an Inner Product

When I say "inner product"[3] I mean an inner product in the space of beams $\mathbb{S}$ (chapter 3). Such an inner product should be based on considerations of physics.[4] Consider physical beams in the spectral basis, that is to say, described in terms of the spectral radiant power density. An overall measure used in radiometry is the radiant power, that is, the integral over the spectrum. This makes perfect physical sense and seems an obvious candidate for a "norm".

However, the setting of colorimetry is quite unlike that of radiometry. In colorimetry you work with *pairs* of beams, or—equivalently—with *virtual* beams such that *negative* spectral radiant power density is permitted (section 3.3.3). In such a setting the integral over the spectrum ("radiant power" of the beam) makes little sense, because many beams different from the empty beam would have zero radiant power. Such situations do not occur in physics, thus you have to be a little creative.

An obvious choice that seems a priori reasonable is to define the norm as the square root of the integral of the squared radiant power density. Thus you obtain the RMS (root mean square) radiant power of a beam. The RMS radiant power is only zero for the empty beam and is positive for any virtual beam. It may be seen as problematic that the RMS radiant power of a real beam is not equal to the "radiant power" of that beam as used in radiometry.[5] However, I think such discrepancies can hardly be avoided.

If you accept such a norm, then it seems evident that the inner product of two beams $\mathbf{p}$ and $\mathbf{q}$ should be defined as

$$\mathbf{p} \cdot \mathbf{q} = \int_{-\infty}^{+\infty} p(\lambda)q(\lambda)\,\mathrm{d}\lambda,$$

for then $\|\mathbf{s}\| = \sqrt{\mathbf{s} \cdot \mathbf{s}}$, exactly the RMS radiant power of $\mathbf{s}$. A point of consideration might be whether you want to introduce a wavelength dependent weighting function at this point, say $\mathrm{d}m(\lambda)$ for some monotonic function $m(\lambda)$ instead of $\mathrm{d}\lambda$. For instance, should you integrate over wavelengths or over photon energies? I will return to this question later. For the moment I suggest that such things do not make so much difference because the visual region is only a narrow part of the electromagnetic spectrum anyway. Thus I will simply proceed and cheerfully accept the definition of the inner product given above.

This particular definition turns out to be the "default" choice of inner product, that is (usually silently) assumed in all of the literature that I have read so far. The most likely reason is that the authors simply didn't give the issue a thought.

8.2   Construction of a Canonical Basis

So how do you do use the nifty "inner product" tool? The main use is based on the following intuitive insight. The space of colors is *much* smaller than the space of beams—only three-dimensional instead of infinitely-dimensional. Thus we (as the human race I mean) are *blind* as a bat to most of the riches of beams. You enjoy only a *very small part* of all there is to enjoy. Thus you may think of beams as being *composed* of a part that you enjoy and a part you are blind to, a rather natural notion. The part you enjoy is a *cause* of your luminous experiences. The part you miss is *causally ineffective* and might as well not exist. Thus it might be phantasized that any beam might be "split" into unique causally effective and causally ineffective parts.[6]

If this fantasy (see figure 8.2) indeed applies, then the causally effective part might rightfully be called the "fundamental component" and the causally ineffective part the "black component" (because it is invisible to you). Then all fundamental components may the supposed to live in "fundamental space" $\mathbb{F}$ or "fundamental visual space" or just "visual space", whereas the causally ineffective parts would live in the "black space" $\mathbb{K}$. These spaces are expected to be mutually orthogonal, linear subspaces of the space of beams ($\mathbb{S} = \mathbb{F} \oplus \mathbb{K}$), because Grassmann's laws are empirical evidence for the linearity of colorimetric equalities.

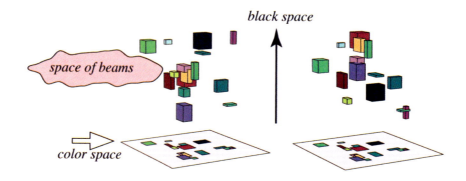

Figure 8.2 This is the very crux of colorimetry from a metrical perspective. The two configurations in color space are identical, yet the configurations in the space of beams are very different. They differ only in the "black dimensions" though. Color vision is very simple in this picture. You see the space of beams itself except for the unfortunate fact that you are blind as a bat to the black dimensions. In this picture color is not different from space proper, where you see the 3D space "you move in" itself, except for the fact that you are blind as a bat (at least monocularly) to the "depth" dimension. For the spatial case you "see" two out of three dimensions, and for the color case three out of infinite dimensions.

Although this is indeed a nice enough phantasy (almost too good *not* to be true), I couldn't make it come true thus far. There is simply *no way* to associate a unique "fundamental component" to a beam in the absence of an inner product. That is because the notion of an "orthogonal projection" requires an inner product.

The idea that any spectrum may be split into a unique "causally effective" beam and a "causally ineffective" (black) beam is originally due to Wyszecki and known as Wyszecki's hypothesis. It dates from the 1950s. This hypothesis is evidently *false*, for there are infinitely many complements to the black space that have equal claim to be called "causally effective" and no way to choose between them. However, in the context of a metric things are very different, for there exists a unique *orthogonal* complement to the black space. Thus, Wyszecki's hypothesis comes true when you decide on a metric. Thus many authors consider the issue to be settled: "*Wyszecki's hypothesis is true, as was shown by Cohen*". I have a rather less cheerful attitude, though. True, given a metric the orthogonal complement of the black space is uniquely determined. But how about the metric itself? There are infinitely many choices at your disposal, all equivalent to each other, and no formal way to choose between them. Each metric yields a different orthogonal complement, so where is the gain? I consider Wyszecki's hypothesis to be false in the sense that you may either select an arbitrary complement by fiat and live with the metric it implies, or you may select an arbitrary metric by fiat and you live with the orthogonal complement that implies. However, either route can be defended, much like the selection of the achromatic beam by fiat turned out to be extremely useful. In order to select a metric you have to use arguments derived from considerations of *physics* (for beams in the black space may well have finite length), whereas in order to select a complement you need to draw arguments from *color science* (for physics has nothing to say concerning the black space and so forth). In this chapter I explore the former, and in the next chapter the latter way.

With the inner product in place you are in good shape. You should look for a projector that, when applied to a beam, "strips off the black fluff", and presents you with the unique, totally causally effective "fundamental component" of that beam. The problem is simply to construct this operator. All the ingredients are already at your disposal. You evidently need an *orthogonal projection* on "fundamental visual space." The problem in finding it is that the color matching matrix is a general, that is an *oblique*, rather than an orthogonal, projection (figure 8.3).

8.2.1 Cohen's Matrix-R

The possibilities offered by the presence of a metric are evidently intriguing. You can immediately add this to the account of colorimetry I have given earlier. All you have to do is introduce a metric. Of course, therein lies the hard problem. I'll discuss choices later.

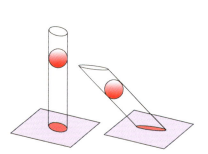 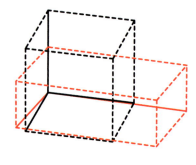

Figure 8.3 Left: Depending upon the projection the image of a sphere can be a circle (lower dimensional sphere) or an ellipse. In the latter case one say the "image is deformed" or that "this is not an honest image;" Right: The CIE basis (red) compared with an orthogonal unit frame (black). Although somewhat skew, the standard CIE xyz-basis is actually close to a rectangular box. Instead of being square the angles are 114°, 85°, and 96°, whereas the edge lengths are in the rations 43:100:115. Of course, these measures are taken from fundamental space, and the CIE representation is a mess.

Remember that the color matching functions $\langle \chi_{1\ldots3} |$ allow you to find a beam

$$|c\rangle = \left(\sum_{i=1}^{3} |p_i\rangle\langle\chi_i| \right) |s\rangle,$$

that is indiscriminable from a given beam **s**, where $|p_i\rangle$ are the spectra of the primaries and $\langle\chi_i|$ the corresponding color matching functions. Thus $\mathbf{c}\,\square\,\mathbf{s}$, whereas $\mathbf{c}$ is in the space $\mathbb{P}^3$ spanned by the primaries $\{\mathbf{p_i}\}$. The colorimetric procedure thus projects any beam **s** on the space $\mathbb{P}^3$, though—since the choice of primaries is arbitrary—the projection is almost certain to be an oblique one.

Classically you define a map $A : \mathbb{S} \mapsto \mathbb{C}$ as

$$\hat{A} = \sum_{i=1}^{3} |f_i\rangle\langle\chi_i|,$$

for a basis $\{\mathbf{f_i}\}$ of "3-dimensional color space $\mathbb{C}$" with

$$\hat{I} = \sum_{i=1}^{3} |f_i\rangle\langle\varphi_i|,$$

$$\langle\varphi_i|f_j\rangle = \delta_{ij}.$$

The matrix elements of this map in the monochromatic basis are

$$A_{ij} = \langle \varphi_i | \hat{A} | \lambda_j \rangle = \sum_k \langle \varphi_i | f_k \rangle \langle \chi_k | \lambda_j \rangle = \sum_k \delta_{ik} \langle \chi_k | \lambda_j \rangle = \langle \chi_i | \lambda_j \rangle.$$

Thus you obtain a matrix with three rows which contain the coordinates of the color matching functions and all other rows filled with zeroes.

Conventionally, the "color matching matrix" is written in table form with the three color matching functions as columns. Thus the matrix of the operator $\hat{A}$ is the transpose of the classical color matching matrix.

In an ideal world it would be the case that whereas e_i $(i = 1 \dots \infty)$ would be an orthonormal basis of the space of beams $\mathbb{S}$, the subbasis e_i $(i = 4 \dots \infty)$ would span the black space $\mathbb{K}$ and the remaining subbasis e_i $(i = 1 \dots 3)$ would span what might be called the "fundamental space $\mathbb{F}$". Thus the space of beams could be written as the direct sum of the black and the fundamental space, these being mutual "orthogonal complements" and the basis would nicely reflect that explicitly.

In such an ideal world the "color matching matrix" would be

$$M = \begin{pmatrix} 1 & 0 & 0 \\ 0 & 1 & 0 \\ 0 & 0 & 1 \\ 0 & 0 & 0 \\ 0 & 0 & 0 \\ \vdots & \vdots & \vdots \end{pmatrix}.$$

In this world, the product $M^T M$ would be the 3×3 unit matrix (the Grammian of the basis of fundamental space), whereas the product MM^T would be the $\infty \times \infty$ matrix

$$R = MM^T = \begin{pmatrix} 1 & 0 & 0 & 0 & \cdots \\ 0 & 1 & 0 & 0 & \cdots \\ 0 & 0 & 1 & 0 & \cdots \\ 0 & 0 & 0 & 0 & \cdots \\ 0 & 0 & 0 & 0 & \cdots \\ \vdots & \vdots & \vdots & \vdots & \ddots \end{pmatrix}.$$

Notice what this matrix does: It is the orthogonal projector on fundamental space. It has rank 3, trace 3, is symmetrical ($R^T = R$) and idempotent ($R^2 = R$).

What we are after is the matrix of this projector in the general case. It will still have rank 3, trace 3, be symmetrical and idempotent. It will be independent of the choice of primaries, though.

Suppose you have an arbitrary basis and an arbitrary matrix A instead of M. A singular values decomposition (see page 8.4) lets you write $A^T = UWV^T$, where W is diagonal with just three singular values, V is an orthonormal basis of a subspace of the space of beams, and U an isometry in color space. The singular values and the isometry in color space depend upon the choice of primaries and you need to get rid of them. The basis V is interesting though, for it is evidently a basis of fundamental space. Thus $R = VV^T$ is the projector that you are after.

In terms of the spectral basis the matrix of this operator R is known as Cohen's matrix-R after Joseph Cohen, who constructed it in the second half of the twentieth century,[7] more than a century after colorimetry became a mature science in the hands of Maxwell, Grassmann, and Helmholtz. Because the rank, trace, and determinant are invariants, Cohen's matrix-R has rank three, trace three, and determinant zero. Its rows are mutually orthogonal, and the matrix is symmetrical.

In practice, you simply run the SVD algorithm on the color matching matrix and keep only the first basis V (say). Its "square," VV^T is Cohen's matrix-R. Because it is the orthogonal projector on the fundamental space, application of the operator on a spectrum "strips of the black fluff" and leaves you with the causally effective component of the spectrum, hence its importance.

Cohen's matrix-R is traditionally defined in terms of the color matching matrix A. The expression is

$$R = A \left(A^T A\right)^{-1} A^T.$$

You easily show that $R^2 = R$, thus R is a projection, and $R^T = R$ (R is symmetric), thus R is an *orthogonal* projection. (Because the transpose R^T depends upon the scalar product, I have used the metric here!) How did Cohen hit on the expression $A \left(A^T A\right)^{-1} A^T$?

Cohen did not use the SVD algorithm, but figured out how to construct the matrix R via algebraic operations. This is a conceptually very opaque procedure. It becomes somewhat transparent in terms of the singular values decomposition, though. From the SVD decomposition $A^T = UWV^T$ only the final factor V^T is going to play a role, thus instead of $R = A \left(A^T A\right)^{-1} A^T$ you have $R = V(V^T V)^{-1} V^T$, but since $(V^T V)^{-1} = I$ (for V has orthonormal columns, thus $V^T V$ is the unit matrix) this simplifies to $R = VV^T$. (You can find the full derivation in this endnote[8] in case you fail to see the irrelevance of U (an isometry, thus $U^T = U^{-1}$) and W (a diagonal matrix,thus $W^T = W$ and W commutes in matrix products) in the SVD decomposition of the color matching matrix.) Apparently $A \left(A^T A\right)^{-1} A^T$ is the same as $V(V^T V)^{-1} V^T$, thus the SVD, followed by discarding W and U serves to strip the color matching matrix A of most of the fluff that depends on your arbitrary choices.

Cohen's matrix-R is the matrix of the desired orthogonal projection operator[9] on fundamental space. This is the form in which Cohen presented his famous matrix-R. After some exercises Cohen's matrix-R, will take on a life of its own and might even appear obvious to you. Just remember my warning; if you really want to *compute* it, go through the SVD algorithm rather than Cohen's magic formula.

A minor generalization, due to Fairman,[10] is sometimes convenient. You may find Cohen's matrix-R from a *pair* of color matching functions, where one pair is obtained from the other through a change of primaries. You easily prove Fairman's formula following the same algebraic manipulations as I did above. I will not use this generalization here, though.

Notice that in Cohen's formula $A \left(A^T A \right)^{-1} A^T$ the first factor is A^T, which maps spectra to colors. Thus the operator

$$\Xi = A \left(A^T A \right)^{-1},$$

maps colors on fundamental spectra. It is like an "inverse" of the projection on the space spanned by the primaries, called its "pseudoinverse"[11]. This operator is very useful in its own right, since it lets you associate a unique beam (alas, usually virtual) to any color.

The fundamental and the black spaces are each other's orthogonal complements in the space of beams, i.e., their direct sum is the space of beams and any fundamental beam is orthogonal to any black beam and vice versa.

Thus, any beam $\mathbf{a}$ (say) can be uniquely decomposed as

$$\mathbf{a} = \mathbf{f_a} + \mathbf{k_a},$$

where $\mathbf{f_a}$ is in fundamental space and beam $\mathbf{k_a}$ in the black space. You evidently have

$$\hat{R}\mathbf{a} = \mathbf{f_a},$$

and

$$\mathbf{k_a} = \mathbf{a} - \mathbf{f_a} = (\hat{I} - \hat{R})\mathbf{a}.$$

Apparently the operator $\hat{I} - \hat{R}$ is the projector on the black space $\mathbb{K}$.

If $\mathbf{a}$ and $\mathbf{b}$ are any beams, then

$\mathbf{a} \square \mathbf{b}$ implies $\mathbf{f_a} = \mathbf{f_b}$, *thus instead of mere colorimetric equivalence an equality of beams.*

Notice that the black components do not matter. Thus when I define the perturbed beams

$$\mathbf{a}' = \mathbf{f_a} + \mathbf{k_b} \text{ and,}$$
$$\mathbf{b}' = \mathbf{f_b} + \mathbf{k_a},$$

The various mappings of colorimetry

It may not be totally superfluous to summarize the important correspondences of colorimetry. The all important spaces are the space of beams $\mathbb{S}$ and the metameric black space $\mathbb{K}$. The black space is a linear subspace of the space of beams. The space of beams is just physics (radiant power spectra), and the black space is a complete description of an observer's discriminability of beams: Any two beams that differ by an element of the black space cannot be distinguished by the observer.

In classical colorimetry you consider the space of affine hyperspaces parallel to the black space.[12] This is the "space of colors" $\mathbb{C}$. To each element of $\mathbb{S}$ corresponds a unique color, to each color corresponds a "metameric suite" of spectra that mutually differ by an element of $\mathbb{K}$. The correspondence is quantified by the color matching matrix and the primaries. This is all there is.

If you admit a scalar product there exists a unique "fundamental space" $\mathbb{F}$ orthogonal to the black space. To each beam corresponds a unique fundamental beam. The fundamental beams and the colors correspond in a one-to-one fashion. Cohens matrix maps metameric suites to fundamental beams, while the pseudoinverse maps colors on fundamental beams. Notice that color space is not even needed when you label colors by their fundamental beams. Then Cohen's matrix-R and fundamental space are all you'll ever need.

you still have $\mathbf{a}' \square \mathbf{b}'$ if $\mathbf{a} \square \mathbf{b}$ *regardless* of the fact that the black components have been arbitrarily interchanged (within the black space of course). Thus you have a wonderful, intuitive image of colors:

> *Colors simply relate to fundamental beams in a one-to-one fashion, then you have black beams which are causally ineffective and may be added to any beam without changing its color.*

It seems to take away most of the mystery that colorimetry has for the layman. Finally, here is something you can easily *explain*.

I show some simple examples (figures 8.4–8.6) using familiar "average daylight" and "incandescent light" spectra. As you split these spectra into their fundamental and black components, you see that the fundamental components look quite different from the original spectra, and that, where the original spectra look indeed qualitatively different, the fundamental spectra are somewhat similar and appear to differ mainly with respect to the emphasis put upon the two prominent modes. The black components are causally ineffective, so you may just as well forget about them.

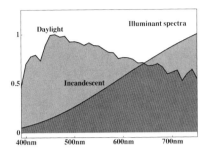

Figure 8.4 The original spectra (average daylight and incandescent light). A major difference is that the light-bulb emits an under-dose of the blue part as compared to daylight.

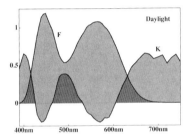
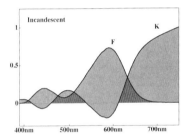

Figure 8.5 The original spectra of daylight and incandescent light (see figure 8.4) decomposed in terms of the fundamental and the black component. Left for the average daylight spectrum, at right for the incandescent spectrum.

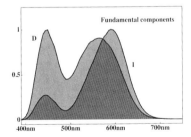
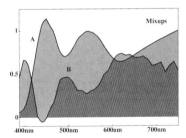

Figure 8.6 Left the fundamentals of the average daylight and incandescent light spectra shown in figure 8.4. Right I have interchanged the black components. Notice that one of the spectra is not entirely real. This can easily be remedied by scaling down the black component a bit. Notice how different these spectra look from the originals, yet their colors remain unchanged.

Colorimetry and the philosophers

When you see examples like those in figure 8.6, it is all too easy to conclude that the human visual observer is totally unable to classify spectra. This is often summarized by the saying, "The eye fails to see by wavelength," an expression that refers to an even more elementary mistaken assumption (the one-to-one relation between color qualia and the refrangibility of homogeneous lights as suggested by Newton) but can be applied equally well to spectroscopic analysis. And if you don't "see by wavelength" many conclude that colors must be (mere) "mental paint."

If you are a hardcore scientist, you should stick to the *objective* facts, and this means *colorimetry*. Notice that I am not saying that psychology is not about "objective" facts. But the crux is that the "objective" in that case has a different ontology from that generally accepted in the sciences.

Then it should be clear to you that color vision (as described colorimetrically) is nothing but *low resolution spectroscopy*. The cheapest spectroscopic devices give you a dozen degrees of freedom; as you pay more this goes up to the hundreds or thousands. The eye is a device from a five-and-ten store. But low as the resolution might be, it still counts as spectroscopy. No secrets here. There is simply no room for philosophy, because on the level of the sciences there is nothing more to explain.

Why do so many philosophers fail get this? I think it is partly due to their lack of understanding of linear geometry (they get fooled by examples such as those in figures 8.4 and 8.6) and partly due to their lack of understanding of the strength (and thus beauty) and obvious limits (no qualia) of the colorimetric paradigm. This is indeed the common source of misunderstanding by laymen (I count many of my colleague physicists and mathematicians among them) and philosophers alike.

This once again shows the incredible genius of Maxwell. Nobody ever explained the point to him; he simply did the right thing. Even as a physicist, I rank his work in color above his monumental achievements in electromagnetic theory and statistical mechanics.

Once the decomposition has been done I can play all kinds of strange games. For instance, I can combine the fundamental component of daylight with the black component of incandescent light and vice versa. This does not change the colors of the beams at all because the black components are causally ineffective. Only the fundamental components are causally effective, but these remain unchanged. Notice how different the resulting spectra appear.

This is an easy recipe to construct as many members of a metameric suite as you want; simply add black beams obtained from various spectra.

In figure 8.7 I summarize the geometrical relationships that make up colorimetry from the metrical perspective. Of course, it is not possible to draw the real thing, since the space of beams is infinitely dimensional. (I could easily gloss over that in a drawing, but it wouldn't be of much service to your intuition.) Since dimension three is as high as human intuition goes, that will be the dimension of the space of beams in the drawing. Since the black space should at least be drawn two-dimensionally in order to be able to show certain relations, that's how it should be in the drawing. That leaves only one dimension for color space, which is really too bad. So the translation has to be:

Space of beams $\mathbb{S}$ infinitely-dimensional, three-dimensional in the picture;

Space of colors $\mathbb{C}$ three-dimensional, one-dimensional in the picture;

Fundamental space $\mathbb{F}$ three-dimensional, one-dimensional in the picture;

Black space $\mathbb{K}$ "infinity minus three" dimensional, two-dimensional in the picture;

Metameric suite $\mathbb{M}$ "infinity minus three" dimensional, in the picture two-dimensional.

Since color space is simply a convenience, an in many ways superfluous *picture* of fundamental space, but definitely not *part* of the space of beams, I draw the space of beams and color space separately side by side. All the other relevant spaces are subspaces of the space of beams. Since color space is simply isomorphic to fundamental space you can indeed do without it. If you do, then the space of beams is the real arena of colorimetry. Instead of "mere mental paint" the (colorimetric) colors are physical entities.

The space of beams contains a subspace called "fundamental space." All points of fundamental space are distinguished by the human observer. No points of the space of beams are ever judged to be distinct from all points of fundamental space. The points of fundamental space are also called "colors." The transpose of the color matching matrix A^T maps a point of fundamental space (a beam) on a color and the pseudoinverse Ξ is the inverse map. The transpose of the color matching matrix also maps points of the space of beams not in fundamental space to color space. All points of the space of beams that map on a single point of color space are metameric, they subtend the "metameric suite" of the representative in the space of beams, the "fundamental component" common to each of these metameric beams.

The space of beams is thus *foliated* by the metameric suites. These form a family of mutually parallel affine hyperplanes, all parallel to the black space, which is the metameric suite of the origin (the "empty beam"). Cohen's projection (matrix-R) maps arbitrary points of the space of beams on the fundamental component, and thus each metameric suite on its representative in fundamental space.

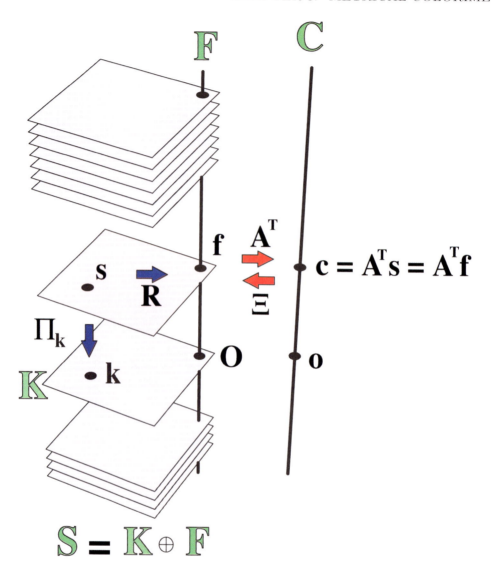

Figure 8.7 The geometry of colorimetry from the metrical viewpoint. Here $\mathbb{S}$ is the space of beams, $\mathbb{F} \subset \mathbb{S}$ fundamental space and $\mathbb{K} \subset \mathbb{S}$ the black space. The space of beams is the direct sum of fundamental space and black space, the space of beams is thus foliated into parallel copies of the black space. Cohen's projector selects a unique fundamental component of each such copy of metameric beams. Notice that "color space" $\mathbb{C}$ stands apart in this scheme and isn't really necessary at all. In this picture you can easily do without it.

Cohen's projector simply "discards the black space". The complement of this mapping (the identity minus Cohen's projection $\Pi_B = I - R$) maps arbitrary points on the black space; it discards the fundamental component.

The space of beams is the direct sum of the black space and fundamental space. Human vision is blind as a bat to the black space and simply discards it. What is left is fundamental space, which—historically—became known as "color space".

Thus you obtain a nice geometrical picture that is easily kept in mind. It will keep you from committing any one of the many silly mistakes so common in discussions of colorimetry.

8.2.2 A semimetric in the space of beams

Given a beam $\mathbf{s} = \mathbf{f} + \mathbf{k}$ you have (by the Pythageorean theorem) $\|\mathbf{s}\|^2 = \|\mathbf{f}\|^2 + \|\mathbf{k}\|^2$. Since

$$\|\mathbf{f}\|^2 = \hat{R}\mathbf{s} \cdot \hat{R}\mathbf{s} = \mathbf{s} \cdot \hat{R}^2\mathbf{s} = \mathbf{s} \cdot \hat{R}\mathbf{s},$$

you see that Cohen's matrix-R can apparently be interpreted as a metric tensor in $\mathbb{S}$. It is not a true metric tensor in $\mathbb{S}$, though, because $\mathbf{s} \cdot \hat{R}\mathbf{s} \neq \|\mathbf{s}\|^2$. Thus $\mathbf{s} \cdot \hat{R}\mathbf{s} = 0$ implies merely $\mathbf{s} \in \mathbf{K}$, but not $\mathbf{s} = \mathbf{o}$. You have only a *semimetric*, not a metric.

In the semimetric all black dimensions are isotropic. The isotropic dimensions are orthogonal to any beam, including themselves. Two beams $\mathbf{p}$ and $\mathbf{q}$ at zero distance from each other, thus $\|\mathbf{p} - \mathbf{q}\|^2 = 0$, need not coincide. In case they don't, I will call such points "mutually parallel". Parallel points evidently differ by a black beam, thus parallel points are members of the same metamer. Apparently the semimetric neatly implements the crux of colorimetry:

> *Parallel points in the semi-metric are indiscriminable and vice versa.*

Using the pseudoinverse you have $\mathbf{f} = \hat{\Xi}\mathbf{c}$, where $\mathbf{c} \in \mathbb{C}$ is the color of the beam $\mathbf{s}$. Thus $\|\mathbf{f}\|^2 = \hat{\Xi}\mathbf{c} \cdot \hat{\Xi}\mathbf{c} = \mathbf{c} \cdot \hat{\Xi}^T\hat{\Xi}\mathbf{c}$, apparently $\hat{\Xi}^T\hat{\Xi}$ is a metric tensor in $\mathbb{C}$. Some algebraic simplification yields

$$\|\mathbf{c}\|^2 = \|\mathbf{f}\|^2 = \mathbf{c} \cdot \hat{M}\mathbf{c} \quad \text{with} \quad \hat{M} = (\hat{A}^T\hat{A})^{-1},$$

thus you can find the metric of $\mathbb{C}$ directly from the color matching matrix $\hat{A}$. You have used the scalar product because the expression contains an adjoint.

In cases where the rows of the color matching matrix are mutually orthonormal, you have $\hat{A}^T\hat{A} = \hat{I}$. For instance, the operator $\hat{V}$ that you obtain from the SVD is of this type. A color space based on $\hat{V}$ thus has the Euclidian metric.

8.2.3   The structure of Cohen's matrix-R for human vision

A straightforward singular values decomposition of the transpose of the CIE color matching matrix yields singular values in the ratios 3.86 : 1.28 : 1; thus the CIE color space is rather skewed. Discarding the irrelevant deformation and rotation, you obtain Cohen's matrix-R as VV^T, where V is the spectrum-side orthonormal basis. For let the basis be denoted V, then Cohen's matrix-R is seen to be

$$A \left(A^T A\right)^{-1} A^T = V(V^T V)^{-1} V^T = VV^T = R,$$

due to the orthonormality of V.

Since Cohen's matrix-R is quite independent on the choice of primaries the route to get here is really not too important. As a first investigation, I plot the values of the coefficients of the matrix (figures 8.8 and 8.9). Much has been made of the fact that Cohen's matrix-R, for human vision has three distinct modes on the diagonal (trichromacy). However, when you compare the (equally trichromatic) matrixes for the Grassmann, Helmholtz, and local models, you see that you have to take this with a grain of salt.

Notice the three major positive modes and a few minor, negative side lobes. The rows (or columns) of the matrix are the fundamental spectra of the monochromatic beams of unit radiant power. The diagonal elements thus indicate the square[13] of the radiant power density of the fundamental of monochromatic beams at the wavelength of these beams.

In figure 8.10 I plot the fundamental spectra of the monochromatic primaries used to obtain the color matching data by Stiles and Burch. Although not monochromatic, these

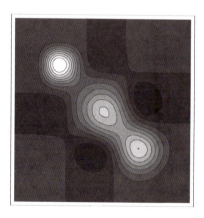 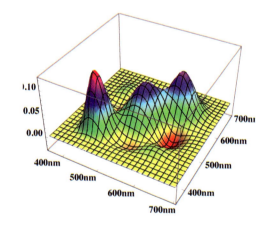

Figure 8.8 A density plot and a 3D surface plot showing the magnitude of the coefficients of Cohen's matrix-R.

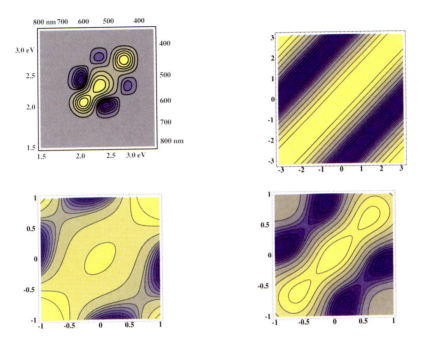

Figure 8.9 The structure of Cohen's matrix-R. From left to right, top to bottom: The human observer, the Grassmann model, the Helmholtz model, the local model.

spectra are rather sharply peaked at the corresponding wavelengths of the monochromatic beams and lie near the extrema seen in the values of the coefficients of Cohen's matrix-R; thus they make indeed good choices as "primaries".

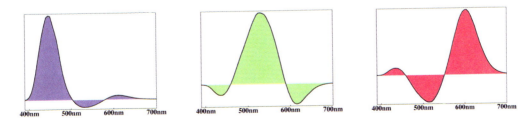

Figure 8.10 The fundamental spectra of the three monochromatic beams with wavelengths 645.16nm, 526.32nm, and 444.44nm, the primaries for the CIE 1960 10° color matching data.

When you add all columns (or, equivalently, rows due to the symmetry of the matrix) together you obtain the fundamental of the sum of all monochromatic components of unit radiant power, that is to say, the fundamental component of the beam with a flat spectrum (figure 8.11 left).

Most of the action is along the diagonal of this (symmetrical) matrix, so I also plot the values of the square root of the diagonal elements of R (the diagonal elements of R are the squares of the lengths of the fundamental components of the monochromatic beams of unit radiant power) (figure 8.11 right).

8.2.4   What the metrical representation does *not* imply

It is important to appreciate that although color space indeed is a "true image" of the space of beams, because it inherits its metric, this does *not* mean that adjacent colors correspond to adjacent beams. The opposite is indeed the case though, adjacent beams generate adjacent colors. That the former does not hold can perhaps be understood best by noticing that the (at first blush puzzling) "additional distance" can easily be generated through an (invisible) black component. This is similar to the case of photographs of your daily environment. Things adjacent in the photograph are not necessarily adjacent in space, though things that are adjacent in space will be adjacent in the photograph.

Just think of Bill taking the obligatory photograph of Alice in front of the leaning tower of Pisa. Alice assumes a special pose, holding her hands in a special way, and Bill looks for exactly the right vantage point. In the final photograph, Alice keeps the leaning tower from

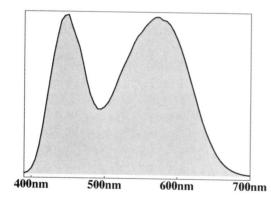 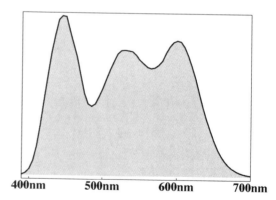

Figure 8.11 Left: The magnitude of the summed columns of Cohen's matrix-R. Right: The magnitude of the square roots of the diagonal coefficients of Cohen's matrix-R (these equal the lengths of the fundamental component of the monochromatic beams of unit radiant power).

toppling over by "pushing" with her hands. In real life, the tower was quite a distance from Alice, for Bill made sure that Alice and the tower were rendered about equally high in the photograph. This is not a fancy simile; geometrically the two cases are almost identical. The "depth" in the case of the photograph corresponds to the black space in the case of colors.

Even orthogonal beams (how different can you get?) may appear virtually indistinguishable. Here is an example. Divide the wavelength domain into narrow bins and label them sequentially. Then the beam that has a fixed radiant power on the even bins and the beam that has the same fixed radiant power on the odd bins, either beam having no radiant power on the other bins, will be as indistinguishable as you want (simply decrease the bin width). However, these are clearly mutually exclusive, hence orthogonal beams. This example shows that colors are somehow more "stable" than beams. In setting up colorimetric standards you have to be keenly aware of this. For example, the aforementioned beams are perhaps not the best candidates for an "achromatic" beam, *even though neither one will look different from a uniform (flat) spectrum*. Such beams with nontrivial microstructure lead to very visible problems when combined with equally structured surfaces, especially when the spectral reflectance of the surface is "synchronized" with that of the beams.

For instance, it is easily possible to work magic tricks like the following. I take two light sources and a surface of my choice. From the surface I cut a little square that I place on a piece of ordinary "gray" paper. Then it is easy to set things up such that the paper will look duly gray under either source, whereas the square may have *any two chosen colors* when placed sequentially under the two sources. Here the gray substrate "proves" that the two sources are "achromatic," and then the fact that the square may have entirely different colors under these (equally) achromatic sources "proves" that color is merely "mental paint." Various philosophers have concluded from such examples that color tells you nothing about the way the world is. Apparently anything is possible.

Such a demo is (at least in principle) possible. You may want to figure out for yourself how to do it (it is not very hard).[14]

Although such formal experiments are entertaining, it will be almost impossible to actually implement such demos with real materials and sources. The reason is that you need highly articulated spectra that are mutually correlated. Finding highly articulated spectra is possible (though hard enough), but finding mutually correlated articulations is next to impossible. Examples of spectrally highly articulated sources are low-pressure gas discharges and examples of spectrally highly articulated materials are rare earth glass powders. Because the articulations are due to different physical systems, they are highly unlikely to be mutually correlated. (Rare examples are used in the optical laboratory, *e.g.*, neodymium glass filters that (by sheer coincidence) absorb the sodium D-doublet.)

Why take color seriously at all? Because

> *spectrally highly articulated illuminants and surfaces are rare. Strong corre-*
> *lations between the spectral structure of source radiance and surface reflectance*
> *virtually never (ever) occur in real life.*

This is something philosophers won't accept as a valid argument, which is one reason why they continue to be puzzled by (and talk nonsense about) color. But it is these facts that make trichromatic color vision a viable proposition. Without them there would not have existed an evolutionary drive. Color vision is really useful in real life, it reveals properties of the world (the philosophers have it wrong there), that is the very reason why it has evolved.

8.2.5   Canonical bases for color space

Since the metric of the space of beams induces a metric in color space, this allows you to draw undistorted pictures of a three-dimensional glimpse of the space of beams—that is, color space. You do not have a unique coordinate frame though, for an SVD of the color matching matrix will yield an orthonormal basis which spatial attitude and orientation depends upon the (incidental) choice of primaries.

It would be advantageous to have a "canonical basis," because then pictures from different laboratories would look the same and coordinate values could be directly compared. Since such a state of affairs is desirable, here is one method to arrive at such a canonical basis.

Assume that you have decided upon the achromatic beam. Then it makes sense to have one basis vector (e_3 say) in the achromatic direction. The other basis vectors may be assigned arbitrarily; thus, you have one arbitrary degree of freedom, a rotation about the achromatic axis. A possible choice is to take the "yellow spectrum generator" (complementary to the short wavelength spectrum limit, subtract its projection upon the achromatic direction, and normalize it to obtain the basis vector e_1. Then the $e_3 \wedge e_1$-plane is the "yellow-blue plane". Now you are left with a binary choice for the basis vector e_2. You may orient it such that a movement from yellow to green turns you from the first to the second basis vector. This settles the frame completely.

Notice the arbitrary choices you have made to arrive at a canonical frame:
— you decided on a particular scalar product;
— you decided on a particular achromatic direction;
— you decided on the yellow-blue plane;
— you decided on the yellow toward green orientation.

Is that all?

At this point I have introduced essentially all of the theory of colorimetry. You are ready to read the literature. You will encounter many new concepts and methods, but you should be able to work your way into this material with your present background. If not, perhaps a technical book like Kang's *Computational Color Technology* might help.[15]

In principle this information suffices to reproduce the canonical frame in any laboratory, you could reproduce it when stranded on some deserted island, for instance.

Many of the illustrations in this book were made in the framework of such a canonical basis. The main advantage is of course the undistorted view of fundamental space.

8.3 Real Problems to Solve

"Real problems" are so diverse that it is impossible to cover them all. What I intend to do here is to discuss just a few "generic problems" that should allow you to feel confident to handle any problem you might meet. There are only a few problems that might be called "generic". The most commonly encountered ones are

— to find a real beam that comes closest to some virtual beam;

— to find a beam with certain general spectral properties that leads to a given color;

— to find surfaces that look the same under more than a single illuminant.

There are many more interesting problems in the area of object colors of course, but I'll come to that later.

Notice that the problems I listed here cannot be solved through the standard colorimetric methods, because these methods are *linear*. Linear methods will not allow you to differentiate between real and virtual beams. For instance, given a color it is easy to find a beam that will generate it, you simply apply the pseudoinverse of the color matching matrix to the color and you obtain the corresponding fundamental spectrum. But the fundamental spectrum is likely to be a virtual beam. Of course, you can generate as many metameric beams as you want, for instance by the addition of random black beams, and at some point—if you're lucky—you may come up with a real beam, but that is hardly a principled method. You need additional tools, and these tools are necessarily of a nonlinear nature.

Most problems can be cast in the following general form:

> To find the point of a convex set that is closest to a given point (typically not in the convex set).

There are a number of common methods that will allow you to solve such problems. The most important ones are *linear programming* and *iterative projection on convex sets*.

8.3.1   Using the primaries

If your primaries are real colors, then any color can immediately be expressed in terms of a linear superposition of the primaries. There is no guarantee that such a combination would also be a real beam. However, you can select primaries in terms of which *the large majority* of colors will come out as real beams. One example is Schopenhauer's partition of daylight mentioned before. It is indeed surprisingly effective. Of the colors you will meet with in typical scenes, only a very small number of extremely saturated colors lead to virtual spectra. For the large majority of colors, you find a unique (because of the fixed nature of the primaries) beam that will evoke it. This is a cheap but effective way to handle many (perhaps the majority of) problems, I will discuss it in more detail in chapter 13.

An example is shown in figure 8.12. The Schopenhauer partition was calculated for average daylight. Notice that any beam which has a color within the Schopenhauer RGB crate (which exhausts the important part of the volume of the color solid) now can be written as a linear superposition of such parts. An example is the incandescent beam. It turns out to be especially rich in red and lacking in blue (in daylight terms). Notice how different the real beam from the RGB parts representation is from the actual incandescent light spectrum. Yet these spectra are metameric by design.

8.3.2   Postulating canonical forms

In some cases you won't need the power tools. For instance, you may already know the general spectral properties of the beams of given colors you are looking for. For suppose you are looking for a beam that has a color near to what most people would be prepared to call WHITE. Such a beam will be close to a uniform spectrum. It will certainly be possible to describe the spectrum as some level, with a certain linear trend and a certain curvature (quadratic trend), for instance

$$s(\lambda) = \exp(a + b\lambda + c\lambda^2/2).$$

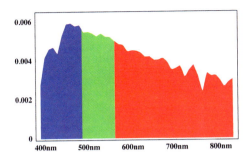 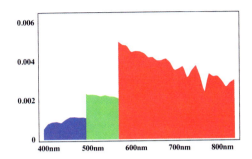

Figure 8.12 At left the daylight spectrum (see figure 8.4) has been partitioned into red, green, and blue (Schopenhauer) parts of daylight. At right the incandescent beam (see figure 8.4) has been represented by its Schopenhauer parts. Notice how "little blue" there is in the light of an ordinary light bulb.

Notice that these beams are automatically real. You simply have to find parameters $\{a, b, c\}$ such that the color comes out right. This is a simple nonlinear estimation problem that is easily solved (iteratively) by any of the standard numerical methods.

The canonical form $s(\lambda) = \exp(a + b\lambda + c\lambda^2/2)$ is just an example of course. In practice you may have to produce a certain color starting from given sources, filters, and so forth. You will then set up a canonical form accordingly.

In figure 8.13 I show some standard illuminant spectra fitted with the slope-curvature template (red curves). For spectra such as these this approximation is rather good and—not

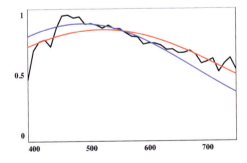 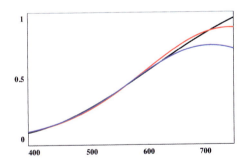

Figure 8.13 At left the daylight spectrum and at right the incandescent beam approximated by the template functions discussed here. First I use some arbitrary fitting procedure to find a reasonable approximation to the spectrum (red curve), then I iterate to find a template function that exactly reproduces the color (the blue curve).

surprisingly—the colors of the approximations are very near the colors of the actual beams. Thus you obtain the perfect position from which to find the actual template metamer through iterative adjustment of level, slope and curvature. Typically a few iterations suffice to obtain the blue curves, which are the perfect metamers.

8.3.3   Linear programming

Linear programming is a standard method to solve optimization problems that are characterized through a (possibly large) number of linear constraints. The constraints can be linear equalities or inequalities, and the optimization involves finding the maximum of a linear functional. This is a common and well understood problem, and methods have been designed that will solve large problems efficiently. The size of the problems encountered in colorimetry (at most hundreds of constraints) is such that you will always be able to solve them in very reasonable time.

The basic concept of "linear programming" is illustrated in figure 8.14. You have a (commonly very large) set of linear equalities and (more important) inequalities. These define a convex polyhedral volume (possibly of infinite extent, perhaps in an affine hyper-

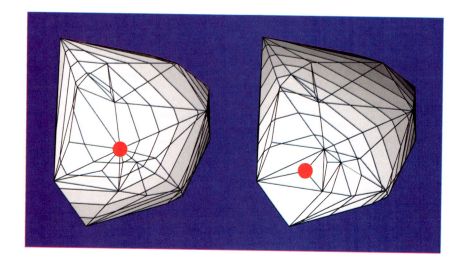

Figure 8.14 The structure of "linear programming". Here the set of linear constraints is geometrically represented by the convex polygon, whereas the objective function to be optimized is shown by the grayscale values. The two figures show a single set of constraints but two different objective functions. You have usually some freedom in selecting an objective function in these problems, it makes a difference to the results, you may find different metamers of the solution.

space if equalities are involved). You also have a linear functional. The objective is to find a point of the aforementioned volume at which the linear functional assumes a maximum (or minimum, it makes no essential difference). It is geometrically evident that the generic solution is a vertex of the boundary of the convex volume. Thus you need only to select one of a finite number of contenders.

However, typically the dimensionality is so high this is not a viable option. Very effective algorithms have been constructed that find the vertex in reasonable time, though. The importance of this derives from the fact that so many problems can be approximated easily via linear constraints, if you can add as many as you want. After all, any nonlinear constraint can be arbitrarily well approximated with a quilt of locally linear constraints.

Colorimetry offers many examples. For instance, although the spectrum cone is all but linear, you can approximate it to any accuracy as the intersection of half-spaces (linear constraints). The only thing you need to do is sample at a finite number of cone generators. A similar remark applies to the Schrödinger color solid.

The colorimetric problems are exactly of the type that fit linear programming because the geometrical entities that play a role are either linear or are convex sets. If you approximate the problem by discretization of the wavelength or photon energy domain (in most cases the problems are already posed in that form) you are dealing with a finite number of linear constraints. In almost all cases you will be able to characterize the solution as the maximum of some linear functional too. In such cases linear programming is likely to be tool of choice. Environments like Mathematica allow you to solve them in a single line of code. The effort is in setting up the constraints, and typically that will be simple enough.

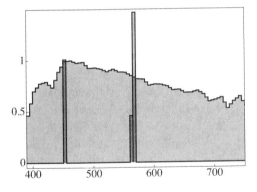 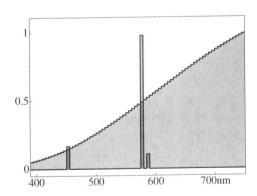

Figure 8.15 At left, the daylight spectrum, and at right, the incandescent beam with a real metamer found by straightforward linear programming. (The peaks have been scaled down relative to the smooth curve by a factor of ten to improve readability of the figure.)

The results are often surprising but perhaps less desirable. Figure 8.15 shows a fairly typical example. The illuminant colors were entered in a linear programming algorithm. Only constraints were nonnegativity of the radiant power density and the color. I minimized the total radiant power. As you see, the linear programming algorithms tend to select solutions that consist of a small number (less or equal than three) of monochromatic components. These are true metamers of the illuminants and they indeed real beams, but the solutions will rarely be useful. Some control is possibly via the function to be extremized, or through the introduction of additional constraints. For instance, you might start to think of linear constraints that might favor more smooth solutions, and so forth.

8.3.4   Iterative projection on convex sets

Suppose I have any number of convex sets and my task is to find a point in their common intersection if such a point exists. The "method of iterative projection on convex sets" allows you to solve such problems effectively and simply. Since many problems in colorimetry can be cast into this form, this algorithm is a very useful tool.

Now suppose you know how to find the point in any convex set that is closest to some given point. Then you may start with an arbitrary point and project it on all of convex sets (in arbitrary sequence). If the result is the same point that you started with, then that point is a solution. If it is not, you repeat the procedure until you have reached a fixed point. If you do, then that's a solution. If it takes forever, it is likely that a solution doesn't exist. That's all. This method is attractive, because all you need to do is implement methods to find a nearest point of a convex set. Since you can always ensure that the convex sets you deal with are planes, half-spaces, spheres, and so on, it really easy to get into business. I find this method attractive because of its simplicity. It solves a host of problems.

In figure 8.16 I show a simple example. A real beam with the color of the daylight color is sought for, starting with some arbitrary guess. In this case there are two convex sets on which to project. One is the positive part of the space of beams. Here the projection is done by simply clipping negative coordinates. The other convex set is the metameric suite of daylight. It is an affine subspace of the space of beams, the black space translated to the daylight color. To set up the projection is an elementary exercise. The projections are each written in just a few lines of code. You iterate in some arbitrary order (I used alternately projection upon the one and on the other set) and check how close the daylight color is approximated. Convergence is obtained in less than a dozen iterations. Starting with a random initial guess (a very rough, noisy spectrum) you obtain a noisy result, starting with a uniform spectrum a smooth solution. Either solution is an accurate metamer of daylight. With this method the initial guess gives you quite a bit of control over the nature of the solution.

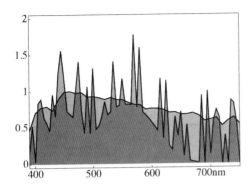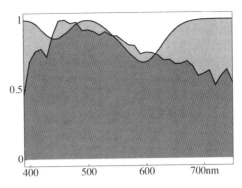

Figure 8.16 A real beam with the color of the daylight spectrum was sought via iterative projection on convex set. Left: Initial beam random with zero mean; Right: Initial beam constant and positive.

8.4   The Difference the Scalar Product Makes

In the literature, one often sees Wyscecki's hypothesis, which means the notion that any beam can conceptually be split in a fundamental component, which is "causally effective," and a black component to which you are blind. It is at least strongly suggested that this split is *unique* for a given beam. This is *wrong*. The nature of the split depends on your choice of inner product. This renders Wyscecki's hypothesis rather less attractive, of course. The "fundamental component" is—at least to some degree—up to you.

In this section I give an example that illustrates this. Thus far I have seen no examples in the literature.

I will consider a description of spectra in terms of radiant spectral density on wavelength basis and an inner product defined as

$$\mathbf{a} \cdot \mathbf{b} = \int_0^\infty a(\lambda)b(\lambda)\,\mathrm{d}\lambda,$$

and an alternative description of spectra in terms of photon number density on photon energy basis with an inner product defined as

$$\mathbf{a} \cdot \mathbf{b} = \int_0^\infty a(\varepsilon)b(\varepsilon)\,\frac{\mathrm{d}\varepsilon}{\varepsilon}.$$

where $\varepsilon = \hbar\omega$ ($\hbar = h/2\pi$ the reduced Planck's constant, $\omega = 2\pi\nu = 2\pi c/\lambda$, c the speed of light) is the photon energy for a photon of angular frequency ω, or wavelength λ. Notice that $a(\lambda)\,\mathrm{d}\lambda$ is a radiant power, whereas $a(\varepsilon)\,\mathrm{d}\varepsilon$ is a photon number density.

These two description are quite different. The first is the conventional description. The second description is for a number of reasons the more attractive one. In this section I will simply look at the differences.

In order to be able to convert the color matching functions from the conventional system to the photon-based system, it is most convenient to refer them to monochromatic primaries of unit radiant power.

The spectrum of daylight looks quite different in the two descriptions of course. This is illustrated in figure 8.17. Remember that $\varepsilon = hc/\lambda$, thus the scales run in opposite directions. Apart from this the shapes of the curves are rather different.

The color matching functions also have to be converted (figure 8.18). This is fairly trivial, and not very much changes. At this point you are all set to use the new metric.

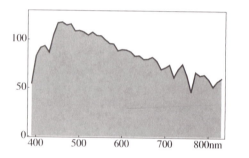 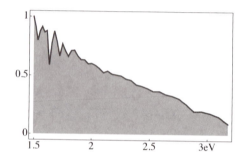

Figure 8.17 The daylight spectrum in the two descriptions.

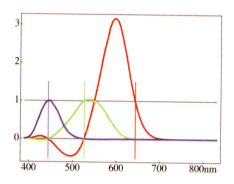 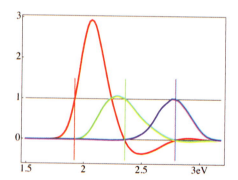

Figure 8.18 At left, the CMFs for the conventional, and at right the CMFs for the photon description. Notice that, at the locations of the wavelengths of the primaries, each time one color matching function is one, whereas the other two are zero.

The first thing to do is to find Cohen's matrix-R in the two descriptions. I show the pattern of matrix coefficients in figure 8.19. Apart from the different orientation the patterns look rather similar. In order to enable a better comparison, I plot the row sums and diagonal elements in figures 8.20 and 8.21.

Next you find the fundamental spectrum and the black component of daylight in the two descriptions (figures 8.22 and 8.23). Notice that it is not at all easy to compare these distributions by eye. You really have to put them in the same framework in order to be able to compare them.

Notice that, except for inessential differences such as the fact that the axes run into different direction, and so forth, these results look indeed very similar. This was to be expected, but it is difficult to decide by eye whether they lead to the same decomposition of fundamental and black components or not. Are these decompositions the same? In order to find out, you must show that the fundamental component in one description has no black component in the other description.

In order to show this more effectively, I look for the black components in the photon description of the basis vectors of an orthonormal basis of the fundamental space of the wavelength description (figure 8.24). Notice that these black components look rather simple, they are "only just black" with no more than three zero crossings.

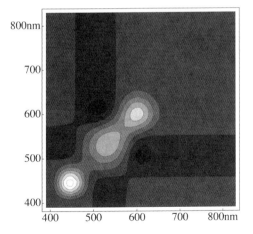 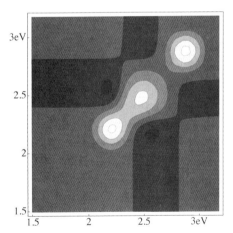

Figure 8.19 Cohen's matrix-R in the two descriptions.

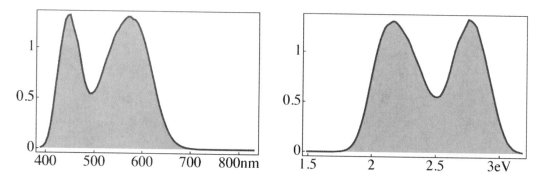

Figure 8.20 Row sums in Cohen's matrix-R in the two descriptions.

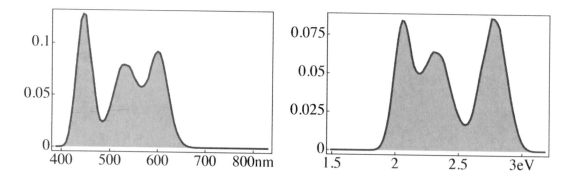

Figure 8.21 Diagonal of Cohen's matrix-R in the two descriptions.

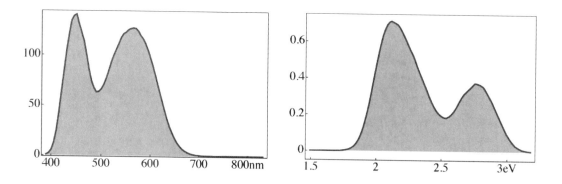

Figure 8.22 The fundamental component of the daylight spectrum in the two descriptions.

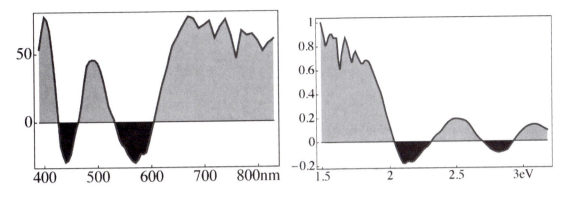

Figure 8.23 The black component of the daylight spectrum in the two descriptions.

This example shows convincingly that the choice of scalar product indeed matters. Wyscecki's hypothesis comes true only if you once and for all decide on a particular scalar product. No spectrum has a unique fundamental (causally effective) partper se. The conclusion is that the two descriptions apparently clash with each other. Yet it is equally clear that the question of which one is right makes no sense. Both descriptions are very reasonable and clearly made in good faith.

What happens here? Clearly the black space has to be the same in either case, for you cannot render a beam visibly by merely changing your definitions. Then how can it be that

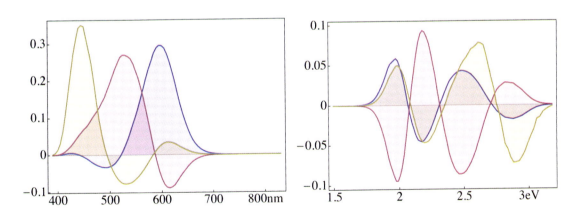

Figure 8.24 Left: orthonormal basis vectors that span the fundamental space in the wavelength description. Right: their black components in terms of the photon description.

Further examples of the importance of the inner product

It is convenient to study the importance of the choice of an inner product in the context of algebraic models. I do this for the case of the discrete model in appendix H.2. An especially simple example is explored in appendix H.3.

---

the fundamental spaces are apparently different? Many professionals with whom I have discussed this matter were puzzled to the extreme and actually doubted the computations. Yet there is nothing mysterious going on here.

The reason is that when the fundamental space is indeed orthogonal to the black space in a given description, it not necessarily orthogonal to the same black space in another description. The very notion of "orthogonality" derives from the inner product; if you change that, then the orthogonal decomposition of the space of beams changes, too. (In appendix H.3 I show a simple, low-dimensional example that should clarify things if you are still puzzled.) A subspace that is orthogonal to the black space in one description need not be orthogonal to the black space in another description. Thus although the black space is indeed an invariant, the "fundamental space" follows your (arbitrary) definitions. Thus "fundamental space" is really a misnomer. If there is anything "fundamental," it surely is the black space. The black space is well defined, even if you do not define a scalar product at all.

Thus the fact that the scalar product is typically *implicitly assumed* in much of the literature should be deeply troubling to you. If you use the method of Cohen's matrix-R, then you should be specific about the precise inner product you are going to use. Some definitions evidently make more sense than others. It is important to consider the options and make a rational choice. The fact that the choice of inner product is up to you is actually an advantage; you can use it to adapt the tools to the problem as well as possible.

8.1 Inner products

8.1.1 Inner product geometry

With the inner product (at least preliminarily) in place, you need to introduce some extensions to the linear algebra as you practiced it so far. First of all, consider the basic properties of the inner product:

- the inner product maps pairs of vectors to the real numbers;

- the inner product is symmetric in its arguments:

$$\mathbf{p} \cdot \mathbf{q} = \mathbf{q} \cdot \mathbf{p};$$

- the inner product is linear in its first argument (and thus—by symmetry—also in its second argument):

$$(\alpha\mathbf{u} + \beta\mathbf{v}) \cdot \mathbf{w} = \alpha\mathbf{u} \cdot \mathbf{w} + \beta\mathbf{v} \cdot \mathbf{w};$$

- the inner product of a vector with itself is nonnegative:

$$\mathbf{p} \cdot \mathbf{p} \geq 0;$$

- equality obtains only for the empty beam:

$$\mathbf{a} \cdot \mathbf{a} = 0 \Rightarrow \mathbf{a} = \mathbf{0}.$$

This is usually summarized as:

> *The inner product is a bilinear, symmetric function of pairs of vectors to the reals.*

The definition I gave clearly is an inner product in this sense.

Geometrically, I think of the inner product as a *slot machine*. The machine has two slots, both accept only vectors, the machine ignores their order. When both slots are filled the machine immediately pops out a number. This is a conceptually nice device, for it induces you to think about the state of the machine if you fill only one of its slots. You obtain something like

$$\mathbf{a} \cdot ?.$$

This is also a machine, a *single* slot machine, waiting for a vector and in response popping out a number. This machine is linear in its single, empty slot. Thus $\mathbf{a} \cdot ?$ is nothing but a *linear functional*. It clearly depends upon the vector $\mathbf{a}$, thus

> *the inner product induces a map that uniquely assigns linear functionals to vectors.*

The action of this linear functional on a vector $\mathbf{b}$ (say) is to produce the value of the inner product $\mathbf{a} \cdot \mathbf{b}$. This is best appreciated in the Dirac notation. Starting with the vector (ket) $|a\rangle$, I use the inner product to find its dual (bra) $\langle a|$ then the inner product $\mathbf{a} \cdot \mathbf{b}$ can be written as the bracket $\langle a|b\rangle$, that is the dual $\langle a|$ of the vector $|a\rangle$ (the dual is a linear functional) operating on the vector $|b\rangle$. Now things are set up in such a way that the bracket $\langle a|b\rangle$ (which stands for the "contraction of a vector on a dual vector" remember?) can also be understood as the *inner product* of the vector $|a\rangle$ with the vector $|b\rangle$. It is easy to get confused here with the bracket $\langle \bullet|\bullet \rangle$ doing double duty (watch it), but once you catch on to what happens, the Dirac notation is a tool you never want to be without. (Figure 8.25.)

That the inner product lets you uniquely associate linear functional with vectors is geometrically evident when you introduce the metric in your drawings. The metric is best represented by the locus

$$\langle a|a\rangle = \|a\|^2 = 1,$$

that is to say, the *unit sphere*. A sphere lets you associate a unique plane to any given point and

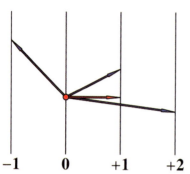

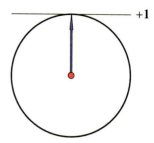

Figure 8.26 Given a unit circle, it is easy to find the unit level line of the dual of a unit vector. It is not hard to draw similar geometrical constructions for vectors that are not unit vectors (look for "construction of the polar line of a point" in geometry books), but it may be easier to use scaling. Remember that if you scale a vector to twice its length, then the level planes of its dual become spaced by *half* the original spacing.

Figure 8.25 The blue vectors make inner products of -1, 0, 1, and 2 with the (unit) red vector. In order for this to happen, it is necessary only that the tips of these vectors should be on the black lines, otherwise they may be arbitrarily inclined with respect to the red vector. The black lines are the *level lines* of the linear functional dual to the red vector. In fact, the black lines (imagine a continuum of them for all functional values) are an intuitive, geometrical representation of the dual. This is the way to think of the scalar product; it blends the "bilinear product of vectors" and the "contraction of vector on linear functional" ideas.

vice versa. Thus, given a point (a vector), you may construct the unit level plane of the associated linear functional. Given the unit plane of a linear functional, you may construct the associated point. Thus you can construct the inner product $\mathbf{a} \cdot \mathbf{b}$ geometrically by constructing the unit level plane of $\mathbf{a}$ and contracting $\mathbf{b}$ on it. (Figure 8.26.)

It is often useful to interpret the inner product geometrically in terms of vectors only, not referring to the associated linear functional of one of the vectors. This is much less attractive from a geometer's perspective, but it is the usual practice among scientists and engineers. You easily see that you obtain the inner product by pro-

jecting $\mathbf{a}$ orthogonally on the direction of $\mathbf{b}$, thus obtaining $\mathbf{a}_{\|\mathbf{b}}$ (say). Then the inner product is the product of the lengths

$$\|\mathbf{a}_{\|\mathbf{b}}\| \, \|\mathbf{b}\| = \|\mathbf{a}\| \, \|\mathbf{b}_{\|\mathbf{a}}\| = \mathbf{a} \cdot \mathbf{b}.$$

The inner product $\mathbf{a} \cdot \mathbf{b}$ is thus always smaller than the product of the lengths of the vectors. The fraction

$$\frac{\mathbf{a} \cdot \mathbf{b}}{\sqrt{(\mathbf{a} \cdot \mathbf{a})(\mathbf{b} \cdot \mathbf{b})}} = \frac{\mathbf{a} \cdot \mathbf{b}}{\|\mathbf{a}\| \, \|\mathbf{b}\|},$$

must lie in the range $[-1, +1]$ and is the cosine of the angle subtended by $\mathbf{a}$ and $\mathbf{b}$, thus

$$\mathbf{a} \cdot \mathbf{b} = \|\mathbf{a}\| \|\mathbf{b}\| \cos \angle(a, b).$$

When it is zero, the vectors are said to be *orthogonal* to each other.

Notice that in this definition of the scalar product mutual spectral exclusiveness implies orthogonality, that is to say, $\langle a|b \rangle = 0$ if the spectra of the beams do not overlap (of course, the reverse does not apply). Although this seems important, it really doesn't imply that much. The reason is that the regions of overlap/exclusion can be arbitrarily narrow and arbitrarily closely spaced. Thus two beams can be almost indistinguishable (even at fine—though finite—spectral resolution) and still be orthogonal.

This geometry puts you in a position to speak of the length of a beam, of the angle subtended by two beams, and so forth. You can speak of "orthogonal beams," and you can project one beam on another. Notice that you gain a measure of the *difference* between beams, for example, the beams **a** and **b** are a distance $\|\mathbf{a} - \mathbf{b}\|$ apart. Thus you can define a "spherical neighborhood" of a beam (figure 8.27), which is very useful in the study of the effect of small perturbations on beams. Using the inner product, you may verify that the spectral basis as defined in a previous chapter is an *orthonormal* basis, as nice as can be. (Remember that there exist infinitely many different orthonormal bases, though. The monochromatic basis is by no means unique.)

Next consider *linear operators*. The bracket

$$\langle a|\hat{\mathrm{T}}|b\rangle,$$

is usually interpreted as the operator $\hat{\mathrm{T}}$ operating on the vector $|b\rangle$ with the result

$$\hat{\mathrm{T}}|b\rangle,$$

being contracted on the dual vector (or "bra" or "linear functional") $\langle a|$. But it is also possible to read the bracket thus; the operator $\hat{\mathrm{T}}$ (but in another guise) works on the bra $|a\rangle$ and the result

$$\langle a|\hat{\mathrm{T}},$$

(a dual vector, or linear functional) operates on the vector $|b\rangle$ to produce a number. The operators $\hat{\mathrm{T}}$ working on vectors ("to the right") and on

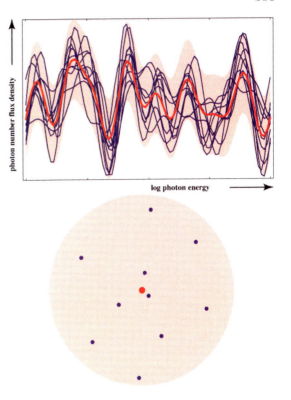

Figure 8.27 Spherical neighborhood of a spectrum (left). The red curve is the fiducial spectrum, the spherical neighborhood the pink region. The blue curves show various random points (that is, spectra) in the spherical neighborhood. Plotted in the space of beams (right), the neighborhood is a hypersphere and the perturbed spectra points. Thus you can think of spectral perturbations "geometrically."

dual vectors ("to the left") are quite distinct, operating on different spaces. The operator working to the left is known as the "adjoint" of the operator working to the right. The adjoint is defined through the condition that $\langle a|\hat{\mathrm{T}}|b\rangle$ should be the same number, irrespective of whether the operator $\hat{\mathrm{T}}$ works to the right, or its adjoint (conventionally written $\hat{T}^{\dagger}$) to the left. In inner product notation you have

$$(\hat{T}^{\dagger}\mathbf{a}) \cdot \mathbf{b} = \mathbf{a} \cdot (\hat{T}\mathbf{b}).$$

It is easy enough to show that the matrix representation of the adjoint is the transpose of the matrix of the operator itself. Notice that in the expression

$$\hat{T}^{\dagger}\mathbf{a} \cdot \mathbf{b},$$

the adjoint works on the *vector* $\mathbf{a}$, whereas in the expression

$$\langle a|\hat{T}|b\rangle,$$

the $\hat{T}$ works on a *dual vector* $\langle a|$ when it is assumed to act to the left.

8.1.2 Projections

You have seen that a "projection" $\hat{E}$ is characterized by

$$\hat{E}^2 = \hat{E}.$$

A projection that is "symmetric," meaning that

$$\hat{E}\mathbf{a} \cdot \mathbf{b} = \mathbf{a} \cdot \hat{E}\mathbf{b},$$

thus that the projection equals its adjoint $\hat{E}^{\dagger} = \hat{E}$, is called an *orthogonal projection*. Remember that a projection $\hat{E}$ and its complement $\hat{I} - \hat{E}$ project "parallel" to each other's kernels and let you split the space as a direct sum $\mathbb{U} \oplus \mathbb{V}$ (say) such that $\mathbb{U} = \ker(\hat{I} - \hat{E})$ and $\mathbb{V} = \ker\hat{E}$. If the projection is *orthogonal*, then

$$\mathbb{V} = \mathbb{U}^{\perp},$$

(and vice versa); thus the two subspaces are each other's *orthogonal complement*. Each vector in the one is orthogonal to each vector in the other. Orthogonal projections are mainly of interest because of the following reason. The projection $\hat{E}$ projects on the subspace $(\ker\hat{E})^{\perp}$. You will have

frequent occasion to use this important property (figures 8.28 and 8.2).

Consider the coordinate representation in a basis $\{|g_i\rangle\}$. You have the dual basis γ_i, such that, for any vector $|v\rangle$ you can write

$$|u\rangle = \sum_i (|g_i\rangle\langle\gamma_i|) |u\rangle = \sum_i u_i|g_i\rangle,$$

with

$$u_i = \langle\gamma_i|u\rangle.$$

Likewise, you have

$$|v\rangle = \sum_i v_i|g_i\rangle.$$

How to compute $\mathbf{u} \cdot \mathbf{v}$? Well, exploiting linearity you have

$$\begin{aligned} \langle u|v\rangle &= \sum_i u_i|g_i\rangle \sum_j v_j|g_j\rangle \\ &= \sum_{i,j} \langle g_i|g_j\rangle u_i v_j = \sum_{ij} G_{ij} u_i v_j, \end{aligned}$$

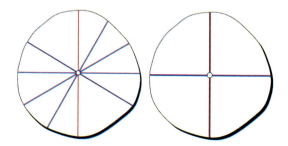

Figure 8.28 Left: There exist infinitely many "complements" (blue) to the black space (red). Each of the "blue" spaces spans the full space of beams together with the black space. Each of these spaces has equal right to the title "fundamental space;" Right: There exists a unique "orthogonal complement" (fundamental space) to the black space. But you require an inner product to construct it.

and thus the number of summands is quadratic in the dimension. You also need the coefficients

$$G_{ij} = \langle g_i | g_j \rangle,$$

of the "Grammian" of the basis $\{|g_i\rangle\}$. This is a computational nightmare.[16] This shows that the smart choice is that of an "orthonormal basis," that is, a basis such that $G_{ij} = \langle g_i | g_j \rangle = \delta_{ij}$, for in that case you simply have

$$\langle u | v \rangle = \sum_i u_i v_i,$$

a computation that is merely *linear*, instead of *quadratic*, in the dimension. When you have an inner product, then it almost never makes any sense to use any general basis. The bases you want (and the only ones I will use) are almost always nice, *orthonormal* bases. For an orthonormal basis the Grammian is simply the identity matrix, and you can forget about it. (Figure 8.29.)

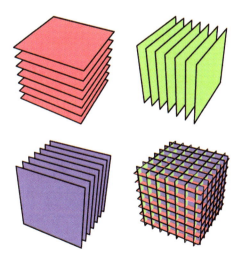

Figure 8.29 The dual basis of an orthonormal basis: 3D Cartesian graph paper.

What you need to do is to construct a nice basis. In fact, you need *two* of them, a basis for the space of beams and a basis for color space. You need to set it up such that the first three dimensions of the space of beams simply coincide with the first three dimensions of color space. Then color space is fundamental space and the remaining $\infty - 3$ dimensions span the black space.

8.1.3 The singular values decomposition (SVD)

The standard tool to do this can be found in any linear algebra toolbox and is known as singular values decomposition, or SVD. I know many people who think of the SVD as of the universal Swiss Army knife of linear algebra and as their secret weapon that is certain to get them out of any situation. I grant you, it is close to being that.[17]

What are "singular values"? The idea is that for any operator $\hat{T}$ you have that $\hat{T}^\dagger \hat{T}$ is symmetric, meaning that it is equal to its adjoint. Even better, it is a *positive* operator in the sense that

$$\hat{T}^\dagger \hat{T} \mathbf{a} \cdot \mathbf{a} = \hat{T} \mathbf{a} \cdot \hat{T} \mathbf{a} > 0,$$

for $\mathbf{a} \neq \mathbf{o}$. This again means that $\hat{T}^\dagger \hat{T}$ is guaranteed to have a complete set of nonnegative eigenvalues. These are the "singular values" of $\hat{T}$, which may itself be completely general. The operator can be written

$$\hat{T} = \hat{S} \sqrt{\hat{T}^\dagger \hat{T}},$$

(its "polar decomposition") in which $\hat{S}$ is an isometry (thus $\hat{S} \mathbf{a} \cdot \hat{S} \mathbf{b} = \mathbf{a} \cdot \mathbf{b}$ for all $\mathbf{a}$, $\mathbf{b}$). The part

$$\sqrt{\hat{T}^\dagger \hat{T}},$$

is an *anisotropic scaling*, where the scaling factors are the square roots of the singular values.

It is most intuitive to think of the polar decomposition as being a higher dimensional analog of the familiar *polar form*

$$z = \varrho \, e^{i\varphi},$$

of the complex number

$$z = x + iy.$$

(Think of z as an *operator* that turns complex numbers w into complex numbers zw; it is easily seen to be linear.) Here the positive number

$$\varrho = \sqrt{\bar{z}z},$$

is the "singular value," a scaling factor, whereas

$$e^{i\varphi},$$

is the "isometry," a rotation over φ.

The idea of the "singular values decomposition" is closely analogous to this. It is possible to find (essentially unique) orthonormal bases such that the operator $\hat{T}$ maps one basis on the other such that the lengths of corresponding basis vectors are scaled by the square roots of the singular values. In the case that a subset of singular values is zero, these dimensions "collapse" and you have effectively a projection on a subspace. It is one of the beauties of the SVD that it lets you handle such degenerate cases without any problem.

Let the SVD of an operator $\hat{T}$ yield bases $\mathbf{u}$ and $\mathbf{v}$ and singular values σ_i. The meaning of the operator is best revealed by its action on an arbitrary vector $\mathbf{a}$: you express the vector $\mathbf{a}$ in terms of the $\mathbf{v}$ basis, scale all components by the respective factors $\sqrt{\sigma_i}$, and perform an isometry (a rotation if you want) that brings the $\mathbf{v}$ basis in coincidence with the $\mathbf{u}$ basis. The "polar decomposition" of $\hat{T}$ evidently plays a major role here. The isometry is the rotation between the bases and the scaling $\sqrt{\hat{T}^\dagger \hat{T}}$ is the projection on the first basis followed by the dimension-wise

scalings by the square roots of the singular values. The geometrical meaning of all this should be evident.

Numerically the SVD is a dream. (Of course I assume that you use a professionally engineered algorithm, don't even try to roll one yourself.) There exist very efficient algorithms that essentially never fail. Writing such an algorithm is best left to the expert, but using it is nothing to be afraid of.[18]

8.2 The discrete model

I have introduced the "discrete model" on page 163.

The discrete model has the color matching matrix

$$A = \begin{pmatrix} 1 & 0 \\ \frac{2}{5} & \frac{2}{5} \\ 0 & 1 \end{pmatrix}.$$

The oblique projection is the transpose of this matrix. Using the standard expression for Cohen's matrix R, you find

$$R = \frac{1}{33} \begin{pmatrix} 29 & 10 & -4 \\ 10 & 8 & 10 \\ -4 & 10 & 29 \end{pmatrix}.$$

Try the monochromatic beam $\{1,0,0\}$. Applying Cohen's matrix R you obtain the result $\{29,10,-4\}/33$. The color of this "fundamental component" (which is seen to be a virtual spectrum) is $\{1,0\}$, the same as the color of the original beam. The difference of the two spectra is $(2/33)\{2,-5,2\}$ which is indeed in the black space. The fundamental component is indeed orthogonal to the black space, so things work out as advertised.

The SVD of the transpose of A can be written $U^T W V$, where W is a diagonal matrix containing the "singular values." Because the rank is only two, you expect two nonzero singular values. In this low-dimensional case you can do the singular values decomposition by inspection:

$$U = \frac{1}{\sqrt{2}} \begin{pmatrix} -1 & -1 \\ 1 & -1 \end{pmatrix},$$

$$W = \begin{pmatrix} \frac{1}{5}\sqrt{33} & 0 \\ 0 & 1 \end{pmatrix},$$

$$V = \frac{1}{\sqrt{2}} \begin{pmatrix} -\frac{5}{\sqrt{66}} & -2\sqrt{\frac{2}{33}} & -\frac{5}{\sqrt{66}} \\ \frac{1}{\sqrt{2}} & 0 & -\frac{1}{\sqrt{2}} \end{pmatrix}.$$

You regain Cohen's matrix R as $V^T V$, thus you may discard the matrixes W and U. The matrix U simply describes a rotation of axes in color space whereas the matrix W describes a nonisotropic scaling. Thus is is a good thing to drop W, since you get rid of a deformation. The matrix V is the perfect basis of fundamental space that lets you get a color space that is an "honest" glimpse of the space of beams, unlike the original color matching matrix, which yields a deformed view.

The pseudoinverse is

$$\frac{1}{33} \begin{pmatrix} 29 & -4 \\ 10 & 10 \\ -4 & 29 \end{pmatrix}.$$

The pseudoinverse, when applied to the color $\{1,0\}$, yields the spectrum $\{29,10,-4\}/33$, which is just the fundamental component of $\{1,0,0\}$ (which has the color $\{1,0\}$) that I found earlier by way of Cohen's matrix R. Thus there is a one-to-one correspondence between colors and fundamental spectra.

In figure 8.30 I show the colors computed for a set of tenthousand random beams. The beams were drawn from a uniform distribution over the unit sphere of the space of beams. (Thus many beams were virtual. That makes no difference here, since you could easily shift the unit sphere so as to have only real colors with the same result.) As you see, the colors lie in an elliptical, not in a circular disk, and thus this color space yields a distorted view of the space of beams. From the value of the singular values you easily predict the shape of the ellipse, meaning that you might as well correct for the distortion. If you use the orthonormal basis of the space of beams that you obtain from the singular values decomposition of the transpose of the classical color matching matrix instead of the color matching

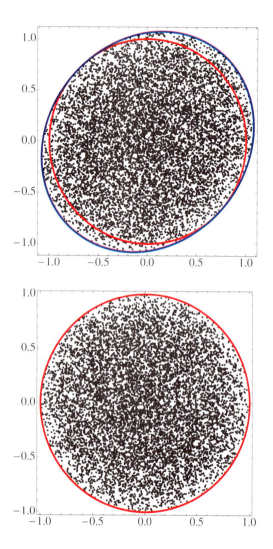

matrix itself, you automatically correct for it. In this basis the black space is orthogonal to fundamental space and the basis of fundamental space is a nice orthonormal subbasis of the space of beams. The image of the unit sphere in the space of beams is now a circle in fundamental space. Since (hyper)spheres thus map on (hyper)spheres (only of lower dimension), it is reasonable to say that this yields an undistorted view of the space of beams.

Figure 8.30 The colors of ten-thousand random beams drawn from a uniform distribution over the unit sphere of the space of beams. Upper: In the original color basis. The region is elliptical, rather than circular; Lower: In a canonical basis. The image of a sphere is a circle. The projection may be said to be without distortion.

8.3 A low-dimensional example

Consider a very simple model (even simpler than the discrete model) of a two-dimensional space of beams and a one-dimensional space of colors, with the simplest color matching matrix possible, that is

$$A = \begin{pmatrix} 1 \\ 0 \end{pmatrix},$$

and the standard inner product. How simple can you get?

Cohen's matrix R is simply

$$R = A \left(A^T A \right)^{-1} A^T =$$

$$= \begin{pmatrix} 1 \\ 0 \end{pmatrix} \left(\begin{pmatrix} 1 & 0 \end{pmatrix} \begin{pmatrix} 1 \\ 0 \end{pmatrix} \right)^{-1} \begin{pmatrix} 1 & 0 \end{pmatrix}$$

$$= \begin{pmatrix} 1 \\ 0 \end{pmatrix} \begin{pmatrix} 1 & 0 \\ 0 & 1 \end{pmatrix}^{-1} \begin{pmatrix} 1 & 0 \end{pmatrix}$$

$$= \begin{pmatrix} 1 & 0 \\ 0 & 0 \end{pmatrix},$$

which is a result you should have guessed without computation at all. Notice that the trace is one as it should be. The black space is evidently spanned by $\{0, 1\}$ and the fundamental space by $\{1, 0\}$.

The geometrical picture is presented in figure 8.31 top. For this inner product the unit circle in terms of the inner product is simply the unit circle as you have it in your head. The conjugate direction of the the black space is perpendicular to the black space in the sense of the inner product, and it is indeed the fundamental space I just found.

If I take an arbitrary spectrum (magenta dot), I can immediately write it in terms of the sum of a fundamental and a black component. A spectrum $\{x, y\}$ (say) has a fundamental component $\{x, 0\}$ (red dot) and a black component $\{0, y\}$ (black dot). This is exactly the generic picture that people entertain when they think of Cohen's theory.

In order to make things perhaps slightly more interesting I define an alternative inner product of two vectors $\mathbf{a}$ and $\mathbf{b}$:

$$\mathbf{a} \circ \mathbf{b} = \mathbf{a} \cdot \begin{pmatrix} 1 & \frac{1}{2} \\ \frac{1}{2} & 1 \end{pmatrix} \cdot \mathbf{b}.$$

Notice that "·" denotes the standard scalar product, whereas "○" denotes the alternative one. First I check whether this is indeed a valid scalar product. It is clearly bilinear and symmetric, no need to check that. You have

$$\begin{pmatrix} x \\ y \end{pmatrix} \cdot \begin{pmatrix} 1 & \frac{1}{2} \\ \frac{1}{2} & 1 \end{pmatrix} \cdot \begin{pmatrix} x \\ y \end{pmatrix} =$$

$$x^2 + xy + y^2,$$

which, in terms of polar coordinates

$$\{\varrho = \sqrt{x^2 + y^2}, \varphi = \arctan y/x\},$$

can also be written

$$\varrho^2 (1 + \frac{1}{2} \sin 2\varphi),$$

from which you see that

$$\mathbf{a} \circ \mathbf{a} \geq 0,$$

whereas

$$\mathbf{a} \circ \mathbf{a} = 0,$$

implies $\mathbf{a} = \mathbf{o}$. Thus the alternative definition indeed specifies a valid inner product.

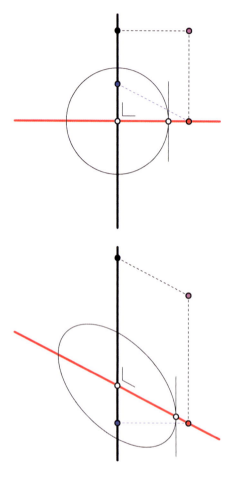

Figure 8.31 On top, the decomposition of a spectrum (magenta dot) into a fundamental (red dot) and black component (black dot) in the standard representation. In this representation the metric is given by the "unit circle" as depicted here. Notice that the fat black line is the black space, the fat red line fundamental space and that the two span the plane, that is the space of beams. At bottom, this configuration is repeated for a different inner product. The blue dots indicate the black components in one description of the "fundamental" component in the other description.

The black space cannot depend upon the (arbitrary) definition of an inner product for it specifies the possible results of colorimetric experiments. I mean *real* experiments here, and no outcome of a real observation can depend upon an arbitrary definition. The world is the way it is, no matter what you care to define. The fundamental space should be orthogonal to the black space *in the sense of the inner product*. It is not hard to see that $\{2, -1\}$ is orthogonal to the black space, for

$$\begin{pmatrix} 2 \\ -1 \end{pmatrix} \circ \begin{pmatrix} 0 \\ 1 \end{pmatrix} = 1 - 1 = 0.$$

This is an algebraic proof, but you can also do this geometrically as shown in figure 8.31 bottom. The only thing to watch here is the perhaps unusual appearance of the "unit circle", which is the "ellipse"

$$\begin{aligned} \left\| \begin{pmatrix} x \\ y \end{pmatrix} \right\|^2 &= \begin{pmatrix} x \\ y \end{pmatrix} \circ \begin{pmatrix} x \\ y \end{pmatrix} \\ &= x^2 + xy + y^2 = 1. \end{aligned}$$

The fundamental space is the conjugate of the black space (thus orthogonal to it) in the sense of this unit circle. Thus the fundamental component of a spectrum $\{x, y\}$ is $\{x, -x/2\}$, whereas the black component is $\{0, y + x/2\}$. These two vectors sum to the original spectrum.

Notice that you have obtained a formally valid decomposition of spectra in fundamental and black components in the sense of the alternative inner product. Notice also that it is evidently *different* from the decomposition I obtained for the standard inner product. "Standard" is, of course, not a good term, since either definition has equal claims to that title.

The black components in one description are also black in the other description, though their magnitudes are clearly different. But the fundamental component in one description has a black component in the other description. I have indicated that with the blue dots in the figure. The conclusion has to be that the decomposition of a spectrum in terms of a "fundamental" and a "black" component is not at all unique and essentially *arbitrary*. It is only the black space that yields an invariant description of color vision. This is something that was already evident to Maxwell. It isn't always understood that well in the modern literature, though.

8.4 The metrical differential geometry of the color solid

I already discussed the structure of the color solid in the affine setting. Given a metric you may start to explore its structure in terms of classical, metrical differential geometry of curves and surfaces. Here I will consider the Euclidian differential geometry.

8.4.1 The spectrum cone and the edge color loci

It is most convenient to parameterize the cool edge color curve by arc length. I will use the notation $\mathbf{c}(s)$ for the cool edge color curve, where $\mathbf{c}(0) = \mathbf{k}$. Then the first derivative is a unit vector $\mathbf{g}(s)$, these are the generators of the normalized spectrum cone. The derivative of the tangent is the curvature $\kappa(s)$ times the normal $\mathbf{h}(s)$ of the edge color curve. Of occasional interest is the binormal $\mathbf{b}(s) = \mathbf{g}(s) \times \mathbf{h}(s)$. The derivative of the normal is $-\kappa(s)\mathbf{h}(s) + \tau(s)\mathbf{b}(s)$, where $\tau(s)$ denotes the torsion of the curve.

The curvature $\kappa(s)$ and torsion $\tau(s)$ determine the curve up to movements. Adding the tangent and normal at the black point fully determines the cool edge color curve.

The curvature is nonnegative throughout (remember that space curves have nonnegative curvature by definition, and a curvature zero is a singularity), the torsion turns out to be positive too. The tangents form the familiar spectrum cone, while the normals form a conical surface that is much less known. It looks like a combination of two planes glued together by a "twist." (Figure 8.32.)

8.4.2 The boundary surface of the color solid

From the tangents to the cool edge color curve you obtain the coefficients E, F and G of the classical "first fundamental form $\mathrm{I}(u, v)$" of the Newtonian patch, that is, the metric ds^2 on that surface. You have, in the classical "line element" notation, that the first form is

$$ds^2 = E\,du^2 + 2F\,du dv + G\,dv^2,$$

with

$$\begin{aligned} E &= \mathbf{g}(u) \cdot \mathbf{g}(u) = 1, \\ F &= -\mathbf{g}(u) \cdot \mathbf{g}(v) = \cos\varphi_{u,v}, \\ G &= \mathbf{g}(v) \cdot \mathbf{g}(v) = 1, \end{aligned}$$

thus

$$ds^2 = u^2 - 2\cos\varphi_{u,v}\,du dv + dv^2.$$

The determinant of the first fundamental form is

$$\begin{aligned} \det \mathrm{I}(u, v) &= EG - F^2 \\ &= \|\mathbf{g}(u)\|^2\|\mathbf{g}(v)\|^2 - (\mathbf{g}(u) \cdot \mathbf{g}(v))^2 \\ &= \sin^2\varphi_{u,v}. \end{aligned}$$

The area element of the surface of the color solid is

$$dA = \sqrt{\det \mathrm{I}(u, v)} = \sin\varphi_{u,v} > 0.$$

The condition that the parameterization is nondegenerate, that is, $\det \mathrm{I}(u, v) \neq 0$, holds throughout the parameter range. The only exceptions may occur near the black and the achromatic points where $\varphi_{u,v}$ may vanish when both parameter curves tend to the same spectrum cone

Figure 8.32 The spectrum cone $\mathbf{g}(s)$ (left) and its derivative $\mathbf{h}(s)$ (right).

generator. The parameterization works near the edge color curves, though you should remember that the one-to-one relation between wavelength and arc length breaks down there: on a wavelength basis these are singular curves of the parameterization.

Because E and G are equal and constant, the parameter net has meshes that are parallelograms of constant edge length and the surface is not only a double *surface of translation*, and also a *Chebyshev surface*.

Since the patch is both a surface of translation and a Chebyshev net, a parameterization by transition loci using the arc-length parameterization of the edge color loci yields a marvelously convenient and beautiful description. (See figure 8.33).

The Chebyshev mesh angle $\varphi_{u,v}$ is easily visualized as the angle subtended by the spectrum generators at $s = u$ and $s = v$. From this you immediately see that the meshes are nearly squarish for the full colors, thus near the color solid "equator," and become increasingly elongated parallelograms when you approach one of the "poles" (the black point). This can be seen clearly in figure 8.33, where the oblique edge of the patch collapses as it maps on the black point. Near the black point the square meshes of the parameter patch degenerate into the spectrum generators.

If you think of the parameter patch as a net of rigid rods (the edges of the parameter mesh), with linkages at the vertices, then you can wrap the net snugly around the color solid simply by adjusting the angles between the rigid rods by pulling in a diagonal direction. Conversely, you can spread out the patch on the color solid in the plane, and you will obtain a regular square net.

This marvelous geometry becomes somewhat more intuitive if you consider the coarsest Chebyshev net that still makes sense (see figure 8.34). As you deform it in the manner I described above you obtain the Newtonian half of the RGB-cube,

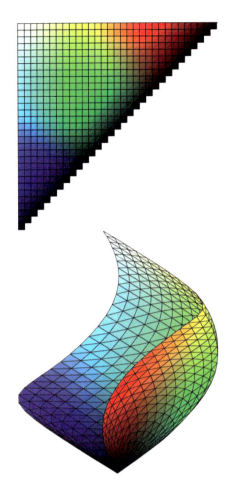

Figure 8.33 At top, the Newtonian optimal colors mapped by edge color locus arc-length, using equal intervals. The vertical and horizontal edges represents the cool and warm edge color loci respectively, the oblique edge the black point(!) At bottom, this patch is shown as the patch in 3D. Notice that the edge lengths of the checks are identical in the two figures (the patch is a Chebyshev net). All parameter lines are translated pieces of edge color loci (the patch is a surface of translation). This representation contains all of the relevant differential geometry if you think hard enough.

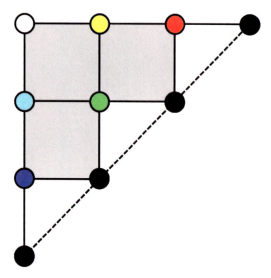

Figure 8.34 This is the coarsest Chebyshev net that makes sense. You can deform it (using the vertices as flexible links, treating the drawn edges as rigid rods) into the Newtonian half of the RGB cube. Notice that "all four black vertices" will merge into the single black vertex.

a surface patch consisting of the three faces of the cube that meet at the green vertex. There seem to be four distinct black vertices in the planar representation of the mesh, but after the deformation these merge into the single black vertex of the RGB-cube. It is not different with the fine (infinitesimal faces) mesh, the whole "diagonal" collapses into the black vertex of the color solid.

An analogous reasoning applies to the non-Newtonian patch, though here the diagonal represents the white, rather than the black vertex. In many cases I use a square parameter patch (for instance, in figure 8.35). This is convenient, but you should understand that it really falls apart into two disjunct triangular patches. Their "common diagonal" represents the black pole for one patch and the white pole for the other patch.

The surface normal is the direction of the cross-product of the two tangents, that is

$$\mathbf{n}(u, v) = -\frac{\mathbf{g}(u) \times \mathbf{g}(v)}{\sin \varphi_{u,v}}.$$

In order to investigate the curvature of the surface you need the *second-order* derivatives of the position. These second-order derivatives of position are first-order derivatives of the tangent to the edge color helix, thus equal to the curvature of the helix times the normal to the curve.

From these second-order derivatives you immediately find the coefficients of the second fundamental form $\mathrm{II}(u, v)$, which is conventionally written as

$$\mathrm{II}(u, v) = L\,\mathrm{d}u^2 + 2M\,\mathrm{d}u\mathrm{d}v + N\,\mathrm{d}v^2.$$

You find

$$
\begin{aligned}
L &= +\frac{\kappa(u)}{\sin \varphi_{u,v}}[\mathbf{h}(u), \mathbf{g}(u), \mathbf{g}(v)], \\
M &= 0, \\
N &= -\frac{\kappa(v)}{\sin \varphi_{u,v}}[\mathbf{h}(v), \mathbf{g}(u), \mathbf{g}(v)].
\end{aligned}
$$

The vanishing of the coefficient M of the mixed term is very convenient because it implies that the tangents of the parameter curves apparently coincide with the directions of principal curvature.

The determinant $\det \mathrm{II}(u, v) = LN - M^2$ is

$$
\begin{aligned}
\det \mathrm{II}(u, v) &= \\
&- \frac{\kappa(u)\kappa(v)}{\sin^2 \varphi_{u,v}} \times \\
&([\mathbf{k}(u), \mathbf{g}(u), \mathbf{g}(v)][\mathbf{k}(v), \mathbf{g}(u), \mathbf{g}(v)]).
\end{aligned}
$$

From these expressions you immediately find the Gaussian curvature as $K = \det \mathrm{II}/\det \mathrm{I}$ and the mean curvature as $2H = (L + N)/\det \mathrm{I}$ (see figure 8.35). The signs of these differential invariants are closely related to the geometry of the

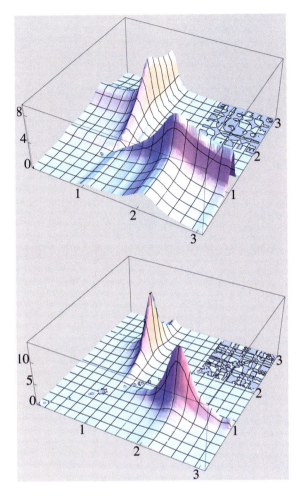

Figure 8.35 The mean curvature H (top) and the Gaussian curvature K (bottom) as a function of the arc length parameters $s_{1,2}$. The mean curvature peaks on the rounded edges of the slightly inflated RGB-cube, whereas the Gaussian curvature peaks on the rounded corners of the slightly inflated RGB-cube. Neither curvature changes sign; the surface is elliptically convex throughout.

cool edge color curve and one easily proves that the surface is elliptically convex, as expected.

The mean curvature is large in the neighborhood of the edges and vertices of the inflated RGB-cube, whereas the Gaussian curvature is large near the vertex neighborhoods. Notice especially the "green" vertex (compare figure 8.34). Otherwise both curvatures are rather minor (the surface being rather flattish). This makes good intuitive sense because most of the area corresponds to the "flat faces" of the RGB-cube. In some areas of the parameter plane, the curvatures are close to zero and the sign of the curvatures is undefined (figure 8.35). The "edges" of the RGB-cube correspond to the lines in the parameter plane $u = 1, 2$, $v = 1, 2$, and the "vertices" correspond to $\{u, v\} = \{1, 2\}$ ("green") and $\{2, 1\}$ ("purple"). The ridges and peaks in figure 8.35 are evidently "explained" through this correspondence; the skeletal structure you need to keep in mind is simply figure 8.34.

Notice that the expression

$$\kappa(u)[\mathbf{h}(u), \mathbf{g}(u), \mathbf{g}(v)]/\sin \varphi_{u,v},$$

is just the projection of $\kappa(u)\mathbf{h}(u)$ on the tangent plane at $\{u, v\}$. This enables you to gain an intuitive grasp of the way the second fundamental form varies over the surface. This is especially useful in finding mutually conjugate directions (for instance, the "meridians" and "latitude circes" of the color solid). The two directions $\{du, dv\}$ and $\{\delta u, \delta v\}$ are mutually conjugated if

$$L\, du\, \delta u + N\, dv\, \delta v = 0,$$

thus almost orthogonal when $L \approx N$. When $|L|$ is very small the conjugated direction of almost any direction becomes the u-parameter direction, when $|N|$ is very small the conjugated direction of almost any direction becomes the v-parameter direction.

The dihedral angle of the creases at the edge color loci can be numerically specified given the metric. In the affine case the dihedral angle could be constructed, but not quantified. Figure 8.36 illustrates the construction. In interpreting this figure you should remember that the planes you see in the figure are oppositely oriented. The "inner" crease angle is always obtuse.

Figure 8.36 The dihedral angle of the crease at the edge color locus equals the dihedral angle subtended by the planes spanned by the spectrum limit generators and the generator at the corresponding wavelength. (The two planes are oppositely oriented.)

Exercises

1. Scalar products I [CONCEPTUAL] Think of different ways to add a scalar product to the space of beams. Does the definition make sense in terms of physics? Of colorimetry? Can you do with less than a scalar product, for example, a semimetric? What will change in the exposition of this chapter?

2. Scalar products II [NOT HARD, BUT INVOLVES SERIOUS CALCULATION] Specifically, work out the metrical colorimetry, using the "naïve" scalar product, in terms of radiative power on wavelength basis and in terms of photon number density on photon energy basis. Do the same for a logarithmically scaled photon energy scale. To what scalar product does the "naïve" choice reduce in these cases? What are the differences? Which choice do you think preferable and for what reasons?

3. Cohen's Matrix-R [NUMERICAL] Construct Cohen's Matrix-R from the color matching functions using the explicit formula. Check its rank and trace. Study it symmetry properties (for instance, transpose, square), row sum and diagonal. Select any three (distinct) rows or columns and reconstruct the matrix from these.

4. Metamers [EASY EXERCISE BUT INSTRUCTIVE] Take two beams of different colors and find their fundamental and black components. (Check the colors of these components!) Take the fundamental component of one beam and add the black component of the other beam to it (if necessary scale it such as to obtain a real beam). The result is a metamer of the beam that yielded the fundamental spectrum. Check the colors and compare the spectra. This is an easy way to construct metamers that look good.

5. Singular values decomposition [ELABORATE] Use singular values decomposition to construct Cohen's Matrix-R and compare with the result of the direct formula. From the singular values decomposition you also obtain an orthonormal basis. Use this to plot undistorted "true" views of the spectrum cone, the color solid and Schopenhauer's RGB crate. How badly distorted is the conventional CIE representation?

6. Find a beam that yields a given color   [EASY] Construct the pseudoinverse. Now find beams that would yield a given color. Check this. Notice that the beams are often virtual.

7. Finding a real beam with the same color as a given virtual beam [COMPUTATIONAL] Use iterative projection on convex sets to implement this. You need to project on the spectrum cone (the intersection of many halfspaces) and on the affine plane parallel to the black space that contains the fundamental spectrum of the given beam. Study the rate of convergence.

8. Fairman's generalization of Cohen's matrix-R   [FOR MATHEMATICIANS] Investigate the geometrical meaning of Fairman's generalization of Cohen's matrix-R: Matrix-S is defined as $S = B(A^T B)^{-1} A^T$, where B replaces the color matching matrix A with another color matching matrix. First prove this (easy if you use the SVD representations). Does it amount to an alternative choice of inner product as has been suggested to me on some occasions? If so, is it a general way to do so?

9. An orthonormal basis of monochromatic beams   [TAKES COMPUTER HACKING] Show (by numerical procedure) that it is possible to construct an orthonormal basis of monochromatic beams and that this basis is unique. How do the wavelengths of these primaries compare to those of the so called the "prime colors" (450nm, 540nm, 605nm)?[19]

10. The concept of "saturation"   [CONCEPTUAL, ELABORATE] By now I have discussed a variety of measures for what might informally be called "saturation": The official (CIE) definition that depends upon V_λ, a measure via the cross-ratio involving the achromatic direction, Ostwald's color content, and so forth. Compare these concepts and explicitly study the various implicit assumptions and dependencies. To what extent are the definitions compatible? How do they depend on the existence of a luminance function, an achromatic direction, a white point, and so forth? The official definition does not depend on the complementary spectrum generators, the others do: Can you delete this dependency? Do you have a preference?

11. The shape of the color solid   [ELABORATE] Study the differential geometry of the boundary of the color solid numerically. Find the Gaussian and the mean curvature, the principal curvatures and the directions of principal curvature. Look for ridges and umbilical points. Construct meridians and find the latitude curves by conjugation.

12. The Chebyshev net [FOR GRAPHICS PEOPLE, ELABORATE] The notion of the surface of the color solid as a Chebyshev net is something many people have difficulty understanding, thus it calls for some fancy demo. Prepare an animation that shows the transformation of a planar patch into the surface of the color solid. Make sure the structure of the Chebyshev net is visible, that is to say, the meshes of the triangulation should show as parallelograms that only change through their internal angles. Vary the coarseness of the triangulation from the RGB-cube to a virtually smooth surface. If well done, this should call a lot attention!

13. The singular values decomposition [TRIVIAL] Solve the the overdetermined system of linear equations $x = 1$, $x = 3$, in the single variable x by writing it as a matrix equation

$$\begin{pmatrix} 1 \\ 1 \end{pmatrix} (x) = \begin{pmatrix} 1 \\ 3 \end{pmatrix},$$

and using the singular values decomposition (which you should be able to do by inspection). Do the same exercise for the underdetermined system $x = 1$, in the two variables x and y, in which case the matrix equation is

$$(1 \quad 0) \begin{pmatrix} x \\ y \end{pmatrix} = (1).$$

Now make the exercise more interesting by solving *large* over- and underdeterminent systems. Study the solutions geometrically.

Chapter Notes

1. Notice that these useful possibilities are not independent from each other.
2. You should indeed be wary of metrical colorimetry, but perhaps not too much. The major problem is that many people are entirely unaware of the prior assumptions. That is indeed bad. But if you remain constantly aware of that, there is no problem whatsoever.
3. The "inner product" also appears as "scalar product" and "dot product." Although the conceptual backgrounds are different, these are formally equivalent in the present setting.
4. The notion that considerations of physics might be of any importance here is quite alien to the colorimetric literature, though. This is not at all a desirable situation.
5. Indeed, it is quite unclear what the physical meaning of the "RMS radiant power" might be. Such a measure only rarely occurs in contexts of optics, for instance in Born and Wolf's treatment of coherence.
6. The idea is simple enough: Any two beams with the same "fundamental component" but different "black components" will look absolutely the same, whereas two beams with different "fundamental components" but the same "black component" will (in principle, that is, if the difference is significant) be discriminable. Thus the fundamental component is causally effective, whereas the black component is causally ineffective. This type of logical reasoning is quite common and may be taken as a definition of causal efficacy. Yet many students are confused when presented by the following example:

> One day Sean drinks a glass of water and bottle of whiskey and wakes up the next morning with a terrible headache. The next day he tries a glass of water with a bottle of gin, with the same outcome. Finally he tries a glass of water with a bottle of vodka but doesn't fare any better. Sean decides never to take water again on the rationale that it happens to be the common factor.

7. J. B. Cohen, "Metameric color stimuli, fundamental metamers, and Wyszecki's metameric blacks," *Am.J.Psychol.* 95, pp. 537–564 (1982); —, and W. E. Kappauf, "Color mixture and fundamental metamers: Theory, algebra, geometry, application," *Am.J.Psychol.* 98, pp. 171–259 (1985); —, "Color and color mixture: Scalar and vector fundamentals," *Color Res.Appl.* 13, pp. 5–39 (1988); —, *Visual Color and Color Mixture. The Fundamental Color Space*, (Urbana: University of Illinois Press, 2001).

8. The full derivation is:

$$A \left(A^T A \right)^{-1} A^T =$$

$$(UWV^T)^T \left((UWV^T)(UWV^T)^T \right)^{-1} (UWV^T) =$$

$$(VWU^T) \left((UWV^T)(VWU^T) \right)^{-1} (UWV^T) =$$

$$(VWU^T) \left(UW(V^T V)WU^T \right)^{-1} (UWV^T) =$$

$$(VWU^T) \left(W^{-2}U(V^T V)^{-1}U^{-1} \right) (UWV^T) =$$

$$V(V^T V)^{-1}V^T =$$

$$VV^T =$$

$$R.$$

Thus the SVD, followed by discarding W and U, serves to strip the color matching matrix A of most of the fluff that depends on your arbitrary choices. Not all of the incidental structure is removed, though, for in the SVD $A^T = UWV^T$ there is still some arbitrariness in the basis V. This will always be a basis of fundamental space, but for two color matching matrixes A and B (say) you will have $V_A^T = SV_B^T$, where S is an arbitrary isometry. This makes no difference to the matrix-R, for

$$\begin{aligned} V_A V_A^T &= (SV_B^T)^T (SV_B^T) \\ &= V_B(S^T S)V_B^T \\ &= V_B V_B^T. \end{aligned}$$

9. Indeed, R is a projection because $R^2 = A^T(AA^T)^{-1}AA^T(AA^T)^{-1}A = A^T(AA^T)^{-1}A = R$ and R is an *orthogonal* projection because $R^T = \left(A^T(AA^T)^{-1}A \right)^T = A^T(AA^T)^{-1}A = R$.
10. H. S. Fairman, "Metameric correction using parametric decomposition," *Color Res.Appl.* 12, pp. 261–265 (1987).
11. The so-called Moore-Penrose pseudoinverse $A^{(-1)}$ of a matrix A satisfies

$$\begin{aligned} AA^T A &= A, \\ A^T AA^T &= A, \\ (AA^T)^T &= AA^T, \\ (A^T A)^T &= A^T A. \end{aligned}$$

It has the useful property that $z = A^T x$ is the *least squares* solution to the equation $Az = x$. In case that $A^T A$ has an inverse the Moore-Penrose pseudoinverse of A is

$$A^{(-1)} = (A^T A)^{-1} A^T.$$

12. The classical authors are not necessarily aware of the fact that they indeed do this.

13. Let $|f_\lambda\rangle$ denote the fundamental of the monochromatic beams of unit radiant power $|m_\lambda\rangle$, thus $|f_\lambda\rangle = \hat{R}|m_\lambda\rangle$. Now you have the computation

$$\begin{aligned}\|\mathbf{f}_\lambda\|^2 &= \langle f_\lambda|f_\lambda\rangle = \langle \hat{R}|m_\lambda\rangle|\hat{R}|m_\lambda\rangle\rangle \\ &= \langle m_\lambda|\hat{R}^2|m_\lambda\rangle = (R^2)_{\lambda\lambda},\end{aligned}$$

that is the diagonal of $R^2 = R$. Thus $\|\mathbf{f}_\lambda\| = \sqrt{R_{\lambda\lambda}}$. Notice that I have used the fact that R is an orthogonal operator; thus $R^2 = R = R^T$

14. Here is how you do it. Divide the wavelength scale into bins in such a way that the sampling density is ample enough. Now divide each bin into a number of sub-bins. Assign each sub-bin to one of your objects, either an illuminant radiant power spectrum or a surface spectral reflectance. Then only objects that have been assigned the same sub-bin interact. Thus you can set up an illuminant that is different for different surfaces. Likewise you can set up surfaces to have different spectral reflectances for different illuminance. There is no constraint at all on the spectra you use, the sub-bins neatly sort everything out. This puts you in a position to set up the most bizarre demonstrations, and there is no limit to what is possible. This is the perfect setting in which to demonstrate in the most convincing manner that *color vision is impossible*. Then why does color vision work, in fact? That is because such highly structured spectra that happen to "mesh" perfectly never occur in the real world. But it is good to remember the fact that they are by no means ruled out by any basic principle of physics. Thus color vision is indeed possible *contingent* upon the generic structure of the human biotope.

15. H. R. Kang, *Computational Color Technology*, (Bellingham Washington USA: SPIEE Press, 2006).

16. In physics it is usual to amend the problem through the introduction of the reciprocal basis. This is conventional in crystallography and solid state physics for instance. It is very convenient, but I will not use it here.

17. The singular values decomposition (SVD) is extremely useful in practical linear algebra. It automatically handles standard problems such as overdetermined or underdetermined sets of linear equations, both cases that are far more common in practice than the "correct" setting (equally many equations as unknowns, independent

equations). In the overdetermined case the SVD finds you a best estimate in the least squares sense, while in the underdetermined case it annuls the projection on the space of indeterminate solutions, both instances of desired behavior.

For instance, consider the overdetermined system of linear equations

$$x = 1, \quad x = 3, \text{ in the variable } x.$$

The initial reaction of my students invariably is "this is impossible" or "how do I know which equation to ignore?" But clearly the desired solution is the *average*, thus

$$x = 2.$$

(In this case you check that

$$(x - 1)^2 + (x - 3)^2,$$

is minimal for

$$x = 2,$$

which is indeed the "solution" advertised before.)

In the case of the underdetermined system

$$x = 1, \text{ in the variables } x \text{ and } y \text{ (!)},$$

the initial reaction of my students is again "this is impossible (not even a problem!)" or "how do I get the missing equation?" But the desired solution is clearly

$$x = 1, \quad y = 0.$$

If you don't agree, then what value for y would you like most? Perhaps 123456 or -10^{137}? In the absence of any prior knowledge Occam's razor says $y = 0$. This is the solution of the least norm. (You easily check that

$$(x - 1)^2 + y^2$$

is minimal for

$$x = 1, \quad y = 0,$$

which is the "solution" mentioned earlier.) The SVD automatically produces such desirable "solutions" (of course it will yield the usual solution—no quotes here—in the regular case too) for any dimension.

18. Numerical linear algebra has become easy now that good libraries of algorithms are readily available. It makes no sense to write your own, unless you want to embark on a new career.

19. See M. H. Brill, G. D. Finlayson, P. M. Hubel, and W. A. Thornton, "Prime colors and color imaging," *Sixth Color Imaging Conference: Color Science, Systems and Applications Scottsdale, Arizona* 6, pp. 33–42 (1998).

Chapter 9

Color Vision as Coarse-grained Spectroscopy

9.1 Wyszecki's Hypothesis Without a Metric

One major advantage of a metric is that it makes Wyszecki's hypothesis come true.[1] However, it is not the only way to achieve this. The simplest way to make Wyszecki's hypothesis come true is to select *any* complement to the black space and bluntly call it "fundamental space." This allows any beam to be split in a unique manner into a "causally effective" fundamental beam and a "causally ineffective" (mere "black fluff" that should be stripped off!) black beam. Everything I said about this split and its many uses carries over without a hitch. The problem is that the choice of complement is not an obvious one. With infinitely many equivalent alternatives to pick from, how do you even start? Your predicament is worse than that of a child in a candy shop.

But wait: In a sense you *already made the choice* when you selected your primaries. In this sense the search for a "good" fundamental space simply boils down to a search for a set of desirable primaries. You just need to make up your mind as to what you would be prepared to call "desirable" in primaries.

Given an achromatic beam, the choices that might make sense are somewhat restricted. All colors dominated by the achromatic beam are contained in the Schrödinger color solid. All colors that can be described through a triple of coordinates in the range [0, 1] lie in a crate subtended by the colors of the primaries. If you want to describe as many colors as possible in this manner (and this is very desirable, as I will explain), this crate must be the crate of maximum volume inscribed within the color solid. Notice that this condition does in no way depend upon a metric, for the ratio of volumes is an affine invariant.

You have already encountered this maximum volume crate; it is the crate subtended by Schopenhauer's parts of daylight (since this sounds so nice, I take the achromatic beam to be daylight; the choice makes little difference in principle). In terms of the Schopenhauer parts of daylight any color in the crate becomes an RGB color with red (R), green (G), and blue (B) coordinates in the range of zero to one. Moreover, almost *any* color can be described in this manner, for colors within the Schrödinger color solid that happen to be exterior to the Schopenhauer crate have—even in the worst case—only minor over or undershoots, but—more importantly—are extremely rare among natural colors. The reason is that the spectral reflectances of such colors are very special (and thus rare) because they are very close to Schrödinger optimal colors. For a black body spectrum of 5700°K (similar to daylight), the Schopenhauer crate exhausts 70 percent of the volume of the color solid. "Overflows" of the coordinate values amount to less than about 17 percent in the red, 19 percent in the green, and 7 percent in the blue. A test using publicly available databases of natural spectral reflectances reveals that over 99 percent of the colors fit comfortably in the Schopenhauer crate (figure 9.1).

This is clearly a very useful choice of complement to the black space. It implements color vision as "coarse-grained spectroscopy," dividing the wavelength range into three bins (the Schopenhauer red, green, and blue parts of daylight), and it allows you to assign RGB labels to colors that are very convenient because intuitive and immediately ready to display on an RGB monitor. Such a construction removes most of the mystery of colorimetry (the CIE XYZ-coordinates and so forth) for the educated layperson. Convenient as it may be, the complement is not an *orthogonal complement* to the black space, as Cohen's fundamental

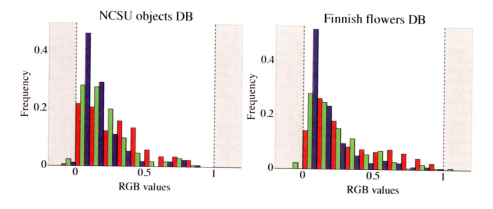

Figure 9.1 Histograms of color coordinates for two well-known databases of object spectral reflectance factors. Apparently cases of underflow or overflow are very rare. This is due to the very construction of Schopenhauer's parts of daylight.

space is. In the present setting, the very notion of "orthogonal complement" is meaningless since I have not introduced a metric to begin with. In order to construct the Schopenhauer complement, you need only affine constructions. Of course this is to be seen as a major *advantage*. After all, since there is no unique way to pick a metric, it might be wise to avoid it altogether.

9.2 The Geometry of the Schopenhauer RGB Crate

The Schopenhauer trisection[2] involves two cut loci on the wavelength axis, say λ_- and λ_+, with $\lambda_- < \lambda_+$. To make this concrete, think of $\lambda_- \approx 483$nm and $\lambda_+ \approx 567$nm for the case of average daylight. Thus you have a tripartition of the wavelength domain into

$$\chi_B : 0 < \lambda < \lambda_- \qquad \text{"blue"}$$
$$\chi_G : \lambda_- < \lambda < \lambda_+ \qquad \text{"green"}$$
$$\chi_R : \lambda > \lambda_+ \qquad \text{"red"}$$

The regions χ_B, χ_G, and χ_R do not overlap. I form the unions

$$\chi_C = \chi_G \cup \chi_B \qquad \text{"cyan"}$$
$$\chi_M = \chi_B \cup \chi_R \qquad \text{"magenta"}$$
$$\chi_Y = \chi_R \cup \chi_G \qquad \text{"yellow"}$$
$$\chi_W = \chi_R \cup \chi_G \cup \chi_B \qquad \text{"white"}$$

and add the empty region $\chi_K = \varnothing$ "black." (See figure 9.2.) Set inclusion induces a partial order on these; the Hasse diagram of inclusion is a lattice that looks much like the projection

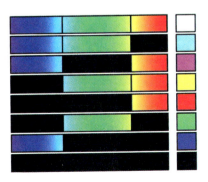

Figure 9.2 Schopenhauer cardinal colors and their spectra formed by set union and/or intersection of the various supports on the wavelength domain.

of a cube (figure 9.3). Thus the two cut loci automatically generate the cardinal colors and using overlap of spectral support as "glue" even the color circle itself.[3] The structure of the RGB cube is a very tight one, in many respects similar to that of the Schrödinger color solid itself. I devote a full chapter to the RGB cube (chapter 14).

The cut loci can be chosen in ∞^2 ways[4], though maximizing the volume of the inscribed crate turns out (by exhaustive search) to yield a unique result. This is specific for the human visual system; for instance, there exist infinitely many solutions in the case of the trichromatic Grassmann model (though one may argue that this is a singular case due to the rotational symmetry of its color solid). The idea of a maximum volume inscribed crate is very natural.[5] (A slight complication is that, because it is natural to conceive of the spectral domain as circular—just think of the spectrum of "purple"—it would be acceptable to have three cut loci: you would consider the two outer spectrum regions as one. I leave this aspect out of the discussion because the final construction turns out to be based on just two cut loci.)

The red, green, and blue colors are either edge colors or linear combinations of edge colors. Thus, red is simply $\mathbf{r} = \underline{\mathbf{e}}_{\lambda_+}$, green is $\mathbf{g} = \underline{\mathbf{e}}_{\lambda_+} - \underline{\mathbf{e}}_{\lambda_-}$, and blue is $\mathbf{b} = \mathbf{w} - \underline{\mathbf{e}}_{\lambda_-}$ (where $\mathbf{w}$ denotes white). The volume of the crate is

$$
\begin{aligned}
V &= [R, G, B] = \\
&= [\underline{\mathbf{e}}_{\lambda_+}, \underline{\mathbf{e}}_{\lambda_+} - \underline{\mathbf{e}}_{\lambda_-}, \mathbf{w} - \underline{\mathbf{e}}_{\lambda_-}] = \\
&= [\underline{\mathbf{e}}_{\lambda_+}, \mathbf{w}, \underline{\mathbf{e}}_{\lambda_-}] = \\
&= \mathbf{w} \cdot (\underline{\mathbf{e}}_{\lambda_-} \times \underline{\mathbf{e}}_{\lambda_+}).
\end{aligned}
$$

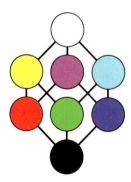

Figure 9.3 The Hasse diagram of the partial order of Schopenhauer cardinal colors as defined by the set inclusion of spectral supports. The partial order is a lattice. Notice that it looks much like the projection of a cube.

Variation of the cut loci yields the condition

$$\delta V = 0 = \mathbf{w} \cdot (\alpha \mathbf{g}_{\lambda_-} \times \underline{\mathbf{e}}_{\lambda_+} + \beta \underline{\mathbf{e}}_{\lambda_-} \times \mathbf{g}_{\lambda_+}),$$

($\mathbf{g}_{\lambda_-}$ the spectrum cone generator at wavelength λ_-) for arbitrary α, β, and thus you obtain the two simultaneous conditions

$$[\mathbf{t}_{\lambda_-}, \mathbf{w}, \mathbf{r}] = 0, \quad [\mathbf{t}_{\lambda_+}, \mathbf{w}, \mathbf{y}] = 0.$$

Here $\mathbf{t}_{\lambda_\pm}$ are the tangents of the edge color locus (which are parallel to the spectrum cone generators $\mathbf{g}_{\lambda_\pm}$). Apparently the warm edge color locus runs parallel to the WRK-plane at the yellow point and parallel to the WYK-plane at the red point. By central symmetry analog conditions apply to the cool edge color locus. (See figure 9.4.) Thus you have a beautiful, very tight, geometrical structure. In figure 9.5 I show the parts of daylight in the spectrum.

That the Schopenhauer parts of daylight representation is *really* nice becomes visible when you plot the maximum volume crate in terms of the CIE XYZ system. Notice that this pathetic squashed crate is *identical* to the unit cube. It only *looks* different because CIE XYZ color space is drawn on an inconvenient (distorted) canvas (see figure 9.6). This looks *much* better if you draw the configuration in a canonical coordinate frame constructed via the Cohen (metrical) method (figure 9.7).

However, you can do even better *without a metric* by Schopenhauer's metric (not really a "metric" of the space of beams at all), you simply promote the crate in color space to the status of "unit cube." This is a mean trick of, course!

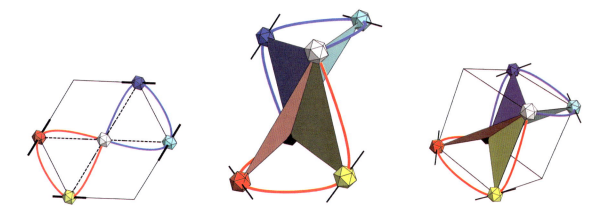

Figure 9.4 The geometry related to the maximum volume crate.

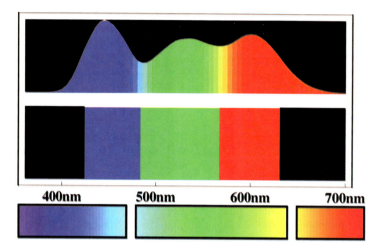

Figure 9.5 The parts of white in the spectrum. The spectral density at top yields a continuous representation. It will be discussed later. The sections have been cut at the correct locations for average daylight.

This indicates that the "natural canvas" of the parts of daylight representation is perhaps more than a mere convenience. Later in this chapter I will propose to take its metric indeed seriously.

9.3   A Special Semimetric

Once you have constructed the RGB crate by way of the Schopenhauer parts of daylight, it makes sense to adapt that as your framework. In the standard Cartesian space $\mathbb{R}^3$, you plot colors using the Schopenhauer RGB coordinates. As a result the Schopenhauer crate appears to the unwary observer as the *unit cube* of $\mathbb{R}^3$. But of course the notion of a "unit cube" implies a *Euclidian metric* for the three basis vectors have "equal length" (one!) and all angles are right angles (90°). Adapting this metric, you have arrived at the standard RGB space that I will discuss in detail in chapter 14. This has many advantages, I discuss the most important ones here.

Notice that this is a major move, for I have introduced a metric. On the other hand, it is only a metric of the "canvas" upon which I draw my picture of color space (that is $\mathbb{R}^3$). For the moment, that is exactly how I will treat this "metric." Thus it has a categorically different ontology than the metric introduced by Cohen, for that is a metric for the space of beams. Thus I am still in "affine colorimetry" here, the metric of the canvas

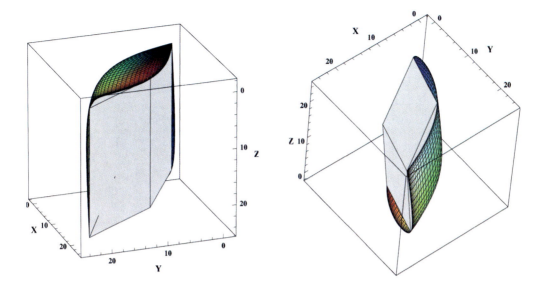

Figure 9.6 The maximum volume crate in the color solid for average daylight, plotted in the CIE XYZ color space. I don't like the deformation induced by this canvas. I prefer my RGB cube to look like the unit cube! Yet this is the "official" picture. You're free to take your pick. Apparently the bulk of colorimetrists enjoy the deformation, or at least they cheerfully live with it.

has no relevance in the space of beams. Below I will change this perspective and define a semimetric on the space of beams that is "induced" (in a natural way) by the metric of the canvas.

One obvious advantage is that RGB colors, in the sense of triples of "RGB values"—or points in the RGB cube, are *labels of colorimetric colors* (equivalence classes of physical spectra) with immediate intuitive content. Anybody with the least experience in computer graphics (any human being in the Western world nowadays?) will simply *know* what color you mean with an RGB color. Compare this with the conventional CIE XYZ or Lab coordinates: only professionals are able to understand these. RGB (or, equivalently CMY) colors actually *mean* something to most people. You can go straight from a reflectance spectrum to an RGB color using straight colorimetric computations; it is not different from doing the standard CIE computations, but the result actually has some *meaning*.

Does this mean that you can proceed from spectra to qualia? Of course not. The RGB colors are still colorimetric entities. The moment you *look* at them they turn into *seen* colors, which are qualia. This is where you enter the mental realm and colorimetry will not get you over the threshold. But exactly the same applies to geometrical entities like

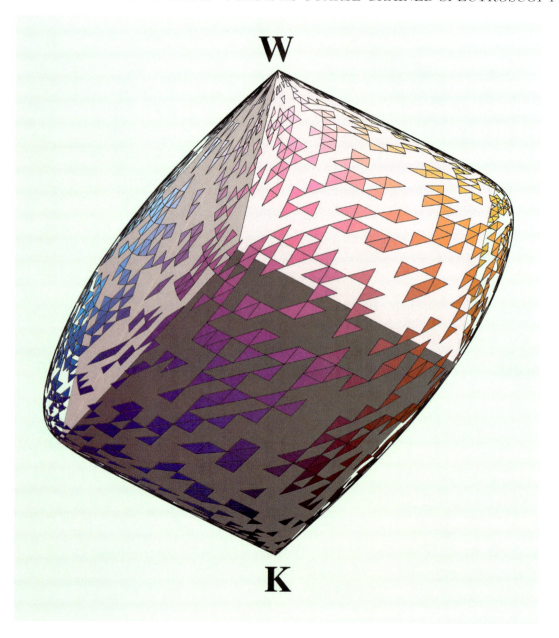

Figure 9.7 The Schopenhauer RGB crate inside the Schrödinger color solid. The surface of the color solid has been triangulated and only a fraction of the faces of the triangulation have been painted, so that the crate can be seen in its interior. It is clearly a tight fit.

lengths or angles (for instance) and nobody complains. A stretch of 10cm is not different in his respect (I'm talking ontology here) from an RGB color $\{1, 1, 0\}$ ("yellow," not YELLOW).

An RGB color from a display monitor is a physical beam, labeled with R, G, B coordinates. When this beam is colorimetrically equivalent to some given beam, the RGB coordinates may be used as a label for that beam. It is a very apt label, because the input to the brain is the same for the arbitrary beam and its RGB equivalent. In discussing the color of the beam you may substitute its label, it makes no difference to your perception. As long as you talk colorimetry there is no problem. The moment you actually *look* at the RGB color psychology kicks in. Talking about the color experience (what it is like to "have" the color in mind) takes you out of the safe haven of colorimetry, and you're out there in the cold. As long as you fully understand such issues, the use of the RGB colors as labels for beams is just fine.

The utility of "RGB labels" is enormous. It works for the overwhelming majority of natural colors: if I may generalize from the databases available on the Internet easily over 99 percent of them.

Another advantage is that you obtain a "free" metric, simply the Euclidian metric in the RGB cube. This has a great many uses; I'll discuss some in the next section. Here I merely discuss the possible extension to a metric in the space of beams. A very simple way to extend the Euclidian metric in the RGB cube to the space of beams is to define the scalar product of any two spectra as the scalar product of the corresponding RGB colors. This has some—at first blush perhaps surprising—consequences though, due to the fact that all dimensions in the black space necessarily become *isotropic* dimensions. Any vector in the direction of an isotropic dimension has length zero! Thus the metric you gain is highly *degenerate*. Yet it conforms to the usual requirements of a "metric," except for the requirement that length zero does not imply a zero vector. Such metrics are often called "semimetrics". They are useful in many settings; for instance, the semi-Euclidean plane is the natural area for one-dimensional classical mechanics (the "Galilean plane").[6]

The semimetric is very natural and useful in colorimetric settings because it effectively strips off the black fluff of beams. It is perhaps less so in physical optics, though, because it "ignores" many non-null beams. It lets you metrically relate beams only in so far as they affect your vision.

Notice that this semimetric is different in kind from Cohen's metric. It is still little more than the metric of the canvas, and there is no implied physical meaning in it. Even with this semimetric you are safely in the realm of affine colorimetry. This makes the semimetric much more attractive than Cohen's metric.

There is no pressing need to look for any *physical roots* of the semimetric as there certainly is for metrics of the Cohen type. I am at a loss to motivate Cohen's choice from first principles.

9.3.1   The semimetrical geometry of color space

Given any two beams, the vanishing of their distance in the semimetric by no means implies their physical identity. In almost all cases such points in color space at zero distance from each other will be *distinct.* This is nothing but the fact of metamerism. The semimetric implements metamerism by treating the black space dimensions as isotropic: any black beam has length zero. In the geometrical context it is natural to call such points *parallel points.* This illustrates the fact that the space of beams, fitted out with the semimetric implied by Schopenhauer's parts of daylight, has a very different geometrical structure from what most people are familiar with.

Another property (or oddity if you want) is that black beams are orthogonal to any other beam, including themselves. This implies that the complement of the black space defined by the Schopenhauer parts of white is an orthogonal complement of the black space in terms of the semimetric. This is not very useful in Cohen's sense, for you might have picked *any* complement of the black space to begin with, not just the Schopenhauer parts of daylight one, and it would still be the case!

It is indeed natural to implement Wyszecki's hypothesis in terms of the Schopenhauer parts of daylight. The "parts" are not just colors, but really beams, the short, medium, and long wavelength parts of the spectrum of the illuminant. An RGB color thus implies a unique beam, namely the parts linearly combined with the RGB coordinates as weights. In the huge majority of cases (I consider samples from the daily environment) the resulting "fundamental component" will have nonnegative radiant spectral power throughout, thus will be actually realizable as a physical beam. The related "black component" will of course be a virtual beam. I show an example in figure 9.8. Starting from the reflectance spectrum of a banana I construct its "fundamental" and "black" parts. The fundamental part has a very simple structure, but appears to approximate the physical reflectance spectrum quite well. The important point is that this simple spectrum serves to send exactly the same signals to your brain as the banana would. The black part is very structured, but this structure is irrelevant in the sense that it doesn't tickle your optic nerves at all.

Since the Schopenhauer parts of white description is so nice, it makes sense to develop metrical colorimetry in terms of its semimetric. Especially the geometry of the edge color locus and the Schrödinger color solid are of much interest. In order to do differential geometry it is most convenient to rectify the edge color loci and to substitute the arc length along the warm edge color locus, starting at zero at the black point, as a substitute for the wavelength scale. This makes good sense because the wavelength scale is essentially arbitrary as far as colorimetry is concerned.

In the map from the wavelength domain to the arc length scale, you may consider the full wavelength range from zero to infinity. Of course the range from zero throughout the

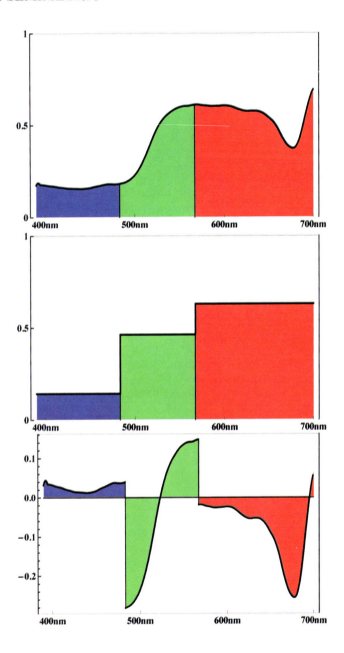

Figure 9.8 From top to bottom the spectral reflectance of a banana, its fundamental component and its black component according to the Schopenhauer parts of daylight description.

ultraviolet is mapped on zero and the range through the infrared to infinity is mapped on the full arc length of the edge curve. Thus you gain in elegance of description. In terms of the edge color arc length, the differential geometry of the color solid takes on its simplest form. This is essentially an implementation of the geometry discussed much earlier, when the (semi)metric was not yet available (section 7.7).

The arc length is also useful to measure the "sizes" of the Schopenhauer parts. I find that the sizes are almost equal, that is to say R:G:B= 1 : 1.02 : 0.97. The median locations of the parts are 445nm, 527nm and 601nm, close to frequently quoted "typical" wavelengths for these hues. The median of the visual range is 527nm. The "extent" of the visual range is 409–657nm if I take the 1–99 percent arclength limits, and 434–616nm if I take the 10–90 percent arclength limits. Thus the range in "octaves" is 0.5 for the 10–90 percent limits, and 0.68 for the 1–99 percent limits. The visual range can thus be characterized as about a half-octave band, about 150–200nm wide, centered at 527nm. (See figures 9.9, 9.10, and 9.11.)

The color matching functions for this description are shown in figure 9.12, plotted on a rectified wavelength scale. Notice that the contributions are mainly from the respective "parts", but that there certainly occurs some spillover into the bins of adjacent parts. These spillovers are the cause of (rare) "undershoots" (negative values) or "overshoots" (values exceeding one) of the RGB coordinates. The color matching functions assume a peculiar "squarish" shape in this representation.

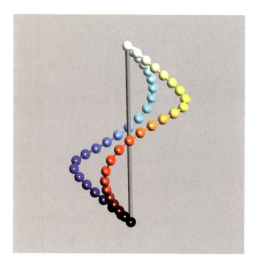

Figure 9.9 The arc length parameterization of the edge color loci.

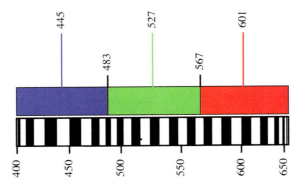

Figure 9.10 A "rectified" wavelength scale (it runs from zero to infinity!) with the Schopenhauer parts of daylight, the cut loci, and the medians of the parts. (Wavelengths in nanometers.)

The color matching functions peak at the location of Thornton's prime colors.[7] The prime colors have found many applications in color engineering, without much of a conceptual motivation. It seems to me that the Schopenhauer parts of daylight serve to demystify the issue

The undershoots and overshoots are fairly mild, as can be seen from the plot of RGB coordinate values for the semichromes (figure 9.13). Notice that these are "worst case" because binary spectra. Most real spectra are smooth, and overshoots or undershoots very rare.

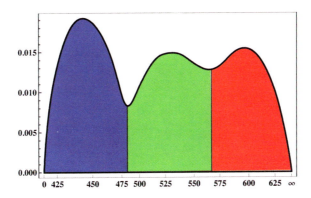

Figure 9.11 The arc length "rate with wavelength" as a function of wavelength. Notice how the rate is maximum near the centers of the parts and "hesitates" on their common boundaries. A behavior like this makes the parts appear "natural". The spectral contributions have been plotted on a rectified scale. The "parts" peak at Thornton's prime colors.

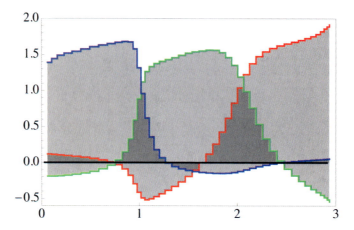

Figure 9.12 The color matching functions for the Schopenhauer parts of white primaries, plotted on a rectified scale. I have mapped the wavelength scale (0–∞ nm!) on the segment $(0, 3)$. The parts are $(0, 1)$, $(1, 2)$, and $(2, 3)$. Notice the occasional undershoots or overshoots.

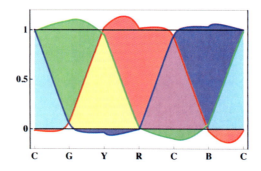

Figure 9.13 The RGB coordinates of the semichromes as a function of hue angle. Except for occasional undershoots or overshoots, the picture is almost identical to what you would have for an ideal RGB cube.

A similar behavior is observed in the Ostwald black, white, and color contents (see figure 9.14). The occasional undershoots and overshoots are mild, though.

The (semi)metric is very useful in graphics, since all renderings come out in the clearest possible manner due to the fact that the "parts" are natural and spread out evenly (that is to say, orthogonally) and with equal magnification. For example, in figure 9.15 I show some views of the spectrum cone. Notice that the representation is very similar to that obtained in the Cohen representation.[8]

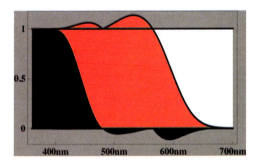 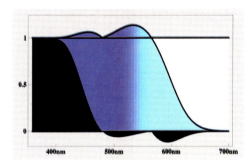

Figure 9.14 The black, color, and white content of an edge color series. Ocasional undershoots or overshoots of the black, white, and color contents occur because the edge color loci run outside of the RGB cube. At left, "black", "white," and "color" contents are indicated with black, white, and red colors; at right, the color content has been filled with the actual hues of the cool edge color series.

The "hue angle" can be defined in the (semi)metric, since the azimuth is a cylindrical coordinate system with the gray axis as cylinder axis. It is often useful as a simple measure of hue. More interesting is the rate of hue angle with wavelength, plotted on a rectified

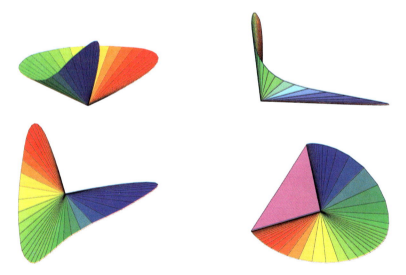

Figure 9.15 The spectrum cone as it appears in a "metrically correct" representation. In the picture at bottom right the generators have been normalized and the plane of purples added.

wavelength scale (see figure 9.16). The rate is very low, except at the boundaries of the parts, and thus the parts can be regarded as being of approximately uniform hue. The switchovers are very sudden indeed. This illustrates once more how natural Schopenhauer's notion of the parts of daylight is.

In this description the Schrödinger color solid becomes very similar to the RGB cube (that is to say, the RGB cube turns out to be an excellent approximation to the color solid). Here are some overall measures:

	Color Solid	RGB cube
Volume	1.43	1
Area	7.22	6
Edge color arc length	3.13	3
Semichrome arc length	5.89	6

Since the RGB cube is *inscribed* into the color solid, most of its measures are slightly smaller than for the corresponding entities of the color solid. Exception is the length of the semichrome locus. Here the curve on the color solid cuts corners where the polygonal arc on the cube cannot.

The Schödinger color solid thus appears as a slightly inflated copy of the RGB cube. The area of the circumscribed sphere of the RGB cube equals $3\pi = 9.42$, thus the area of the color solid is closer to that of the RGB cube than that of its circumscribed sphere (hence "slightly" inflated). (See figure 9.17.)

In figure 9.18 I show the crease angle between the Newtonian and non-Newtonian patches on the surface of the color solid, plotted as a function of the edge color arc length.

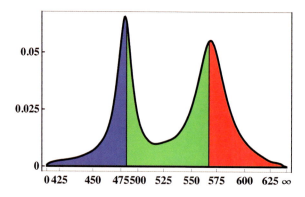

Figure 9.16 The hue angle rate with wavelength on a rectified scale. Notice that the angle "suddenly switches to the next value" as you cross a "parts boundary."

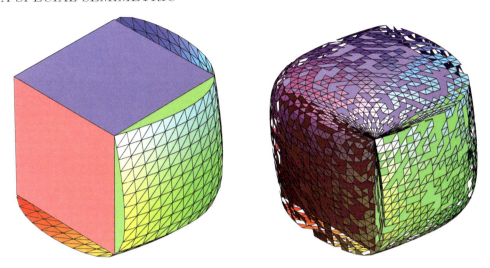

Figure 9.17 The RGB cube inside the color solid. The color solid appears as a slightly inflated cube. This is by far the most "natural" representation of the Schrödinger color solid.

The crease angle is defined such that small values imply that the crease is almost flattened out. The inner angle is always obtuse, though it is close to a right angle (a little larger) in the green and red parts, whereas the crease increases from almost flat to markedly creased in the blue part. The "crease angle" of the RGB cube is just the dihedral angle between

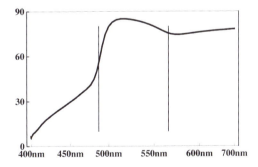

Figure 9.18 The dihedral angle subtended by the two normals at either side of the color edge color locus. The angle is plotted as a function of the edge color arc length. A small dihedral angle implies that the crease is almost flattened out, thus the inner crease angle is always obtuse. The vertical lines indicate the Schopenhauer cuts for the parts of daylight.

adjacent faces, thus 90°. Apparently the crease angle of the actual color solid surfaces approximates this rather closely in the Schopenhauer metric.

The semichrome locus is similar to the YGCBMR (periodic) edge sequence of the RGB cube. That is to say, it approximates a twisted hexagonal polygon. (See figures 9.19 and 9.20.) Seen from the direction of the gray axis it is the "color circle". Thus the color circle is actually a twisted, slightly rounded hexagonal arc.

The R, G, and B parts of daylight very naturally generate the color circle (figure 9.21). This is yet another way to conceive of the semichrome locus that relates it closely to the RGB cube.

The semichromes (or full colors) and edge colors form the skeleton of the color solid and the RGB cube alike. These loci closely approximate the corresponding edge sequences of the color cube. (See figure 9.22.) This is already implicit in the edge color locus alone, since all other loci can be constructed from it.

The geometry also carries over to the chromaticity domain. A convenient way to define the chromaticity domain is by way of the RGB triangle which is simply an equilateral triangle (thus treating R, G, and B on the same terms as barycentric coordinates). (See figure 9.23.) Notice how closely the chromatic semichromes locus approximates the RGB triangle. The spectrum locus runs far outside of the triangle, but the overwhelming majority of actual chromaticities comfortably fits with the interior of the triangle. In almost no case it would make sense to plot the spectrum locus at all; it is too far out. Notice how different this RGB triangle is from the conventional CIE xy diagram.

Chromaticity diagrams are projective planes, not affine planes. Thus you cannot expect area ratios to be meaningful. In figure 9.23 the area outside of the RGB triangle appears very appreciable as compared to the area of the interior of that triangle. If you believe your eyes here, you're in for a surprise, though. The exterior area stands for a minor part of the

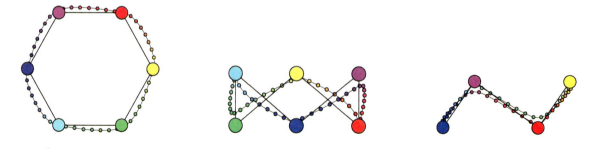

Figure 9.19 The semichrome locus with the six cardinal colors and subdivided by way of Ostwald's principle of internal symmetry. The semichrome locus closely approximates the YGCBMR twisted polygon of the RGB cube.

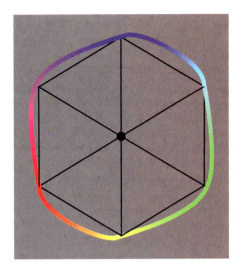

Figure 9.20 The semichrome locus, as seen from the direction of the gray axis, is the color circle. At left, it has been mensurated according to Ostwald's principle of internal symmetry. At right, a view from the direction of the gray axis allows you to compare it with the YCBMR-hexagon of the RGB cube.

volume of the color solid, whereas the interior stands for the bulk of the volume. Moreover, the exterior area maps colors that are very rare in real scenes, close to actual optimal colors. Thus the exterior area is largely irrelevant; almost all the action is in the interior part. This is a typical problem you meet in the naive perusal of chromaticity diagrams. As I explained before, I'm not a fan. The present example is yet another reason.

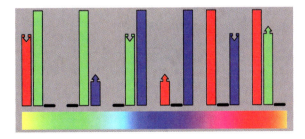

Figure 9.21 The RGB crate colors automatically generate the color circle.

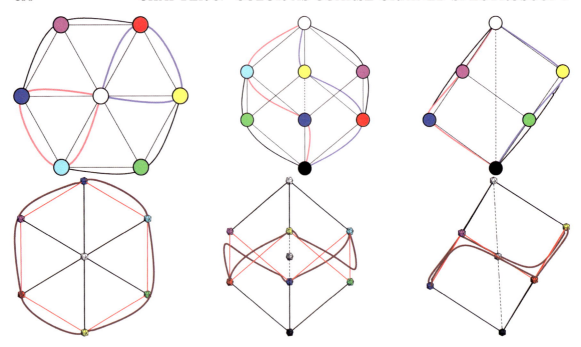

Figure 9.22 Top row: the edge color loci, and bottom row: the semichrome locus, plotted in the context of he RGB cube. Notice how closely the corresponding edge sequences of the RGB cube approximate the true loci. The Schopenhauer RGB crate is really a very good approximation to the Schrödinger color solid.

9.4   Comparison with the Cohen Metric

Since the description in terms of the Schopenhauer parts of daylight is so very different from Cohen's approach, yet yields in many respects very similar results, a comparison is in order.

In figure 9.24 I show a list plot of the coefficients of the matrix of the projector on RGB space. Notice that it is evidently not at all symmetric. This is no cause for surprise for the projection is of course not an orthogonal one, the very notion of "orthogonal projection" is undefined in the absence of metric. The matrix thus is in most respects very different from Cohen's Matrix-R.

If you define the metric as the Cartesian metric of the canvas of RGB space, you obtain a metric tensor that, *mirabile dictu*, is almost identical to Cohen's matrix R (Cohen's matrix R is the matrix of the projection on fundamental space, but doubles as the metric tensor in

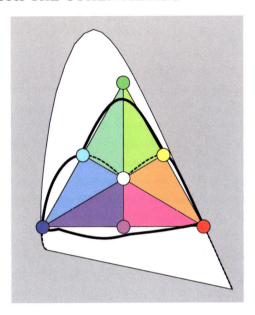

Figure 9.23 The RGB triangle with the semichromes locus (drawn thick curvilinear polygon). The edge color loci are dashed. The vertices of the triangle and the mid-points of the edges are the Schopenhauer cardinal colors. Notice how closely the triangle approximates the semichrome locus. Apparently the most vivid colors are located on the edges of the triangle.

Cohen's description). (See figure 9.25.) Indeed, the coefficients (as flat lists) have a linear correlation coefficient of more than 0.99.

The reason for this is simply that the Schopenhauer parts are mutually orthogonal in virtually *any* metric because their supports—by design—hardly overlap. This renders this description especially robust and elegant.

In the Cohen metric, the RGB basis has basis vectors that have lengths in the ratios $0.95 : 0.92 : 1$, subtending mutual angles of $89.2°$, $90.6°$, and $93.2°$; thus the RGB crate is almost cubical in the Cohen metric.

There isn't really anything that you can do in the Cohen paradigm that you cannot do in terms of the Schopenhauer parts of daylight. The advantage is of the latter is that you completely avoid the hairy issue of a metric in the space of beams. The Cohen approach is really problematic in this respect, that is to say, except when you have valid reasons (necessarily based upon considerations of physics) to select a specific metric a priori.

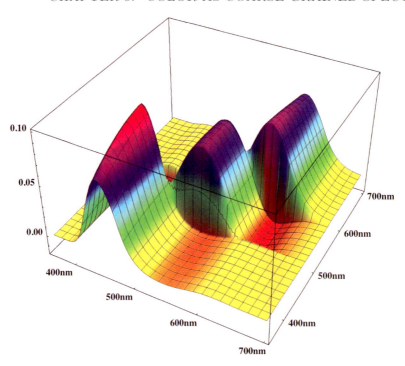

Figure 9.24 A list plot of the coefficients of the matrix of the projector on RGB space. Since the projection is not orthogonal (the very notion makes no sense in this description!), the matrix is not symmetric.

9.5   Change of Illuminant

What if you change the illuminant? After all, the Schopenhauer parts of daylight have been defined with respect to a *special* illuminant, namely daylight. One possibility would be to recompute the Schopenhauer cut loci for each illuminant. However, then you would need to redesign the visual system for each new illuminant, which evidently makes little sense. The alternative is to stick to the cut loci that are optimal for daylight and simply live with the fact that these will necessarily define a suboptimal crate for other illuminants. Since most illuminants encountered under natural conditions are fairly similar, this actually works out quite well. For each illuminant the crate may be normalized to the unit cube; this involves just a linear transformation of the RGB coordinates. The color solids, transformed in the same way, will be quite different. Such a transformation indeed transforms the crates into each other, but in no way implements an isomorphism between the gamuts inside the

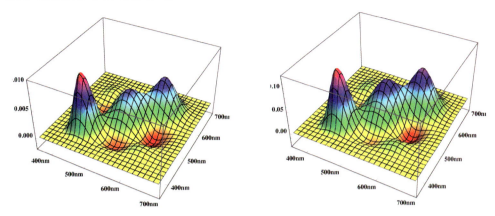

Figure 9.25 At left, a list plot of the coefficients of the metric tensor of the RGB space. At right, a list plot of the coefficients of Cohen's matrix-R. These matrixes are very similar. The flat lists of coefficients correlate better than 99 percent.

crates[9]. Two colors that are distinct under one illuminant may be indistinguishable under another and vice versa. However, in realistic cases such effects of metamerism are expected to be minor.

9.6 Opponent Color Systems

Instead of the RGB basis $\{\{1, 0, 0\}, \{0, 1, 0\}, \{0, 0, 1\}\}$ it is often convenient to refer colors to an "opponent" (overcomplete) basis. This can be done in various ways, but a representation that fits the RGB cube naturally is:

$$\mathbf{e}_{KW} = \frac{1}{\sqrt{3}}\{1, 1, 1\},$$

$$\mathbf{e}_{YB} = \frac{1}{\sqrt{6}}\{1, 1, -2\},$$

$$\mathbf{e}_{GM} = \frac{1}{\sqrt{6}}\{-1, 2, -1\},$$

$$\mathbf{e}_{RC} = \frac{1}{\sqrt{6}}\{2, -1, -1\},$$

(see figure 9.26.) This differs from Hering's original proposal in that it represents an overcomplete basis. In Hering's system, purple and turquoise play no role.

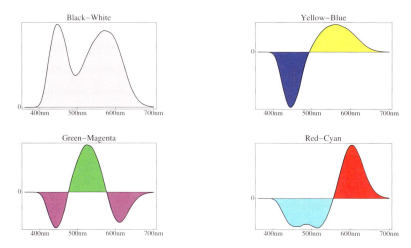

Figure 9.26 "Opponent systems": From left to right, top to bottom a KW, YB, GM, and RC system.

Except for the KW-channel the YB, GM, and RG-channels have both positive and negative coefficients. Their signals can conveniently be decoded via unsigned push-pull pairs; for instance, the YB-channel can be decoded as the ordered pair of projection on $\mathbf{e}_{\mathrm{YB}}$ and $-\mathbf{e}_{\mathrm{YB}}$, clipped so as to retain only the positive parts. Thus you need six color-opponent channels and one intensity channel. An encoding like this is easily implemented in biological "wetware," and your actual brain code is something like this. (The actual system is even more overcomplete, and the push-pull pairs occur in a variety of spatial "center-surround" organizations.) In such a system the hue is coarsely coded via a "labeled lines" scheme, although more precise analog data are available.

The system is elegant because it retains the natural hexagonal symmetry of the RGB system. This representation is fully equivalent to the RGB one, and you can easily convert from the one to the other. This holds true for the opponent color systems in general; in colorimetry proper they are mere changes of bases. Their significance relates to psychology and physiology, rather than colorimetry.

9.7   RGB Colors and Color Appearances

Isolated RGB colors can easily be associated with color appearances. For most purposes this is entirely practical, even though the rationale must remain in the dark. In order to be able to recall RGB colors to your visual imagination, it is best to think of colors as made up of white, black, and (pure) color parts. The pure colors are mixtures of pairs of analog colors, like YG, GC, CB, BM, MR, and RY. Notice that each pair involves a primary and a secondary color. These pairs are special because they can be mixed without graying the mixture: the mixture is again a pure color (semichrome).

You can visualize such decompositions in various ways. The mixtures can be mapped upon the Hasse diagram, each color involving one item from each level in the lattice. In figure 9.27 I have indicated the contributions of the parts by area. Another useful representation is as a Maxwell color top. When you would spin these disks they would appear

Figure 9.27 Some random colors. In each case the color has been split into black, white and two related colors, one primary (R, G, B), and one secondary (C, M, Y). On the left-hand side of each figure, the color has been mapped on the Hasse diagram, while on the right-hand side the colors are shown as such (center), as mixtures of pure color with gray (inner annulus) and fully analyzed into their white, black, primary, and secondary components (outer annulus).

as uniform areas of the fiducial color. As an alternative to the full split into white, black, primary, and secondary colors, you may split the fiducial color into "color" (a semichrome, the mixture of the primary and secondary components) and "gray" (the mixture of the white and black components). It is useful to train yourself to actually *see* these various components in any color you encounter.

There exist three types of optimal colors, the natural shades, the natural tints, and the natural pure colors.

The natural shades are the colors on the boundary of the RGB-cube that lie on a face containing the black point, among them the mixtures RK, GK, and BK, the "reds", "greens," and "blues". These are "natural shades" because they cannot be mixed with

---

Colorimetry in terms of Schopenhauer's parts of white

In my view, the Schopenhauer parts of white description easily yields the optimal way to do colorimetry or explain it to newcomers. In this description

— the primaries have immediate meaning as the "bins" of a low resolution spectroscope;

— the color matching functions have meaning as picking out Thornton's prime colors;

— the color coordinates have meaning as the red, green, and blue contents;

— color space has meaning as the RGB cube;

— the color solid effectively coincides with the RGB cube;

— the edge color and semichrome loci effectively coincide with the corresponding edge-sequences of the RGB cube;

— the chromaticity diagram effectively reduces to the RGB triangle;

— the colors can be immediately intuited.

In terms of the physics colorimetry reduces to mere coarse-grained spectroscopy, the spectrum being sampled in three bins. It is easy to guestimate the colors of given spectra, you do it in terms of their "fundamental spectra". In terms of graphics you obtain the most effective images, much better than the standard pictures you see in the literature, which invariably look "distorted" (sometimes beyond recognition). And yet, you're still in the realm of affine colorimetry. You have no need for Cohen's (hard to justify) metric in the space of physical beams.

I find it hard to understand why the world has decided to do colorimetry the hard way.

white (producing "tints") without graying them out, because the white will combine with the black component to form gray. Thus the natural shades have no tints among the gamut of optimal colors. The reds and blues are much more frequent than the greens in natural scenes because they involve just a single spectral transition (they are edge colors), whereas green involves a pair of such transitions.[10] The tints of the natural shades (if they involve not too much black) look quite different from the pure colors. "Pink" (a tint of red) and "baby blue" (a tint of blue) are the most common.

The natural tints are the colors on the boundary of the RGB-cube that lie on a face containing the white point, among them the mixtures CW, MW, and YW, the "cyans", "magentas," and "yellows". These are "natural tints" because they cannot be mixed with black (producing "shades") without graying them out, because the black will combine with the white component to form gray. Thus the natural tints have no shades among the gamut of optimal colors. The cyans and yellows are much more frequent than the magentas in natural scenes because they involve just a single spectral transition (they are edge colors), whereas magenta involves a pair of such transitions. The shades of the natural tints (if they involve not too much white) look quite different from the pure colors. "Olive" (a shade of cyan) and "brown" (a shade of yellow) are the most common.

The natural pure colors are the mixtures YG, GC, CB, BM, MR, and RY. Some of the 50–50 percent cases have special names, like "orange" (YR), but it is generally more intuitive to consider the environments of R, G, B, C, M, and Y. For instance the "environment of yellow" will include the YG mixtures with less than 50 percent G (the "greenish-yellows") and the YR mixtures with less than 50 percent R (the "reddish-yellows"). The "natural pure colors" allow both tints and shades among the gamut of optimal colors. Mixing in either white or black will not gray out the naturally pure colors. The RY and BC colors are most frequent because they are edge colors with a single spectral transition. The environments of G and M are less frequent, since they involve a pair of such transitions.

I will return to the topic of the structure of RGB colors in a later chapter.

9.1 Canonical angles

In order to be able to compare various choices of subspaces, you need a metric in the space of beams. This is awkward, the nicest thing about the Schopenhauer's parts of daylight is exactly that you *need not adopt such a metric*. However, it is of some interest to be able to compare subspaces and in order to make this possible I will adopt Cohen's metric in this appendix.

Suppose you have any two (linear) subspaces of the space of beams that have only the origin in common, how can you compare them? It is intuitively clear that you will need to specify some kind of angle in order to measure the difference. The way to do this is by way of "canonical angles". The notion is closely related to that of "canonical correlations" in statistics.[11]

Suppose you have $\mathbb{A}, \mathbb{B} \subset \mathbb{F}$, and let

$$\dim \mathbb{A} \geq \dim \mathbb{B} \geq 1.$$

The smallest angle $\vartheta_1(\mathbb{A}, \mathbb{B}) \in [0, \pi/2]$ between $\mathbb{A}$ and $\mathbb{B}$ is defined as

$$\cos \vartheta_1 = \max_{u \in \mathbb{A}} \max_{v \in \mathbb{B}} u^T v, \quad \|u\| = 1, \|v\| = 1.$$

You may go on and iteratively define $\dim \mathbb{B}$ "principal angles," but I will stick to the smallest angle ϑ_1 here.

A simple way to compute the angle is to use the SVD to find orthogonal bases A, B for $\mathbb{A}$, $\mathbb{B}$, then apply the SVD once more to find the largest singular value of $A^T B$, which is the cosine of the angle ϑ_1.

In order to compare primaries I first find Cohen's fundamental space $\mathbb{F}$ and the black space $\mathbb{K}$. You find that $\vartheta_1(\mathbb{F}, \mathbb{K}) = 90°$ as expected, for fundamental space is the orthogonal complement of the black space.

One set of primaries of interest are those of Stiles and Burch used in the definition of the CIE 1964 rgb color matching functions. These primaries are monochromatic beams of 645.16nm, 526.32nm and 444.44nm. Another set that I will compare is Schopenhauer's parts of daylight. Schopenhauer's parts can be adopted as primaries, and these span very different three-dimensional linear subspaces of the space of beams. The angle subtended by the two subspaces is 75.9°, and thus the spaces are indeed *very* different.

The Stiles and Burch primaries subtend an angle of 69.3° with fundamental space and an angle of 5.3° with the black space.

The parts of daylight primaries subtend an angle of 21.3° with fundamental space and an angle of 50.7° with the black space.

If you believe that Cohen's metric has any significance, then a "good" set of primaries should span a space that subtends an angle of 0° with fundamental space and an angle of 90° with the black space. Then the parts of daylight are apparently much to be preferred over the Stiles and Burch primaries. Remember that such measures depend upon the adopted metric.

Exercises

1. Find the cut loci   [HARD LABOUR] Find the cut loci for the Schopenhauer parts of daylight by exhaustive search. Repeat for the case of incandescent light. Interpret the differences.

2. Fancy graphics   [FUN] Render the important colorimetric structures in the Schopenhauer metric. It will be worth your while to devote some effort to this exercise, because you will find frequent opportunity to use the results for illustrations.

3. Differential geometry   [IF YOU KNOW DIFFERENTIAL GEOMETRY] Rectify the edge color loci and develop the full differential geometry of the color solid in terms of this parameterization. Compare the (numerical) results with the geometry of the RGB cube.

4. Relation to the Cohen metric   [RESEARCH PROBLEM] Why does the Cohen metric work so well? Is there any obvious relation between the semimetric discussed in this chapter and Cohen's metric?

5. Vision as coarse-grained spectroscopy   [CONCEPTUAL] Flesh out the notion of vision as coarse-grained spectroscopy. Reinterpret current ideas on color in philosophy. Are colors "in the world"?

6. Measures of difference   [CONCEPTUAL] Use the notion of canonical angles between linear subspaces to measure the difference between various sets of primaries in terms of the first canonical angle.

7. Sanity checks   [GET THINGS IN PERSPECTIVE] What fraction of the volume of the solid falls *outside* the Schopenhauer crate? In the representation of chromaticities where the RGB triangle appears equilateral, how much of the apparent area within the spectrum locus falls *outside* that triangle? Find databases of object spectra on the internet and find the fraction that falls outside the crate and/or outside the triangle. Compare with typical (mostly chromaticity) representations. Conclusions?

8. Another basis for the space of beams [CONCEPTUAL WITH POSSI-BLE APPLICATIONS] The Schopenhauer RGB representation can be seen as the start of a complete basis of color space. (This is best done if you use edge color arc length instead of wavelength.) One very elegant construction is to use bases of Walsh functions[12] (see figure 9.28) for each of the three (R, G, and B) spectrum regions. Use this to construct random black spectra, taking into account that the probability of finding highly articulated spectra falls off fast with the number of wriggles. (In figure 9.29 I show an example.) This construction has many potential applications.

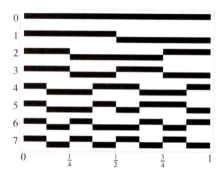

Figure 9.28 A basis of Walsh functions for a spectral region consisting of eight bins. Notice that the Walsh functions are an orthonormal basis of functions that are piecewise constant over the bins.

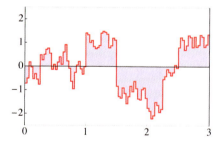

Figure 9.29 A black beam constructed for a Schopenhauer representation (blue region $(0,1)$, green region $(1,2)$, and red region $(2,3)$) where each region was subdivided in 256 bins. The probability for a certain order drops as the second power of the order, leading to fairly smooth spectra.

Chapter Notes

1. This is the standard main stream opinion, at least implicitly. In virtually no instance is the fact that Cohen's construction involves a choice of metric (scalar product) explicitly acknowledged.

2. *A part from Schopenhauer's "Über das Sehn und die Farben" in which he gives the relative sizes of the "parts" of daylight. looks like this:*

3. What I mean by "glue" is the following: Consider the relations between green, yellow, and red. The spectral support of yellow overlaps with both that of red and green, whereas red and green do not mutually overlap. Thus, yellow has to be positioned in between red and green. Likewise, cyan is positioned between green and blue, and magenta is positioned between blue and red. Thus you obtain the cyclical order red-yellow-green-cyan-blue-magenta and red again, that is to say, the *color circle.*

   Thus the partial order by set inclusion of spectral support generates the topology of the color circle for the Schopenhauer cardinal colors. This is no minor feat in view of the fact that the different topologies of the color circle $\mathbb{S}^1$ and the spectrum $\mathbb{I}^1$ have bothered scientists for centuries.

   The standard solution has been to declare the spectrum "scientific" and the color circle "artist's lore." This is seen to be bad judgment.

4. Actually in ∞^3 ways. The point is that the spectral extremes should be conceived of as mutually connected, for otherwise one could not have purples, for instance. Thus you need three cuts in order to obtain three parts. However, the third cut turns out to be the IR-UV-connection itself (as might have been expected); thus you only have to worry about the locations of the other two cuts, hence "∞^2 ways".

5. H. E. J. Neugebauer, "Über den Körper der optimalen Pigmente", *Z.f.wiss.Photographie* 36, pp. 18–24 (1937).

6. Consider the xy-plane with metric $ds^2 = dx^2$, thus with "isotropic" y-dimension. You obtain a geometry (the "dual number plane") with the group of congruences,

or "movements":

$$x' = x + t_x$$
$$y' = \gamma x + y + t_y$$

where $\{t_x, t_y\}$ is a translation and γ an isotropic rotation. The plane is different from the Euclidian plane in that the angle metric is parabolic (like the distance metric) instead of elliptic. You cannot turn around in this plane! This is a good model to obtain some intuition for isotropic dimensions.

A particularly nice text that will give you a feeling for isotropic dimensions in a geometrical setting is I. M. Yaglom, *A simple non-Euclidean geometry and its physical basis: an elementary account of Galilean geometry and the Galilean principle of relativity, (transl., A. Shenitzer)*, (New York: Springer-Verlag, 1979).

7. William A. Thornton Jr. was 1979 U.S. National Inventor of the Year, and he discovered the prime colors of human vision. He founded the company PCI (Prime Color Inc.), whose main work was the application of prime colors in illumination and display engineering (choice of phosphors, and the like). Papers by Thornton are: W. A. Thornton, "Spectral sensitivities of the normal human visual system, color-matching functions and their principles, and how and why the two sets should coincide," *Color Res.Appl.* 24, pp. 139–156 (1999); —, "A system of photometry and colorimetry based directly on the visual response," *J. Illum. Eng. Soc.* 3, pp. 99–111 (1973); —, "Toward a more accurate and extensible colorimetry. Part I. Introduction. The visual colorimeter-spectroradiometer. Experimental results," *Color Res.Appl.* 17, pp. 79–122 (1992); —, "Luminosity and color-rendering capability of white light," *J. Opt. Soc. Am.* 61, pp. 1155–1163 (1971). In an interesting paper Brill et al. offer a "maximum volume proof" which is closely connected to the Schopenhauer parts of daylight concept, M. H. Brill, G. D. Finlayson, P. M. Hubel, and W. A. Thornton, "Prime colors and color imaging," *Proc. of IS&T and SID's Sixth Color Imaging Conference: Color Science, Systems, and Applications*, pp. 31–42 (1998).

8. The Cohen representation is indeed very similar to the one obtained by way of the Schopenhauer construction. This is largely a lucky strike, since many different choices of metric would spoil the resemblance. It turns out that Cohen's choice of metric happens to be a convenient one.

9. Transformations that map RGB crates on each other are (generalized) von Kries transformations, named after Helmholtz's famous pupil von Kries. They implement a simple and fairly automatic (one needs only an estimate of the white point to implement it) approximate form of "color constancy." More about that in chapter 13.

10. A statistics of the number of "transitions" (suitably defined) in the visual range for large data bases of reflectance spectra of "natural" materials reveals that the average number of such transitions is small, such that a single transition is most frequent, higher numbers of transitions implying lower frequency of occurrence.

11. A useful introduction is S. N. Afriat, "Orthogonal and oblique projectors and the characteristics of pairs of vector spaces," *Proc. Cambridge Philos. Soc.* 53, pp. 800–816 (1957).

12. J. L. Walsh, "A closed set of normal orthogonal functions," *Am.J.Math* 45, pp. 5–24 (1923).

Chapter 10

Riemann Metrics

10.1 Affinely Invariant Distance Measures

A "distance" d_{12} between points is a nonnegative function of pairs of points such that

$d_{12} = 0$ for coincident points only,

$d_{12} = d_{21}$ ("symmetry"),

$d_{12} + d_{23} \geq d_{31}$ ("triangle inequality").

Is it possible to find such a distance function that is invariant with respect to affine transformations? (Here I consider only transformations that leave the origin fixed.) It is indeed possible, but not for the *whole* space. It can be done for a *half-space*, though, and this is good enough for colorimetry, since a half-space easily accommodates the spectrum cone, and thus all real colors.

Take $\{\mathbf{q_1}, \mathbf{q_2}, \mathbf{q_3}\}$ such that these "special" three points span a plane such that the spectrum cone is on one side of it. It makes sense to consider the geometry of *four* points, an arbitrary point $\mathbf{p}$, and the three "special" points $\{\mathbf{q_1}, \mathbf{q_2}, \mathbf{q_3}\}$. The reason is that any four points define a *tetrahedron* and thus a *volume*. Since affinities conserve the *ratio of volumes*, you can easily construct *invariants*. It makes sense to look for such an invariant that can be used as the distance between two points. Such a distance function evidently will depend on the "special" points. However, it should not depend on the order of these. Thus, you are led to construct the invariant D_{12}^2 involving two arbitrary fiducial points $\mathbf{p_1}$, $\mathbf{p_2}$ and the three special points $\{\mathbf{q_1}, \mathbf{q_2}, \mathbf{q_3}\}$:

$$D_{12}^2 = \frac{[\mathbf{p_1}\mathbf{p_2}\mathbf{q_2}\mathbf{q_3}]^2 + [\mathbf{p_1}\mathbf{p_2}\mathbf{q_3}\mathbf{q_1}]^2 + [\mathbf{p_1}\mathbf{p_2}\mathbf{q_1}\mathbf{q_2}]^2}{[\mathbf{p_1}\mathbf{q_1}\mathbf{q_2}\mathbf{q_3}]^2[\mathbf{p_2}\mathbf{q_1}\mathbf{q_2}\mathbf{q_3}]^2}.$$

Here I use [...] to denote the volume of a tetrahedron defined by a four-tuple of points. Notice the symmetry. D_{12}^2 vanishes only if $\mathbf{p_1} = \mathbf{p_2}$ and is invariant against permutations of both the special and the fiducial points. You do not yet have a distance function, though; you need to make sure the triangle inequality works out. It can be shown (do it!) that

$$d_{12} = \sinh^{-1} D_{12},$$

is indeed a distance function. By construction it is invariant against affinities.

This distance explodes as one point approaches the $\mathbf{q_1 q_2 q_3}$-plane. Apparently the metric works only for a half-space at one side of this plane. For color space this is fine, though. The *alychne* is an excellent candidate for the $\mathbf{q_1 q_2 q_3}$-plane, or you could use the plane spanned by the opponent directions (red-green and yellow-blue) of a Hering basis, for instance. Then all pairs of real beams will generate colors at finite distance from each other. The empty beam (black point) will be infinitely distant from all generic real colors, though. This gives the metric a decidedly "logarithmic" nature. I will come back to that soon.

Thus it is indeed possible to define a distance function on classical color space that is invariant against changes of the primaries (that is, affinities). The construction is due to von Schelling.[1] It can be shown that the metric treats color space as a space of constant negative curvature. It is equivalent to a certain Riemann space treated later on in this chapter; thus I will postpone further analysis of the metric here. But it is important to see that nothing keeps you from grafting a metrical structure on classical color space. I have not seen this used in the modern literature though, important as it is.

10.2   Riemann Metrics Proper

So far I have considered only the simplest structures: color space as a projection of the space of beams, either considered as a general linear space or as a Hilbert space. This makes it impossible to implement certain general psychophysical "laws" such as the Weber-Fechner law (see the following). In order to increase the flexibility, you need to loosen the straitjacket of colorimetric response reduction, something that should not be attempted lightly. After all, the colorimetric response reduction is the very backbone that keeps colorimetry upright as an objective science. To my mind, the only acceptable "generalization" of colorimetric response reduction was proposed by Erwin Schrödinger in the 1920s.[2] For reasons that I fail to understand or appreciate, Schrödinger's proposal has been largely ignored by the colorimetric community. This is such a key topic that I will introduce it in some detail.

10.3 Schrödinger's Generalization of Matches

Schrödinger's proposal was framed in the context of *heterochromatic photometry*. This is a generic example of a property that cannot be operationalized in the setting of standard colorimetric response reduction (see appendix J.2). If you have two beams of different hues, then what could "equal intensity" possibly mean? Certainly *not* that the beams are indistinguishable, for they are—by hypothesis—distinguishable by *hue*, whatever that may mean. Any "meaning" cannot be a *colorimetric* meaning, so much is evident. Clearly the notion of "heterochromatic brightness" requires a nontrivial *extension* of the colorimetric method of response reduction.

The colorimetric response reduction asks for a single judgment concerning the relation between a pair of beams:

> *The beams are or are not distinguishable.*

Schrödinger proposes to replace this by a relation between a beam (I refer to it as "fiducial beam") and a *set of* beams, singling out a certain *special member* of the set (I refer to it as "most similar beam"), as follows

> *Among all members of the set of beams the special beam is* **most similar** *to the fiducial beam.*

If the set contains a beam that is metameric to the fiducial beam, then that beam is certain to be the "most similar" one (being indistinguishable), and Schrödinger's criterion reduces to the colorimetric paradigm. Otherwise the criterion identifies a pair of beams (the fiducial and the most similar beam) relative to the given set that have no special colorimetric relation (of indistinguishability).

How does this help to define a "heterochromatic brightness match"? Consider a beam **a** and the set $\mu\mathbf{b}$ for $\mu \geq 0$, where for no value of μ a colorimetric equality $\mathbf{a}\,\square\,\mu\mathbf{b}$ pertains. Then the beam $\mu_0\mathbf{b}$ *most similar* to **a** can be *defined* (a mean trick to be sure) as "equally bright" to **a** (figure 10.1). This may make some intuitive sense, but it is certainly not colorimetry. It indeed makes intuitive sense, since the members of the set $\mu\mathbf{b}$ differ only by intensity (radiant power to be sure), which is supposed to be the "physical equivalent" of the "subjective" notion of "brightness."

The problem is that to consider the set $\mu\mathbf{b}$ as a set of colors differing only by "intensity" is an implicit *definition* of "intensity." In this sense, you *force* the result.

Notice that Schrödinger manages to avoid the subjective evaluation of the various visual qualities of a beam ("brightness," "hue," "saturation," or what have you). Thus he *almost* manages to avoid the hairy psychological issues altogether. What I like about Schrödinger's proposition is that:

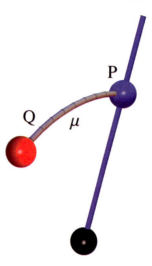

Figure 10.1 Schrödingers proposal. Let the observer indicate a point in a given set (here: the "blue" chromaticity) that is "most like" the fiducial color (here the "red color"). Since the blue colors differ only by intensity, the "most like" color might be said to be "equally bright" as the fiducial color, although it is not assumed that the observer has any concept of "brightness." (The curve μ suggests a "geodesic plumb line.")

> You indeed don't know what the observer is doing, but at least you know what the experimenter is doing.[3]

Not quite, though. Schrödinger's criterion somehow lies in a gray area between colorimetry (full response reduction, colorimetry proper) and judgment of quality ("relaxed" response reduction, not colorimetry proper). I consider this a monumental achievement, and I have no doubt that this is the best you can do. I have some doubts whether such a task is a reasonable one, though. It may well be that observers might experience difficulty in singling out *any* member of a given set as "most similar." Such problems cannot be expected with the colorimetric criterion of equality. Only in cases where the "most similar" element is actually "almost identical" can you be fairly sure that the task is a reasonable one; if the "most similar" element is still "very different," the observer may be hard put to select such an item. This is indeed what is found in practice.

Although Schrödinger's method was proposed in the setting of heterochromatic photometry, it is of much more general applicability (as Schrödinger himself was well aware). For instance, it can just as well be used to "define" such qualities as "saturation," "hue," and so on, in an operational sense.

10.3.1 Consequences

Suppose I take a fiducial color $\mathbf{a}$ and a slightly different color $\mathbf{a} + \mathbf{da}$ (figure 10.2). You may suppose that the "perceptual distance" of the colors is a smooth function of the difference $\mathbf{da}$ and can be represented with sufficient accuracy by the second order Taylor term $\langle \mathbf{da} | \hat{G}(\mathbf{a}) | \mathbf{da} \rangle$, where the operator $\hat{G}(\mathbf{a})$ (the "metric") is self adjoint. Schrödinger writes in coordinate language

$$\mathrm{d}s^2 = \sum_{i,j} G_{ij} \, \mathrm{d}a_i \, \mathrm{d}a_j,$$

where the "infinitesimal distance squared" $\mathrm{d}s^2$ denotes the "line element" in the sense of Riemann[4]. Notice that the geometrical locus $\mathrm{d}s^2 = \varepsilon^2$ for some constant ε is a triaxial ellipsoid centered on $\mathbf{a}$. The "most similar" color among the colors $\mu\mathbf{b}$ (where $\mathbf{b}$ is very similar to $\mathbf{a}$) is thus the point of tangency of the ellipsoid that touches the ray $\mu\mathbf{b}$. This again means that the colors "most like $\mathbf{a}$" in an infinitesimal neighborhood of $\mathbf{a}$ lie on a

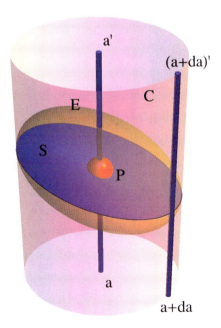

Figure 10.2 Consider a point $\mathbf{P}$ and a chromaticity $\mathbf{aa'}$. The chromaticity $(\mathbf{a+da})(\mathbf{a+da})'$ is infinitesimally close. The equidistance locus given by the ellipsoid E then defines a unique plane S (the ellipsoid touches the cylinder with generators parallel to the chromaticity along a planar ellipse) that might be regarded as an infinitesimal part of an "equibrightness surface."

planar element with an orientation that is the *conjugate plane* (figure 10.2)[5] of the direction **a** for the quadric $\mathrm{d}s^2 = \varepsilon^2$. If you perform this construction throughout color space you obtain a *field of contact planes* that may or may not be integrable.[6]   *If* the field is integrable there exists a (nonnegative real valued) "brightness function" $L(\mathbf{a})$ (say) such that $L(\mathbf{a}) = L(\mathbf{b})$ implies that **b** is "most like" **a** among all colors $\mu\mathbf{b}$ andvice versa, whereas $L(\mathbf{a}) = \|\mathbf{a}\|$ for some (arbitrarily chosen) color **a**. If the field is not integrable, then such a brightness function doesn't exist. Thus the existence of a brightness function is up to empirical investigation.

10.3.2   Just noticeable differences (JNDs)

Schrödinger's construction of a Riemann metric in color space is an improvement on a proposal by Hermann von Helmholtz from the late nineteenth century. Helmholtz simply proposed that the "perceptual metric" would be described by a Riemann line element, whereas Schrödinger improves on this by his ingenious generalization of the colorimetric response reduction, thus defining the metric in a true operational sense (already implicit in Helmholtz though not explicated).

In psychophysical practice the "line element" is *measured* through the determination of the just noticeable difference or JND. Two colors **a** and **b** are said to "differ by a just noticeable difference" if they can be just discriminated. This is operationalized by measuring the probability of detecting the difference between a beam **a** and another beam

$$\mathbf{c}(\mu) = \mathbf{a} + \mu(\mathbf{b} - \mathbf{a}),$$

for $\mu > 0$, as a function of the parameter μ. This may be done (for instance) by having an observer decide "equal" or "different" in a sequence of trials in which either the pair $\{\mathbf{a}, \mathbf{a}\}$ or the pair $\{\mathbf{a}, \mathbf{c}(\mu)\}$ is presented. For very small values of μ the probability will tend to 50 percent (pure guessing), whereas for large values of μ the probability will tend to 100 percent (the observer is dead sure). Then the "JND" can be defined as $\mu(\mathbf{b} - \mathbf{a})$ for the value of μ corresponding to a probability of 75 percent for instance.

The definition has to be a statistical one because observers turn out to give (on some sufficiently fine grained scale at least) unpredictable responses that vary from instance to instance for identical stimuli. Thus the response domain does not have a metric in the formal sense. For instance, it is quite possible to have "intransitive triples" such as $A = B$, $B = C$ and $C > A$ among your observations. The response domain is a "tolerance space" (see appendix J.1).

In practice you find that the one JND surface for a given color is close (though sometimes significantly different from) a triaxial ellipsoid. Much as you would like to have an empirical determination of a dense field of such JND surfaces, all over color space, such a task is too

Herculean to be feasible. Moreover, very significant interobserver differences and differences between laboratories exist, and JNDs for very reduced settings may be totally uninteresting in practice. Thus

> *the conventionally accepted "metrics" are for a large part based on make belief, rather than empirical fact.*

Although it is true that the perceptual metric is known in (very) rough outline, it certainly isn't in sufficient detail. Neither is this possible at all, since metrical relations depend critically upon so many details that any "complete description" is totally beyond reach. Since such a situation is not acceptable for the needs of industry, people substitute convention for facts.[7] This is of course not acceptable to the scientist, and I will regard the problem as essentially open to speculation until sufficient material is available, which is not likely to happen in the foreseeable future.

In the absence of a sufficient body of data, you make do with theoretical proposals. Helmholtz was the first who ventured a perceptual Riemann metric on a conceptual basis, and Schrödinger also proposed such a conceptual model.[8] Later in the twentieth century various modifications were proposed, most of them in bad taste from a scientific perspective (due to the overfitting of sparse and heterogeneous data). Here I will concentrate singularly on simple, conceptual models. If you are mainly interested in applications you should probably use the CIE Lab system[9], which has become something of an industry standard. Although convenient, its basis in empirical fact is not impressive, and the system has little or no scientific appeal. (Are my personal convictions showing through here?) There is a wealth of literature available to guide you here.[10] You will need a strong stomach to digest it, though.

10.3.3 Formal Metrics

Helmholtz's Metric

The metric originally proposed by Helmholtz[11] was based on the Weber-Fechner law. In the mid nineteenth century Ernst Heinrich Weber discovered that the JND for some nonnegative scalar variable (weight, loudness, brightness, and so on) is generally proportional with the average level, that is,

$$\Delta L = w\, L,$$

where w denotes some constant, a relation now generally known as Weber's law.[12] The constant w is known as the Weber fraction and is circa one percent for light intensity. Later in the nineteenth century Gustav Theodor Fechner "integrated" Weber's law such as to define a "psychophysical function" known as the Weber-Fechner law".

Fechner's idea is simple enough. An infinitesimal interval $\mathrm{d}L$ contains $\mathrm{d}L/\Delta L$ JNDs, thus the number of just noticeable levels between L_1 and L_2 is

$$n = \int_{L_1}^{L_2} \frac{\mathrm{d}L}{\Delta L} = \int_{L_1}^{L_2} \frac{1}{wL}\,\mathrm{d}L = \frac{1}{w}\log\frac{L_2}{L_1} = \Psi(L_2) - \Psi(L_1),$$

where

$$\Psi(L) = \frac{1}{w}\log\frac{L}{L_0}.$$

Thus Fechner's psychophysical law says that

> the "psychical (perceptual) magnitude" is the logarithm of the physical magnitude.

This "law" (at least in Weber's incremental formulation, I fail to see how you might verify Fechner's Psychophysical law empirically) holds empirically over several decades of radiant power for the "brightness" of beams. Whereas Fechner derived his law for a scalar variable, Helmholtz attempted to generalize this to vector magnitudes such as color.

In its simplest formulation,[13] Helmholtz's proposed "line element" is

$$\mathrm{d}s^2 = \frac{1}{w_S^2}\frac{\mathrm{d}S^2}{S^2} + \frac{1}{w_M^2}\frac{\mathrm{d}M^2}{M^2} + \frac{1}{w_L^2}\frac{\mathrm{d}L^2}{L^2},$$

where S, M, and L (for "short," "medium," and "long" wavelength "mechanisms") denote three suitable, orthogonal color coordinates. (Figure 10.3.) Thus Helmholtz simply combined three separate Weber relations in the best Pythagorean tradition. It is simple enough to show that this metric leads to a brightness function. In order to see that, note that in the space with coordinates

$$\{u, v, w\} = \{\log S^{\frac{1}{w_S}}, \log M^{\frac{1}{w_M}}, \log L^{\frac{1}{w_L}}\},$$

the line element is simply the Euclidean line element. The chromaticities (lines through the origin of color space) map to lines in the $\{1, 1, 1\}$ direction of $\{u, v, w\}$ space; thus the equibrightness surfaces are planes orthogonal to that direction. Translated to color space, this implies that the loci of equibrightness are

$$S^{\frac{1}{w_S}} M^{\frac{1}{w_M}} L^{\frac{1}{w_L}} = \Lambda,$$

for constants $\Lambda > 0$. As Helmholtz noticed to his dismay, these surfaces fail to fit the facts even *qualitatively*. (Figure 10.4.) The surfaces are curved in such a way that the

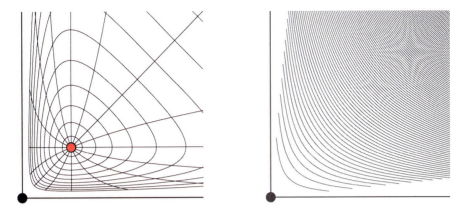

Figure 10.3 Left: Geodesics radiating out from a point and equidistance circles around a point according to the Helmholtz metric. Right: Equibrightness loci according to the Helmholtz metric.

equal mixture of two equibright colors is predicted to be *brighter* than each of its components, whereas experiment reveals that the opposite pertains. Thus Helmholtz's method to generalize the Weber-Fechner law fails miserably. The edition of Helmholtz's *Handbook of Physiological Optics* that appeared after his death was edited by his former pupil Johannes von Kries, who deleted the section on the Riemann metric from the text.

Despite the fact that the Helmholtz metric is evidently not the way to go, people have attempted to "patch the holes" via various small changes and additions to Helmholtz's basic metric. Such attempts have even found their way into modern textbooks. Needless to say, I consider such attempts ill-motivated (because bound to strand) and in generally bad taste (because you should strive to get the foundation right instead of patching holes).

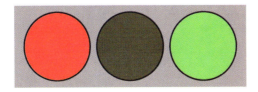

Figure 10.4 Equibrightness prediction from the Helmholtz metric.

Schrödinger's metric

Schrödinger was led to believe (due to sparse and spurious psychophysical data, known as Abney's law"[14]), that the surfaces of equibrightness should be a family of parallel planes; thus he looked for a line element that would lead to this while simultaneously respecting Weber's law for any individual color. He mentions that a remark by Wolfgang Pauli (like Schrödinger of quantum-mechanical fame, but this happened *before* the formulation of quantum mechanics) induced him to set

$$ds^2 = \frac{1}{4(\sigma S + \mu M + \lambda L)} \left(\sigma \frac{dS^2}{S} + \mu \frac{dM^2}{M} + \lambda \frac{dL^2}{L} \right).$$

This is indeed a shrewd choice, because

$$\frac{1}{4} \left(\sigma \frac{dS^2}{S} + \mu \frac{dM^2}{M} + \lambda \frac{dL^2}{L} \right) = (d\sqrt{\sigma S})^2 + (d\sqrt{\mu M})^2 + (d\sqrt{\lambda L})^2.$$

Since the $\{\sqrt{\sigma S}, \sqrt{\sigma M}, \sqrt{\sigma L}\}$-space is Euclidean, the equibrightness surfaces are spheres with center at the origin in this space. Translated back to $\{S, M, L\}$ space these become the planes

$$\sigma S + \mu M + \lambda L = \Lambda,$$

for constants $\Lambda > 0$. The factor

$$\frac{1}{\sigma S + \mu M + \lambda L},$$

doesn't change this, but serves to implement Weber's law for the multiples of any given color.

Because Schrödinger's proposal implements the *luminance* functional, it is something of an engineer's dream.

Too bad the dream doesn't fit reality. (Figures 10.5 and 10.6.) Whereas Schrödinger predicts that the equal mixture of two equibright colors should be just as bright as either of these, the empirical fact is that such a mixture appears *less* bright than each of the components.

The Fechner metric

It is not difficult to set up a line element that fits the data at least qualitatively. Consider

$$ds^2 = \frac{dS^2 + dM^2 + dL^2}{(S + M + L)^2},$$

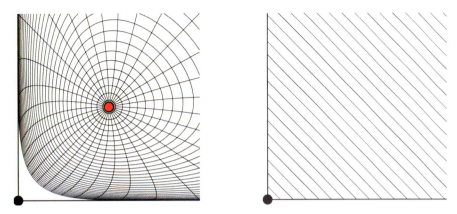

Figure 10.5 Left: Geodesics radiating out from a point and equidistance circles around a point according to the Schrödinger metric. Right: Equibrightness loci according to the Schrödinger metric.

where I have refrained from inserting various inessential constants for the sake of simplicity.[15] Clearly Weber's law is implemented for any single color. Since the JND ellipsoids are degenerated into spheres, the conjugate planes are orthogonal to rays through the origin, and you can "integrate by eye" to arrive at the fact that the surfaces of equal brightness must be concentric spheres about the black point. Thus the "brightness function" is

$$\sqrt{S^2 + M^2 + L^2} = \Lambda,$$

for constant $\Lambda > 0$. For this brightness function you predict that the equal mixture of two equibright colors would appear *less bright* than either color separately, which is (at least qualitatively) the empirical finding. (Figures 10.3–10.8.)

Notice that this metric is exactly of the type that I introduced at the beginning of this chapter (you want to check this!). This makes it a very serious proposition in the context

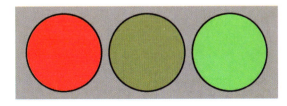

Figure 10.6 Equibrightness prediction from the Schrödinger metric.

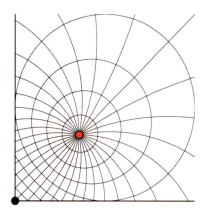 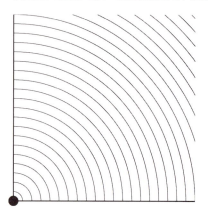

Figure 10.7 Left: Geodesics radiating out from a point and equidistance circles around a point according to the Fechner metric. Right: Equibrightness loci according to the Fechner metric.

of colorimetry. If any of the proposed analytical line elements comes somewhat close to the truth, then this one would be my bet. The special $(\mathbf{q_1 q_2 q_3})$ plane is

$$L + M + S = 0,$$

in this case. I consider this metric to be the most natural generalization of the (one-dimensional) Weber-Fechner law to three-dimensional color space. In a way, it appears to be exactly what Helmholtz was after, more than a century ago.

What I consider the best data set on heterochromatic brightness available today is that of Guth.[16] Guth proposes to describe his (very extensive) data phenomenologically by way of his "vector model," by which he means a brightness function (except for inessential constants) $\sqrt{S^2 + M^2 + L^2}$. Thus this line element implements the Guth vector model.[17]

I will refer to this metric as the "Fechner metric," not because Gustav Fechner ever considered it, but because I consider it to be the most natural generalization of Fechner's

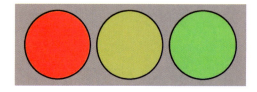

Figure 10.8 Equibrightness prediction from the "Guth vector model".

psychophysical law to higher dimensions. This makes the metric a very elegant one, and in my view it certainly *deserves* to fit the facts!

Whether it actually does is a matter of confrontation with the empirical facts. This is not an easy matter, because the psychophysical data are scarce to begin with, and hard to interpret and estimate on their true value, because usually reported in terms of models or conceptual thought patterns. In almost all cases this makes it very difficult to decide on the empirical "truth" of the matter.

10.4 The Actual Equibrightness Function

Is there such a thing as an "equibrightness function" in the first place? Not necessarily, for even if the Riemann space model works, there still exist many metrics for which such a global brightness function cannot be constructed. Although there always exist *local* pieces of equibrightness surface (the planes conjugated to the rays through the origin according to the metric), it is not at all guaranteed that these will globally "mesh," that is to say, the field of local planar elements need not be an integrable. In fact, generically it will not. Schrödinger was fully aware of that. For the metrics I discussed here, such equibrightness functions do exist. (Compared in figure 10.9.)

When you are confronted with two beams of various hues and are asked whether these appear equally bright, then the answer is immediate if one is much more intense than the other. However, if the beams are at least somewhat matched, the answer is surprisingly difficult to decide on.[18] Many people consider the question an impossible one to deal with and consider it senseless.

It is easier if the beams are close in hue; thus the idea to consider sequences of similar hues that range between rather different limits is a natural one. I show some examples in figure 10.10. At least for me, the top row seems to pass through a dip, that is to say, the patches grow gradually darker, then become lighter again, as I go from left to right through the series. Likewise, the series at the bottom seems to pass through a bump. I'm not so sure which series is the most "uniform" one, though.

Figure 10.9 The brightness functions of the Helmholtz, Schrödinger and Fechner metrics compared.

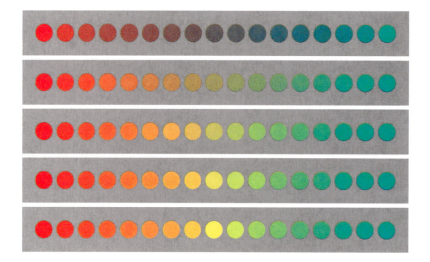

Figure 10.10 In which of these continuous series do successive patches appear equally bright to you?

These series are located in the red-green plane, and I can plot them that way (figure 10.11). The linear path is the prediction for the Schrödinger metric (and Grassmann's fourth law). At least for me it is uneven, for it clearly takes a dip at the equal mixture, which looks much too dark. The Helmholtz metric would be even worse than that, so is totally out. These are subjective judgments, you should look for yourself.

The pure addition of red and green yields a nice, very bright yellow, which clearly outshines the red and green. If there is an equibright part it must be somewhere in between. The circular arc looks at least reasonable to me. It is the Guth vector model. Thus the Fechner metric seems to be at least in the ball park with respect to the equibrightness issue. It is clearly to be preferred over the other contenders.

It is possible to make some observations concerning "brightness" that clearly bear upon the topic, yet have no relation to any metrical description. I allude to the topic[19] of "spectral domination" illustrated in figure 10.12.

If you subscribe to the belief that

>   *A beam cannot be made to "look darker" by adding any other beam to it.*

the partial order of domination becomes very relevant to the brightness issue. I myself strongly believe in this statement, and I am not aware of any experiment that has ever demonstrated its violation. However, I grant you that the issue is an empirical one and

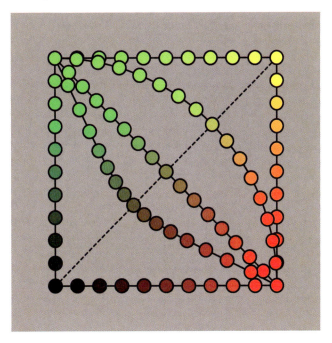

Figure 10.11 Some paths between red and green.

that it is perhaps the case that psychophysicists have spent insufficient effort in trying to construct counterexamples (I have tried myself but have not been able to construct any). The statement has at least a ring of likely truth in it. I have never seen it quoted as "such and so's conjecture" or "such and so's law," so in all likelihood most psychophysicists knowingly or unknowingly accept the statement on face value. If you do, too, then

— if **a** dominates **b**, then **b** does not look lighter than **a**;

— if **a** is dominated by **b**, then **a** does not look lighter than **b**;

— if **a** neither dominates **b** nor is dominated by **b**, then their brightness relation remains undecided.

In the latter case you may speculate that the question "is **a** lighter or darker than **b**?" would make no sense to the human observer, something that is perhaps not totally strange in view of the known facts.

If this were all there is, then you should perhaps discard the notion of equibrightness and replace it with the notion of domination. Since dominance is not a full, but only a

Figure 10.12 A medium gray color, some colors it dominates, and some colors that dominate it.

partial order, the psychophysical investigation of dominance (of *patches*, thus different from the spectral dominance of *beams*) is not a simple matter.[20]

10.5   Consequences of a Metric

Schrödinger suggests additional applications of the Riemann metric that are of much interest. For instance, he suggests that loci of equal hue in the chromaticity diagram may be given by geodesics through the achromatic point in equibrightness surfaces. This is indeed of considerable interest, since such equal hue loci are empirically found to be significantly curved, a fact for which colorimetry (by design) cannot account. It is not so much that the prediction is wrong as that there simply cannot *be* such a prediction. The idea is reasonable, since the geodesics through the achromatic points are the "shortest routes" to the (fully saturated) spectral colors. Such loci may be constructed for each of the three metrics introduced here. You find qualitatively very similar (though quantitatively different) results, in at least semiquantitative agreement with the empirical facts. The possible importance of the geodesics was already hinted at by Helmholtz.[21]

A prediction that does allow you to differentiate between these metrics is the so-called Bezold-Brücke effect. According to von Bezold and Brücke, the hue of a beam changes with intensity in a very characteristic manner. You may account for this by way of Schrödinger's proposal. Pick any fiducial color **a** and a set of colors of equal brightness different from that of **a**. The most similar color among this set might be considered the color of "the same hue as **a**" of that different brightness. The most similar color is geometrically constructed by dropping a geodesic plumb line on the equibrightness surface from the fiducial point. You easily show that the chromaticities (lines through the black point) are geodesics for the Helmholtz and Schrödinger metrics, from which it is immediately evident that these metrics do not predict any "Bezold-Brücke shift." For the Fechner metric, things work out differently, for here the geodesics are circles in planes perpendicular to the

$$S + M + L = 0$$

plane, with centers in that plane. Thus the chromaticities are not geodesics, and you predict a finite Bezold-Brücke shift. This shift is largely in saturation, brightening a beam leading to less saturated hue. At least to my eye this is indeed the major effect. However, there appears to be no formal empirical evidence on this issue in the literature. You also find a hue shift and the metric can be "tuned" such as to "predict" ("fit" might perhaps be the more apt term here) the empirical function quantitatively. Thus this metric, at least qualitatively and semiquantitatively, can be made to represent the (unfortunately sparse) empirical facts. This is interesting, because the metric implies that the geometry is that of Lobatchevsky's "hyperbolic space."

Because the metric describes a space of constant negative curvature (a "homogeneous space") it is possible to derive an explicit expression for the distance between any two points,[22] even points that are remote, instead of the infinitesimally close points implied in the Riemannian "line element." This distance function is equivalent to the affinely invariant distance function that I introduced at the beginning of this chapter. This makes the present metric rather special. It is indeed the closest to any higher-dimensional form of Fechner's (one-dimensional) "psychophysical function" that I know of, a three-dimensional version of the Weber-Fechner law. It is rewarding to see how elegant geometry, Schrödinger's proposal for the relaxation of colorimetric reduction and Guth's empirical "vector model" converge on the same structure.

10.6   Ontological Status of the Schrödinger Conjecture

Schrödinger's conjecture leads to predictions of perceptual judgments and thus can be viewed as a kind of *model* of vision. Of course, there exists a plethora of contenders. There are many models that purport to predict the same perceptual judgments as Schrödinger's conjecture allows you to do. However, virtually all of these models are of an ontologically completely different kind.

The bulk of models attempts to model the *physiology* at some level of coarse graining. That is to say, the various variables that occur in the model are interpreted as levels of electrochemical activity in human nervous tissue, the various subparts that make up the model are interpreted as various types of neurons (for example, the various receptor types in the retina) or neuron assemblies, and the various connections that define the structure of the model are interpreted as actual connections of the neural nexus that is the human brain. This is fine as far as it goes, that is to say, as long as the model is regarded as a model of the physiology. However, such modest claims are very rare. The typical modeler claims that the model predicts aspects of *visual perception*. The various variables that occur in the model are then immediately compared with *quale* of the luminous variety. For instance, there might occur a variable labeled "brightness." If this variable assumed a certain value, then the prediction would be that the human observer would experience a corresponding brightness. This is the most common structure of such models.

To my mind, these models are in utterly bad taste because they silently assume some psychophysical correspondence of a—scientifically speaking—*magical* kind. In essence, such models[23] are of the photometer type, or inner screen type". In the photometer paradigm the model contains a photosensitive device connected to various signal processing boxes. At some level in the model the output from such a box is reified as a luminous experience "brightness." In the inner screen paradigm, the model produces some image, that is to say, a two or even three ("inner stage") line array of intensities or vector quantities that might be likened to a *movie screen*. The model is supposed to predict the experience an observer would have when actually looking at a movie screen featuring these images. People sometimes speak (not you or me, of course) of the *homunculus*,[24] a little man in the head, enjoying the movie. Needless to say, such primitive notions should be repulsive to the scientist. Yet most existent models are exactly of this kind.

Because nobody has even the faintest idea (I have to be honest here) how physiological processes give rise to experiences, models of the physiology are very limited in their predictive power when applied to perceptual judgments.

The only thing that might be inferred with certainty is that if the model removes a certain structure of the optical input, then the prediction is that the experiences cannot causally depend on that structure. Or, in other words, models of the physiology let you predict *selective blindnesses* of various kinds. This is a very powerful tool and certainly not to be despised. Colorimetry is a key example of the power of this principle. But you have to accept the limitations of such theories. For instance, you cannot say anything concerning subjective brightness for the very reason that it is not an instance of such a selective blindness.

In order to model such perceptual judgments as relative subjective brightness, you need an ontologically distinct type of models. Let me call them "cognitive models" as distinct from "physiological models." Schrödinger's conjecture clearly falls in this class of cognitive models.

Not only are the physiological models of doubtful status, but they are also rather less powerful than cognitive models can be. For instance, Schrödinger's conjecture leads to the prediction that anything like a "brightness function" need not but might exist. It is an empirical matter whether such a function can be said to exist or not. Most of the data would suggest that it does not, but that observers can rank certain colors with respect to brightness whereas they are unable to do so for other colors. Such a possibility does not even arise in the physiological models, though. If there is some neural activity labeled "brightness," then you thereby define a brightness function. There is no way to account for the fact that observers cannot reach a conclusion as to the relative brightness of some colors. After all, all they have to do is compare the value of the "brightness variable" in their brains. The class of physiological models (the bulk of the literature), quite apart from their doubtful ontological status, appears too restricted to be worth serious consideration. Such models should be used for the purpose of understanding the neurophysiology, but not the understanding of perception.

10.7 Isochromes

I mentioned Schrödinger's speculations concerning the nexus of geodesics. A geodesic can be regarded as the path that connects two locations in such a way that you need a minimum number of JND's to go from one end to the other. Thus all points on the geodesic are in Schrödinger's sense "most like" each other under constraint of the end points. This makes the geodesics very special and it is indeed reasonable to speculate upon their correlation with "eye measure" in many settings. These ideas can already be found with Helmholtz, thirty years before Schrödinger.[25]

One interesting idea due to Schrödinger concerns the so-called Abney shift.[26] Abney observed that admixture of white (desaturation) to a spectrum color not only lets the color appear paler (a "pastel color" or "tint"), it also tends to shift the apparent hue of the color. This happens at equibrightness, for instance (approximate interpolations between the achromatic beam and monochromatic beams of the same luminance). This means that the Helmholtz coordinates of a color have less immediate appeal, because constant dominant wavelength fails to imply invariant apparent hue, as is often suggested. It must be said that the latter suggestion is not altogether ludicrous though, for the Abney shift is not large. It is also the case that the Abney shift is rather ill defined. It is a pure "eye measure" entity, much as brightness is. Thus apparent hue has no place in lower colorimetry. (Figures 10.13–10.15.)

Figure 10.13 Do these patches look the same hue to you? Of course their saturations are different. They are mixtures of a single red with various amounts of white. Whether the "hues" look different is a matter of "eye measure".

It is not easy to define "apparent hue" formally. Again, it was Schrödinger who first came up with a sensible notion. Schrödinger's idea was to consider the surface of constant brightness through the achromatic point and construct geodesics in this surface (using the induced metric from color space in which the surface is embedded) fanning out from the white point. Such radial geodesics would then be the loci of constant apparent hue as defined by the metric. Why? Well, it certainly looks like an intuitively reasonable notion.

Although utterly *ad hoc*, such a definition makes some intuitive sense and certainly extends the realm of higher colorimetry appreciably. Of course, it is very hard to justify, or rather, *impossible* to justify from a colorimetric perspective. In the case of heterochromatic photometry it seems possible to come up with a more or less (depending upon your scientific conscience) convincing argument, but in the case of the Abney shift (like in the case of the so called Bezold-Brücke shift to be discussed later) it seems at first blush more problematic.

> *In a way, the Schrödinger definition can be regarded as a definition of the subjective attribute "hue."*

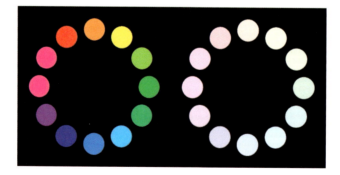

Figure 10.14 The Abney effect. When you desaturate colors, they change hue. In the desaturated hue circle they appear to crowd toward the blue. To me they do, at least; decide for yourself! In order to be in a position to make such a judgment, you should present the stimuli on a linearized CRT monitor in a darkened room.

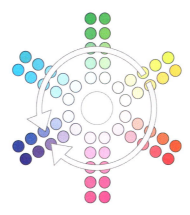

Figure 10.15 Another shot at the Abney effect. When you desaturate colors, they change hue. Yellow is an unstable, blue a stable fixed point. Red and green are not invariant against admixture of white.

When you have to find a patch constrained to move on a locus of equal colorimetric saturation (say a spectrum color) and luminance that looks "most like" a monochromatic patch of the same luminance, it is perhaps reasonable to say that what the closest patch has in common with the monochromatic patch must be hue. As a definition, it is as good as any other. After all the operational definition of "hue" is up for grabs. (The colorimetric definition is dominant wavelength and essentially meaningless.) When observers are able to perform such a task reliably it may be said to define *something*, and why not call that something "hue"? Once you grant the definition you're all set to start empirical and theoretical work on the topic of hue. As such, it is a constructive move.

In retrospect, I think Schrödinger's approach is a reasonable one for which exists no real alternative. I view it as an interesting way to boot up a field of experimental psychology. The next thing to do is to see how far the idea carries you. The approach is certainly much more attractive than what is usually done (even today) where people seem to think it reasonable to assume that observers know what they mean when the topic of "hue" is mentioned. Such is not the case at all. You can talk sensibly about dominant wavelength (which has nothing to do with how beams look), and you can talk about hue (which has no relation to colorimetric quantities) with people who act like they understand what you're talking about. But in the latter case, who is to say whether you're talking "sense"? Science has certainly nothing to say about that. At least Schrödinger has made an interesting opening, and it may well pay off to explore that. The best possible outcome would be a novel piece of psychology.

Schrödinger calculates the "isochromes" (curves of constant apparent hue) for his metric[27] (figure 10.16) and suggests that the result closely resembles the Abney shifts as empirically determined. It is not easy to judge to what extent this cheerful suggestion applies. Certainly the trend seems to be about right. Schrödinger's calculations have—to the best of my knowledge—not been pursued for other hypothetical metrics, nor have they been numerically determined for empirically determined metrics.

In the case of the *Helmholtz metric*, it is very easy to determine the isochromes. In this case you can map the space on the log-space in which the metric becomes the Euclidean one. Thus the isochromes are simply straight lines through the white point in the isobrightness plane. When you remap to color space you find that the isochromes are curved lines in the chromaticity diagram. The result is a strong Abney shift (figure 10.17), not at all unlike the shift predicted by Schrödinger for his metric.

In the case of the *Fechner metric*, the isochromes can be found from simple symmetry arguments. The equibrightness surfaces are spheres and the isochromes meridians of these spheres with the chromaticity defined through equality of all three cone excitations as pole. Since these curves lie in planes through the chromaticity, I conclude right away that there seems to be no Abney shift in the Fechner metric. However, you have to reason a little more subtly here.

The argument as it stands is correct, but it doesn't really describe the *actual experiments* very well. In the experiments people fix all colors to a *constant luminance plane*. Thus

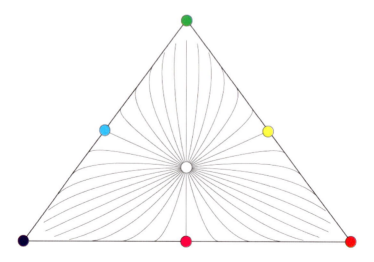

Figure 10.16 The "isochromes" in the RGB triangle in case of the Schrödinger metric. Notice that the isochromes diverge away from CMY and converge toward RGB.

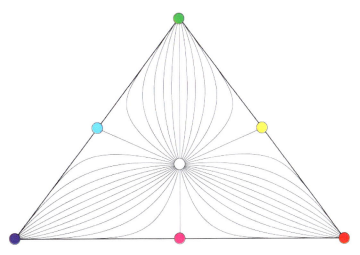

Figure 10.17 The "isochromes" in the RGB triangle in case of the Helmholtz metric. Notice that the isochromes diverge away from CMY and converge toward RGB like in the Schrödinger metric, though there exist significant quantitative differences between these metrics.

you need to find the geodesics in such planes, and these are in general different from the geodesics in color space. For the Fechner metric there are two possibilities for the metrics of planar subsets. When the plane is very special, namely such that the scaling function of the Fechner metric is constant, the metric is Euclidean. For all other planes it is hyperbolic. In the planes with a hyperbolic metric, the geodesics are curved; thus you do expect an Abney shift. The nature of the shift depends upon the tilt of the planes of constant luminance. When you do the calculations you find that the Fechner metric also "predicts" the Abney shift. (Figure 10.18.)

However, whereas the qualitative structure of the isochrome congruence is similar for the Helmholtz and Schrödinger metrics, it is different for the Fechner metric. In the case of the Helmholtz and Schrödinger metrics the isochromes diverge from the yellow, magenta, and cyan and converge toward red, green and blue. Thus a slightly reddish "yellow" of low saturation will be matched to a more reddish (chromaticity-wise) yellow of a higher saturation, and so forth. In the case of the Fechner metric, the isochromes diverge from the yellow and converge to the blue. The "unique green" and "unique red" will shift when you change the saturation, whereas they are invariant in the cases of the Helmholtz and Schrödinger metrics. This issue is evidently decidable by experiment. For instance, you may compare predictions with a pure "eye measure" scale as for instance the Munsell color atlas. In the Munsell atlas the isochromes diverge from the yellow and converge to the blue,

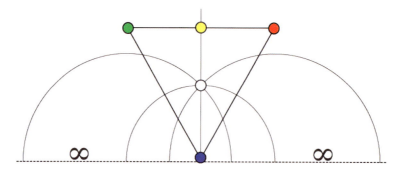

Figure 10.18 The "isochromes" in the RGB triangle in case of the Fechner metric. Notice that the isochromes diverge away from Y and converge toward B, whereas R and G are not invariant. In this (extreme) example I assume that the ideal blue has vanishing brightness.

whereas red and green are not stable with respect to desaturation. However, I am not sure that the data are clean enough that you could use them to decide the issue. The Munsell atlas doesn't come with an indication of tolerances.

So you end up with a partial draw here. The Abney shift (probably) yields no empirical data precise enough to help decide between the relative validity of the Schrödinger, the Helmholtz, and the Fechner metrics. If I had to guess, I would surely put my money on the Fechner metric, though.

10.7.1   The Bezold-Brücke effect

Another quirk of apparent chromaticity is the so-called Bezold-Brücke effect (figure 10.19). When an observer is asked to find a color of given brightness that closest matches a given color of another (given) brightness, most observers don't stick to the same chromaticity[28]. Thus *apparent chromaticity shifts with intensity*. ("Apparent chromaticity" is really awkward, since colorimetric chromaticity is by definition invariant with intensity.)

There are quite a number of open questions here. One problem concerns the actual task of the observer. Here I have formulated a task in the Schrödinger style, but most often experimenters have asked different questions such as "*find me the same hue*" or something like that. Then there is the problem of the exact freedom granted to the subject. You may keep brightness or luminance fixed for instance. You might limit the set of possibilities to monochromatic beams. Finally, there is the question of whether a shift (if one exists) depends on the *ratio* of brightnesses or on their *absolute magnitudes*. When I analyze the literature (which happens to be rather scarce to begin with) I fail to arrive at clear-cut

Figure 10.19 The Bezold-Brücke effect. When you darken colors they change hue. In the darkened hue circle they crowd toward the red, green, and blue. At least for me; again, decide for yourself.

answers. Experiments have typically been conducted with specific preconceived ideas in mind, and the results tend to be almost impossible to interpret in a general framework.

The least I can do is to relate the canonical story, which is in my opinion not that much different from "myth," though. When the observer has to find a wavelength of a monochromatic beam at one luminance that matches a monochromatic beam of a fiducial wavelength at another luminance, the observer selects wavelengths that systematically depend on all parameters and are different from the fiducial wavelength. The results may be represented as a function $\nu(\lambda, L)$, where $\nu(\lambda, L_0) = \lambda$ for some fiducial luminance L_0. The function $\nu(\lambda, L)$ or (more often) $\nu(\lambda, L_1)$ for some luminance L_1 (say ten times L_0) are what you find in the literature. Especially the latter curve is often identified with the Bezold-Brücke effect. I think there are good reasons to mistrust this way of putting the matter, which is really bound up with a specific mechanistic model that has never been up to independent investigation as far as I know. The model assumes that the activities of the three cone channels are individually subject to nonlinear transfer functions, essentially saturations for increasing luminance. The outputs of these nonlinear transforms would then be fed into the standard colorimetric expressions. It is just one random model someone pulled out of a hat. You would prefer data presentations to be put in a less dedicated form.

The conventional mechanistic interpretation seems (to me at least) to be an example of the (common) "photometer fallacy" (figure 10.20). You assume that somewhere in the head you have an indicator device (the place where the buck stops as far as vision is concerned) which indication (whoever reads it) determines "what you see". The whole idea seems preposterous to me, and it does probably to many people, because few come right out and

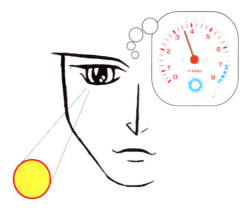

Figure 10.20 The "observer as photometer" notion.

say that this is what they mean. But when you look carefully, you will notice that the majority of scientists use the idea implicitly. Whether they are are abashed to say so or simply don't know what they're talking about, I don't know. I suspect the latter. It may be that it is the lack of a good answer to the question "what it is that we see" that makes people go for such strange notions. I don't have the answer to the question either, but I am of the opinion that an open question is better than a ludicrous answer.

Notice that there is no pressing need for the arbitrary assumption of nonlinear mechanisms (and of course almost anything can be fitted with such a "model") because the metrics actually predict Bezold-Brücke shifts from Schrödinger's principle. The observer always tries to find the nearest color to the fiducial one in the set of colors at his or hers disposal. It is clearly preferable to explore such predictions before other arbitrary elements are introduced to account for the facts, especially when—as I believe to be the case here—the "facts" have been massaged into a form that is suited to fit the assumptions to begin with.

When you apply Schrödinger's method, you have to find the geodesic plumb line on the set of colors at the disposal of the observer, starting from the fiducial color passed to him or her. When the set is constrained you need to find geodesics for the induced metric. This becomes necessary if you limit the set to monochromatic beams (the spectrum cone), for instance. This is important, but it complicates matters substantially. Here I will commence with the former case.

When the fiducial color is a point not on the given surface of equibrightness, you simply drop a geodesic plumb line from the fiducial point on that surface. Candidate plumb lines are geodesics through the fiducial point. The surfaces of equibrightness are defined as surfaces that are everywhere orthogonal to the chromaticities (lines through the origin) in the sense of the metric. An immediate consequence is that if lines through the origin can also be geodesics they must coincide with the sought for plumb lines. This actually solves the problem for the Helmholtz and Schrödinger metrics right away without any calculation at all, because in these metrics *all lines through the origin happen to be geodesics*. Thus you predict no Bezold-Brücke shifts for these metrics. This is evidently contrary to the facts.

The case of the Fechner metric is different, though. (See figure 10.21.) Here the only line through the origin that happens to be a geodesic is the chromaticity of equal cone excitations (lets agree to call it the achromatic chromaticity in this discussion). Thus there is no shift for the achromatic chromaticity, but you expect shifts for all other chromaticities. The direction of the shift is very simple, mainly a *desaturation*. It is because the geodesics are circles in planes that contain the achromatic chromaticity. Thus the plumb lines have the same dominant wavelength, but different saturation. The higher the saturation of the fiducial chromaticity, the stronger the desaturation effect.

The prediction that colors of high brightness look more desaturated than colors of the same chromaticity at lower brightnesses is fully brought out in *my* experience, but I have never seen formal empirical results. I am in good company here, because Helmholtz (in the *Handbook*) also remarks that as you increase the intensity, colors become more whitish, such that—at very high intensities—the entire spectrum becomes colorless. The effect is

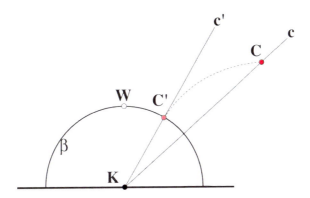

Figure 10.21 The Bezold-Brücke effect in the Fechner model. (In this simple figure I show only a section through the achromatic chromaticity.) The color C is much brighter than the colors on the surface of heterochromatic brightness β. The task is to find a color C' on β that looks most like C. The solution is to drop a geodesic plumb line on β. The foot point C' is the dimmer color most like C, and its chromaticity c' is less saturated (W is an achromatic point) than c. In this metric, colors desaturate when made brighter.

expected to depend on the ratio of the brightness of the fiducial color and the fiducial brightness of the set of colors to pick from.

Whether this is in accord with empirical data hasn't been decided so far. I personally believe that the data strongly indicate it, but the "official" reading is quite different, I'm sure mainly because a dependence upon absolute levels fits the a priori model. Purdy, in his 1931 paper, actually remarks

> *so far as one can judge from these limited data, the amount of shift appears to depend more closely upon the ratio between the two intensities than upon their arithmetic difference. In other words, it depends upon the 'psychological' rather than the 'physical' intensity difference.*

I take this as evidence against photometer models and as an indication that Helmholtz-Schrödinger style metrical models are to be preferred.

The actual data on the Bezold-Brücke shift have been obtained in much more constrained experiments. Typically, the beams were confined to monochromatic beams. The finding is that with increasing luminance reddish and greenish hues tend to become more yellowish, whereas cyan and deep blue hues tend to become more average bluish. Yellow, some average blue, and (roughly) the far green point are "fixed points"; their hues don't seem to be affected by an increase of luminance. Yellow and the average blue point are both "stable" fixed points (or "attractors") because all hues in its neighborhood are attracted by them when you increase the luminance. The green fixed point is *unstable* because all hues in its neighborhood are pushed away from it when you increase the luminance. Because of topological reasons, there has to be another unstable fixed point on the purple line.

In this case you have to use as the geodesics the curves of shortest distance that run completely within the spectrum cone surface, and these are most likely different from the general geodesics in color space. Since you expect the effect to be fairly small, it is sufficient to approximate the spectrum cone mantle in the neighborhood of any generator by the tangent plane at that generator, though, which simplifies the problem somewhat. In more technical terms, you need to find the metric (called the "induced metric") in the planes through the origin.

In the case of the *Helmholtz metric* you may transform into a Euclidean space, whereas chromaticities map to parallel lines. The spectrum cone becomes a cylinder and the generators are geodesics that are also plumb lines on the constant brightness loci, which are section of the cylinder with orthogonal planes. Thus you cannot expect a Bezold-Brücke shift even when the choice is confined to monochromatic beams. In the Helmholtz metric there simply is no Bezold-Brücke hue shift, contrary to experience.

In the case of the *Schrödinger metric* you can show in a similar way (it is a little more involved) that you don't expect a Bezold-Brücke effect, again contrary to experience.

Color experiences

With Schrödinger's loosening of the strict colorimetric reduction paradigm you get somewhat farther on the way of a formal account of color experiences. However, it should be understood that the visual realm is enormously richer than what can you can handle with these methods. In order to obtain a notion of the complexity, you should consult sources of experimental psychology and the visual arts. To give you some glimpse of the variety of visual experiences you may read appendix J.3. I show only the top of the iceberg. For many of these "effects," "illusions," and so forth, people have framed theories in the context of physiological, cognitive, or plainly phenomenological models. The relation to (even much relaxed) colorimetry of is the most tenuous nature though. If you set your aims to promote at least some of these phenomena to the colorimetric realm, you're in for a lifetime's endeavor.

The case of the *Fechner metric* is (again) different. In the planes through the origin the metric is that of the hyperbolic plane. You do expect shifts because the relevant geometry will be very similar to that shown in figure 10.21. The Bezold-Brücke shift (shift of apparent dominant wavelength) will depend on wavelength and brightness ratio, the actual functional dependence strongly depending upon the exact shape of the spectrum locus. The spectrum locus varies its distance from the achromatic axis quite sharply (see figure10.22).

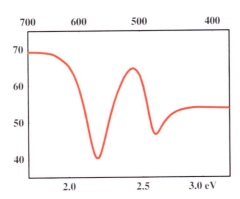

Figure 10.22 The angles (in degrees) between the achromatic direction (for a uniform spectrum) and the spectrum cone generators. (I used a canonical basis; otherwise the angular distance would make no sense.)

Thus the tangent planes to the spectrum cone mantle will make rather varying angles with the dominant wavelength planes. When such an angle is small you expect a shift similar to that in the case discussed earlier (see figure10.21), if the angle is almost a right one, you don't expect any shift for reasons of symmetry. (For suppose the cone were a right circular cone with the achromatic axis as its axis of symmetry. Then all wavelengths play equal roles. Thus you don't expect any shift.) The tangent planes are perpendicular to the dominant wavelength planes where the curve figure 10.22 has extrema, thus at the yellow, the far green point, and some average blue. These will be the predicted fixed points. A little reasoning shows that the yellow and blue fixed points are stable, the green one unstable. This is indeed what you find when a precise numerical calculation is done. The Fechner metric predicts the observed shift in detail.

However, the power of this prediction should not be overestimated, for you expect qualitatively similar results for the predictions of the "saturation models," simply because the functional dependence will mainly depend on the exact shape of the spectrum locus. A critical experiment to distinguish between the metrical descriptions and the saturation models would be to distinguish between the influences of absolute magnitudes versus ratios. Such experiments have—to the best of my knowledge—not been performed yet. The saturation models predict that the shift will depend upon the *absolute* levels, whereas the prediction for the metrics is that it depends only upon the *ratio* of the radiation levels involved in the task. When I read the major source of empirical data (almost exclusively Purdy's work[29]) closely, I seem to detect that this author favors the relative view. However, I cannot be sure, especially in view of the fact that the color vision community as a whole seems firmly committed to the absolute view for reasons that I fail to understand. Could so many people be wrong? Although it is perhaps unlikely, my guess is that they are.

My personal choice would be to apply Occam's razor and simply ignore the saturation models. From the available metric descriptions I would not hesitate to pick the Fechner model as the one that accounts best for the facts as I know them. This would make a good platform for consequent empirical researches. Time will tell.

10.1 Tolerance spaces

The metrics usually proposed in "higher colorimetry" are of various kinds. Often they refer to the psychophysical notion of JND, a topic that perhaps stands in need for some explanation. (Skip this appendix if you happen to do psychophysics yourself.)

This concerns problems of finite precision, that is to say, the variability of the observer. As I remarked before, this introduces certain problems that involve the very *topology* of color space.

The essential problem is already evident for the case of a simple scalar variable. Suppose you have a physical parameter—I will call it the "intensity"—that can be specified by some nonnegative real number. (I will take its physical dimension for granted here.)

Suppose an observer is confronted with ordered pairs of intensities and is forced to make a choice as to which of the two is "more intense." When both intensities give rise to a perception (they are "supraliminal," as one says), this choice apparently should make sense to the observer. In reality the observer may well be in doubt when the difference or ratio is close (in some sense) to either zero (in case of differences) or one (in case of ratios). When you repeat such forced choices many times, you obtain an estimate of the probability that the first will be rated as more intense than the second. When the first is held fixed and the second is increased from very low to very high values, you expect this probability to rise monotonically from zero to one, at least statistically. You expect the second *never* to be rated as more intense than the first when its actual intensity is *much* lower and *always* to appear more intense than the first to the observer when it is *much* higher. When the two are actually *equal*, you expect the probability to be one-half, at least when the observer

has an "honest" guessing strategy (has no "response bias"). The resulting curve is known as a "psychometric curve" (See figure 10.23). For the sake of argument I will assume that it can be described with sufficient accuracy by an error curve (integrated Gaussian) with the midpoint at physical equality. The steepness of the psychometric curve then reveals the certainty of the observer for deviations from equality.

No one knows what goes on in the head of the observer in performing such a task. But notice that it is *as if* some normal perturbation were added to the intensities and the decision based upon the comparison of the perturbed val-

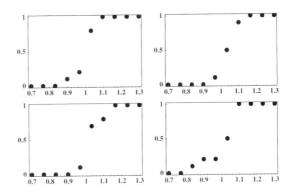

Figure 10.23 Here are four psychometric curves of a discrimination experiment. In each trial the observer had to discriminate two levels ("say whether the first level was higher than the other"). The fraction of "higher" answers for a session of ten trials was obtained for a number of different level differences (on a ten-log scale). The experiment was repeated four times. As you see, the results are different in detail, though similar in kind. From such measurements you obtain an insight into the degree to which the psychometric continuum is "locally scrambled".

ues. Thus people often assume some "internal parameter" (say) that is supposed to correspond to the physical parameter such that the internal parameter has an intensity equal to the actual intensity of the stimulus *plus* a perturbation, where the perturbation is some normal deviate with zero mean whose variance is some function of the value of the intensity (figure 10.24). In such a model the data will be described in terms of the "internal noise" and its dependence on the intensity. This type of description is the one typically invoked in communication and information theoretical settings.

Another way to think of what you have here is to conceive of transformations from the "stimulus domain" to some "internal domain." You may think of a large set of stimulus values and their corresponding "internal" values. In the ideal (or trivial) case the internal values would simply equal the stimulus values, or be a monotonic transformation of them (that is irrelevant to the arguments). Because of the inherent uncertainty of the observer's judgments, the internal domain has become locally disorderly, while the monotonic transformation is still *globally* monotonic, but not *locally* so, for it has the character of a "local shuffle". Moreover, because of the stochastic character of the observer's judge-

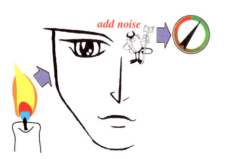

Figure 10.24 The observer troubled by "internal noise."

ments, you must assume that the local shuffles will differ on any repeat. You simply have no idea where the shuffling comes from, but just have to take it for granted.

> *You have to conceive of the internal domain as an ensemble of locally shuffled copies of the stimulus domain.*

Suppose you present the observer with a pair stimuli. At any moment you are in a particular member of the ensemble of internal domains and the stimulus pair maps on a certain pair of internal values say. Assume that the observer will consider the first stimulus as more intense than the second one when it turns out that the first internal value exceeds the second internal value andvice versa. At a repeat of the trial you have another member of the ensemble; thus the answer will appear to be a stochastic process. For a given input set, the output set will fluctuate from trial to trial. You may consider the ensemble average, which is simply a blurred or smoothed copy of the input pattern. The blurring kernel is a Gaussian whose width depends on the stimulus value as determined via the psychometric functions. Notice that *this ensemble average is a somewhat artificial construct, for at one trial you really have an instance of the ensemble, not the average.* This is what I will call a "tolerance space". (See figures 10.25 and 10.26.)

Notice that $A = B$ and $B = C$ by no means implies $A = C$ in a tolerance space.[30] These are important facts that the psychophysicist has to take for granted. (Figure 10.27.)

In a tolerance space it may well happen that when three stimuli are compared pairwise you will obtain responses that are evidently *inconsistent*. For instance, the first might exceed the second, and the second exceed the third whereas the third again exceeds the first (an "intransitive triangle"). Clearly this is an intolerable state of affairs. Of course, you expect that such incon-

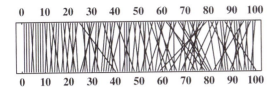

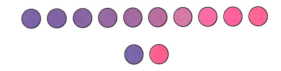

Figure 10.25 Here a set of levels (0...100) has been observed through a system that suffers from Gaussian noise disturbances with an amplitude of 10 percent of the mean level ("Weber fraction of ten percent). Notice how the topology is lost on a small scale, though it is retained on a rather coarse scale.

Figure 10.27 You probably have a hard time to discriminate between any two adjacent blobs in the top row. The bottom row shows the two endpoints of the top row. You easily discriminate them.

Figure 10.26 Left: A nice, sharp scene. Right: What a little bit of local scrambling will do.

scales, everything works out fine. You have to be very careful in interpreting "equality" when you deal with a tolerance space, as you invariably do in psychophysics.

sistencies will not systematically repeat on many trials.

Due to the task reduction (forced choice) you *never* encounter true equality, for this is not a possible response. Notice that—by design—you will find that the first exceeds the second or the second the first even if the second actually *equals* the first (same stimuli). In retrospect, that is after you have obtained the psychometric curve, you may assign—essentially arbitrary—fiducial levels (say 25 percent and 75 percent) and *define* equality via the 25 to 75 percent interval and so forth. This is an operational definition of equality and of greater than and lesser than. This type of "equality" fails to be "transitive" by design, at least on a fine-grained scale. On coarser

10.2 Varieties of judgement

What *can* be asked of an observer? This depends on the setting as well as on the observer. Trained observers (especially when paid for their efforts) can be induced to perform tasks that anyone in their right mind would refuse. This can be of interest in some cases. For instance, professionals in some particular trade can often make discriminations that require training. Well-known examples are wine tasting and the assessment of perfumes and cosmetic products. Trained painters are often able to see what particular pigments were used on a canvas. This is often of use in art historical problems (for instance, certain pigments are characteristic for periods and regions). In such cases there is an objective criterion as to the degree of success of the professional.

After tasting the wine and arriving at a conclusion, the label may be produced and the pro be revealed as right or wrong. After smelling a perfume and coming up with a list of ingredients, you may consult the chemist who prepared the concoction and judge the pro right or wrong. After pronouncing zinc white, a probe may be taken off the painting and an physicochemical analysis run.

In the some cases an analysis after the fact may be impossible, then you may try to reach some modicum of objectivity by statistical comparison of a large sample of pros on the task, or by a study of repeatability of the same pro on the same task over time.

In many other examples it may be impossible to arrive at any reasonable level of objectivity though.

The pro who mixes a novel, trendy perfume for the modern urban professional woman cannot be called right or wrong (marketing success is the only metric here); the colorist who judges "lightness" or "apparent hue" also cannot be called right or wrong.

When somebody asks you "what is the temperature outside?" it is clearly expected that you report what a thermometer would read, regardless of whether it feels chilly or hot to you when you stick your head out of the window. The question invites an *objective response*. The same is the case when you are asked what time of the day it is. The expected answer is what a clock would read, not whether it feels early or late to you. When you are asked what color that apple is? you can often get away with "it looks reddish to me," but I think that the *expected* answer is also in the objective mode, say what a colorimetric reading would be.

> *You aren't being asked about your feelings, you're being asked about the apple.*

The problem with feelings is that they are idiosyncratic. At the same room temperature (thermometer reading, I mean) I may wonder why you strip while I want to put on another sweater. It tells me something about you and me I guess, but your behavior simply doesn't seem to correlate with room temperature to me. No doubt we would never agree on feelings such as "awfully hot" or "very chilly indeed" if you act like that, whereas we of course would easily enough agree on a temperature reading (the height of a certain column of mercury against a scale). In answering the question about the color of the apple, I think I try to see what the color of the apple really is. At sunset I would (consciously or unconsciously) correct for the overdose of low energetic photons hitting the apple and thus the (relative) overdose scattered into my eye. Of course I may err in this, as I may err in reporting temperatures, times, and lengths.

The problem of "eye measure" is so problematic because it falls squarely in the class of tasks whose success or failure cannot be ascertained in any objective way except through extensive lateral or longitudinal statistical studies. In the final instance, only the pure colorimetric task of judging the (in)discriminability of beams is a fully objective one. Even Schrödinger's very gentle relaxation treads on dangerous grounds because the task is to select something "most like" something else. In principle there is no way to call the result right or wrong. Yet Schrödinger's extension of colorimetry is generally considered *far too conservative*, especially by psychologists.

Certainly you are interested in *much more*, apparently because you, like most people, intuitively feel that it is not complete madness to require such more far reaching judgments from even naive observers (figure 10.28). Indeed, you know that the painter can look at a scene and without further ado mix the right pigments to put a tolerable likeness (colorwise, that is) on the canvas. Apparently the painter can take a look and come up with a recipe for paint, a kind

Figure 10.28 Left: What color would you say this apple is is? What if I say it's blue? Right: Which blob looks most like yellow to you? Are you likely to change your mind tomorrow?

of absolute judgment of the colorimetric aspect of beams. You have to be careful in such conclusions, though. Although I described how the painter might start, the painting typically requires iterations in which the painter alternately views the canvas and the scene. Moreover, the reality for the painter is the canvas, not the scene. The painter produces a canvas such as to evoke certain feelings in spectators of the canvas, *in the absence of the scene*. Even so, the stubborn fact remains that most people will readily agree that the fire truck (close to us, in broad daylight, and so forth) is *red*. That is more than colorimetry or Schrödinger's extension of it will ever give you.

So what types of tasks would it be reasonable to ask from a naive observer (remember I'm talking color here)? Here are a few examples:

— given a pile of chips, name them;

— given a pile of chips, order them;

— given two pairs of chips, say which pair is made up of more similar chips;

— given a fiducial chip and a pile of chips, select a chip from the pile that is most like the fiducial in some aspect (white content, hue, you name it);

— given three chips, judge whether they are collinear (one an interpolation of the other two);

— given four chips, judge whether they are coplanar (one an interpolation or extrapolation of the other three);

— given a pile of chips, pick the one that represents the purest yellow;

— given a pile of chips, select the gray ones;

— given a pile of chips, pair them off in complementary pairs;

— given a linear sequence of chips (that is to say, adjacent chips look very similar), pick out a subsequence such that successive chips are equally spaced;

— and so on, and so forth, . . .

Few people, when confronted with such a task, would need long explanations. Yet few of the tasks are respectable from the standpoint of colorimetry or Schrödinger's extension of it.

10.3 Color Experiences

Color experiences depend on a great many factors and the study of color experiences properly belongs to experimental psychology. Of course, you use colorimetry as a tool in psychophysics, but colorimetric measures—by themselves—have limited predictive value, most certainly in the cases of most interest to the psychophysicist. Thus a discussion of these matters doesn't properly belong to this book. However, because the topic is raised so frequently in discussions (especially with those unaware of the limitations of colorimetry proper) I present some interesting examples in this appendix, limiting the discussion to phenomenology. Much of psychophysics aims at a connection with physiology, I will skip this entirely.

The effects that are perhaps most often mentioned are those of (simultaneous) contrast (figure 10.29) and assimilation (figure 10.30) described by Chevreul in the early nineteenth century but known to painters for ages.

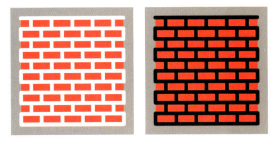

Figure 10.30 Classical "assimilation." The red of the bricks depends upon the color of the thin white or black stripes. Assimilation was especially important for Chevreul who worked with colored threads.

Other well-known effects have to do with boundaries and edges. Cornsweet has a nice demonstration (figure 10.31) that shows the importance of the edge and the "filling in" at either side of the edge. This filling in makes that irregular blobs tend to be seen as bounded by regular boundaries (figure 10.32), an effect that makes children's coloring books work out so well and makes that coloring near edges suffices to (subtly) color areas (figure 10.33) an effect often seen on ancient, hand-colored maps.

In many cases it seems that the eye[31] automatically posits objects that are not there. Vision apparently has its own "logic" quite independent of the *ratio* of the (conscious) intellect. A good example is "Kanizsa's triangle" (figure 10.34 left). Notice that the triangle is "whiter than white," without any application of a superwhite paint! As figure 10.34 right demonstrates, it is also easy enough to darken the paper without any application of gray paint. The so-called neon effect (figure 10.35) is an especially striking instance of essentially the same phenomenon.

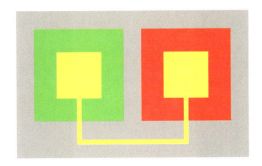

Figure 10.29 Classical simultaneous "contrast." The yellow patches at left and right look stubbornly different despite the fact that the bridge between them reveals them (visually!) to be identical.

Figure 10.31 The "Cornsweet illusion." The graph shows the grayness of the printed strip. At the left and right outskirts the grayness is the same. Yet the left and right sides look quite different in the upper figure. Occlusion of the edge (lower) figure removes the effect.

Figure 10.34 The "Kanizsa triangle" is induced through three strategically placed pacmen. Notice that the "triangle" looks whiter than the white of the paper upon it has not been printed. Likewise the paper base within the square at right looks grayish.

Figure 10.32 An irregular blob (left) "regularized" through a black "key line" (right).

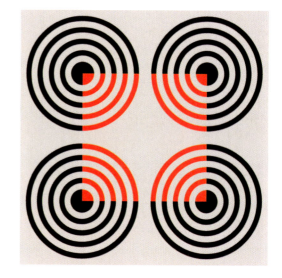

Figure 10.35 The "luminous red square" is an instance of the neon illusion.

Figure 10.33 The "watercolor effect." Notice that the paper inside the left and right ovals looks subtly different in color. This "filling in" was frequently used by the early cartographers who colored engravings by hand, using colored ribbons of dilute watercolor.

The spontaneous organization of the visual field was especially stressed by the Gestalt school of psychology. A major issue is the process of "grouping," whereby "Gestalts" are formed.

Color plays a very important role in grouping. A demonstration is shown in figure 10.36. These figures are geometrically identical, but the spontaneous organization depends upon the coloring. Using suitable coloring, I can render certain interpretations immediately obvious or almost impossible to see. This is very important in daily life, for it is all too easy to design maps, graphs, slides, and so forth such that they are thoroughly incomprehensible. This unfortunate effect is very common with people who indiscriminately apply the great color schemes freely available in their computer programs. Even professional cartographers often manage to render their maps visually incoherent. Bad examples include the indiscriminate use of the "rainbow scale" in the rendering of scientific images. Often the resulting Gestalt grouping visually fragments the relevant objects.

The eye is really very sophisticated and performs all kinds of optical reasoning automatically, that is to say, quite outside your consciousness. The demonstration shown in figure 10.37 makes this point. With some exercise you can "invert the depth" in this picture, and as a result

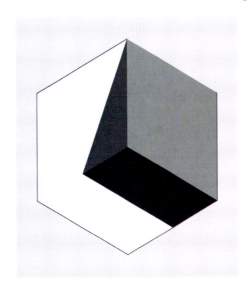

Figure 10.37 Try to see the vertex at the center as either close to you or far away. In the former case you see the exterior of a cube, and the gray areas look like "paint," and thus the faces are varicolored. in the latter case you look into a room and the gray areas appear as shade and shadow. Now the three faces appear as uniformly white (though shaded in various ways).

Figure 10.36 The figures are geometrically identical, though I tend to "see" (left to right, top to bottom) a hollow red/green star, a blue hexagon, a yellow parallelogram, and a red lozenge. With some effort I can see two interlocking triangles, and so forth.

you will see the colors of the "cube faces" change at the moment of the reversal. This demonstrates vividly that it will not be possible to build simple models that predict the "experience colors" of patches in the visual field.[32] Such models will (at best) be limited to abstract "Mondrian paintings."

In figure 10.38 I show a photograph of a gate at St. James park, London. The decoration appears "gilded," but is "golden" really a color? Analysis reveals that it is simply a pattern of white, yellow, and brown patches. Yet the impression of a "golden" color is hard to resist. Here the eye (quite automatically) interprets a pattern of colored patches in terms of the mate-

Figure 10.38 A gate at St. James park, London. The decoration appears "gilded," but is "golden" really a color? It is not in the colorimetric sense, the "gilding" being rendered through browns, yellows, and whites. The blobs at right are samples.

rial constitution of some pictorial object. Pictorial objects live in three-dimensional space and possess physical properties (of course, neither the space nor the properties need really exist). Colorimetry has nothing to say about that.

Edwin Land is the author of a number of very striking demonstrations. The example I show here is inspired by his example. Consider the array (figure 10.39) of colored squares that represent binary mixtures of red and green in sorted order.

As you randomly scramble either columns or row in the previous figure its appearance changes in a striking fashion (figures 10.40 and 10.41). Notice how different the results of the column and row scrambles look. Yet both images contain the same set of colored little squares.

If I scramble the rows and then the columns, the effect is even stronger (figure 10.42).

By far the strongest effect is obtained if you scramble the squares individually, independent of their row or column (figure 10.43). In this figure I spot colors that I do not see in the sorted array, despite the fact that it is exactly the same collection of colored squares. Apparently the order in which the squares appear in the array has a strong influence on their colors. This is something for which a pure colorimetric basis is entirely lacking.

Cutting the squares loose and dithering the array does strengthen the effect, as was perhaps to be expected, though not a whole lot.

These examples perhaps suffice to demonstrate that colorimetry has only a limited power in predicting color experiences. But of course colorimetry was never designed for that in the first place.

The psychology of color experiences is probably more interesting and certainly more difficult a topic than mere colorimetry. Small wonder that many people have very little interest in the topic of colorimetry. But a lack of interest is no excuse for ignorance. A sound understanding of colorimetry is a sine qua non for any study of color experiences.

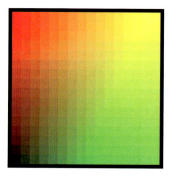

Figure 10.39 All binary (additive) mixtures for a given discretization in lexicographical sorted order.

Figure 10.40 Rows randomly scrambled. Notice overall difference in hue.

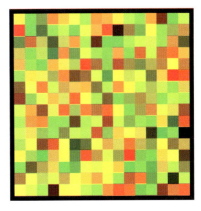

Figure 10.43 Here the individual squares have been shuffled without any reference to their row-column origins. Notice the huge difference to the previous case. In this pattern I see olives and browns that I fail to see in the sorted case (figure 10.39). Yet these examples contain exactly the same set of squares!

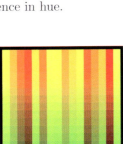

Figure 10.41 Columns randomly scrambled. Notice overall difference in hue.

Figure 10.42 Random scrambling of rows and columns in succession yields a pattern that might be used for a fabric (easy to weave). Compare colors to the case of sorted order.

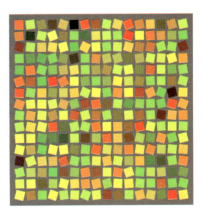

Figure 10.44 Cutting the squares loose and dithering them increases the effect, though by no means dramatically so.

Exercises

1. Geodesics [FOR MATHEMATICIANS] Find explicit expressions for the geodesics in the various metrics. Are they planar curves? Are lines through the origin (chromaticities) among the geodesics? Is it possible to write explicit expressions for the distance between any two points?

2. Equibrightness surfaces [NEED SOME FORMAL BACKGROUND] Formulate the conditions (in terms of constraints on the coefficients of the line element) for the existence of global equibrightness surfaces. For this to be the case the field of local surface elements as discussed in the text has to be an integrable field of contact elements. The condition simplifies for specific forms of the metric, for instance if there are no mixed terms or when you have a multiple (with an arbitrary function as factor) of the Euclidian line element. Now apply your condition to the Helmholtz, Schrödinger, and Fechner metrics. Construct an example of a metric that does not allow equibrightness surfaces (easy).

3. Isochromes [ELABORATE] Draw the "isochromes" in the chromaticity diagram for the various methods. (Use numerical methods if the going gets rough.)

4. The Bezold-Brücke effect [ELABORATE] Compute the Bezold-Brücke shift for a doubling of the intensity for monochromatic beams for the Fechner metric. You need to construct the induced metric in the spectrum cone surface (or in the tangent planes to the spectrum cone). Compare the result with the the curve (a psychophysical result by Purdy) reprinted in many textbooks. (When you suitably adjust various free parameters it is not hard to achieve an almost "perfect fit," which is very impressive.)

5. The CIE-Lab System [INSTRUCTIVE IF YOU HAVE APPLICATIONS IN MIND] Look up the definition of the CIE-Lab system on the Internet. Trace some of its consequences. Why is CIE-Lab space unfit as the arena for much of colorimetry (for instance, is it a linear space)? Does the question of whether a brightness function exists arise in this system? Does the system depend on the "photometer fallacy"? Where do the nonlinear functions and magical numbers come from?

6. Added sophistication [CONCEPTUAL] Color experiences are codetermined by the spatiotemporal structure of the retinal irradiation patterns. The CIE attempts to define sophisticated systems that take such influences (at least approximately and partly) into account. Find definitions on the Internet and consider their relations to colorimetry proper. Would you agree with my decision to leave a discussion of such systems out of this book? How would *you* define the boundary of "colorimetry proper" and why?

7. Actual discrimination thresholds [OF INTEREST TO PHYSIOLOGY AND PSYCHOPHYSICS] Look up the wavelength discrimination as a function of wavelength and luminance (and many other—for instance, spatiotemporal—parameters) for normal trichromats and dichromatic observers. Find the experimentally determined regions of confusion in the chromaticity diagram (so-called Macadam ellipses) and color space (Brown-Macadam ellipsoids)[33]. Compare with the metric in the space of beams and the various speculative metrics discussed in this chapter.

8. Predictions for the normal trichromatic observer [VERY ELABORATE] Find explicit predictions by carrying out all calculations (numerically) for the standard observer. In order to be able to do this you need to fix the metrics with the necessary bells and whistles in order to make them fit the available psychophysical data. Finding and evaluating such (often conflicting and lacunary) data is not easy. In many cases you will have to supply educated guesses. Is it possible to eliminate certain metrics, or, even better, certain classes of metrics? Of the metrics that cannot be refuted are some more desirable than others? (Criteria could be predictive power, mathematical elegance, number of free parameters, and so forth.) Don't get too carried away, for this is a research project that could easily take you a lifetime!

9. The curvature of color space [FOR MATHEMATICIANS] From the covariant components of the metric tensor $g^{\lambda\mu}$ compute the Christoffel symbols (affine connection) $\Gamma^{\lambda}_{\mu\nu}$ and the Riemann curvature tensor $R^{\lambda}_{\mu\nu\sigma}$. Contract to find the Ricci tensor $R_{\mu\nu} = R^{\lambda}_{\mu\lambda\nu}$ and finally the scalar curvature $R = g^{\mu\nu}R_{\mu\nu}$ (notice that I use the Einstein summation convention) for the Helmholtz, the Schrödinger, and the hyperbolic metric. Show that the Helmholtz space has zero scalar curvature ($R = 0$), the Schrödinger space is elliptic ($R > 0$), and the hyperbolic metric—indeed—hyperbolic ($R < 0$). For all three metrics the scalar curvature is constant.

10. Geodesic planes [FOR MATHEMATICIANS] Consider the (geodesic) midpoint AB of a geodesic arc from A to B. In general it differs from the colorimetric equal mixture, but in the sense of the metric it is the closest point to both A and B, it is the "subjective average" say. Now consider three points A, B, and C, not on a common geodesic. Construct the subjective averages AC and BC. Now consider the geodesics from A to BC and from B to AC: Do they meet in a common point? (A drawing might help.) If not (the geodesics are "askew"), the triangle ABC apparently does not exist as an all geodesic plane. Check this for various metrics and discuss the implications.

11. Line elements in the literature [TAKES SOME DEDICATION] Find some of the Riemann metrics (they usually go under the name of "line elements") proposed for color space in the literature (use the Internet). A generic example is by J. J. Vos and P. L. Walraven (1972), "An analytical description of the line element in the zone-fluctuation model of colour vision—II. The derivation of the line element", Vision Research **12**, pp. 1345-1365. Do they offer useful additional insights? Do you deplore the fact that I allotted no space to these things?

Chapter Notes

1. H. von Schelling, "Concept of distance in affine geometry and its applications in theories of vision," *J.Opt.Soc.Am.* 46, pp. 309 (1956).

2. E. Schrödinger, "Grundlinien einer Theorie der Farbenmetrik im Tagessehen," *Ann.Phys.* 63, pp. 397–447 (1920).

3. There exist numerous instances in the literature of experimental psychology in which I haven't the faintest notion of what the experimenter is doing.

4. An empirical check on the validity of this description is A. B. Poirson, B. A. Wandell, D. C. Varner, and D. H. Brainard, "Surface characterizations of color thresholds," *J.OptSoc.Am.* A 7, pp. 783–789 (1990). The authors conclude that "the best-fitting ellipsoid approximates our threshold data with the same precision as replications of the data."

5. Consider a family of concentric quadrics $(a_{20}x^2 + 2a_{11}xy + a_{02}y^2)/2 = $ constant in the plane and a direction $\{\alpha, \beta\}$. The direction is tangent to the quadrics when $\{a_{20}x + a_{11}y, a_{11}x + a_{02}y\} \cdot \{\alpha, \beta\} = 0$, that is on the line $(\alpha a_{20} + \beta a_{11})x + (\alpha a_{11} + \beta a_{02})y = 0$. This line is the direction "conjugated" to $\{\alpha, \beta\}$. It is orthogonal in the special case that $a_{11} = 0$, that is, when the quadric's principal axes coincide with the coordinate frame. It coincides with the $\{\alpha, \beta\}$ direction when $a_{20} = a_{02} = 0$, $a_{11} \neq 0$. This works out essentially the same in higher dimensions, except that the conjugate of a direction has dimension $n - 1$; thus in 3D the conjugate of a direction is a plane.

6. Frobenius's theorem gives a necessary and sufficient condition for the "Pfaffian equation" (the field of surface elements) to allow of an integral manifold. J. F. Pfaff, "Methodus generalis, aequationes differentiarum particularum, necnon aequationes differentiates vulgares, utrasque primi ordinis inter quotcumque variabiles, complete integrandi," *Abh. K. Preuss. Akad. Wiss. Berlin*, pp. 76–135 (1814–15); G. Frobenius, "Über das Pfaffsche problem," *J. für Reine und Angew. Math.* 82, pp. 230–315 (1877).

7. The interested reader should study the construction of the present industry standard, the Lab-system. This is widely used as a substitute for actual human observers, something that is evidently desirable in the industrial setting.

8. Helmholtz's metric is found in the second edition of his handbook on physiological optics; the last edition he edited himself. It was removed from the third edition (the one available in English translation) by his pupil von Kries.

9. The CIE-Lab system, the coordinates are known as $L^\star$, $a^\star$, and $b^\star$. The "lightness" $L^\star$ represents the difference between light (where $L^\star = 100$) and dark (where $L^\star = 0$). The coordinate $a^\star$ represents the difference between green and red, and $b^\star$ represents the difference between yellow and blue. Let $\{X, Y, Z\}$ denote the color coordinates in the CIE 1931 XYZ color space, then the transformation is:

$$
\begin{aligned}
L^\star &= 116 f(Y/Y_n) - 16, \\
a^\star &= 500[f(X/X_n) - f(Y/Y_n)], \\
b^\star &= 200[f(Y/Y_n) - f(Z/Z_n)],
\end{aligned}
$$

where

$$f(t) = t^{1/3},$$

for $t > 0.008856$,

and $7.787t + 16/116$ otherwise.

The coordinates of the white point are $\{X_n, Y_n, Z_n\}$. See K. McLaren, "The Development of the CIE 1976 $(L^\star a^\star b^\star)$ Uniform Colour Space and Colour-difference Formula," *J. Soc. of Dyers and Colourists* 92, pp. 338–341 (1976).

10. The official site of the CIE is: http://www.cie.co.at/cie/.

11. H. von Helmholtz, "Kürzeste Linien im Farbensystem," *Sitz. Ber. Akad. d. Wiss. z. Berlin* 17 Dec. 1892, pp. 1071–1083 (1891). and H. von Helmholtz, "Versuch einer erweiterte Anwendung des Fechnerschen Gesetzes im Farbensystem," *Z. f. Psychol. n. Physiol. d. Sinnesorgane* 2, pp. 1–30 (1891).

12. In practice one takes various limiting cases into account through minor alterations of the basic expression. It is evidently problematic that ΔL approaches zero as L approaches zero, for the threshold is expected to bottom out. When L grows beyond bounds bad things are bound to happen to the eye, so one expects a "saturation," perhaps. There is much literature concerning the "correct" form of Weber's law in such limiting cases. It is of little interest to the topic here.

13. Helmholtz himself included the patches necessary to make things work out okay in limiting situations. Such patches are obvious, and since they merely serve to detract from the argument I leave them out.

14. W. Abney, and E. R. Festing, "Colour photometry," *Phil.Trans.R. Soc.London* 177/II, pp. 423–456 (1886).

15. You would certainly do this if you would use the metric to predict actual psychophysical data quantitatively of course.

16. S. L. Guth, and H. R. Lodge, "Heterochromatic additivity, foveal spectral sensitivity, and a new color model," *J.Opt.Soc.Am.* 63, pp. 450–462 (1973);

S. L. Guth, R. W. Massof, and T. Benzschawel, "Vector model for normal and dichromatic color vision," *J.Opt.Soc.Am.* 70, pp. 197–212 (1980).

17. The line element introduced here is by no means the only one that would yield Guth's vector model as well as conforming to Weber's Law in any dimension. Another instance of such a metric is

$$ds^2 = \frac{dS^2 + dM^2 + dL^2}{S^2 + M^2 + L^2},$$

which is perhaps even more in the spirit of Guth's model. The geodesics for this metric are the same as for Schrödinger's metric, that is to say, planar logarithmic spirals. This is the reason I mention this (very pretty) metric only briefly, for it turns out to be the case (see below) that this metric fails to predict the Bezold-Brücke hue shift, and thus is ruled out on empirical grounds.

18. In M. Ikeda, and Y. Nakano, "Spectral luminous-efficiency functions obtained by direct hetrochromatic brightness matching for point sources and for 2° and 10° fields," *J. Opt Soc.Am. A* 3, pp. 2105–2108 (1986), you find actual data for many observers with the empirical spread. Total spread may exceed a factor of five, heterochromatic matching of rather different hues is *very* variable.

19. Formally this is the theory of partially ordered linear spaces and conical order.

20. The psychophysical investigation of dominance should take dominance to be a (possible) attribute of colors, not of beams. In fact, the "spectral dominance of beams" is a hairy topic, since it depends upon spectral resolution. However, this need not impede progress in the study of dominance among colors.

21. H. von Helmholtz, "Kürzeste Linien im Farbensystem," *Sitz. Ber. Akad. d. Wiss. z. Berlin* 17 Dec. 1892, pp. 1071–1083 (1891).

22. Since the geodesics are circles in planes perpendicular to the plane $L + M + S = 0$, the distance follows from a simple integration. The distance of a point at angular distance ϑ from the top-point (relative to the $L + M + S = 0$-plane) to that top-point is

$$d(\vartheta) = \log \frac{1 - \tan \frac{\vartheta}{2}}{1 + \tan \frac{\vartheta}{2}}.$$

From this you can simply find the distance of any two points by subtraction, after you have constructed their common circular arc. This is a kind of three-dimensional generalization of the Weber-Fechner law.

23. An example is described in C. von Campenhausen, C. Pfaff, and J. Schramme, "Three-dimensional interpretation of the color system of Aguilonius/Rubens 1613," *Color Res.Appl.* 26, pp. 17–19 (2001).

24. In alchemy, a homunculus is a little human created by mixing different substances, including sperm. It will then protect and help its creator, much like a golem. The first person said to have created a homunculus was the famous alchemist Paracelsus (1493–1541). I think an attempt to reproduce his experimental method would prove to be a waste of your time, though.

25. Ibid.

26. S. A. Burns, A. E. Elsner, J. Pokorny, and V. C. Smith, "The Abney effect: chromaticity coordinates of unique and other constant hues," *Vision Res.* 24, pp. 479–489 (1984).

27. In figure 10.16 I show a simple example where I only consider the $R + G + B = 1$ plane.

28. On the Bezold-Brücke effect see D. McL. Purdy, "Spectral hue as a function of intensity," *J. of Psychol.* XLIII, pp. 541–559 (1931); —, "The Bezold–Brücke phenomenon and contours for constant hue," *Am.J.Psychol.* XLIX, pp. 313–315 (1937); R. M. Boynton, and J. Gordon, "Bezold-Brücke hue shift measured by color-naming technique," *J.Opt.Soc.Am.* 55, pp. 78–86 (1965); J. Lillo, L. Aguado, H. Moreira, and I. Davies, "Lightness and hue perception: The Bezold-Brücke effect and colour basic categories," *Psicológia* 25, pp. 23–43 (2004); R. W. Pridmore, "Bezold-Brücke hue-shift as functions of luminance level, luminance ratio, interstimulus interval and adapting white for aperture and object colors," *Vision Res.* 39, pp. 3873–3891 (1999).

29. D. M. Purdy, "The Bezold-Brücke phenomenon and contours of constant hue," *Amer. J. Psych.* 49, pp. 314 (1937).

30. $A = B$ in no way implies $A = B$ (on the next trial) either (nor does it imply $A \neq B$).

31. When I say "eye" I of course mean the "visual system", though not in the physiological sense. Something like "psychogenetic process" catches the meaning, though the term is unfortunately already in use for quite different purposes. The old (German) term from the Gestalt school "Aktualgenese" is close to the "microgenesis" as used in modern psychiatry and may be the most apt to use. Since all these terms have various unintended fringe meanings, I prefer the term "eye". It is at least simple, but you should not take it too literally. I'm sure you know what I mean though.

32. N. Moroney, M. D. Fairchild, R. W. G. Hunt, C. Li, M. R. Luo, and T. Newman, "The CIECAM02 Color Appearance Model," *IS& T/SID CIC Scottsdale* 10, pp. 23–27 (2002).

33. D. L. MacAdam, "Visual sensitivities to color differences in daylight," *J.Opt.Soc.Am.* 32, pp. 247 (1942); W. R. J. Brown, and D. L. MacAdam, "Visual sensitivities to combined chromaticity and luminance differences," *J.Opt.Soc.Am.* 39, pp. 808–834

(1949); G. Wyszecki, and W. S. Stiles, *Color Science—Concepts and Methods, Quantitative Data and Formula*, (New York: Wiley-Interscience, 2000).

Part IV

The Space of Object Colors

Chapter 11

Object Colors

What is the "color of an object"? In many cases there is an obvious answer. For instance, (I cover RGB here), the consensus is that fire trucks are RED, meadows GREEN, and the clear sky BLUE. "Conceptual apples" come GREEN (sour), RED (sweet), or BROWN (rotten). But how about a *real* apple (figure 11.1)? Most objects are not of a uniform COLOR, nor are they uniformly illuminated. They show the colors of the environment through reflexes and specularities or glosses, though their own varicolored nature also shows through. If

Figure 11.1 What is "the COLOR of an apple"? Or even of THIS apple? Few objects are of a uniform color, their parts are hardly ever uniformly illuminated, and so forth. Does an apple "have a COLOR"? Does *any* object?

you are a painter, trained in the realistic tradition (that is, if you know how to "see"), you know that "the" COLOR of any actual apple is a nonentity. What do I mean by "object color" then? You will have to understand this term in a formal, technical sense.

Color scientists conventionally distinguish between *aperture colors* and *object colors*. The difference is categorical.

Suppose you experience a "patch of color." Except for its "color," a patch may (and most probably *will*) possess numerous additional properties. For instance, *any* patch appears to have some size and/or shape and persists for some given time slice. Except for these basic chronogeometrical properties, a patch may be perceived as having some spatial location and/or attitude ("frontal," "slanted," and so forth), as well as any number of material properties ("metallic," "wet," "smooth," "soft," "translucent," and so forth) or radiant properties ("glowing," "fluorescent," "glossy," "translucent," "being illuminated in a certain way", and so on).[1]

A patch has perceived spatial location and/or attitude or material properties only if *seen in context*, usually as *part of a scene*. (Figure 11.2.) An isolated patch (necessarily seen *out of context*) has only size, shape and duration, but has no specific position or orientation (it doesn't look "surfacelike") or material or radiant properties. Isolated patches tend to appear "self-luminous", but in an eerie sort of way, not like a regular source of light. Such isolated patches are rarely experienced in normal circumstances (see figure 11.3), although they appear as run-of-the-mill items in the laboratory.

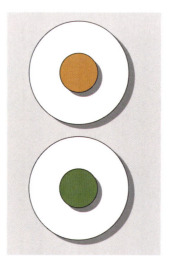

Figure 11.2 Brown and olive cannot be encountered as aperture colors. They occur only in context.

Figure 11.3 The apple is a varicolored object, but through the aperture you see a uniform reddish patch, "the color of the apple", especially if you focus on the aperture as if it were an object rather than trying to see the apple through the aperture.

An isolated patch is much like a "point at the fund of the eye" in terms of Berkeley's "New Theory of Vision."[2] It stands for a single "visual ray" (or a slender pencil of these) and is unable to reveal any properties of the environment but the local color. Other properties must be due to "cues," and the cues are only to be found in the context, that is to say, the retinal image as a whole. Thus a patch has any qualities (apart from mere color) only because it is part of the full optical structure available at the eye.

Isolated patches can be produced through the isolation of uniform patches from the environment, although in the laboratory they are usually produced from scratch. A generic way of generating an isolated patch from the environment is to punch a small aperture in a screen and look *at* (not *through*) the aperture. Then the aperture is perceived as filled with a uniform color that is due to what is behind the aperture. Thus an aperture (for example, a hole punched in a business card with a standard office punch held at normal reading distance) can be used to *isolate* small patches from the scene in front of you. Doing so turns a small part of the scene in front of you into an *aperture color*; that's where the name comes from. Painters in the impressionistic style often use the aperture trick to turn colors of a scene into aperture colors. This allows them to judge which paint to use to render the patch in the "impressionistic" mode. (Figure 11.3.)

Aperture colors are quite distinct from "object colors," which are the "colors" of patches of the scene in front of you as *seen in context*. For instance,

- they cannot look *metallic*, thus "gold" is an object but not an aperture color;

- they cannot look *glossy* or *matte*, thus the difference between paper and smooth plastic vanishes;

- they cannot look *gray* or *white*;

- neither can they look *brown* or *olive* (figure 11.2), or like any color that "contains black";

- they cannot appear to have a well-defined orientation with respect to a light source, in fact, they don't appear "illuminated" at all.

The aperture colors are what elementary colorimetry (chapters 4, 5, and 8) is all about. Up to this point in the book I have singularly regarded aperture colors, not object colors.

The differences between the colors of beams and the colors of materials have been very strikingly illustrated in paint by Philip Otto Runge, a Danish painter who turned himself into something of a color scientist. In his *Morning* (figure 11.4) you see the colors of beams in the air and in the "Light Lily," whereas the colors of materials are present on the earth. Notice that no GREENS and BROWNS are present in the air. It is suggested that the additive mixture of YELLOW and BLUE beams does not yield GREEN (as when you stir earthy paints together). In the border areas you see scales from LIGHT to DARK (subterranean to heavenly). On the bottom, darkness is illustrated by an eclipse of the sun; on the top you have the "realm of pure light". Runge was a "light mystic," and he perceived the Trinity in the primary triple RED, BLUE, and YELLOW. He clearly worked in a prescientific way. However, he wrote a booklet (*Color Sphere*) in which he clearly outlined the structure of the space of object colors in essentially modern form. He should be counted as one of the heroes of color science. He is certainly one of mine. Goethe printed a letter by Runge in his *Farbenlehre*, but he (unfortunately but perhaps not unexpectedly) failed to grasp the full scope of the painter's ideas.

All surface colors (figure 11.5) for a given "illuminant" (this is the conventional term, and I will use it, but of course you should really talk of "irradiant") lie in a certain convex, fusiform body that is smooth throughout except at two points (the black and white points), and two creases (the edge color loci); this is just the Schrödinger color solid. The "illuminant" must be understood as the spectral normal irradiance of the surfaces. Once you decide upon the illuminant the "color solid" is well defined. In the neighborhood of the black point the color solid is identical (in the limit) with color space, that is, the space of patches. It is augmented color space because the achromatic beam has been decided upon. It is the illuminant. Thus

> *the space of patches is simply the tangent space of the color solid at the black point.*

It would indeed have been much more rational to *start* with the surface colors and *derive* the colors of beams instead of the other way around. I merely decided to follow the tortuous

Figure 11.4 Philipp Otto Runge, in his *The small morning*, treats the difference between object colors and aperture colors from his perspective as a painter. There is a lot of light symbolism that it would not be proper to discuss in this book, but the LIGHT-DARK, WHITE-BLACK scales should be fairly obvious.

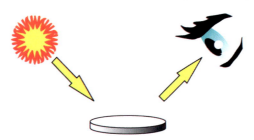

Figure 11.5 In the case of object colors you are aware of the context of a patch, and you see the patch as being illuminated in such and such a way. Object colors are not "self-luminous" as mere patches are. As Goethe says, they are witness to the fate of light (he uses "suffering") in the world.

path in this book because people are so used to it. Maybe when somebody writes a book on color a century from now that would be the rational way to structure it. But it is important to remember that

> *surface colors are of primary importance in color science, while aperture colors (or patches) are derivative.*

Notice that it is *very dark* in the immediate vicinity of the black point. That is to say, entities such as the "monochromatic beams" that exist only in the tangent space cannot be really said to exist; they are all varieties of "blacks". The "homogeneous rays" of Newton are indeed "spectral" in the sense of resembling a specter or ghost. The Goethean concept of colors as "parts of sunlight" or "shadowlike beings" (page 262) makes far more sense. All surface colors are dominated by the illuminant, thus they can be considered as proper "parts" of the illuminant. The colors on the boundary of the color solid have a close affinity to the Goethean edge colors (*Kantenfarben*, chapter 6), to Ostwald's full colors or semichromes (section 7.3.1) and Schopenhauer's parts of daylight.

The colors close to the boundary must be due to spectral reflectances that are similar to Schrödinger optimal colors (chapter 7). Since rapid variations are rarely encountered in typical materials, you find many instances of the edge color type. Moreover, colors can be bright and vivid only if the spectrum is not highly structured. For instance, the YELLOW or RED colors of flowers are typically due to carotenes and are very close to the warm edge color family. Vivid GREENS and PURPLES are close to Ostwald full colors (combinations of two edge colors).

Object colors look as part of the scene in front of you (figure 11.5), typically as surface patches belonging to the boundaries of extended objects (hence the name). Many "colors" are indeed named after generic objects; think of "orange" or "ice-blue," for instance. In many cases the color is used to judge material properties; for instance, green, red, and brown signal "unripe", "ripe" or "rotten," for various fruits.

Because object colors are (small) parts of the scene in front of you, they are generically not uniform patches but are almost certainly *textured*. Thus they do not map on single points of color space, but on a *cluster of points*, a color gamut. This adds additional dimensions to object colors that I will explore at a later stage. For the moment, I assume that object colors are at least approximately uniform patches.

Object colors arise when material objects interact with incident radiation. Typical interactions include surface scattering, a process exemplified by a piece of paper. No matter how the paper is irradiated, you typically are able to see it from any viewing direction due to the fact that the surface of the paper scatters the incident radiation into all directions. Many other types of interaction generically occur in typical scenes. In general the beam scattered toward your eye by an object depends in an intricate fashion on the geometry of the incident beam, the nature of the object and the viewing geometry (figure 11.6). This extreme richness is important (indeed, necessary) to daily-life vision, where you routinely perceive geometrical layout, light field structure ("luminous atmosphere"), and surface properties (both geometrical and physical), all at the same time. This is too much to deal with right now, so I will start to discuss the type of "stimulus reduction" that will allow you to get an initial grip on object colors.

11.1 Formal Notion of Object Color

When you look at a surface the beam that reaches your eye depends both on the beam that irradiates the surface and on the (spectral) scattering properties of that surface. What you see will also depend upon the background on which the chips are laid out (figure 11.7), because the background (especially when large and uniform) is automatically accepted as the reference (the "illuminant"). Various additional circumstances may (and usually do!) complicate things even further. For instance, you may happen to wear tinted glasses, or a veiling illuminance (perhaps diasclerally hence probably unnoticed) may influence the retinal illumination pattern. Also the geometry may play a role (figure 11.8). In realistic settings you need to be aware of all such possible effects. I will skip most of these complicating factors here.

When you place a set of chips on various backgrounds (under the same illuminant) (see figure 11.9) you see apparent color changes due to contrast. I have illustrated this before in the introductory part on visual field effects. The effect is a strong one as you can judge for yourself. In some settings you may actually get the illuminant wrong and apply (quite unconsciously) an uncalled-for correction.

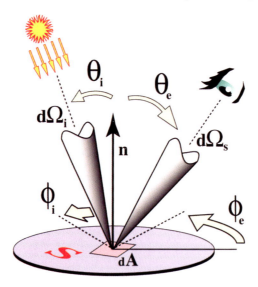

Figure 11.6 The concept of the BRDF. A surface element dA of the surface $\mathcal{S}$ has unit normal $\mathbf{n}$. I define the "vector surface element $\mathbf{dA}$ as $dA\,\mathbf{n}$. The surface is irradiated from the direction parameterized by the angles $\{\theta_i, \phi_i\}$, causing an irradiance E (say). The surface is viewed from an another direction, parameterized by the angles $\{\theta_e, \phi_e\}$, causing a radiance N (say). The bidirectional reflectance distribution function (BRDF) is defined as the ratio N/E of the radiance in the viewing direction to the irradiance caused by the beam from the direction of irradiation. It is a function of two directions, or four angles, that characterizes the surface scattering. In many calculations it proves to be convenient to use the "vector solid angles," as indicated in the figure.

When you place the chips on a single background but change the mutual illuminant (see figure 11.10), you also see apparent color changes, but these are of a different nature. The figure is somewhat ambiguous as you are led to believe in a single illuminant for the

Figure 11.7 Context matters even if it is of the simplest kind. Here the three inmost disks are of the same color, but they are put in slightly different contexts.

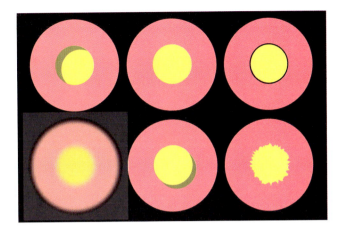

Figure 11.8 The actual physical implementation of the chips matters, too. I will simply ignore this aspect here, but in real life you cannot do that.

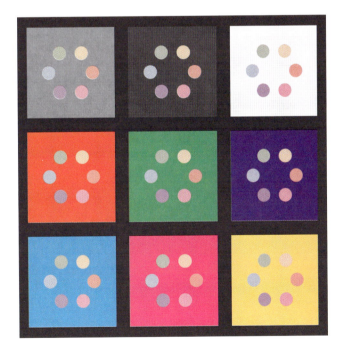

Figure 11.9 The same set of six colors put upon different backgrounds. The illuminant is the same in all cases. So are the six samples. Notice the different appearances.

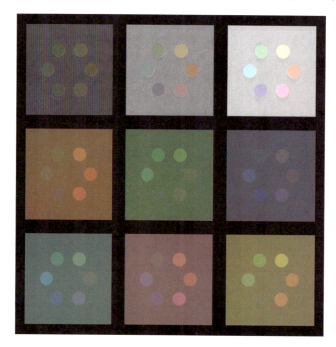

Figure 11.10 The set of six colors on a (single) median gray background with different illuminations from a variety of sources. In assessing the colors in the six cases, the color of the background should appear as "not different," that is to say, the only differences are due to the illuminance. The concensus in vision science is that human observers "discount" such differences.

picture as a whole. Imagine that you are confronted with the subpictures one at a time in isolation. Compare these two pictures carefully. In the first picture the six chips are objectively identical (printed with the same inks in the same manner), only the backgrounds differ, but in the second the disks themselves (like the backgrounds) are different.

Again, when you put a number of colors on a single background under a single illuminant and then add various veiling illuminations (see figure 11.11) you experience different apparent colors. Both the disks and the backgrounds are different for the subpictures. However, the effect is quite different from that in the previous case.

Exactly *what* you will see in these various circumstances is still a matter of hot scientific debate. The second case is the "color constancy" situation. "Color constancy" would pertain when the set of colors would appear the same in all cases. Of course, it doesn't.

In real life, you are constantly confronted with arbitrary combinations of all these factors (contrast, illumination, veiling), and the layout of the real world is more complicated than

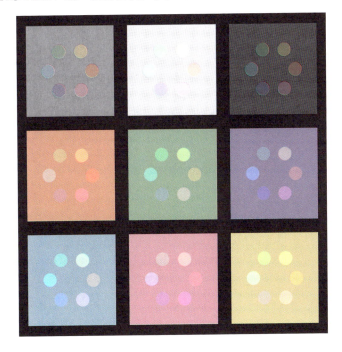

Figure 11.11 The set of six colors on a (single) median gray background with different veiling illuminations added. In assessing the colors in the six cases, the color of the background should appear as "not different", that is to say, the only differences are due to the veil. Human observers are supposed to "discount" veils too.

six round disks in a hexagonal pattern on a uniform square background. I used this simple layout in the example in order to enable you to compare corresponding disks in the three pictures.

Real life is different. No wonder you are prone to make frequent mistakes when you have to identify objects on the basis of subtle color differences. Yet even in real life the world is not a *total* chaos, though I am the first to acknowledge that it often comes close. There are many materials that you may count on as convenient anchors, for instance, human skin tones. Indeed, most of the complaints photo labs get from naive customers have to do with "impossible" skin tones in automatically processed snapshots. A human operator in a photo lab will almost automatically apply filtering to get at least the skin tones right. Another family of anchors are the achromatic objects. Examples include asphalt, car tires and concrete. Exceptions are frequent. A European might assume the notepaper in a picture to be WHITE, whereas (in the United States) it is often YELLOW.

In this part of the book I concentrate upon a very simple, but important, special case: the colors of "simple objects" under various illuminants. Simple objects are objects that are (from a colorimetric perspective) completely specified through their spectral albedo. That is to say, if you know the power spectrum of the irradiance of the surface of your simple object, you can simply find the power spectrum of the beam that gets to your eye from the knowledge of the spectral albedo. Convenient as this may be, it is—sadly but true—*not your typical reality.* Most objects are rather remote from being simple objects in the sense that the beam that gets into your eye depends not simply on the irradiance of the surface of the simple object, but also on the precise way of *how* the surface is being irradiated, and not simply on the fact that you happen to look at the simple object, but also on precisely *how* you look at it. This is important, because the "irradiance" of a surface is an average over all the directions by which the surface happens to be irradiated. Even when the irradiance is the same the actual scene can differ widely (sunlight, overcast sky, and so on). And there are many ways you can look at a surface, for instance the slant and tilt under which you view the surface may differ widely (looking straight on, obliquely, at a grazing angle, and the like). Most objects look different when you vary the lighting set up or the viewing angles. Only certain *ideal objects*, called "Lambertian," are special in that they *look exactly the same* no matter how they are irradiated and no matter how you happen to look at them. This is so special that *simple surfaces like that don't exist.* They are mere fancy, although many authors actually make believe they *do* exist. Don't fall for it. They don't.

However, it must be admitted that many real surfaces come somewhat close to Lambert's ideal. In this chapter I will join in the general make-believe and do *as if* the real objects we're interested in (apples and so forth) are simple objects. I will make up for this in later chapters, don't worry. What this buys you is that the topic of object colors (at least from a colorimetric perspective) becomes quite manageable.

In the simplest setting a uniform, planar surface is irradiated by a uniform beam, whereas the surface acts as an ideal Lambertian[3] surface. A Lambertian surface scatters radiation in all directions, such that the radiance of the scattered beam in any direction depends only on the irradiance of the surface, not at all on the specific viewing or illumination geometry. Although true Lambertian surfaces do not exist, the concept of such a surface is a convenient idealization. I assume that the only interaction of the irradiating beam with the surface is such Lambertian scattering. This implies that if the irradiating beam is monochromatic, then the scattered beam is, too (and at the same wavelength of course). In such cases you can define the "reflection factor" as the ratio of the radiance of the scattered beam (notice that I don't have to specify the viewing direction here) to the radiance scattered by a perfectly conservative Lambertian surface (a conservative surface is defined as a surface that scatters all incident radiation, absorbing none).

The reflection factor evidently lies in the range of zero to one. It may be different for different wavelengths; thus you deal with a "spectral reflectance factor" or "spectral reflectance" (short for "spectral reflectance factor") or "reflectance" for short (though this is often confusing).

This is an extreme simplification of the actual physics that goes on when beams of radiation hit your eye when you happen to look at a typical scene in front of you. It is a setting that can be approximated quite well under (rather simple) laboratory conditions though. It is in this setting that I will discuss "object colors" in this chapter. It is important to appreciate the enormous amount of *stimulus reduction* going on here. Thus it is quite possible to experience "object colors" in actual scenes that are not to be found under the "object colors" generated in this canonical setting (figure 11.12). Many textbooks on colorimetry may leave you with the suggestion that the spectral reflectance factor is all you need to describe an object color. Don't fall for it. The world is much more interesting than that.

The *conservative* Lambertian surface may be denoted "white." This is not so much a subjective rating of what such a surface looks like as a *definition*

> A **white** surface is a surface that scatters all radiation incident upon it and that is Lambertian, that is to say, the scattered radiance is the same for all directions.

Notice that this is a *physical* definition, thus being "white" has nothing to do with color as experience. Whether such a surface would actually *look* WHITE when present in a certain scene would depend on the context. In all probability it would not.

Notice that a *perfect mirror* also reflects all radiation, but that it cannot be said to look white (figure 11.13). It is essential that the viewing direction with respect to the surface should not matter. Thus, the restriction to Lambertian surfaces is indeed *necessary*.

Figure 11.12 This looks like a "uniformy black surface" although the tones span the gray scale.

Figure 11.13 Some samples of colored paper (close to Lambertian) on my desk, photographed from only slightly different directions. The oval object at the center is a mirror. Notice that the mirror is not "white;" it isn't *any* (single) color! The mirror is an object that—as a piece of white paper—remits all radiation that impinges on it. But—unlike paper—the mirror looks different from any direction; it doesn't have a color of its own. Notice that my desktop—an average gray—is glossy, that is a little bit mirrorlike, too. This is the case for many smooth surfaces.

Carrying on the stimulus reduction, I define an *object color in the formal sense* in terms of a very simple scene, providing the minimal *context* necessary to move from an aperture color to an object color (in the subjective sense). Consider a "scene" consisting of an isolated, extended white surface, on which is placed a smaller Lambertian surface with some arbitrary spectral reflectance factor. In order to fully specify the scene, you would have to specify the shapes and sizes of these surface patches, their surround, the way the smaller surface is overlaid on the white surface (for instance, is there a visible "drop shadow"?), and so forth. I will refrain from that here, but you get the drift. Notice that the smaller surface (that is what I am really interested in) appears in the context of the white ("reference") surface. I will refer to the smaller surface as a "chip" and consider "chip" and "object color" (in the formal sense) as synonymous. Although I usually refer to the smaller surface as the "chip," the white reference surface is invariably implied.

In certain cases the white reference can be left fully implicit. This is the case, for instance, when you have a large gamut of widely different colors. If "for any wavelength" there is a color in the gamut with reflection near to one at that wavelength, then the "white" object is effectively implied by the totality of beams in the gamut[4]. This is important in practice, for in actual situations there will rarely be a true "white reference." Instead, the context of the global scene must be used (at least in so far as the scene is more or less uniformly illuminated).

11.2 Two Important Perspectives on Object Colors

An object color is not specified through a radiant power spectrum, but through a spectral reflectance factor. Formally the spectral reflectance factor is an operator $\hat{R}$ that transforms the incident irradiant power spectrum $|i\rangle$ into the radiant power spectrum of the scattered beam **s**. To keep things simple, I specify the incident irradiant power spectrum $|i\rangle$ in terms of the radiance scattered by a white surface. Notice that the spectral reflectance factor is a *diagonal* operator $\hat{R}$ (say). This is due to the assumption of a Lambertian surface. In real scenes there could arise off-diagonal terms due to fluorescence for instance. This makes it possible to write

$$|s\rangle = \hat{R}|i\rangle = \left(\sum_\lambda \varrho_\lambda |\lambda\rangle\langle\lambda| \right) |i\rangle,$$

where $|i\rangle$ denotes the beam as it would be scattered from a white Lambertian surface, $|s\rangle$ the beam as it is scattered from the actual surface, and ϱ_λ the spectral reflectance factor of that surface. Notice that ϱ_λ is a pure number between zero and one.

Now consider the color matching matrix (section 4.2)

$$\hat{\Pi} = \sum_k \mathbf{e_k} \langle \pi_k |.$$

The color of the beam scattered to the eye is

$$
\begin{aligned}
\mathbf{c} &= \hat{\Pi}|s\rangle = \hat{\Pi}\hat{R}|i\rangle = \\
&= \left(\sum_k \mathbf{e_k}\langle\pi_k|\right)\left[\left(\sum_\lambda \varrho_\lambda|\lambda\rangle\langle\lambda|\right)\right]|i\rangle = \\
&= \sum_{k,\lambda}(\varrho_\lambda i_\lambda\langle\pi_k|\lambda\rangle)\,\mathbf{e_k}.
\end{aligned}
$$

Thus the color coordinates of the scattered beam are simply

$$c_k = \sum_\lambda \varrho_\lambda i_\lambda \langle\pi_k|\lambda\rangle.$$

In this description "the eye" in the form of the projection operator $\hat{\Pi}$ operates upon the beam $\hat{R}|i\rangle$ scattered to the eye. This is just standard colorimetry, and it describes the physical situation in a direct manner.

Notice that $\sum_\lambda \varrho_\lambda i_\lambda \langle\pi_k|\lambda\rangle$ could be factored in various ways. In one interpretation you have the standard colorimetric sequence. The "eye" operates on the scattered beam that indeed hits the eye in the physical sense. In an alternative interpretation things are quite different, though. Here the "eye" is represented by an operator that operates on the *reflectance factor* ϱ_λ. The reflectance factor doesn't physically "hit the eye"; here the eye "seeks out" the reflectance factor, whereas the eye itself has been "changed" by the incident beam. This is less strange than it may sound, for in the case of object colors the eye indeed looks for "objects" rather than "beams" (the reflectance factor rather than the scattered beam) taking the context into account (thus "the eye is changed by the context"—here the incident beam in the form of the white substrate). Although these interpretations are quite different (figure 11.14), they formally lead to exactly the same result, thus either interpretation is (at least formally) valid in the sense of not being false.

The latter interpretation is the *common sense* interpretation. A banana looks yellow, either at noon (when the daylight is bluish) or at sunset (when the daylight is reddish). The banana is always the same, whereas it is the eye that varies. The former interpretation is that of the hardcore scientist. You never see a banana in the first place; all that happens is that certain beams happen to hit your eye. In such a mindset it is *paradoxical* that the banana should look the same at noon and at sunset (how can things that are

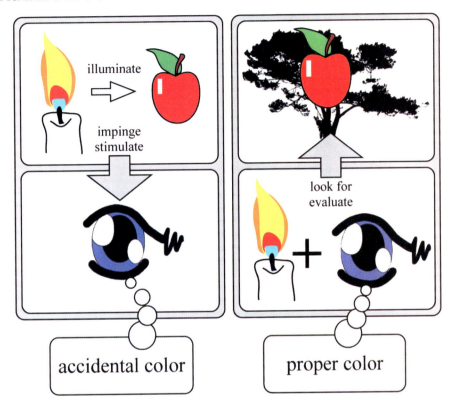

Figure 11.14 Left, the account of the hardcore scientist; right, that of "folk psychology". "Accidental colors" are as close to colorimetric colors as it gets. To see these takes a genius like (perhaps) Monet. "Proper colors" are the colors of things. The eye discounts the illuminant and uses all contextual constraints it can get. Anyone can see these.

different appear the same?) a strange fact ("color constancy") which requires an explanation (such as Helmholtz's "unconscious inference"). In this chapter I will stress the former (commonsense) view of the matter.

11.3 Formal Description of Object Color

In order to describe the situation I will represent the reflectance factor not as a diagonal operator

$$\hat{R} = \sum_\lambda \varrho_\lambda |\lambda\rangle \langle\lambda|,$$

but as a vector

$$|r\rangle = \sum_\lambda \varrho_\lambda |\lambda\rangle,$$

and the eye not through the projection operator

$$\hat{\Pi} = \sum_k \mathbf{e_k} \langle \pi_k |,$$

but the related operator (changed by the irradiance)

$$\hat{\Pi}^\star = \hat{\Pi}\,\hat{I},$$

where

$$\hat{I} = \sum_\lambda i_\lambda |\lambda\rangle\langle\lambda|,$$

(a diagonal operator), thus

$$\hat{\Pi}^\star = \sum_k \mathbf{e_k}\langle \pi_k| \sum_\lambda i_\lambda |\lambda\rangle\langle\lambda| = \sum_{k,\lambda} i_\lambda \langle \pi_k|\lambda\rangle \mathbf{e_k}\langle\lambda|.$$

Notice that this involves a change of the physical dimensions. The vector $|r\rangle$ is dimensionless, rather than a radiant power density, whereas the operator $\hat{\Pi}^\star$ is not dimensionless, but a radiant power density. This simply reflects the change of viewpoint. Now you have

$$|c\rangle = \hat{\Pi}^\star|r\rangle = \sum_{k,\lambda} i_\lambda\langle\pi_k|\lambda\rangle \mathbf{e_k}\langle\lambda| \sum_\mu \varrho_\lambda|\mu\rangle = \sum_{k,\lambda}\varrho_\lambda i_\lambda\langle\pi_k|\lambda\rangle\mathbf{e_k},$$

which is (of course) the same as I found before. This is the formal description of the intuition drawn in figure 11.14:

In

$$|c\rangle = \hat{\Pi}^\star|r\rangle \text{ with } \hat{\Pi}^\star = \hat{\Pi}\hat{I},$$

the "adapted eye $\hat{\Pi}^{\star}$" works on the "object $|r\rangle$", whereas in

$$|c\rangle = \hat{\Pi}|i^\star\rangle \text{ with } |i^\star\rangle = \hat{R}|i\rangle,$$

the "colorimetric eye $\hat{\Pi}$" (that is the color matching matrix) works on the "beam scattered to the eye $|i^\star\rangle$". Although the descriptions are completely different from a conceptual perspective, they are formally equivalent.

11.3.1 "White" versus "bright" and "black" versus "dark"

Because of the definition of the reflectance factor the scattered beam is always dominated by the beam scattered by the white substrate. As a consequence all object colors are necessarily dominated by the white color, irrespective the nature of the incident beam.

It makes sense to define the *achromatic* beam as the beam scattered by the white substrate. In fact, any other choice would be irrational. In this context it evidently makes sense to define an achromatic *beam*, rather than an achromatic *color*. That is indeed the reason why I insisted on achromatic beams instead of colors in the context of simple (aperture color) colorimetry (chapter 5), although this might have seemed arbitrary and unmotivated in that context. It is indeed more logical to develop the colorimetry of object colors out of the theory of the colorimetry of object colors than vice versa, and only didactical reasons made me follow the conventional path. In the setting of object colors the achromatic beam is given *ab initio* instead of being introduced in a *deus ex machina* fashion. This brings a coherence to the topic that is sadly lacking in the conventional "bottom up" (aperture colors prior to object colors) development.

Because all achromatic colors are dominated by the white color, the set of achromatic colors consists of the finite segment bounded by the white color and the origin, that is, the black color. Since they are mixtures of white and black, I will refer to these colors as the "grays" and refer to the segment as the "gray axis." Notice that the set of achromatic aperture colors is the half-line at the origin in the direction of the achromatic beam; thus achromatic aperture colors can be of arbitrarily high intensity. Such colors run between "pitch dark" (the origin) and "dazzling bright" (arbitrarily far from the origin). The "brightness" resides in $\mathbb{R}^+$. On the other hand, the grays run from "black" to "white". The "lightness" resides in $\mathbb{I}^1$. Notice the difference between "dark" (said of an aperture color) and "black" (said of an object color) and "bright" (said of an aperture color) and "light", or even "white" (said of an object color).

> Black and white are *surface properties*, whereas dark and bright are *properties of patches of light*.

The distinction is of cardinal importance. Many unfortunate confusions can be traced back to a confusion of this ontology.

The full intuitive understanding of these matters dates from the early nineteenth century and is due to the Danish painter Philipp Otto Runge (figure 11.15). Even in modern work you often see confusions though. In appendix K.2 I offer some thoughts on the nature of BLACK and WHITE.

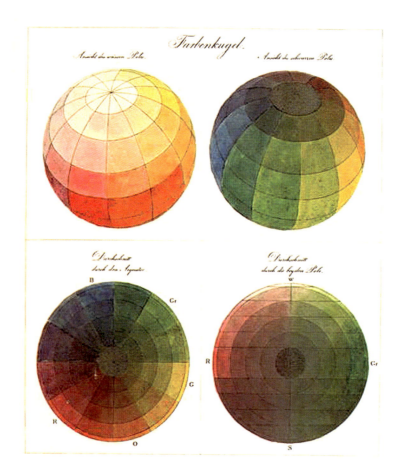

Figure 11.15 Philip Otto Runge's *Color Sphere*. In the early nineteenth century this Danish artist was the first person to get the topology of object colors exactly right.

11.3.2   The set of object colors as a subset of color space

As you have seen, the brightest colors for a given chromaticity, granted the fact that the beam is dominated by the given achromatic beam, are the Schrödinger optimal colors. It follows that the most luminous object colors are generated through spectral reflectance factors that are either zero or one, with at most two transitions throughout the spectral range. I will refer to such brightest object colors as "optimal" (in Schrödinger's sense), too. You may conclude that all object colors for a given incident beam are contained in an "object color solid" whose boundary is made up of optimal object colors.

When you vary the spectrum of the incident beam the object color solid will change *even though the white color is kept invariant.* For a given dominant wavelength the limiting case is reached if the incident beam is composed of a monochromatic beam of that dominant wavelength mixed with a monochromatic beam of the complementary wavelength such as to keep the white fixed (if necessary substitute the appropriate purple). Since these two beams are generators of the spectrum cone, it has to be the case that all color solids for a fixed white color are contained within the double cone that is generated as the intersection of the (filled) spectrum cone and the (filled) inverted spectrum cone at the white point. This double cone is the common envelope of all possible color solids for the given white point. Thus you obtain a complete overview of the *gamuts of possible object colors.*

Just to give an example of the intrinsic difference between patches and chips, let me consider the case of YELLOW paints again (as mentioned in earlier chapters). What is the best (colorwise that is) YELLOW paint? I mean from a fundamental, scientific perspective, not the best paint available in your local paint shop. When I ask students who have followed a course in colorimetry, the answer is usually:

> *The paint should reflect a monochromatic beam (false) for monochromatic beams have the highest saturation (true), thus the spectral albedo should be 100 percent at 580 nm (or whatever the wavelength of "yellow" is) and zero for all other wavelengths.* (WRONG!)

The mistake they make is to confuse chips with beams. A chip that reflects only at a very narrow spectral range ("zero" makes no sense, of course) will be much darker than the white substrate of the chip, and the more "monochromatic" you make the paint, the more it will look like black paint. In order to be *brightly* colored the chip must reflect much radiation and thus have substantial reflection over a wide range of the spectrum. When you go out to buy a pot of best yellow paint and put it before the spectroscope, you will notice that it reflects most radiation from the cyan to the deep red end of the spectrum, that is to say, roughly *half* of the spectral range (figure 11.16). There is indeed hardly anything "monochromatic" about YELLOW paint. When you take an almost monochromatic "yellow paint" (which looks BLACK to most of us) and put it in a dark room, illuminate it with an intense white beam, but without white substrate—that is to say, when you turn it into a patch—the paint will indeed look BRIGHT YELLOW. The moment you put the painted chip on a white substrate (turning it from a "patch" into a "chip" in the formal sense), it becomes BLACK again. This forcefully illustrates the importance of my definition of what a "chip" is.

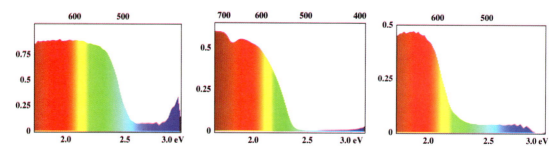

Figure 11.16 Spectral albedo of a YELLOW flower, an ORANGE (marigold) flower, and of a RED berry. If the colors are fairly saturated, your guess at the spectral albedo is likely to be in the right ball park.

What if an object color is "impossible", that is to say, when an "object color" is not dominated by the achromatic beam and thus cannot be "explained" through a spectral reflectance factor less than one? Such a color might still be *possible* for some achromatic beam of the same color as the actual achromatic beam, or it might be *impossible for any achromatic beam of that color*. In the latter case it will lie outside the spectrum double cone for the given achromatic color. In such cases it turns out that observers often perceive not an "object color" in the sense of some "illuminated surface color", but as the color of a "luminescent" surface, or even a veritable "light source". This is but one example of a color "seen as" some physical phenomenon.

Other examples include colors looking like translucent or transparent overlays or colors that seem to belong to some world behind an aperture (window), and so forth. A certain scene, or part of a scene, in which all color patches look like "object colors" appears as an "illuminated" scene made up of (typically opaque) surfaces that (spectrally selective) scatter radiation from some "source" or "illuminant" (not necessarily present in the scene) towards your eye. Such (parts of) scenes are also perceived as existing in a well-defined "luminous atmosphere". In the simplest case the luminous atmosphere is just due to an illuminant, but often it includes some luminous medium, such as when you look at a distant landscape through a bluish haze.

Seeing a scene invariably involves parsing it in terms of luminous atmosphere and illuminated objects. The objects seem to possess a "proper color" and appear in a color that is somehow due to the proper color modified by the luminous atmosphere. These are evidently difficult matters and fall largely outside the proper scope of the colorimetry of object colors although the literature is rife with ontological confusions. I will return to the subject in due course.

When you vary the illuminant, the white point moves through color space. (There is no way to vary the black point at this time. For that you would have to add "air light". I will discuss that topic in chapter 17.) The color solid "moves with the white point," because it is invariably a fusiform body (though differently shaped in detail) spanned up between the black and white points. This perhaps suggests, in a very loose way, that you might map object colors for different illuminants by somehow "normalizing" with respect to the location of the white point. This is more or less what von Kries suggested and it is generally known as a "von Kries transform." I discuss the topic in appendix K.1. The von Kries transformations are done for you on the fly if you set the white point of your electronic camera to automatic. It is often suggested that the human observer works a bit like that and that it is one step on the way to color constancy. Color constancy can be defined in various ways, but it has to do with the phenomenon that bananas look just as YELLOW at sunset as they do in the shade at noon. This is puzzling to the physicist, because the beams scattered to the eye by the bananas are quite different on these occasions. The commonsense explanation is simply that "bananas are YELLOW" of course. Whether this is to be considered naive or a notable wisdom is not something scientists agree upon.

In figure 11.17 I indicate the important regions of color space that separate qualitatively different cases. Notice that the regions E and F are well defined, even if there is no context at all, Regions C, D and the union of B and A are defined only if the achromatic color is defined. Regions A and B can only be differentiated once the spectral composition of the achromatic beam is known. In real scenes even the position of an "achromatic point" is only a fuzzy notion that moreover may differ from region to region in the scene. As the observer parses the scene lacking data has to be provided somehow. Perception is indeed largely "controlled hallucination" in which the observer hunts for evidence that might conflict with the current best guess. Thus one should not be amazed when colored patches are classified ("object color", "light source", "glowing", "glossy", "metallic", "wet", "shaded", 'in shadow", and so forth) dynamically, being updated or adjusted as the "hallucination" ("psychogenesis" perhaps sounds more positive but indicates the same happening) progresses.

11.4 Same Looks Under Different Illuminants

In many applications you would like to have different surfaces (characterized by different spectral reflectances) look the same under different illuminants. Examples are objects that are composed of different materials, for instance, plastic buttons on a cotton shirt. It is not possible to use identical pigments for these materials. But then, when you match the color of the buttons under daylight conditions, they are unlikely to match also under incandescent light, and vice versa.

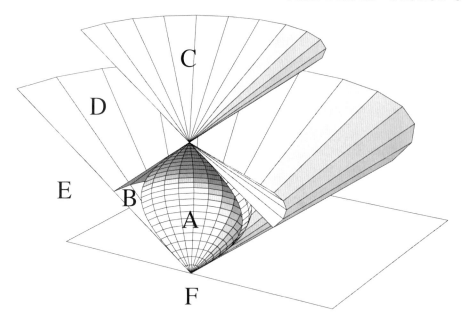

Figure 11.17 Various regions in color space shown in a planar section through the gray axis: A: The region of object colors; B: The region of colors that might occur as object colors for different illuminants of the same color, C, andD: The region of colors that might occur as object colors if the white point were different. For region C the white point would have to dominate the actual white point, E and F: The region of colors that could not possibly occur, no matter what the position of the white point were. (The figure is based on the Grassmann model, but it represents the generic layout.)

This is not a desirable state of affairs if you are in the business of designing objects that are supposed to sell primarily because of their looks.

It is easy enough to set up a scheme that lets you handle such cases. Notice that since the eye is changed by the light source, you have two different color matching matrixes, one for daylight and one for incandescent light for instance. When you stack these matrixes, you obtain a composite color matching matrix composed of six color matching functions. It is as if you are dealing with a hexachromatic color vision system instead of a trichromatic one.

Since there is nothing really special about trichromacy, you can apply all colorimetric methods to the hexachromatic case without further ado. The hexachromatic color you want is simply the trichromatic color with its coordinates repeated once. Applying the hexachromatic pseudoinverse then yields a hexachromatic fundamental component, which you can split into two trichromatic components, which are the required reflectances up to additions of metameric blacks. You can use any of the standard methods to find physically realizable reflectance spectra.

Notice that you can use the same method to find surfaces that look one (specified) color under one illuminant and another (specified) color under another illuminant. The method is not limited to mere two illuminants either. The real problems are problems of *praxis*, not of *theory*. In practice, it is very hard (most often simply impossible) to realize spectral reflectance that satisfy a given prescription.

Ontology of colorimetries

By now you may well be slightly confused by the sheer number of "colorimetries" that have been discussed. It is important to have a clear notion of their ontologies and interrelations. So far I have discussed:

basic (or "austere") colorimetry (chapter 4),
colorimetry with achromatic beam (chapter 5),
metrical colorimetry (chapter 8),
Schopenhauer's "Parts of Daylight" (chapter 9),
"higher" colorimetry (involving Schrödinger's criterion, chapter 10),
colorimetry of object colors (chapter 11).

Most texts on colorimetry follow a similar sequence as I have done in this book, which is—very roughly speaking—the historical sequence. The conceptual relations are rather distinct from this historical sequence, though.

The "higher" colorimetry clearly is a separate discipline since it does not keep to the basic colorimetric response reduction. It draws on the "lower" colorimetries, but stands well apart from them. The "metric" derives from psychophysics, that is, judgments concerning relations between colors.

Metrical colorimetry is an extension of mere "affine" colorimetry that depends on an inner product in the space of beams, that is, an additional structure in our physical description of radiation. This has nothing to do with human judgments (psychophysics). To all affine colorimetries you may construct parallel metrical colorimetries. All these are "true" colorimetries in the sense that the colorimetric response reduction is fundamental. Thus there are no simple relations between "metrical" colorimetries and the Riemann "metrics" of the higher colorimetries. These notions of "metrical" have distinct ontologies.

Thus you are left with the relations between

basic colorimetry,
colorimetry with achromatic beam,
colorimetry of object colors.

Here there exists a simple hierarchical relation in the sense that the colorimetry of object colors includes the colorimetry with achromatic beam as a limiting case (very dark object colors, the environment of the black point) whereas the latter again includes basic colorimetry as a special case (simply forget about the achromatic beam). Thus the rational sequence of introducing these disciplines is exactly the *reverse* of the historical sequence. I did not follow this route for didactical reasons, but at this point you should rethink matters and make sure that you understand the ontology of the various colorimetries well. This is important, for failure to understand these relations often leads to unfortunate statements, even in the modern literature.

11.1 Von Kries transformations

Von Kries was the first to come up with a transformation in which the human fundamental response curves are scaled.[5] The idea derives from physiology and is simply that the "channels" devoted to the signals carrying the information for the long, medium, and short wavelength sensitive retinal photoreceptors might be subject to independent "gain control", somewhere in the system, possibly already in the retina. This would be termed "chromatic adaptation". Then the idea is that a similar gain control or scaling would take place in the space of perceived colors, with the result that color impressions would be subject to such scalings. If the illuminant changes the gain control would set the white point to standard white and all perceived colors would change accordingly. The concept is very similar to the way the white point is set in modern electronic cameras.

In the figures 11.18 through 11.22 I show how simple von Kries transformations applied to an RGB image can be used to move its color balance or white point. Von Kries transformations roughly allow you to simulate the effects of changes of the slope of the illuminant spectrum (common in digital cameras, and present in all image processing programs, often called "change of color temperature") and the effects of changes of the curvature of the illuminant spectrum (rare in digital cameras,[6] and present in most image processing programs).

Notice how strange an idea this really is. Any linear transformation of color space would change nothing as far as colorimetry is concerned, because none of the colorimetric equations would be affected at all. The idea is not about (colorimetric) colors at all, but about *qualia*. Hidden in the concept is some kind of "psychophysical bridging hypothesis" that is not made explicit.

Figure 11.18 A colored scene.

Figure 11.19 A von Kries transformation has changed the spectral slope so as to favor the blues.

Apparently one assumes that the *quale* is somehow causally connected to a set of physiological signal strengths. The idea forces you to subscribe to the "photometer fallacy"; its ontology is alien to that of "hardcore colorimetry."

If this idea (in some general, abstract formulation) is to be viable at all, you have to reformulate it as a "cognitive model" (in the parlance of the turn of the millennium). This lifts it to the correct ontological level, namely the mind, because that is where the qualia belong. The physiology is on a different ontological level and is irrelevant to the issue.

The Von Kries transformation is usually introduced in the context of the "color constancy problem". It is hard to say whether it counts as an "explanation" of color constancy at all though, even in the simplest setting. It depends upon your philosophical stance, I guess.

I define a *generalized* Von Kries transformation as a linear transformation in color space that has three real and positive eigenvalues. Such a transformation is only slightly more general than the true von Kries transformations, but it has the advantage of being independent of physiological notions. The special von Kries transformations are based on assumptions concerning the physiological implementation that I deem not particularly relevant.

One way to reformulate the color constancy problem is to turn it into a problem of gamut equivalence. Suppose I have a gamut that is due to a scene illuminated in a certain way, and another gamut that is due to the same scene illuminated in another way. Either gamut is a cloud of points in (colorimetric!) color space. You might say that the color constancy problem is solved (for this instance) if you can establish a point-by-point correspondence between these clouds. Notice that this involves a type of *response reduction* that is quite similar to that used in colorimetry proper. A few immediate consequences are worth noticing:

Figure 11.20 A von Kries transformation has changed the spectral slope so as to favor the reds.

Figure 11.21 A von Kries transformation has changed the spectral curvature so as to favor the greens.

Figure 11.22 A von Kries transformation has changed the spectral curvature so as to favor the purples.

— The notion is easily psychophysically operationalized for you merely require observers to indicate correspondences. A correspondence is a fully objective fact. This is science;

— In order to establish correspondences, the clouds need not be geometrically identical. A Procrustean metric would do fine.[7] This is important, because changing the illuminant will change even the topology, and points may split or merge;

— The color constancy problem might be *partially* solved; it is not all or nothing. Some correspondences may be in doubt or even totally unresolved, whereas others might be clear-cut. Thus the color constancy problem framed in this way is *robust* and *degrades gracefully*.

In this formulation, the von Kries transformations fits in very well. Since the physiology is irrelevant you may as well consider general linear transformations. Given two gamuts you subject one or both[8] to linear transformations in order to make them fit as well as possible (this is the Procrustean method). For instance, given two gamuts with different white points you might transform the two (different) Schopenhauer RGB crates to unit cubes. Then you can compare the clouds—which are both contained in the unit cube—directly, for instance using Euclidian distances between points in the unit cube. You can use any standard method to establish correspondences. (See chapter 13 for an example.)

It is by no means easy to analyze such a procedure. The reason is that the gamut changes under changes of illuminants are typically not at all simple. (For instance, they are by no means linear transformations, not even general isomorphisms.) As I indicated earlier, it is quite possible for points to split or merge; thus even the topology can only be preserved on a suitably coarse scale.

11.2 On the nature of "black" and "white"

The colorimetric "black point" is well defined, even in the absence of any knowledge of the illuminant. In contradistinction, the colorimetric white point has to be assigned by fiat. Given the white point, the gray segment is defined, too.

"Black" and "white" are evidently object colors in the colorimetric sense. They are special in many ways though, one reason being that their hue remains undefined. Black does not even *have* a chromaticity. White has a well-defined chromaticity, but one that cannot be assigned a hue, much for the same reason that the north pole has no well-defined geographical longitude.

In any (infinitesimal) neighborhood of the black point (or the white point for that matter) you find colors of any possible hue. In that sense "black" may be said to represent *any* possible hue, and so does white. This is well understood by the visual artist. They avoid a "dead black." Any black has some hue; it is either cool or warm. Thus you avoid black from the tube (for example, "lamp black"), but instead you mix your blacks from pairs of complementary colors. It is somewhat easier to stand a pure white, but most whites are either cool or warm. The "color" of white, blacks, and grays is also acknowledged by photographers in black and white, who often make a cult out of the slight hues of their prints.

"Black" and "white" are somewhat ambiguous because different from, though closely related to, "dark" and "bright." The "shades" of a pure color are again pure colors, merely shaded, but they are also "mixtures with black". Thus "mixing with black" is similar to "darkening" (by decreasing the illumination). Mixing with white is different in kind because it is not at all similar to "brightening" (in the sense of increasing the illumination). Mixing in white pro-

duces "turbidity", giving a pure color a "material" look. This is the reason why "black" looks "deep" whereas "white" looks "opaque". A uniform white of large extent appears as a luminous mist whereas a uniform black of large extent looks like deep space. This is equally true for colors. A pure color of large extent looks distant and deep (think of a dark blue sky), whereas a "pastel color" of large extent looks like a material surface or a not too distant colored and luminous misty volume.

This is no doubt due to generic phenomena of ecological optics. A colored, transparent medium can be seen only when in front of some distant light, and it will transmit a pure color to the eye when the layer is a thick one. The red of a glass of wine as seen against a candle is a key example. On the other hand, "white" must be due to multiple scattering, thus a turbid medium of which you mainly see the nearest layers because the medium is opaque in thick layers. A glass of milk, frontally illuminated is the key example here. Sometimes a white area may appear as a diffuse source of light; in that case, it looks superficial and opaque.

A gray always looks turbid and can never appear as an extended source. Hence grays generally have a "muddy" appearance. This applies equally to mixtures of colors with grays. The only way for grays to avoid the "muddy" look is to appear "silvery," that is as a somewhat low reflective metal surface.

A convenient way to study such effects is to build yourself a "Hering box."[9] (See figure 11.23. I constructed mine from an empty container, some black cardboard, and a chopstick.) This lets you study the effect of adding BLACK to an object COLOR. It is most instructive to see a piece of "WHITE paper" turn GRAY or to see a piece of "ORANGE paper" turn BROWN.

Figure 11.23 The "Hering box" lets you compare a piece of paper (on the top of the box)
with a piece of identical paper (on the stick in
the box) under different illumination. You look
from the top and see both pieces simultaneously,
one through a hole in the top. By rotating the
paper seen through the hole with respect to the
light source, you can add black to it. Then
"WHITE paper" becomes GRAY and "ORANGE paper" BROWN.

The box has been cleverly designed to hide
the fact that one piece of paper is in shadow. In
real life this is typically obvious, and the shadow
is (quite automatically) discounted. A fuzzy darkish blob is often discounted this way (figure 11.24)
and is not seen as a stain. Encircling it with a
black line causes the blob to look like a stain,
while shifting the black line makes it look like a
shadow again. This experiment (best done with
a real shadow and a black marker) is also due
to Hering. It makes for an instructive demo for
small groups. You use a black felt marker pen
to draw the outline after having first pointed out
the shadow.

Figure 11.24 Hering's experiment with the
shadow.

Exercises

1. The BRDF [FOR PHYSICISTS] Prove that the BRDF of a Lambertian surface of unit albedo is $1/\pi$.

2. Construct "the eye" [EASY] Numerically construct the operator that represents "the eye" for average daylight. The operator should yield the color of a surface when applied to its spectral reflectance function. Repeat this for various illuminants.

3. Yellow paints [EASY] Study the colors of "yellow paints," that is to say, spectral reflectances that yield a color of dominant wavelength 580nm (say). You will have to think about how to generate (parameterized) instances first. Check that you cannot do better than the semichrome.

4. Von Kries transformations [EASY] Construct the (parameterized) von Kries transformations in the strict sense. (You need the spectral sensitivities of the fundamental mechanisms. These can be downloaded from the Internet.)

5. Metameric object colors [NOT TRIVIAL] Suppose you need to design two spectral reflectance functions that will evoke equal object colors under two different illuminants. (Think of a lipstick and matching nail polish concocted from different materials under daylight and incandescent illumination.) Solve the problem by setting up a six-dimensional fundamental space and building the corresponding matrix-R. (This is really an easy problem, but you may have to think it over.[10])

6. Von Kries transformations [EASY] Can you map spectrum double cones (intersection of spectrum cone at the black point with inverted spectrum cone at the white point) for different illuminants on each other through general von Kries transformations?

7. The illuminant spectrum [NOT TRIVIAL] Discuss choices to make in the definition of a "nice" achromatic beam given some set of spectra that are supposed to derive from it. The spectrum of the achromatic beam should dominate all spectra from the set, of course. Compare the upper envelope, the convex upper envelope and spectra of some fiducial shape (Planckean spectra, for instance).

8. Revealing hidden dimensions [CONCEPTUAL, DIFFICULT IF YOU WANT] Remember Riemann's multiply extended manifolds, the space we move in, and the space of colors. In the space we move it the "depth" dimension is hidden, but you can reveal it by taking a side-step. For instance, seeing someone straight in the face the length of the nose is in doubt, though it is revealed in a side-view. Consider a change of illuminant as "taking a side-view" in color space. Is it possible to see all dimensions of spectral reflectances? Obviously, *yes*, for if you illuminate with many almost monochromatic sources you effectively perform a spectral analysis. How does this work for arbitrary illuminants? What if you don't know their spectrum? Work out the comparison with depth and shape into some detail.

Chapter Notes

1. Colors may appear in various other aspects. They need not be of either the "aperture" or "object" (in the sense of surface-based) variety. For instance, they can appear as volume colors (the "red" seen in a glass of wine), and so forth. This is an important topic, but it properly belongs to psychology, so I skip it here. Good sources to start are D. Katz, "Die Erscheinungsweisen der Farben und ihre Beeinflussung durch die individuelle Erfahrung," *Z.Psychol.* Ergänzungsband 7, pp. 1–425 (1911); R. Hunter, *The Measurement of Appearance,* (New York: Wiley, 1975); J. L. Caivano, "Appearance (Cesia): Construction of scales by means of spinning disks," *Color Res.Appl.* 19, pp. 351–362 (1994); —, "Cesia: A system of visual signs complementing color," *Color Res.Appl.* 16, pp. 258–268 (1991).

2. Berkeley remarks:

> *II. It is, I think, agreed by all that Distance of it self, and immediately cannot be seen. For Distance being a Line directed end-wise to the Eye, it projects only one Point in the Fund of the Eye. Which Point remains invariably the same, whether the Distance be longer or shorter.*

G. Berkeley, *An Essay Towards a New Theory of Vision,* (Dublin: Aaron Rhames, MDCCIX).

3. Two famous books mark the maturation of photometry as a science. Remarkably, both date from 1760. Most modern radiometric concepts can be traced to these remarkable texts. P. Bouguer, *Optical Treatise on The Gradation of Light, translated with an introduction and notes by W. Middleton,* (Toronto: University of Toronto Press, 1961); J. H. Lambert, *Photometry,* (New York: Illum.Eng.Soc. North Am., 2001).

4. The white surface is due to a spectrum that is the upper envelope of the spectra of all possible scattered beams. Thus any color such that the inverted spectrum cone at that color contains all colors in the scene is the color of a possible white object. The (generically unique) color with that property that is closest to the origin may be used as an estimate of the white object. If the scene is sufficiently varied such an estimate is likely to be quite close.

5. J. von Kries, "Chromatic Adaptation," in *Sources of Color Vision,* edited by D. L. MacAdam (Cambridge, MA: MIT Press, 1970), pp. 109–119; Good accounts are to be found in many texts, see for example M. Richter, *Einführung in die Farbmetrik,* (Berlin: Walter de Gruyter, 1976).

6. Typical digital cameras let you set the "color temperature," that is, the spectral slope. There isn't even an accepted term for something like spectral curvature. Some of the more recent cameras (for example, the Olympus-500) let you control spectral curvature, though. No doubt this is to be applauded, though it would be a shame if all your photographs would end up looking like made in "average daylight".

7. D. G. Kendall, "Shape manifolds, Procrustean metrics, and complex projective spaces," *Bull.London.Math.Soc.* 16, pp. 75–80 (1984).

8. Either you transform one cloud such as to fit the other, or you transform both clouds to some canonical form. It makes little difference.

9. E. Hering, *Zur Lehre vom Lichtsinne,* (Vienna: Gerold's Sohn, 1878).

10. S. A. Burns, J. B. Cohen, and E. N. Kuznetsov, "Multiple metamers: Preserving color matches under diverse illuminants," *Color Res.Appl.* 14, pp. 16–22 (1989).

Chapter 12

Color Atlases

A simple way to "measure" colors is to have a human observer *name* them. In many cases this works perfectly fine. When people mention RED fire trucks, YELLOW bananas, or GREEN leaves to you, you are likely to understand what they mean, no problem there. But in other cases it doesn't work at all. When you need to order a pot of paint that matches the color of your car, it is rather unlikely that this type of "measurement" will fit the bill. In all likelihood the painted spot will stick out like a sore thumb. A much better (even surefire) way to get what you need is to order a paint that (physically) matches a *sample* taken from your car. For a match is a match (is a *match*), and you can fully depend on that. Two physically identical objects will have the same color for all observers and under all light sources. This may not count as a "measurement" in the conventional sense, but it it is certain to *work*. The problem is that you will need a physical sample for any color you need to indicate. Moreover, you depend upon the manufacturer to be able to repeat the paint recipe. In typical cases, the most you may hope for is a mere *visual* match under "standard" (daylight, say) conditions. Paint manufacturers are aware of the problem and distribute sets of colored "chips" that demonstrate visually what colors they sell (figure 12.1). Such sets are often fitted with a color naming system that allows you to to specify a color by phone or email (you won't have to mail an actual chip).

Such systems of physical chips augmented by a "denotation system" are the simplest type of color atlases. Thus color atlases are very practical things and—as I intend to show— are also *conceptually* very interesting entities. That is why I have devoted a full chapter to them.

Naming is not as simple as it may seem. There are 5,411 color names in the NBS/ISCC Dictionary of Color Names,[1] which stands in stark contrast to the set of "eleven colors that are never confused" as proposed by Boynton[2](figure 12.2).

Figure 12.1 A color swatch book may be in arbitrary order. It is the most primitive type of color atlas.

If you need to match something, then there is no better way than to compare actual samples. If they look the same to you they look the same to you, period. But notice that:

- they may look different to you under another illuminant;

- they may look different to other observers;

- they may look different to you in different spatial attitudes.

However, things could be much worse, for

- instead of an actual sample, you may have to use a "chip" from some collection;

- you may not be able to see both items simultaneously under the same illuminant;

- you may be forced to make the judgment under an illuminant that is not of your choice;

- you may be forced to leave the judgment to another person;

- and so forth.

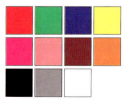

Figure 12.2 The "eleven colors that are never confused," according to Boynton.

Standardizing viewing conditions is obviously important. Knowing something about the samples, light sources, and observers is also desirable. When you use the comparison method you often have to act on incomplete information. For instance, when you use the color sample book of some paint company you will probably be unaware of the spectral reflectances of the samples, whereas—most likely—these are different from those of the paints themselves. Thus you are not in the standard colorimetric paradigm. It makes some sense to wonder whether some color atlases[3] might be more desirable than others and what the nature of possible differences might be.

I will start with making a distinction between the *conceptual structure* of color atlases and their *physical implementation*.

There exist two *categorically different* conceptual structures:

- color atlases based on *eye measure*, and

- color atlases based on *colorimetric principles*.

Since these types are indeed categorically different it is conceptually impossible to compare them. Perhaps unfortunately, a practice has developed in which people actually "mix" these types to produce mongrel structures of indefinite ontological nature.[4] As an honest scientist, I consider this to be in particularly doubtful taste. Thus I will *have* to consider "relations" between these types. The literature on this issue is very muddled to the extent that authors often (even commonly) fail to make the distinction at all, with (conceptually) disastrous consequences. This is a warning that you would do well to mind.

The *physical implementation* of color atlases is a nontrivial issue, but since it is more of an engineering nature I will skip the details. There are many issues to consider:

- whether to define the chips by spectral reflectance or by color;

- the material and surface finish of the chips;

- the layout of the chips in the atlas;

- issues of aging, weathering, and the like;

- instructions on how to use the atlas (how to define the illuminant, how to constrain the viewing conditions, and so forth);

- how to check calibration of the atlas over time and use.

None of these issues has a unique, obvious solution. For colorimetrically defined atlases the calibration yields no special problems, since you can use the familiar methods of spectroradiometry and/or colorimetry. Spectroradiometry is preferred, since it guarantees a unique atlas, whereas colorimetry only specifies (infinitely many different) metameric copies. It is different for the atlases based on eye measure. Such an atlas is (at least *conceptually*) only made *once*, yielding a "master" from which all other copies have to be "cloned." Making a clone again involves only colorimetry, one merely has to specify whether the cloning is done spectrally or by color. In the latter cases the clones are possibly different from the master, and this will show up when the illuminant is varied or different observers use the atlas. The master is *unique*, at least in principle, being the result of a series of judgments by a (single) human observer. Praxis is different, of course. Typically, people somehow average attempts by any number of observers, measure the result colorimetrically, and "smooth" the result in color space in an attempt to remove the effects of random judgment variations that naturally occur in such tasks. At some point you have to "freeze" the definition of the atlas, though. The resulting master is not necessarily the reflection of the eye measure of any single observer, and the difference need not to be confined to a smoothing of some random variation. Often people inject theoretical notions in the final definition of the master. In such (most frequent) cases you obtain a mongrel result of indefinite ontology. Even the early edition of the Munsell atlas was essentially of this type. It is probably best to consider such atlases as being of the eye measure type. They are evidently quite distinct from purely colorimetrically based atlases. You should understand that the literature often judges differently on this issue. Once the result of eye measure has been colorimetrically measured (in retrospect I mean), such an atlas is often considered an object that belongs squarely to colorimetry proper. An example is the Munsell atlas,[5] which is in common use (for mainly historical reasons, its use is pushed by the CIE) and is typically accepted as gospel. Such practice is a source of considerable misunderstanding in the literature.

12.1   Pure Eye Measure Atlases

Eye measure is not only concerned with qualia, although the basic categorical judgment into colors as being RED, GREEN, BLUE, YELLOW, PURPLE, ROSE, BROWN, ORANGE, BLACK, GRAY, WHITE, and so on is the most basic form of *absolute* eye measure. But eye measure also relates to *relative* judgments, that is, color relations. In the *judgment of relation*, the problem of qualia becomes less important or may even vanish (as in the judgment of colorimetric equality). There is a continuous spectrum here. An example of such a relative eye measure judgment is the judgment of a linear color series to be "continuous", or "equally spaced". Consider the series shown in figure 12.3. These look all at least roughly equally spaced, especially when you compare with an obviously unequally spaced series, or a permuted series as in figures 12.4. The permuted series looks "scrambled", the other series obviously "unequally spaced".

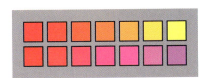 

Figure 12.3 Left: An analog color series, from one cardinal color to the next one (in this case R to Y and R to M), thus staying on the full color locus. You have a gradual hue change, but no admixture of either white or black. Right: Mixtures with achromatic colors are often of interest. Mixing with white "veils" colors and yields them a "pastel" appearance. This "veiling" is what you see when you add milk to your coffee. Mixing with black yields a color "depth", the opposite of mixing with white which yields a more substantial, surfacelike (literally "superficial", that is, not "deep") quality. Mixtures with gray produce "muddy" colors. The subjective appreciation of such effects appears to be closely correlated with generic optical interactions in the daily environment.

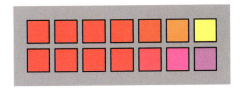 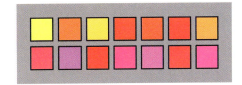

Figure 12.4 The "analog color series" of figure 12.3, but either unequally spaced or permuted.

Although the series of figure 12.3 look "obviously" equally spaced, this is less so for the series in figure 12.5 and perhaps even more less so for the series in figure 12.6. In the latter cases the series "stops in the middle" as you read it from left to right, and there is a sense that you could have continued it in various ways. In figure 12.7 these series have been mapped in the color circle. The "obvious" series are between "analog colors"; any path passing through (or near) the gray point is a questionable one. Such rules are indeed familiar to practicing colorists.

In figure 12.4 I have applied arbitrary permutations. Such scales look very obviously uneven, demonstrating that the impression of "even scale" is not entirely unfounded. Apparently not *every* scale looks "even" (which is not an logical impossibility is it?). There is some property of a series that *causes* it to "look even."

From experiences like this it is clear that you "carry a scale in your eye". You are able to directly judge geometrical relations in color space. Eye measure-based color atlases exploit this fact.

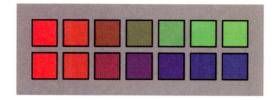

Figure 12.5 An "oblique color series" from one cardinal to the next (in this case, R to G and R to B), thus straying a bit off the full color locus. You have a gradual hue change, but it is accompanied by an admixture of gray. Painters generally avoid such trajectories.

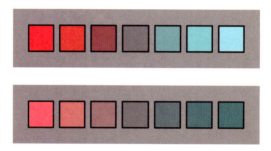

Figure 12.6 Top: A complementary color series, from one cardinal to the complementary one (in this case, R to C), thus leaving the full color locus and passing right through the median gray center of the color solid. You have a sudden hue change as you pass gray, and a gradual graying on either side. Painters use such trajectories to produce "lively grays", generally avoiding black/white mixtures. Bottom: Another "complementary color series", but this one is out of the level of full colors. It starts with a color near to white and ends with a color near to black, passing through the median gray center in between.

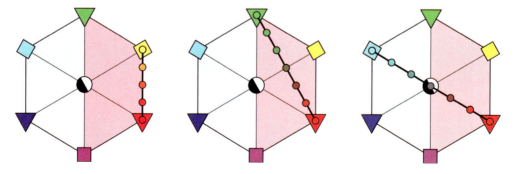

Figure 12.7 Paths from red, from analogous (left) to complementary (right) and in between (center), drawn within the color circle.

An *eye measure* atlas is typically based on some prior notions concerning the "perceptual attributes" of object colors. Since the object color solid is a three-dimensional volume, you need three independent attributes, such that any object color can be identified through a unique triple of attributes. The various eye measure attempts vary qualitatively in the choice of these attributes.

You also have to decide on the "grain" or "resolution" of the desired result. One solution would be to find chips that would differ in one JND of one of the attributes. However, such a method (apart from the sheer magnitude of the task) would yield an atlas of so many chips that it would defy its very purpose, just as a geographical map the size of the actual landscape is not particularly useful. Thus people generally decide on a "grain size" of several or (usually) many JNDs. The aim is to arrive at a grid of attribute values that is fairly regular in some sense.

Typical layouts are either cylindrical or spherical coordinate systems, although many different "systems" have been proposed or are in use. To arrive at such a regular grid is difficult and needs much iteration and smoothing. Once such a grid has been arrived at, the chips can be denoted by their (unique) triple of attribute values. Needless to say, this task is very difficult. Although the grid is likely to be fairly "perceptually regular" in any local neighborhood, regions that are far apart are likely to be less strictly related for the simple reason that it is near to impossible for an observer to compare "distances" that occur in mutually remote regions of color space.

It is here that most "systems" rely on prior theoretical notions to "smooth" the result. This is a tricky business, since any "theories" in this area are necessarily highly speculative, that is to say, psychology rather than colorimetry.

The structure of an eye measure atlas need not take any account of colorimetric facts. A good example is the Munsell atlas, which is probably the atlas in most common use today.[6] Munsell was a painter by profession who attempted to put some order in the gamut of object colors.[7] He used the gray axis as the backbone of his system, correctly perceiving that the grays can be linearly ordered on a finite linear segment. In drawings he put a sun at the white point and a rabbit's hole at the black point, thus showing that he did confuse the black-white with the dark-bright scale. From the gray axis Munsell placed colors in order of increasing saturation on "branches" at right angles to the gray axis, each hue branching off in a different direction, thus obtaining the "Munsell color tree".

The color tree is evidently a cylindrical coordinate system, thus the hues are ordered periodically on a color circle. Whereas this is an order that makes colorimetric sense, Munsell put no limit on the length of the branches. When you find and even more saturated color of a given hue, you simply attach it farther away from the trunk. Thus Munsell had no notion of the fact that the color solid is limited in volume (the most saturated colors are the Schrödinger optimal colors), even though he was familiar with Runge's color sphere.

The Munsell system is a pure eye measure system (with some conceptually rooted backbone structure) and takes no account of the colorimetric facts. This is perfectly fine, were it not for the fact that nowadays the Munsell system has been so integrated in colorimetric *praxis* that many color practitioners (and many text book authors) have lost track altogether.

12.2   Purely Colorimetrically Defined Atlases

A *colorimetrically* defined atlas that may serve as the paradigmatic example of such an entity was developed by Wilhelm Ostwald (a professor of chemistry who earned a Nobel prize on catalysis and worked on the theory of colors during his retirement) almost contemporaneously with Munsell's work. Unfortunately, the nature of this construct has been generally misunderstood. In many textbooks the Ostwald atlas (if mentioned at all) is presented as a minor German emulation of the Munsell atlas. It is true that Ostwald had visited Munsell early on and was aware of his attempts—and vice versa. Yet is is blatantly wrong to accuse Ostwald of plagiarism, as is often done. Oswald's atlas is *categorically different* from the Munsell one. It is important to distinguish between the *concept* of the Ostwald atlas and its *physical implementation* though. The conceptual structure is quite independent of any implementation. Indeed, the Ostwald atlas can be constructed from scratch from purely colorimetric principles[8] (you can easily program it on your computer); thus any implementation is merely incidental. This is a major difference from eye measure atlases, which are necessarily clones from some unique master.

Ostwald defined the chips by their reflectance factors and came up with a denotation system that is colorimetrically defined. In order to understand the structure of the Ostwald atlas, you need to regard both its physical (the spectral reflectances) and the colorimetric aspects. Both are intimately interwoven.

The spectral reflectances of the chips are based on Ostwald's notion of the "semichromes" (section 7.3.1). A semichrome is a Schrödinger optimal color of which the transitions are located at complementary wavelengths. Notice that this implies that the color of the achromatic beam (the illuminant) is known, otherwise the notion of complementarity is not defined. The semichromes are as different from achromatic as possible for any given hue, hence Ostwald also refers to them as "full colors" (in German,Vollfarben). The full colors make up the *color circle*.

Notice that the complementarity of chips does *not* require the prior adoption of an achromatic beam, though. The complementarity of chips is purely based upon their spectral reflectance factors and holds up for any illuminant and any observer. Such is indeed a major advantage over eye measure methods.

The spectral reflectances of the achromatic colors are defined as *physical* mixtures of black and white. The black and white chips are defined in the obvious way, that is to say, the black chip doesn't reflect anything and the white chip is a Lambertian surface of unit albedo. Ostwald suggests soot and flour as physical implementations. Then the grays are defined through mixtures of soot and flour, that is to say, implementations of Lambertian surfaces of albedos less than unity. Thus a gray chip can be defined by its white and black content, for instance 20 percent soot (black) and 80 percent white (flour).

Ostwald proceeds to define the colored chips as mixtures of white, black, and full color. Notice that this describes their *spectral reflectance factor*, and thus the definition is a physical one. Any chip in the Ostwald atlas is thus parameterized by its corresponding full color (a location on the color circle) and its color, white, and black content. The color, white and black contents add to one (or 100 percent), thus you have three independent parameters. (Figure 12.8.)

In order to complete the parameterization, you need to mensurate the color circle. Ostwald achieves this through his ingenious principle of internal symmetry (section 7.4). He then places the origin at the yellow point, a point that is uniquely defined as the full

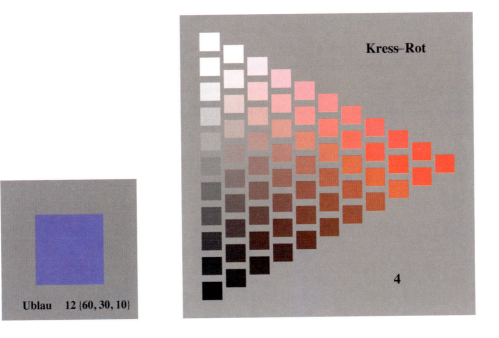

Figure 12.8 An Ostwald chip "U-Blau 12 (60,30,10)" (left) and the Ostwald page for "Kress-Rot". These examples were simply computed from scratch.

color that passes the long wavelength range of the spectrum, and has one transition at the complementary of the infrared limit. The orientation of the color circle is defined as going from yellow toward green. In order to obtain a numerical value for any full color, the only thing left to do is to define the range.

Ostwald originally divided the color circle into a hundred parts[9] (see figure 12.9), which is really inconvenient (Ostwald was a strong believer in the metrical system). Later Ostwald used the more convenient (as well as conceptually proper) 12-, 24- or 48-step scales. For most purposes a 24-step scale amply suffices, in practice a 12-step scale is even more convenient. (Figure 12.10.)

Now any chip can be classified (figure 12.8) by its hue (or hue number), white content, black content, and color content. This is a color specification with an immediate intuitive appeal, for with hardly any training you can learn to "see" the hue and color, white, and black contents. Since the specifications are fully colorimetric you can also measure them. The Ostwald atlas is popular with designers because the various linear series contained

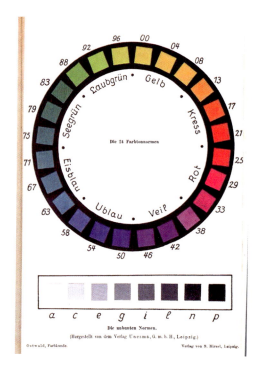

Figure 12.9 An Ostwald color circle with decimal subdivision 00–100(=00) and 24 displayed hues. This is a copy from one of Ostwald's (many) books. Ostwald personally hand-painted all samples.

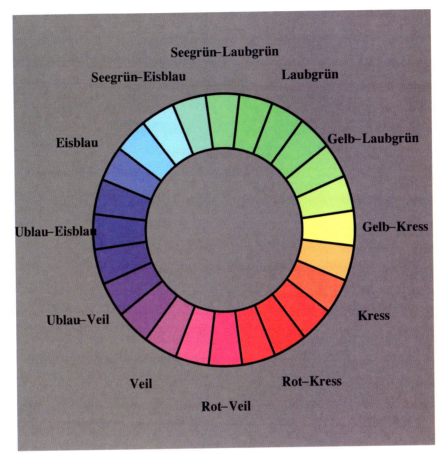

Figure 12.10 A color circle of 24 steps, measured according to Ostwald's principle of internal symmetry. (I computed this one, examples from Ostwald's books look very similar.) The German color names are of some historical interest: Due to hard feelings against the French, Ostwald avoided color names that had French roots (such as "orange") and came up with freely invented (generally charming) German names, often inspired by flower colors.

in it (figure 12.11) have immediate visual appeal and can be used as the basis for "color harmony" systems.

With these definitions, the Ostwald atlas can be constructed numerically from the basic colorimetric data (the color matching matrix and the spectrum of the achromatic beam). Thus the atlas can easily be constructed from scratch in any laboratory. Nevertheless, Ostwald spent years of hard labor to produce a physical copy. (In appendix L.3 I discuss

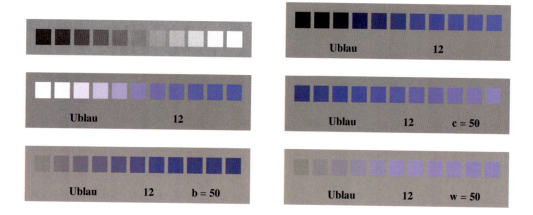

Figure 12.11 Some straight line sections through the Ostwald atlas. From left to right, top to bottom you have: Gray series, shades, tints, constant color content, constant black content, and constant white content.

some of Ostwald's laboratory methods.) This is understandable, because in his time computers and the Internet were nonexistent. Unfortunately, the fact that he undertook this heroic task (although quite irrelevant with respect to his *conceptual* atlas) led to many misunderstandings in (even the recent) literature. The (physical) atlas was confused with eye measure atlases (such as the Munsell atlas), which looked similar, and the chips were measured colorimetrically and showed various (inevitable) inaccuracies that discredited the work. All this is rather unfortunate, since the truly unique nature of Ostwald's achievements remained almost completely ignored. Sometimes (I hope) history simply ain't fair.

There are a few minor problems with Ostwald's construction, though these are easily amended.

One problem is the principle of internal symmetry. Although crucial and of considerable conceptual interest, it is extremely inconvenient in practice and somewhat ill defined in the formal sense.[10] In the 1940s and 1950s, Bouma used the principle to compute a "perfect" Ostwald scale, using an iterative numerical procedure[11]. Ostwald himself iterated through the mixture of pigments, a virtually impossible task that led to various inaccuracies.

It is much more convenient, and still formally correct, to redefine slightly the principle and divide the color circle by equal volume sectors of the color solid. This is formally unambiguous and leads to a very simple numerical procedure[12]. In figure 7.26 I showed a color circle for average daylight that has been measured by this method. In retrospect, the "honest pie slice" method makes more sense than Ostwald's original principle, though the results are virtually the same.

Another problem concerns the fact that the convex hull of white, black, and full colors fails to exhaust the color solid. Thus Ostwald "missed" certain possible object colors. (Figure 12.12.) Ostwald failed to notice or even understand this, but the fact that certain object colors could be found that cannot be placed in Ostwald's system (especially in the red region) was remarked upon by a number of authors even in his time. With the present understanding of the structure of the color solid the problem is obvious enough. It is also clear how to amend the problem. There are essentially two options:

- admit *negative* white or black content;

- admit optimal colors that are not semichromes as "landmarks" (redefine "full colors");

either option will solve the problem. Such changes are only necessary for very special object colors, such as hardly ever occur in real scenes. Since "negative white or black contents" are awkward, the best option seems to be to admit full colors as distinct from semichromes. Fortunately, this can be done in a neat, continuous way. Thus the problem is easily solved, hardly changing anything to Ostwald's original scheme.

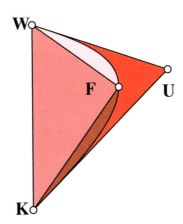

Figure 12.12 The problem with the Ostwald KWF (black, white, full color) triangles (the Ostwald double cone) is that the color solid extends beyond the double cone. There exist optimal colors that cannot be captured in Ostwald's scheme. The problem is amended when you substitute the KWU (black, white, ultimate color) triangles for Ostwald's KWF triangles. This essentially replaces Ostwald's double with the spectrum double cone. Although this contains colors *outside* the color solid, these colors can be reached for the same white point but a different spectral composition of the illuminant. In this scheme the full color *F* is no longer "full", it has finite white and black content.

Perhaps surprisingly, the real difficulties with Ostwald's scheme are hardly mentioned in the literature (the book by Bouma being one of the few exceptions), whereas all kinds of relative irrelevancies (often due to unfortunate misunderstandings) have been widely discussed in the relevant (though largely irrelevant) literature.[13]

12.2.1   The most stable chips and the "non-Ostwaldian" colors

By "non-Ostwaldian" colors I mean real object colors that fail to be caught in Ostwald's net, that is, those that would have negative white or black content. In figure 12.13 the color **f** is an Ostwaldian color. It has color, white and black contents in the range from 0 percent to 100 percent in terms of the full color **F**. This is the generic case that applies to the overwhelming majority of object colors you encounter in normal praxis. In contradistinction, the color **g** might be called "non-Ostwaldian" because it does not have color, white and black contents in the range from 0 percent to 100 percent in terms of the full color **G**. You would need a *negative black content*, thus the color has an special "luminous" quality to it.

In order to solve cases like this, it helps to reconsider the status of the "full colors". Ostwald selected semichromes for the full colors and was convinced that these were indeed "full" colors in the sense of "most colorful" colors. For the generic colors (like **f**) this appears to be the case. If you want to describe the color in terms of an optimal color, white, and black, then the optimal colors at your disposal lie on the arc indicated in figure 12.13. For the full color you obtain a description in which the color content of the color **f** is a minimum (you may want to prove this). This evidently indicates the "full color" status of the semichrome.[14]

For the non-Ostwaldian color **g** the range of possible optimal colors is an arc (indicated in the figure) that does not contain the semi-chrome **G**. This is indeed why the color **g** is "non-Ostwaldian".

The optimal color for which the color content reaches a minimum is the color **G'**, that is the possible optimal color nearest to the semi-chrome **G**. It works the same way for colors with negative white content. Thus you can retain the Ostwald description and simply substitute **G'** for **G**. The rule is

> *The "full color" is the optimal color for which the color, white, and black contents are in the range 0–100 percent and for which the color content has a minimum value. For the Ostwaldian colors (almost all colors) this means that the full color will be the semichrome. For non-Ostwaldian colors the full color is not a semichrome, but some generic optimal color.*

Notice that you need only one rule for all cases and that the Ostwaldian and non-Ostwaldian descriptions smoothly run into each other.

There is another way to regard the situation. If the full color is the color for which the color content is minimal, it means that you have a chip that has as little spectral variation as possible, that is a chip that is as *stable* against perturbations of the illuminant as possible. In that sense the Ostwald chips are truly optimal; they are the most stable choice. Remember that the white and black chip are perfectly stable: They are "black" and "white" under *any* illuminant. The colors of the optimal color chips (those with an optimal spectral reflectance) change when the illuminant changes, though. Thus you want your chips to be "as white or black (or gray)" as possible and thus to have minimal color content. This is an important property that Ostwald perhaps intuited, but never explicitly mentions. I haven't heard it mentioned in the literature either, but who knows? It may be buried in the literature of theoretical colorimetry of the early twentieth century. Much of that has hardly been mined and is generally forgotten.

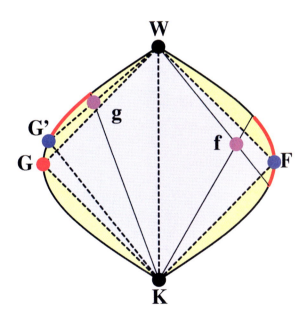

Figure 12.13 Color **f** is generic, color **g** is "non-Ostwaldian". The red arcs are the ranges of optimal colors that could be used in an Ostwald-style description in terms of color, white, and black contents in the ranges 0–100%. The colors that give rise to minimal color content are the semichrome **F** for the color **f** and the optimal color **G′** for the color **g**. Thus in the former case the "full color" is the semichrome, in the latter case the optimal color closest to the semichrome (thus not a semichrome!).

12.3   Physical Realization of Chips

The formal notion of "chips" is far more powerful than might appear on first blush. In a number of cases it even *fully determines the physical properties of the surface.* The easiest case is that of the *black chip.*

The black chip is—by definition—a pair of beams, where the background beam is any beam with positive spectral density throughout and the beam of the chip proper is the empty beam. Because the radiant power spectrum of the background beam is positive throughout, it follows that

> *the black chip must have vanishing albedo throughout the visual region.*

The black chip is simply a surface that scatters *none* of the incident radiation. All the incident radiation is thus absorbed by the black chip. What is remarkable here is that this is actually a *physical* definition of the optical properties. Any black chip must absorb all incident radiation and scatter none. Good approximations are such substances as soot or black velvet.[15] Notice that it will not be necessary to produce a "black standard" and keep it in a vault for reference by later generations of colorimetrists because the definition is fully self sufficient. Any laboratory will be able to design a suitable black chip and check its validity through radiometric measurements.

A similar reasoning applies to the *white chip*, though here the situation is not quite as simple. A thing that is certain is that the white chip should scatter *all* incident radiation and absorb none. For otherwise it would be easy to conceive of an *even whiter* white chip (simply decrease the absorbed fraction of the incident radiation), which contradicts the assumption that you were dealing with a white chip in the first place. However, this is not a *sufficient* condition because the radiation might be scattered in quite different ways (see appendix L.1). For instance, a perfect mirror absorbs no radiation and scatters (or rather "reflects" as people say) all. Yet a perfect mirror can hardly be called "white". When you illuminate it with a candle, the perfect mirror looks *black* instead of white from most vantage points. And when you see the reflection of the candle flame it is still the case that most of the surface of the mirror looks black (except for the image of the candle flame that is). How to handle this complication? The only acceptable solution as I see it is to define the white chip as a Lambertian surface of unit albedo:

> *The white chip is a surface that scatters all of the incident radiation and that looks the same from all conceivable vantage points irrespective the location of he light source.*

Such a surface is not unlike a piece of paper or white chalk, a matte white surface conventionally known as Lambertian. Lambertian surfaces are singled out among all other surfaces through the property of scattering equal radiance in all directions *irrespective* of the incident beam. This definition of the white chip as a Lambertian surface of unit albedo is again a perfectly complete *physical* definition. Good approximations are white chalk, blotting paper, or (the traditional laboratory method) smoked magnesium oxide on a glass substrate. As in the case of the black chip, there is no need to guard an international golden standard. Every laboratory can construct its own white chip and check its validity through elementary radiometric measurements on real chips.

Any other definition would imply that you have to specify more details of the viewing geometry. Clearly this would be most objectionable. In talking of "the color of a thing,"

> *you need to be able to abstract from the multifarious other relations of the thing with respect to its environment, otherwise the thing simply has no color of its own.*

Thus the white chip should be Lambertian. You should not deduce from this that any actual piece of white writing paper is Lambertian of course. Real objects may well change their color when you change the viewing geometry (think of peacock tails for instance). But chips are ideal entities and are *by definition* objects that don't change their color when you look at them differently.

Thus I have reached very satisfactory conclusions in the case of the white and black chip.

> *The difference with the colorimetry of patches is striking in the sense that you have arrived at a colorimetric definition that actually constrains the physical properties of the surfaces.*

The situation is far less simple in the case of all other chips though. For instance, consider the *grays*. By suitably mixing flour and soot you may prepare a surface that has spectral albedo 50 percent throughout the visual spectrum. Such a surface appears as a light gray, I call it the "central gray". (This is not to be confused with "medium gray", I'll come to that later.) But is this the only candidate for this surface color? No way. There exist *infinitely many* surfaces that look exactly like the central gray *under the same illuminant* ("metameric grays"). These surfaces will look *distinct* when you change the illuminant though, *even if you keep the color of the illuminant constant*. The central gray is the colorimetric mean of black and white, and it will be that under *any* illuminant. The other chips (looking the same under the fiducial illuminant) will fail to remain the average of white and black when you vary the spectrum of the illuminant. Such surfaces are *unstable* grays.

> *If you require the achromatic colors to be stable there is no choice, the spectral albedo should be constant throughout the spectrum.*

Since this is the only rational choice, I will always mean the stable variety when I use the term "gray chip". This definition of the gray chips again fully determines the optical properties. There is no need to guard any standard set of grays.

In the case of the chromatic chips, the situation is much less simple. But one thing is certain, recalling Schrödinger's notion of "optimal colors" you have

> *the most colorful chips can only be implemented in a single way, thus their colorimetric properties fully determine their spectral properties.*

In that respect the most colorful chips are much like the white chip. The other chips are less well defined. By introducing a notion of stability under variation of the illuminant you may arrive at a unique definition. However, the situation is slightly more complex than in the case of the grays. Moreover, even modern technology doesn't allow you to obtain the precise chips you would like in real life. You simply have to make do with what can actually be produced. That doesn't mean you should lose sight of the ideal, though. After all, the ideal is not *forbidden* by any of the laws of physics. It's about time that technology catch up.

The fact that

> *in many cases of practical interest the colorimetric specification of chips (in the formal sense as ordered pairs of patches with some additional constraints) fully determines their optical (that is physical) properties*

is highly remarkable. Compare this with the case of the patches where the colorimetric specification only manages to single out an *infinite* set of beams. This highly important observation is conventionally downplayed in the textbooks for reasons that I fail to understand. The distinction of having noticed this remarkable state of affairs belongs without any doubt to Wilhelm Ostwald. Ostwald, as a chemist, thought of colors as *material* properties rather than properties of beams of radiation, as physicists were wont to do. This enabled him to develop the theory of object colors far beyond the state of the art of his times. It is something of a scandal (at least that is how I read the facts) that Ostwald got as raw a deal as he did (posthumously, that is). Even modern methods in common use are in many respects inferior to some of Ostwald's innovations. More of that later.

If you could produce an atlas with chips whose spectral reflectances would be defined as Schrödinger optimal colors, you would have a very interesting object, a color atlas representing the brightest colors (boundary of the color solid) that would work for *any* illuminant and *any* (even anomalous) trichromatic observer!

This is the ideal limit of the Ostwald color atlas concept.

12.4 Relations in Color Atlases

The geometrical relations in a color atlas are much like these in the object color solid, apart from the specific metric and sampling density of the atlas. All relations are induced by the configuration relative to the special elements:

— the white and black points;

— the gray axis and lines parallel to it;

— the spectrum cone, the inverted spectrum cone, and cones parallel to these;

— the WGKM plane and planes parallel to it.

Most of these elements have already been discussed in some detail, the WGKM plane plays a special role because it divides the atlas into "warm" and "cool" colors.

Apart from these special elements you may consider a number of special transformations, for instance,

— scaling with respect to the origin corresponds to the physical effect of *shading*;

— a translation (adding a real color) corresponds to the physical effect of *air light* or *atmospheric perspective*;[16]

— a von Kries transformation[17] approximately corresponds to the physical effect of changing the color of the illuminant.

Take an arbitrary fiducial color. Any other color in the atlas automatically assumes a relation relative to this fiducial color. By constructing the spectrum and inverted spectrum cone and, finally, the WGKM plane, all suitably shifted to the fiducial color, you obtain a division of the full gamut into six sets. These are

— the colors that dominate (thus are lighter than) the fiducial color and are cooler than it;

— the colors that dominate (thus are lighter than) the fiducial color and are warmer than it;

— the colors that are dominated by (thus are darker than) the fiducial color and are cooler than it;

— the colors that are dominated by (thus are darker than) the fiducial color and are warmer than it;

Are chips "samples," "labels," or "qualia"?

In most cases, chips are not meant to be "samples." Even the "samples" distributed by paint manufacturers tend to be printed papers instead of actual paint (*the* paint, I mean) layers. Most chips are "labels" in a certain sense; they "represent" a given color but are not clones of it. Typically, one uses the color denotation rather than the actual chip as a label, though. This makes it easier to communicate by email and so forth.

A chip that really fits your fiducial color sends the same signal to your brain as the fiducial color would. For instance, in case you use RGB colors, the RGB color emitted by your computer screen tickles your brain just as the fiducial color (a beam with a very different spectrum) would. Is the chip a *quale*, then?

The chip in your experience is; the chip as a (physical) thing isn't. Modern people who (as is all too often the case) confuse the mind with the brain are not going to understand this though.

— the colors that do not dominate, nor are dominated by the fiducial color and are cooler than it;

— the colors that do not dominate, nor are dominated by the fiducial color and are warmer than it.

(See figures 12.14–12.18.) You can also consider the various *shades* of the fiducial color, which are the colors that are collinear with the fiducial color and black, the *tints* of the fiducial color, which are the colors that are collinear with the fiducial color and white, and the colors which lie on the line through the fiducial color parallel to the gray axis. (See figure 12.19)

If the latter dominate the fiducial color they are "foggy" versions[18] of it, obtained by adding air light of the same color of the illuminant; if they are dominated by the fiducial color they are "clearer" versions of it, obtained by taking away air light of the same color of the illuminant. Colors that are coplanar with the fiducial color and the gray axis are also special because they come from the same pair of (complementary) ultimate colors as the fiducial color itself. (See figures 12.19-12.18.)

Take a pair of arbitrary fiducial colors, such that one of the pair dominates the other. Such a pair generates a gamut, the volume of the intersection of the spectrum cone centered on the dimmer color and the inverted spectrum cone centered on the bright color. Such a gamut can be considered a veiled and (von Kries) scaled version of the atlas as a whole.

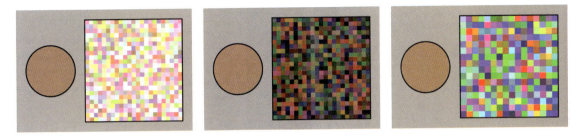

Figure 12.14 A color with the gamut of colors that dominate (are lighter than) it (left), the gamut of colors that are dominated by (are darker than) it (center), and the gamut of colors that are unrelated to it (right).

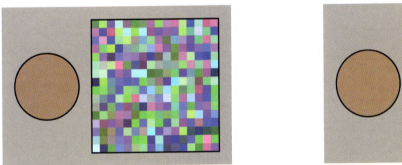
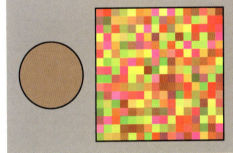

Figure 12.15 A color with the gamut of colors that are cooler than it (left), and the gamut of colors that are warmer than it (right).

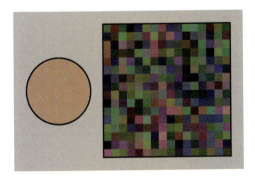
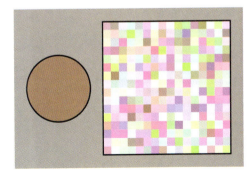

Figure 12.16 A color with the gamut of colors that are cooler and darker than it (left), and the gamut of colors that are cooler and lighter than it (right).

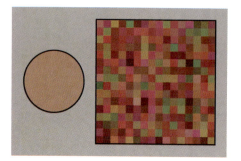 

Figure 12.17 A color with the gamut of colors that are warmer and darker than it (left), and the gamut of colors that are warmer and lighter than it (right).

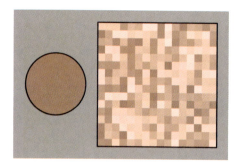 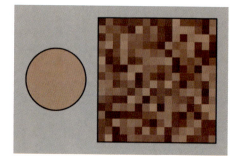

Figure 12.18 A color with the gamut of colors that are foggier than it (left), and the gamut of colors that are clearer than it (right).

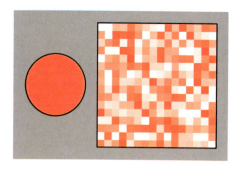 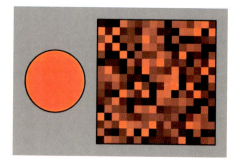

Figure 12.19 A color with its "tints" (left) and "shades" (right).

Likewise, take a pair of unrelated (domination undefined) fiducial colors that are mutually collinear with the gray axis. Such a pair also generates a gamut by intersection of the spectrum and inverted spectrum cones centered at the LUB (least upper bound) and GLB (greatest lower bound) of the pair. The colors in the gamut are unrelated to either fiducial, but if they are close enough the gamut appears as a set of "interpolating" colors. Again, by veiling and (von Kries) scaling the gamut is a transformed version of the corresponding planar section of the atlas though the pair of fiducial colors and the achromatic axis. (Figure 12.20.)

Numerous additional relations between sets of points or between one set of points and another set of points can be defined. Such relations are (often consciously) used by "colorists," *users* of color in the visual arts and design (see appendix L.2).

The distinctions used here to generate specific gamuts related to a given fiducial color indeed exactly fit the terminology you will encounter in books on how to paint.[19]

Of the related psychophysics very little is available. Yet this is an interesting field of endeavor because it enables objective psychophysics rather than the mere recording of subjective impressions or idiosyncratic preferences.

The relations considered here depend only on the divisions induced by the spectrum cones and the WGKM plane. In many cases you would also differentiate by hue, distinguishing (at least) analog colors and complementary analog colors from others. For instance, the gamut of unrelated colors for a given fiducial color is not all that interesting, exactly because the colors are *unrelated*. Selecting the analog and complementary analog colors from the gamut would certainly produce interesting subsets, and so forth.

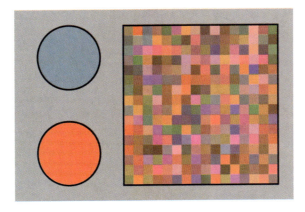

Figure 12.20 Two colors in the same plane through the gray axis with the gamut of colors that are unrelated to both.

12.1 The BRDF and Lambertian surfaces

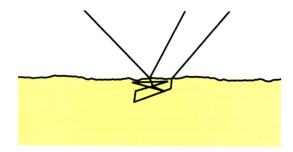

The so-called bidirectional reflectance distribution function (BRDF) measures the ratio of the *radiance* scattered by a surface into a certain direction (for example, the viewing direction) to the *irradiance* caused by an incident collimated beam from a second direction (the direction of the light source). (See chapter 17.) The BRDF is a property of the surface that sums up its scattering characteristics. It is by no means a *complete* description of the phenomenology. For instance, the BRDF description fails to account for the fact that a photon that is incident on the surface at some point might enter the bulk material and emerge at some different point (figure 12.21). Anyway, you have to conceive of the BRDF as a phenomenological description that summarizes complicated optical interactions at some scale. Thus the BRDF is not a "point property" in the usual sense; for instance, in order to describe the "BRDF of foliage," the "points" should include many leaves.

The BRDF is of course a nonnegative, dimensionless quantity, but otherwise there are few constraints on its form. An obvious constraint is energy conservation, evidently the scattered flux should be less or equal to the irradiant flux. A more confining constraint is "Helmholtz reciprocity," which states that the BRDF does not change when you interchange the viewing and illumination directions. This leaves much room for different functional forms, and indeed, very different BRDFs are encountered in real scenes.

The simplest BRDF imaginable is the *constant* one. If such a surface has unit albedo, that is to say, if it scatters all incident radiation and absorbs none, the BRDF is easily shown to be $1/\pi$. This is the Lambertian BRDF.

The importance of the Lambertian BRDF for colorimetry is that it formalizes the concept

Figure 12.21 The BRDF is by no means the full story. Here a ray incident on a surface splits into a reflected ray (Fresnel reflection, somewhat random due to surface irregularities) and a refracted ray. The refractive ray diffuses through the turbid medium (the material), and if it exits it does so at a different location from the location of incidence. Thus the BRDF description is incomplete; it does not account for the spatial shift. On exit, the direction of the diffused ray is largely random.

of the "fully diffuse surface" or Lambertian surface, for which both the source and the viewing geometries are irrelevant. The radiance of the beam received by the eye is simply the irradiance divided by π. Restricting the surfaces to Lambertian ones thus allows you to develop the colorimetry of object colors without awkward references to scene geometry.

Pleasant as Lambertian surfaces are from a formal, colorimetric perspective, they are enigmatic from the perspective of the physicist. Attempts to model Lambertian surface scattering on the micro-scale have been going on since the late eighteenth century, without apparent success. Moreover, there do not seem to exist actual Lambertian surfaces within standard radiometric accuracies. Thus the Lambertian surface remains an ideal object. Many real surfaces come close, though, especially for not too oblique viewing and/or illumination directions. For many purposes, white blotting paper does just fine. The classical laboratory implementation has been magnesium oxide "smoked unto" a glass carrier. Nowadays one buys "white standards" that are more convenient to handle and easily cleaned.

Surfaces with non-Lambertian BRDF have spectral reflectance factors that depend on both the viewing and illumination geometries. Such surfaces have different colors in different circumstances, a bit like a peacock's tail. Most surfaces are like that, though often the changes are minor enough that they may be ignored.

12.2 The structure of the Ostwald color atlas in design

The Ostwald atlas has been popular in art, architecture, and design to arrive at color gamuts with particular properties, with the ultimate aim to arrive at pleasing or striking color combinations in a principled manner. Ostwald himself was much interested in that topic.[20] As Ostwald remarked, it is often striking how linear and planar cuts through the double cone (with mensurated color circle) yield color combinations that look well composed as if they fitted a certain "style". The stronger the constraint the more striking this effect is. Thus an array of random colors (figure 12.22 left) does not look particularly "unified" and an array of random full colors (figure 12.22 right) looks perhaps blatantly garish.

Of course, you may sample all hues but use them in less than full color vividness. If you mix in white you obtain "tints" (figure 12.23 left) and if you mix in black (figure 12.23 left) you obtain "shades."[21] Instead of white or black you may also "tone colors down" by having a fixed gray (fixed both white and black contents; see figure 12.24) content. The tints look like "pastel

Figure 12.23 Random samples from the Ostwald double cone with 50 percent color content, 50 percent white content ("tints", at left) and random samples from the Ostwald double cone with 50 percent color content, 50 percent black content ("shades", at right.)

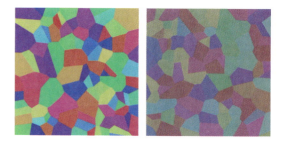

Figure 12.24 Random samples from the Ostwald double cone with fixed gray contents.

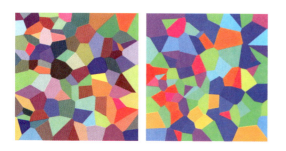

Figure 12.22 Random samples from the Ostwald double cone (left) and random samples from the Ostwald full colors (right).

colors" and have a somewhat "chalky" appearance, whereas the shades look "clear" and appear to have "depth". Graying down colors typically yields a "muddy" look.

You can introduce discrete scales at this point. Since I have already discussed the concents of chromatic "scale" and "key" elsewhere, I will not repeat that here.

If you sample randomly from a KWF triangle, that is to say, if you fix the full color, you obtain already much more "unified" patterns (figure 12.25).

Much stronger constraint is imposed by a linear series. For instance, random patterns of colors of the same hue and constant white content (figure 12.26) look as if the pattern were seen though a luminous "fog", which is why Ostwald refers to such series as *Nebelreihen* (fog series).

If you keep the black content constant (figure 12.27), the colors become a bit muddy or turbid. If you keep the color content constant (figure 12.28), you obtain results that often look like a uniform color that was somewhat loosened up, especially in cases where the color content is high.

Figure 12.27 Random samples with constant black content for a fixed full color (red).

Figure 12.28 Random samples with constant color content for a fixed full color (red).

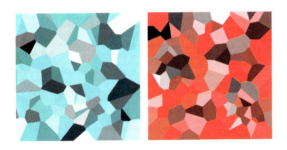

Figure 12.25 Random samples from the Ostwald KWF triangle for cyan (left) and the KWF triangle for red (right).

Figure 12.26 Random samples with constant white content for a fixed full color (yellow).

Series in which the ratio of two parameters is kept fixed are also of interest. For instance, in case you keep the white-black ratio constant (figure 12.29), the series runs through the full color vertex of the KWF triangle. The result often has a somewhat muddy appearance.

In case you keep the white-color ratio constant (figure 12.30) the series runs through the black color vertex of the KWF triangle. The result often looks like it has "depth" and if seen through a darkish, transparent filter.

In case you keep the black-color ratio constant (figure 12.31) the series runs through the white color vertex of the KWF triangle. The result looks subdued, desaturated, but without looking "muddy".

There need be little doubt that the various subjective reactions to such color schemes derive rather directly from automatic inferences as to the physical nature of the scene in front of you, even if you doubtless look at an abstract array of colored patches. For instance, the "tints" look like pastel colors or chalky because pastels are dry powders that scatter quite a bit of the illuminant spectrum to the eye. Thus pastel colors are invariably mixtures with white. This also applies to the fog series, here you assume a certain addition of air light. That the "shades" (the term itself is derived from a physical cause) look "deep" is because there is no indication of scattered illuminant, thus no indication of air light or of turbidity. Such layers are typically transpar-

Figure 12.31 Random samples with constant black-color ratio for a fixed full color (red).

ent absorbers in front of a light background; thus the "shades" are much like the colors of a glass of red wine. From such "ecological" considerations it is no doubt possible to build a more principled theory of these color themes. This is not something that I have seen attempted in texts on color in design, though. That Ostwald was aware of these causes of subjective feelings is evident from the terms he used, such as, *Nebelreihen*, which refer to ecological optics rather than emotional states.

Although Ostwald understood that his colorimetric parameters were defined on continuous domains, his atlas had to be composed of a finite (hopefully *very* finite) number of instances. Thus it was necessary to sample or discretize the continuum. For Ostwald this was at first a mere necessity. For instance, being a strong believer in the advantages of the metric system, he initially divided the color circle into a hundred discrete hues. Later it occurred to him that is is not such a good idea and that a "natural scale" should use a multiple of six. (Albert Munsell initially used a five-based scale, which is equally awkward.) The magical number six is due to trichromacy (the total number has to be a multiple of three) and the requirement that each hue should have a complementary made (the total number has to be even). Similarly, at first Ostwald di-

Figure 12.29 Random samples with constant white-black ratio for a fixed full color (red).

Figure 12.30 Random samples with constant white-color ratio for a fixed full color (red).

vided the "white content" scale linearly. When he (partly mis)understood that "white content" is related to "brightness," he substituted the logarithmic scale, following Fechner's lead with his psychophysical function: The logarithmic scale was supposed to be "well tempered" and thus more "natural" than the linear scale.[22]

In this way Ostwald was led to the definition of discrete scales, much like the scales in music. In a "well tempered" scale, the distances between neighboring instances should be equal and the partition "natural" (like the multiple of six requirement for the hue scale). Now suppose I have a colored image and I use only instances from the scale for its representation. Then the choice of scale evidently has an influence on how the image looks. This is much like the use of scales in music. There are two aspects here. One is the issue of grain, or stepsize. The other is the issue of "phase" or "key."

A piece of music rendered in a continuous scale (fortunately rare because most musical instruments tend to produce a discrete set of tones) tends to sound unpleasant. It is no different with images. The draftsman and painter likewise tend to apply discrete scales to tones, colors, line orientations, line thicknesses, hatching styles, and so forth. This is supposed to prevent a "licked" appearance. These scales are typically very coarse, for the discrete steps should be *apparent*.

Ostwald at first tried to pick scales that were so fine that they could effectively double for continuous scales. Thus he used a hundred-step hue circle because it was next to impossible to distinguish neighboring hues in such a scale. Perceiving this as overkill, he went down to 48, then 24, and for many purposes 12, which is about the finest scale considered by visual artists. You may try for yourself. Whereas you probably have little problem recognizing (on a six-step scale you can even *name* them) all steps on a 12-step scale, this is already hard on a 24-step scale. (See fig-

ure 12.32; you may well have trouble distinguishing the greens.)

Apart from the graininess, you have to deal with the phase. The scale has to start somewhere. Ostwald anchored his scale on yellow (a well considered choice, because yellow can be precisely reproduced by most people) and fixed the orientation (sense) of the scale as initially (from yellow) turning toward green. On a six-step hue scale you then get (figure 12.33)

Figure 12.32 A continuous and a 24-step color circle.

Yellow
Green
Cyan
Blue
Magenta
Red

and (of course) yellow again. If you shift the scale by half a step, you get another scale:

Greenish-yellowish
Bluish-green
Greenish-blue
Reddish-blue
Bluish-red
Orange,

which is in a different "key." Like in music, you may transpose an image into a different key, even sticking to the graininess of the scale. Famous examples of "off-key palettes" in painting include the *Deposition* by Pontormo at Florence (keyed on ORANGE), but most painters use a scale keyed on YELLOW.

A "change of key" is easily obtained through computer image processing. It yields an interesting and powerful handle on the emotional impact of a picture.

Apart from graininess and key, scales often have a variety of symmetry properties. The hue scale, for instance, is generally supposed to naturally divide into two parts, the family of "warm colors" and the family of "cool" colors. As I have argued earlier, this artistic notion is most likely related to the Goethe edge color series. The quintessential "WARM" colors RED and YELLOW belong to one family of edge colors, whereas the quintessential "COOL" colors CYAN and BLUE belong to the other family. GREEN and PURPLE are "neutral" with respect to the WARM-COOL dichotomy (their classification by artists is very variable). Another important aspect of hues is their intrinsic brightness. YELLOW is the BRIGHTEST COLOR, even in its purest state, whereas BLUE is the DARKEST. Because artists need to use "VALUE" to modulate pictorial re-

lief, these are extremely important attributes to them. In figure 7.42 I showed the effect of a reflection in the yellow-blue plane. Since this symmetry[23] conserves brightness relations, the result looks indeed very "natural."

Color harmony is a moot problem for the scientist, though it is of the highest importance to the artist.[24] Harmony in music has been studied since the Pythagoreans (I confine myself to my native Europe here), and harmony in spatial patterns has been the topic of both mathematical research (symmetry groups) and architectural and drawing praxis. In order to make any formal headway, you need a system of measurement and comparison.

When Ostwald had set up his color atlas structurally, it was only natural that he proceeded to use it as a basis for the definition of color harmonies.[25] His system was (in some places still is) widely used in design, architec-

Figure 12.33 A change of key on a 6-step color circle.

ture, and even photography.[26] The basic notion is a simple one:

> *Symmetric configurations that are positioned in a rational manner with respect to the skeleton of the Ostwald double cone are likely to evoke harmonious impressions.*

The eye appreciates the rational structure as it were, just as in musical harmony.

I show a few typical harmonious color schemes in figures 12.34-12.38.

The "monochrome" scheme (figure 12.34) is the simplest. In a sense it isn't a color harmony at all, since there is only a single hue. Such a scheme is indeed harmonious, but (like a single tone played at various loudnesses) perhaps not very exciting.

The "analogous" scheme (figure 12.35) is also quite simple, it includes minor variations of a single hue. Such a scheme is obviously harmonious since the hues are so close; however, it is not too exciting either.

The "monochrome with complementary accent" scheme (figure 12.36) is also quite simple. It introduces the complementary hue. Such a scheme turns out to be harmonious because of the complementary relation of the hues and the fact that one clearly dominates. The scheme is

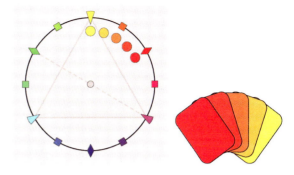

Figure 12.35 An "analogous color scheme."

more exciting because the complementary accent adds spice.

The "analogous with complementary accent" scheme (figure 12.37) is still quite simple. It also introduces the complementary hue. Such a scheme is again harmonious because of the complementary relation of the hues and the fact that one clearly dominates. The scheme is more exciting because the complementary accent adds spice and there is variety in the key hue.

Although this is not generally acknowledged, the areas covered by the various components (as

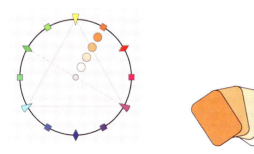

Figure 12.34 A "monochrome color scheme."

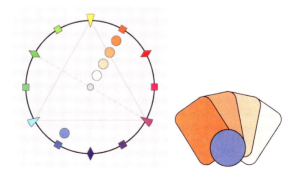

Figure 12.36 A "monochrome with complementary accent color scheme."

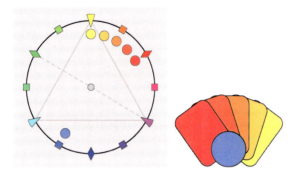

Figure 12.37 An "analogous with complementary accent color scheme."

ered a fairly complex color scheme, probably as complicated as one should ever get. Going into more detail would throw you into chaos, at least visually.

You may split a color into components that yield the original color through partitive mixture. The components may be chosen in some harmonious relation. Thus a painter may "divide the touches" and introduce "chords" of his own choice. This was done by the pointillists and divisionists in a variety of different ways. The effect is sometimes forced by the technique used, for instance, in weaving or mosaic when a limited gamut of colored materials is at the artist's disposal.

well as, obviously, the spatial relations) are very important.

A small but strong accent may "balance" a much larger area. This is well known to visual artists, but there exists only scant material in the visual sciences.

The "triadic" scheme (figure 12.38) is more complicated.[27] Such a scheme is harmonious because of the relation of the three hues. In many applications one component clearly dominates, whereas the others are used as accent that add spice. In most design applications this is consid-

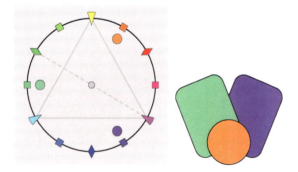

Figure 12.38 A "triadic color scheme."

12.3 Ostwald's workbench

Ostwald's color atlas is not the product of eye measure. As I said before, you can compute the atlas and generate it computer-graphically from the conventional CIE tables. Ostwald did not have access to such convenient tools, though. He literally built the necessary apparatus with his own hands. These tools are fairly simple, and you may enjoy constructing them yourself in order to get a hands-on taste of what practical colorimetry used to be.

None of Ostwald's tools is completely original. For instance, his color synthesizer (see below) is in some respects similar to Maxwell's "color box", the HASCH is similar to Hering's "shadow box," and the POMI is similar to a machine proposed by Dove. But Ostwald showed a remarkable ingenuity in building and employing these instruments. It is worthwhile to consider these self-built tools in some detail.

The first instrument used by Ostwald was his "HASCH" (an acronym derived from the German *Halb-Schatten*, half-shadow, figure 12.39). This instrument allowed him to "scale" colors, that is to say, to change their distance from the origin. This was done by varying the strength of the illumination of a sample, using the geometry of the étendue of the illuminating beam to measure the scale. The sample is placed at one side of a long tube, blackened on the inside in order to avoid internal reflection and scattering. At the other side of the tube is an aperture whose area can be precisely controlled. The étendue is proportional to this area, and so is the illumination of the sample, and thus the radiance scattered by the sample toward the eye. The sample is seen through another tube (again blackened on the inside) placed at an angle with respect to the first one. The instrument is used for relative measurements and thus is executed double-barrelled.

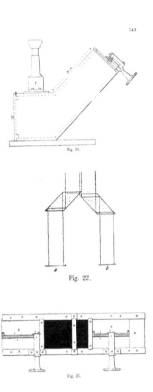

Figure 12.39 Ostwald's "HASCH (*Halbschatten-photometer*). A profile view (top) shows the tilted illumination tunnels, with apertures at top. The viewing tunnels are vertical, with the oculars at top. The viewing geometry (middle figures) uses two offset prisms that achieve two goals: To separate the samples spatially and to obtain a sharply delineated splitfield view. The illumination and viewing pathways are screened from each other and internally blackened. One of the illuminating apertures is made variable (bottom figure) so the illumination can be precisely and quantitatively varied. In use, one sample will be standard white, the other the sample being measured. The principle is clearly based on established photometric practice.

On one side a standard specimen is placed, on the other the fiducial specimen. Much ingenuity is devoted to the optical system used to enable as precise a comparison as possible. You want the two patches to be compared to be close together, meeting at a sharply defined boundary. This is very critical and the various solutions to this problem all have their specific pro and cons. The major part of the HASCH can easily be assembled from cardboard, but you will need some ingenuity to get the splitfield view right. I used the Ostwald design (I had the prisms especially made), which indeed works very well, but many other designs are possible (depending on the materials at your disposition) and should work, too.

The HASCH can be used in a number of ways, the most important probably being in the construction of well-tempered shadow series (*Schattenreihen*), that are series of colors of the same chromaticity.

It is not difficult (and a lot of fun) to build this important instrument. It is also very instructive to use it. For instance, you can use it to produce a precisely calibrated gray scale.

The second instrument of great utility used by Ostwald was his "POMI" (an acronym from the German *Polarizations Mischer*, polarization mixer, figure 12.40). This machine allowed Ostwald to mix two samples in a precise, additive way. The idea for such an instrument is older; you find it described in Ogden Rood's *Modern Chromatics*, where it is ascribed to Dove (figure 12.41). At the heart is a Wollaston prism. This prism is conventionally used to split a beam of natural light into two spatially separated beams of mutually orthogonal linear polarization. Ostwald used the instrument in reverse, combining two beams instead of splitting a single one. A Nicol prism mounted such as to be rotatable about the optical axis is then used to control the mixing ratio very precisely and measurable via its angular orientation. Although this instrument is very simple to assemble, it is

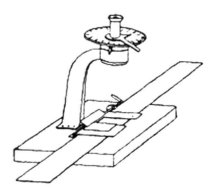

Figure 12.40 Ostwald's "POMI" (*Polarizations Mischer*). A Wollaston prism is used to obtain two beams of different direction that are combined in the ocular. Thus the observer sees two samples that are spatially separated on the groundplane optically superimposed. The two beams are oppositely polarized, so an analyzer can be used to set the mixing ratio. A circular scale lets you set this ratio accurately and quantitatively. The instrument looks like a development of the Dove apparatus (figure 12.41).

rather expensive if you have to buy the Wollaston and Nicol prisms, because these have to be of comparatively large size. I picked up discarded specimens that lay forgotten in a closet of an optics laboratory, which is probably what Ostwald did.

Using the POMI is very instructive. It allows you to mix various colored papers in a precise manner. For instance, you can easily construct a calibrated scale of tints by mixing in white. Some playing with the POMI and the HASCH will allow you to form an opinion concerning the precision of Ostwald's methods. I was impressed.

A third instrument of considerable use is an old-fashioned spectrograph used in reverse (see

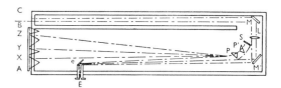

Figure 12.41 Ogden Rood's legend is: "Mode of using Dove's apparatus. R G is cardboard with red and green glass; at P is a prism of calc spar."

Figure 12.43 Maxwell's "color box." There are three entrance slits X, Y, and Z. They will be illuminated by a white paper receiving sunlight, beyond AB. Thus the instrument is a triple monochromator. The intensities of the three beams are controlled by way of the slit widths. The comparison field—white—is due to light entering the aperture CB.

figure 12.42). Typically, you would have illuminated an "entrance slit" of such an apparatus to find a "spectrum" projected at the other side. The spectrum used to be projected upon a photographic plate; thus an old-fashioned spectrograph has a means to insert such a plate, often interchangeable with a ground glass used for focussing. Nowadays the spectrum will be caught by a CCD-array and is likely to be unaccessible. The old-fashioned type is ideal, especially

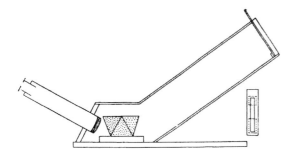

Figure 12.42 Ostwald's color synthesizer. This is simply an old-fashioned spectrograph used in reverse. At right a mask used to define a semichrome.

if equipped with a ground glass. When you use the apparatus in reverse you simply illuminate the ground glass and you look through the entrance slit (which is thus being promoted to exit slit). You can calibrate the machine by using it in its intended forward direction. Almost any old-fashioned spectrograph is easily converted[28].

This instrument is similar in its basic design to Maxwell's "color box" (figure 12.43), and it is interesting to compare the concepts. Maxwell's color box is a triple-barreled monochromator that allows mixtures of three monochromatic beams, whereas Ostwald's apparatus recombines *all* spectral components. Ostwald's design evidently has a much improved throughput, thus yield colors of much higher brightness. Maxwell's design is unnecessarily "monochromatic" in design.

In using this instrument, you place various masks in the plane of the spectrum. If you use no mask at all, the beam you receive through the slit has the color of the illuminant, for the spectrograph in reverse simply recombines the monochromatic components of all wavelength (it is simply Newton's experimentum crucis). Using simple masks you can produce the optimal colors exactly and precisely. It is rather spectacular to generate precise semichromes in this manner; they are the full colors that are the most vivid colors imaginable for the given illuminant. The machine is a veritable color synthesizer.[29]

With these instruments at your disposal you should be able to construct the Ostwald atlas from scratch. The really hard part is the production of the chips, since ideal pigments are not available. This must have been Ostwald's major struggle during the years he spent on the production of the atlas. According to his children, the colors of their father's beard told them daily on which page of the atlas he was working.

It is amazing how little infrastructure suffices for the task of constructing a color atlas from first principles. You can almost do it with your bare hands. Your eye is only necessary to judge colorimetric equalities. Notice how different this is from the Munsell method, in which literally *everything* depends upon eye measure. I am unable to understand how it can come to pass that the scientific color community (at least judging from the literature) fails to be impressed by the difference. Both methods are of interest and use, but they are ontologically different.

Exercises

1. Color series [INSTANT PSYCHOPHYSICS] Plot various chords in the color circle (using COLOR, of course!). Comment on the nature of these series: Are all samples "vivid" or "muted"? Does the series look "logical," or could you change hue in midcourse (for example, at a gray sample)?

2. The Ostwald atlas [TAKES A LITTLE EFFORT, IS LOTS OF FUN] Implement the Ostwald atlas (in RGB colors) on your computer. Study linear chords of the Ostwald double cone.

3. The Munsell atlas [CONCEPTUAL] Would you agree with my assessment of the ontological status of the Munsell atlas? You may have to read some of the literature from various periods in order to be able to weigh the various arguments.

4. Chromatic scale and key [EXPERIMENTAL] Construct a "Mondrian" consisting of random colors, perhaps imposing certain constraints. Study the effect of scale and key.

5. Color atlases [HISTORICAL] Check whether your library has a copy of the Munsell atlas, and if so take a look at it. You may check for other color atlases, too. In rare cases you may be able to see an Ostwald atlas with chips originally hand-colored by Ostwald himself.

6. The most stable chips and the definition of the "full colors" [ELABORATE] Analyze the non-Ostwaldian case and check that the full colors are not semichromes in this case. Use the Grassmann model, then also consider the general case. Study the stability of the chips. What assumptions do you have to make? Prepare examples of perturbations of the illuminant and check the stability numerically.

7. "Colorimetry done right" [REQUIRED IF YOU TEACH A COURSE] At this point I have introduced essentially the whole structure of colorimetry. I have used a variety of formats and conventions, often arbitrary choices, in order to illustrate the various issues. This is the time to take stock, to start from the beginning and *do things right* (your way). Decide on a representation of beams (my choice would be photon number density on a photon energy basis), a scalar product (my choice would be the usual definition, but taken on the log-photon energy axis), a number of generic achromatic beams (my choice would be the black body spectra), and so forth. You may also decide for a particular chromaticity diagram, a way of measuring the color circle (my choice would be based on the "honest pie slice" division of the spectrum double cone), and so forth. Now build the whole of colorimetry from scratch, starting with the color matching functions (I would convert the $10°$ Stiles and Burch data) and perhaps ending with a color atlas. Make sure you write fairly generic algorithms and data structures, so it will be easy to switch over when you reconsider your preliminary decisions. Don't be satisfied with just numbers, but spend some effort on fancy visualization (of course, you will make *true* pictures). It is worth the effort, for this will be your final product. All this takes a bit of doing, but you will be rewarded with a coherent set of pictures, something not to be found in the literature or in any textbooks.

Chapter Notes

1. K. L. Kelly, and D. B. Judd, *The ISCC-NBS Method of Designating Colors and a Dictionary of Color Names*, (Washington, DC: Nat.Bur.Standards, 1955).

2. R. M. Boynton, "Eleven colors that are almost never confused," in *Human Vision, Visual Processing, and Digital Display*, edited by B. E. Rogowitz (Proc.SPIE: 1077, 1989), pp. 322–332. These colors may be thought of as a cloud of points in the color solid that have maximally wide spacing. You may construct such point sets for any cardinality.

   With eleven points the spacing is sufficiently large that the probability of confusion is low, otherwise there is nothing magical about the number eleven. An interesting reference is L. D. Griffin, "Optimality of the Basic Colour Categories," *J. Roy. Soc.* 3, pp. 71–85 (2006).

3. Apart from a theoretical discussion, you might be interested in a color atlas for convenient personal use. The one I like best is published by the U.S. Bureau of Standards, *ISCC-NBS Color-Name Charts Illustrated with Centroid Colors, Standard Sample #2106.*

   These charts are easily taken with you, they are sufficiently cheap that you may consider them replaceable, they have about the right sampling density for informal use, and they give you color names that are easily understood and remembered.

4. Albert A. Munsell was an artist. He thought the mere use of color names too primitive and often misleading and sought to establish a rational way to describe color using decimal notation.

   Starting in 1898, he published his *Color Notation* in 1905, the forerunner of the *Munsell Book of Color* (of 1929 in modern form) in general use today. Munsell used a cylindrical coordinate system with "height" denoting *value*, that is the light-dark scale, "azimuth" denoting *hue* and "radial distance" denoting *chroma*. Value, hue, and chroma were measured on 1 to 10 scales. The coordinates were assigned by eye measure such as to derive at a well-tempered distribution.

   In the 1940s the Optical Society of America overhauled the system through extensive experiments and sample definitions (S. M. Newhall, D. Nickerson, and D. B. Judd, "Final report of the O.S.A. subcommittee on the spacing of the Munsell colors," *J.Opt.Soc.Am.* 33, pp. 385–418 (1943)). The system is arguably one of the best available examples of an eye measure system.

5. The "Munsell Book of Color, Glossy Edition" has more than 1,600 chips. It is sold by Gretagmecbeth. See `http://usa.gretagmacbethstore.com/index.cfm/act/catalog.cfm/`.

6. Ibid.

7. On the history of the Munsell system see R. G. Kuehni, "The early development of the Munsell System," *Color Res.Appl.* 27, pp. 20–27 (2002).

8. An excellent text on colorimetry that consequently follows the Ostwald methodology is S. Rösch, *Darstellung der Farbenlehre für die Zwecke des Minerologen, in: Fortschritte der Mineralogie, Kristallographie und Petrographie*, (Berlin: Selbstverlag der Deutschen Mineralogischen Gesellschaft, 1929), pp. 753–904.

9. At least the lure of the decimal system was some *reason* to take ten as a base, though an irrelevant reason as it turns out. Munsell's early atempts include color circles divided into five cardinal colors, for which I can find no rationale, not even an irrelevant one.

10. Bouma pokes fun at Ostwald on this account, it makes amusing reading if you you are interested in axiomatics. P. J. Bouma, "Zur Einteilung des Ostwald'schen Farbtonkreises," *Experientia* 2, pp. 99 (1946).

11. The Bouma calculation can be found in: ibid. This corrects an abortive attempt by Richter: M. Richter, "Zur Einteilung des Ostwald'schen Farbtonkreises," *Licht* 13, pp. 12 (1943).

12. Indeed, numerically simple and stable. This definition would have been useless to Ostwald, though, since he needed to approximate the scale by actual physical mixture.

13. For example, this is generally considered something of a bible by the colorimetrist: G. Wyszecki, and W. S. Stiles, *Color Science, Concepts and Methods, Quantitative Data and Formulas*, (New York: Wiley, 1967).) The index mentions "Ostwald color system, p. 475". On this page you find the sentence, "Examples include the *Ridgway Color System* and the *Ostwald Color System*." This refers to color-order systems "based primarily on the principles of additive color mixture. Systematic variations of the settings of a Maxwell disk can be used to generate color scales which in turn can be duplicated by actual material standards." This is not altogether wrong, but it is very uninformative, and hardly in tune with the extensive discussions of other topics. The majority of the literature even gets it wrong, which is worse. A (rare) balanced review of the Ostwald system can be found in: P. J. Bouma, *Physical aspects of colour*, (Eindhoven: N.V. Philips Gloeilampenfabrieken, 1947).

14. People are often confused when they learn that "full colors are semichromes." Why two terms if there is only one object? Historically Ostwald defined the "full colors" as the most colorful colors and in retrospect noticed that the full colors are semichromes. Here I define the full colors as colors that lead to minimum color content in an Ostwald-style cwk description, this seems to capture Ostwald's former notion. Then you easily prove that the full color is a

semichrome in the generic case, that is to say, when $((0 \leq c \leq 1) \wedge (0 \leq w \leq 1) \wedge (0 \leq k \leq 1))$, but that it isn't if the condition is violated. However, the full color is as close to the semichrome as the conditions allow. This yields an important novel perspective on the relation between full colors and semichromes. I have not encountered this in the literature (which tends to be rather confused on the topic anyway).

15. Easily the best implementation is as a "black hole." You punch a (relatively) small hole in an internally blackened box. Looked at from the outside, the hole is about the "blackest" thing you will ever encounter. This is the preferred implementation for the laboratory. It is also easy to produce for informal use.

16. The "air light" is an additive component on the radiance of the beam that hits your eye due to radiation of the sun that has been scattered by the air column between your eye and the nearest terrestrial object. The air light is dominant in the case of foggy conditions during daytime; otherwise it becomes mostly noticeable if the scattering air column is very thick, thus when the nearest object is at a large distance from you. The "blue" of the blue mountains at the horizon is almost pure air light.

17. Von Kries transformations are scalings of the dimensions that correspond to the retinal cone action spectra. The idea is simply that you adjust the sensitivities of your long, medium, and short wavelength sensitive mechanisms such as to make the color coordinates of the nominal "achromatic beam" roughly similar.

18. Ostwald named these "foggy" gamuts *Nebelreihen*.

19. Here are some (in my view) exceptionally nice examples to get you started: S. Quiller, *Color Choices*, (New York: Watson-Guptill, 1989); J. Sloan, *Gist of Art: Principles and Practice Expounded in the Classroom and Studio*, (New York: Dover, 1977); T. S. Jacobs, *Light for the Artist*, (New York: Watson-Guptill, 1988); J. Dobie, *Making Color Sing*, (New York: Watson-Guptill, 1986); L. Cateura, *Oil Painting Secrets from a Master*, (New York: Watson-Guptill, 1995); A. Stern, *How to See Color and Paint it*, (New York: Watson-Guptill, 1984); H. Windisch, *Schule der Farben Fotografie*, (Harzburg: Heering-Verlag, 1939).

20. W. Ostwald, *Lebenslinien: Eine Selbstbiographie*, 3 Vols., (Berlin: Klasing, 1926–7); —, *Die Harmonie der Farben*, (Leipzig: Unesma, 1923); —, *Die Maltechnik jetzt und künftig*, (Leipzig: Akademische Verlagsgesellschaft, 1930).

21. The origin of the term "shade" is of course obvious: You obtain shades if you "shade" (that is diminish the illuminance) on a patch. This is easily done by slanting the illuminated surface with respect to the incident beam. On nonplanar surfaces, there are parts of various spatial attitudes, thus slants with respect to the incident beam. Such surfaces are "shaded" to various degrees. If the surface is of constant Lambertian albedo, the colors of the surface will lie on a line through the origin of color space, that is, a series of "shades" of the colors obtained for normal incidence.

22. Since "lightness" and 'brightness' are categorically different, the choice of the Fechner law was not a particularly fortunate one, though, given Ostwald's perspective a rational one. The nonlinearity used in the definition of the CIE-Lab lightness would be a better alternative (though also hardly more than a crutch).

23. Unfortunately, the psychophysical properties of symmetries have hardly been investigated. A rare exception is L. D. Griffin, "Similarity of psychological and physical colour space shown by symmetry analysis," *Color Res.Appl.* 26, pp. 151–157 (2001).

24. Early scientific works include M. E. Chevreul, *De la loi du contraste simultané des couleurs et de l'assortiment des objets colorés, considéré d'après cette loi. Dans ses rapports avec la peinture, les tapisseries des Gobelins, les tapisseries de Beauvais pour meubles, les tapis, la mosaïque, les vitraux colorés, l'impression des étoffes, l'imprimerie, l'enluminure, la décoration des édifices, l'habillement et l'horticulture*, (Paris: Pitois-Levrault, 1839); W. von Bezold, *Die Farbenlehre in Hinblick auf Kunst und Kunstgewerbe*, (Brunswick: Vieweg, 1921); E. Brücke, *Bruckstücke aus der Theorie der Bildenden Künste*, (Leipzig: Brockhaus, 1877). From the artistic and design background I mention H. A. Bühler, *Das innere Gesetz der Farbe, eine künstlerische Farbenlehre*, (Berlin-Grunewald: Horen, 1930); A. Stokes, *Colour and Form*, (London: Faber, 1937); H. B. Carpenter, *Colour, a manual of its theory and practice*, (London: Baltsford, 1915); E. Rijgersberg, *Beknopte Kleurenleer*, (Amsterdam: Ahrend, 1948); J. Albers, *Interaction of Color*, (New Haven: Yale U.P., 1963); H. Chijiiwa, *Color Harmony, a guide to creative color combinations*, (Rockport, MA: Rockport Publ., 1987); S. Kobayashi, *Colorist. A practical handbook for personal and professional use*, (Tokyo: Kodansha Int., 1998); S. Kobayashi, *Color Image Scale*, (Tokyo: Kodansha Int., 1990); J. Pawlik, *Theorie der Farbe*, (Cologne: Dumont, 1969); J. Pawlik, *Praxis der Farbe*, (Cologne: Dumont, 1969); M. Graves, *The Art of Color and Design*, (New York: McGraw-Hill, 1951).

25. Ostwald's main thesis is given on the first page of his *Die Harmonie der Farben*:

> *harmonisch oder zusammengehörig können nur solche Farben erscheinen, deren Eigenschaften in bestimmten einfachen Beziehungen stehen.*

Ostwald then proceeds to enumerate the possible lawful relations. W. Ostwald, *Die Harmonie der Farben*, (Leipzig: Unesma, 1918). It is instructive to compare this with Ostwald's ideas concerning the harmony in spatial patterns —, *Die Harmonie der Formen*, (Leipzig: Verlag Unesma, 1922).

26. At least initially, artists were not all that positive about Ostwald's exercises though. See P. Ball and M. Ruben, "Color theory in science and art: Ostwald and the Bauhaus," *Angew.Chem.Int.Ed.* 43, pp. 4842–4846 (2004).

27. A color circle with opaque overlays having various apertures is often used to select such "chords". A nice one is *The Color Star* by Johannes Itten, published by Van Nostrand-Reinhold (New York, 1986).

28. If you can't find an abandoned spectrograph, the cheapest solution is to get a diffraction grating. Photographic gratings are good enough for the purpose and can often be bought cheaply at science museum stores. Gratings do have the advantage over prisms that the scaling of wavelenghs in the spectrum is lawful rather than arbitrarily nonlinear, but it is more difficult to obtain a bright image because not all of the beam will be diffracted into your spectrum.

29. An ingeneous arrangement was described by Rösch, this is a veritable color synthesizer that allows you to generate any color from the color solid. I have never played with (or even seen) such an instrument, but it should not be particularly difficult to construct. This would make a great science project. Here is the design:

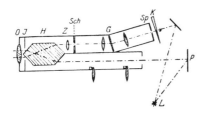

The meaning of the symbols is:

O Ocular

I Field aperture (*Blende*)

H Hüfner prism, a glass block that serves to combine two beams in a half-field view

Z Cylinder lens that combines the spectral components (*Zylinderlinse*)

Sch Variable occluder at location of spectrum (*Schieber*)

G Diffraction grating (*Gitter*)

Sp Entrance slit of spectroscope (*Spalt*)

K Neutral density wedge (*Keil*)

L Light source

P Test object (*Prüfling*)

The lenses in the lower beam allow the apparatus to be used as either a loupe or a telescope. In practice it is convenient to swap the locations of Sch and Sp (it makes no difference to the principle). In a measurement you start by putting a piece of white paper at P and adjust the neutral density wedge such that both haf-fields show the same achromatic color (the illuminant). Then any color can be produced by varying the occluder Sch. Rösch suggests a variety of convenient ways to achieve this. Rösch mentions that Zeiss (Jena) was actually constructing such an instrument, but I have never encountered one.

Chapter 13

Spectra from Colors

Things are really simple, according to Newton's perspective. There was supposed to exist a perfect one-to-one relation between the wavelengths of "homogeneous lights" (monochromatic beams) and the perceived (experienced) hues. Albeit in a somewhat "fuzzy" sense, there indeed exists such a relation (figure 13.1). For instance, most "color normal" human observers agree pretty much on the following rough indication of the wavelengths regions associated with common hues:

Wavelength	Hue
380–436nm	Violet
436–495nm	Blue
495–566nm	Green
566–589nm	Yellow
589–627nm	Orange
627–780nm	Red

Figure 13.1 illustrates this. It is hard to say what this means. The "experienced hues" are *qualia*, whereas the wavelengths are *pointer readings* in Eddington's sense. Thus Newton suggested a connection between ontologically disparate levels "conscious experience" and "the physical world" that is totally incomprehensible from a scientific perspective. Nowadays one would say that Newton suggested a psychophysical bridging hypothesis. There is no need to believe in strict psychophysical parallelism though, because this table is by no means exact. Hue estimates vary quite a bit from trial to trial, and there is at best some statistical correlation in case the circumstances are reasonably well controlled. Newton himself was somewhat wary of strict psychophysical parallelism, and he certainly was aware of the ontological chasm: *"the rays are not colored."*

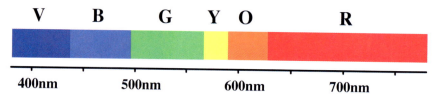

Figure 13.1 Schematic representation of the sequence of spectral HUES on a linear wavelength scale. In reality, the transitions are quite gradual. The capital letters stand for VIOLET, BLUE, GREEN, YELLOW, ORANGE, and RED.

As you proceed from monochromatic beams to more complicated spectra, the Newtonian suggestion is perceived to be almost totally void. You simply don't "perceive color by wavelength". The color science of the Goethe, Schopenhauer, Hering, and Ostwald tradition tends to be rather more to the point.

Suppose you view a scene, experiencing myriads of colors. What can you say concerning the spectral composition of the beam that causes a patch in your visual awareness, or even concerning both the spectral reflectance (or BRDF) and the beam that irradiates the surface element out there that scattered the beam toward your eye that again caused the patch to appear in your visual field? The short answer, no doubt, is: *not much!*

This is "inverse colorimetry," as it were. Even in the simplest settings, colors correspond to infinitely many spectral compositions. Thus you indeed guess that "inverse colorimetry is impossible." This type of (no doubt correct, but perhaps not entirely relevant) analysis has caused philosophers to declare that "colors do not correspond to physical properties in the world" or that "color is (mere) mental paint." Yet such conclusions are immediately perceived to be nonsense by the layman, which should induce you to think it over once more. And, come to think of it, from the perspective of experimental psychology and cognitive science a similar analysis applies equally well to position, length, shape, and so forth, all visual experiences that are commonly believed (also by many philosophers) to reflect physical properties of the world in a rather direct way. The Lockean distinction between "primary and secondary qualities" does not hold up against the evidence from experimental psychology.

That individual colors are "explained" by infinitely many different spectra is not something that should bother you *per se*, as it is true for virtually *any* observed entity. For one thing the infinitely many spectra might be so similar that you wouldn't even bother to distinguish them as a physicist. In order to let differences count, you need to be more quantitative. Even then things don't look so bright though. It is definitely the case that a "monochromatic yellow" looks pretty much the same as a YELLOW due to a beam that is composed of half of the visual spectrum. Clearly these spectra are both qualitatively and quantitatively widely different, yet they evoke identical colors. On the other hand, if you know the achromatic point (and that is reasonably well revealed in a richly differentiated scene), then at least the very vivid colors must be due to reflectance spectra that are close to the optimal (or even full) colors. Thus "inverse colorimetry" may be an impossibility in some cases and at least a somewhat (or even reasonably well) possible proposition in other cases. This is evidently an interesting topic that warrants closer investigation.

13.1 Inverse Colorimetry in Suitably Restricted Settings

Generic scenes are extremely complicated and are more properly considered by "machine vision," which (necessarily) is a much wider field than colorimetry. I cannot even start to tackle such complexity in this book. A useful start can be made by suitable restriction of the admitted scenes, a kind of (extreme) "stimulus reduction," as it were. Later on you may hope to relax one or more of the restrictions, but this will involve methods of machine vision and will take you out of the realm of colorimetry in the proper sense.

Useful restrictions are very similar to the kind of restrictions I considered in the case of object colors. You try to remove dependencies on scene, illumination, and viewing geometries, and you try to remove dependencies on material properties. Moreover, you restrict the scene to a single illuminant. When you do this systematically, you are left with a very narrow range of admissible scenes. Their structure can be described in the following way:

- all surfaces are frontoparallel (restriction on scene geometry, aimed at removal of dependency on illumination geometry);

- the illumination is by a collimated, uniform beam (together with the former this removes dependency on illumination geometry; it also enforces the single illuminant);

- all surfaces are Lambertian (restriction on material properties, renders viewing geometry immaterial);

- there are no photometric interactions between parts of the scene such as mutual shadowing or interreflection. This forces all surfaces to be coplanar;

- for the purpose of simple exposition I assume a pattern of piecewise uniform reflectance (no smooth gradients, like a quilt). This restriction is not essential, merely convenient.

These constraints are very restrictive, they are often referred to as the "Mondrian world" ("quilt world" would be more appropriate given the nature of Piet Mondrian's oeuvre). Yet it is certainly possible to realize such scenes physically, at least approximately so. Looking at paintings in an ideal museum (perfect illumination, no gloss) comes close.

In this setting the problem can be stated precisely. You have a (large) number of colored patches in the visual field. You assume that these are due to an illuminated, frontoparallel quilt. From this you attempt to deduce the illumination (spectrum of the source) and the spectral reflectances of the individual patches of the quilt. Since the problem is highly underdetermined, you will have to be satisfied with some "reasonable" solution plus the space of all possible solutions (of which many will no doubt be quite "unreasonable"). This will allow you to (at least approximately) solve many interesting and important problems. For instance, suppose I take the quilt apart and resew the patches together in another pattern. I show you this new quilt under another (unknown) illuminant. Can you establish the correspondence between pieces of the first and the second quilt? If you can (and a "reasonable" solution should enable you to do it to *some* extent), you have made progress toward the solution of the "color constancy" problem. Since being able to find apples and bananas in various settings is highly important from the perspective of bare survival, this problem counts for something!

How might you approach the problem? In specific cases the problem may be attacked without any spectral estimation at all. The quilt maps to a point configuration in color space and when you change the illuminant you will get another point configuration. Often these will contain equally many points, but in principle points may "split" or "merge". You may find the linear transformation that brings the point configuration into best approximate coincidence in some convenient metric. This establishes a mapping between the point sets that in many cases may be expected to yield a high probability of getting the correspondences right. This is somewhat of a mean trick of course. More generally, I suggest a three-tiered approach:

1. estimate the "white" color;

2. estimate the spectrum of the illumination;

3. estimate the reflectance spectra of the patches of the quilt.

This is reasonable because you need the spectrum of the illumination in order to be in a position to estimate the reflectances and you need the white point in order to be able to estimate the illuminant. At each step you will find infinitely many solutions, and these ambiguities propagate throughout the process. At each step you also attempt to find a "reasonable" solutions, or perhaps even the "best possible one" (or "most likely one") in the given circumstances. This will have to involve a priori knowledge about the likely structure of the scene, a prior understanding of "ecological optics".

13.1.1 Estimation of the white point

For a given white point all object colors are guaranteed to lie within the spectrum double cone with apex at the white point. Since, regardless of the white point, all colors lie within the spectrum cone anyway, the relevant constraint is that that all object colors lie inside the inverted spectrum cone with apex at the white point.

Notice that this essentially solves the problem. Any color such that the inverted spectrum cone with apex at that color contains the full gamut of the quilt is a possible white point. Many of such "solutions" can hardly be called "reasonable," though. For instance, a white point of any chromaticity will do if you make its radiant power sufficiently large. Perhaps it makes sense to look for the solution with the lowest radiant power.

If the colors of the gamut are in a generic configuration, then the lowest position of the inverted cone has been reached if three of the colors of the gamut lie on the inverted cone surface and all others inside it.[1] For then the apex of the inverted cone cannot be moved to a lower position without losing or or more of these three points from its interior. This makes the general solution more precise. All possible white points lie in the spectrum cone with apex at the position of the apex of the lowest inverted spectrum cone. This is the complete solution to the problem. The lowest inverted spectrum cone can be found as the solution of a linear programming problem; thus you also have an effective method to construct the full solution in any practical case.

In order to find a reasonable solution, you must indicate some point of the full solution. Here you have to rely on heuristics and prior knowledge, for colorimetry will not take you beyond the general solution. One obvious solution is to take the point of lowest radiant power. If the quilt is sufficiently varied, you are certainly close to the true white point, though the true white point is likely to have somewhat higher radiant power because the color solid is contained in the spectrum double cone. In experiments I indeed find that this estimate is typically about ten percent off in intensity. You may also wish for a white point of not too extreme a chromaticity; for instance, you might want it to be close to average daylight. You can easily accommodate this by selecting the intersection of the daylight chromaticity with the boundary of the region of solutions. If you want to be sophisticated, you set up formal prior probability densities and do a Bayesian estimate. Any of these methods will yield a reasonable solution.

13.1.2   Estimation of the spectrum of the illuminant

Once you have an estimate of the white point, you are in a position to estimate the spectrum of the illuminant.

One simple way to solve the problem is to use the pseudoinverse to find the fundamental and add the black space to it so as to obtain the complete solution. Since these are purely linear methods, you will obtain many solutions that cannot be physically realized; thus you have to discard all spectra that are not throughout nonnegative from this set. In order to construct a "reasonable solution" you need to specify what you mean by "reasonable". If you are able to do so using linear constraints, and this is always at least approximately possible, you can use linear programming or iterative projection on convex sets to find an instance.

This method is not necessarily the best way to proceed though. It would be much more convenient to shoot for a specific solution that would *already* be reasonable, and add the black space to that for the full solution. In order to be able to do that, you need to supply some prior information right at the start.

In almost all cases of practical interest, the set of expected illuminants is very limited indeed. In most cases it is close to a Planckean spectrum. As for the case of daylight and incandescent sources, it is more involved, but still fairly smooth and not that far from uniform. Only in extreme cases, such as lurid bar illuminations or cheap low pressure gas discharges, will this be different. In such extreme cases, "color constancy" breaks down anyway. You will rarely need to consider them. In the absence of more specific prior knowledge, the best way to proceed is simply to assume a smooth radiant spectral density, characterized by level, slope, and curvature. Varieties of daylight and most artificial sources are rather well described by this approximation. If you do this the problem becomes trivial, for the illuminant you look for is parameterized by three degrees of freedom, whereas the white point gives you three equations. You merely have to solve for the parameters and you are done.

In almost all cases this estimate will automatically be the reasonable solution. In order to obtain the complete solution you only need to add those elements of the black space that will not render the solution negative though certain spectral ranges.

This solves the problem in principle, and also for (almost) all practical purposes. If you need to add a black component at all, the best way to proceed depends on the precise setting of the problem.

The freedom in the estimate of the white point and the freedom in the estimate of the illuminant given the white point have to be compounded. In some (very few) cases it might be of interest to iteratively refine your estimate. From the illuminant spectrum you can find the color solid, which is contained in the spectrum double cone. This lets you refine the white point estimate from which you can update the illuminant spectrum and so forth. I doubt that this will ever be of practical interest.

13.1.3 Estimation of spectral reflectances

Given the estimate of the illuminant you can find the color solid explicitly. The quilt colors now are points in the interior of the color solid. For those near the boundary, the spectral reflectances must be similar to the Schrödinger optimal colors. Thus you have obtained a handle on the reflectance spectra.[2]

The first one to attempt to estimate reflectance spectra from colors was Wilhelm Ostwald, even (a few years) before Schrödinger's publications. Ostwald conceived of colors as mixtures of full colors, white and black. He knew the reflectance spectra of all these components, black uniformly zero, white, uniformly unity, and the full colors either zero or one with at most two transitions in the spectrum, the transitions at complementary wavelengths. (For full colors are semichromes.) Ostwald then conceived of reflectance spectra as linear combinations of these simple spectra with weights according to the color, white, and black content of the corresponding color. In this way you certainly obtain a reasonable estimate, though the complete solution of course contains numerous reflectance spectra that are not of this type. It is unclear whether Ostwald believed in a strict relation between colors and reflectance spectra. From his writings this seems to be the case, but Ostwald often used a spectroscope and most certainly had seen numerous actual spectral reflectances. From his writings it is evident that he was aware of the structure of actual spectral reflectances and absorbances. My best guess is that Ostwald must have considered his procedure as yielding a good approximation but defended his "ideal" position rather too strongly. It would certainly fit his character. Many followers of Ostwald took the correspondence for a real one-to-one relation, though. Small wonder that the "Ostwald color science" became an object of derision in the physics community.

With the historical polemics being (long) over, it makes sense to take stock. The Ostwald method evidently selects only a reasonable solution, by no means a unique or the complete solution. It is quite a nice estimate though because guaranteed to be physically realizable and as smooth[3] as can be; in other words, it is a solution that respects the fact that you are in complete ignorance except for the illuminant and the color. Any further articulation would violate this. The only real problem with the Ostwald solution is that it does not always work, but fails in the cases where the color is outside of the Ostwald double cone. Such cases are (very) rare, but they sometimes occur.

The problem with the Ostwald solution can be amended in a number of ways. One possibility is to admit negative white or black content. This would only happen in rare cases, and all colors can be represented in this way. In cases where the white content has to be taken negative the reflectance must be negative for some wavelength ranges and for negative black content it has to exceed unity in certain ranges. This "solution" (which has been suggested in the past) clearly leads to physically unrealizable solutions. A better solution is to describe any color in terms of a Schrödinger optimal color and the median gray color. The reflectance is $(1 \pm \mu)/2$, for $0 \leq \mu \leq 1$, with at most two transitions in the spectrum. The transitions will (in general) not be at complementary wavelengths, as in the case of the semichromes (Ostwald's solution). This solution can be regarded as the "maximum entropy solution" of the problem and is clearly attractive.

The solution can be completed to the full solution by adding elements of the black space, subject to the nonnegativity constraint. Thus you have indeed obtained the complete solution to the problem, based on an attractive reasonable solution. Of course, the ambiguity has again to be compounded with that in the illuminant estimate.

13.2   Color Constancy

With the reasonable solution you are in a position to predict the quilt colors for different illuminants, and with two reasonable solutions (for the same quilt patches in different configuration under different—unknown—illuminants) you are in a position to identify the patches through comparison of the estimated reflectances. Thus these methods carry you a long way toward the pragmatic solution (there only *are* pragmatic "solutions" as the problem has no *general* solution) of the color constancy problem.

13.3   Relaxing the Constraints

The constraints of the "quilt world" are very restrictive indeed. Is it possible to relax them significantly? To some extent, yes. For instance, consider relaxing the frontoparallel condition. As long as interreflections can be neglected, this will not pose much of a problem. Each quilt color now appears in various shadings. If sufficiently many instances are available there will be a frontoparallel instance, and this will dominate the white point estimate, which is just fine. Problems occur if the surfaces fail to be Lambertian, for then shading is not the only factor that determines the radiance scattered to the eye. Worse still, specularities may be (much) brighter than a piece of white Lambertian material in the same spatial attitude. In many cases you will still be able to obtain reasonable estimates if you discount "outliers due to specularity," though. This is problematic. Otherwise you will treat the same surface patch in different attitudes as different, as the patches on a peacock's tail are generally considered to "change color" as the illumination and/or viewing geometry changes. Different illuminations will be typical in real scenes. In order to be able to proceed, you will have to segment the scene into piecewise uniformly illuminated regions.

All these factors are really problematic and change the problem significantly, namely from a pure colorimetric problem into a problem of "scene interpretation," that is, "machine vision." This is only natural. Indeed, in real life you hardly ever "see colors" per se, you see geometrical layout, shape, luminous atmosphere (or optical space), and material properties. Only painters in the impressionistic tradition really "see colors" and they need a long training period in order to be able to do that. Even then, it is somewhat questionable whether they really "see colors" or merely acquire the ability to *apply colors* (dabs of paint I mean) in such a way as to suggest aspects of the scene in front of them to the cognoscenti. Remember that people were shocked when the early impressionistic paintings came on display. These paintings were by no means considered to be realistic renderings, and pregnant women were kept from visiting because of the fear of miscarriages.

"Colors" in the colorimetric sense are meaningless, whereas patches in the visual field when consciously "seen" are seen in terms of their meaning.

13.4   An Example

In this section I discuss a simple but interesting example. I load a database of object spectral reflectances from the Internet, compute the object colors for average daylight and incandescent light, and study the results of various methods to handle color constancy.

I selected the NCSU database[4] of spectral reflectances of 170 materials from the daily environment. The description has

> *The reflectance spectra of 170 natural and man-made objects. Organized by material order, 171 wavelengths per material. Thus the first 171 numbers are the spectra for the 390–730nm range (in 2nm intervals) for material ♮1, the next 171 numbers are the spectra for material ♮2, and so on.*

This database contains an odd selection of samples, which makes it useful for this study. Whether it is at all "typical" of the daily environment is a moot question. Certainly a set of 171 instances severely undersamples the world. That is not too important, though.

In figure 13.2, I show the gamut of object colors in the data base. Notice that most of them are rather muted, grayish, or very dark. There are a few vivid colors and even a few near-whites. The latter point is important if you want to estimate the illuminant. The daylight illuminated objects look yellowish in the "daylight interpretation" (RGB values obtained via Schopenhauer RGB crate for daylight) and bluish in the "artificial light interpretation" (RGB values obtained via Schopenhauer RGB crate for incandescent light),

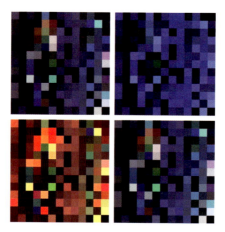

Figure 13.2 The gamut of object colors in the NCSU data base under daylight (top row) and incandescent (bottom row) illumination. The Schopenhauer RGB crates for daylight (first column) and incandescent light (right column) were used to do the translation to RGB.

though they look almost indistinguishable when interpreted in the right frame. This already indicates, before any calculation, that color constancy might not be too difficult to establish.

I load the average daylight spectrum (CIE illuminant D65), the spectrum of a light bulb (CIE illuminant A), and the color matching functions from the Internet and do some work to bring all these table in sync. As a preliminary calculation I find the edge colors, full colors, and optimal colors for both illuminants. I provide some means to find the corresponding full color and/or optical color for any given color. I also compute the colors of the objects in the data base for each illuminant. At this point I'm ready to begin.

13.4.1   Using a (generalized) von Kries transform

The first thing to do is to simply compare the object colors under the two illuminants. In figure 13.3 left, I show a plot of color space with the shifts in color indicated by straight line segments. It is clear that these shifts are very significant indeed; the colors are quite different under the two illuminants. On the other hand, you see that there is a lot of regularity in the shifts, they are by no means totally random. Thus there is some hope to account for the shifts somehow.

Since the shifts clearly depend on the intensity (for instance, black must be invariant under change of illuminant), you need to take that into account. A simple way to do this is

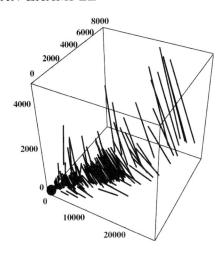 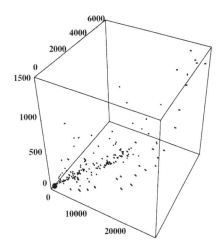

Figure 13.3 Left: A comparison of the object colors. The line segments indicate the shift in color space due to the change of illuminant. Right: A comparison of the object colors after an optimal (generalized) von Kries transformation. The line segments indicate the shift in color space due to the change of illuminant.

to calculate the relative shifts, the coordinatewise ratio of the difference of colors to their sum.

A scatterplot reveals that these are by no means the same for all items; thus there is an appreciable random component. However, the spatial median of the relative shifts is very significant. The median is $\{-0.209, 0.071, 0.526\}$, which is very similar to the relative difference of the illuminants, which is $\{-0.207, 0.092, 0.529\}$. Of course, you should assume that the illuminants are unknown for the constancy problem, but it is of interest to notice the fact.

Apparently the differences can be accounted for to a large extent by some linear transformation. The best linear transformation that maps the daylight colors on the incandescent light colors is

$$\begin{pmatrix} 1.62545 & -0.50477 & 0.58581 \\ 0.04017 & 0.80811 & -0.10814 \\ -0.00164 & 0.00497 & 0.30877 \end{pmatrix}.$$

It is a "generalized von Kries transformation" (see appendix K.1), its eigenvalues are 1.60, 0.83 and 0.31. Not surprisingly, these values are close to the ratio of the colors of the sources, which are $1.52 : 0.83 : 0.31$. After I apply the transformation, the color differences are indeed much reduced,[5] as you can see in figure 13.3 right.

This is a good start to try to find correspondences between the colors under the two illuminants. I simply let each color under daylight illumination correspond to the nearest color under incandescent illumination, taking the von Kries scaling into account. Of the 170 objects in the database, 32 turn out to have an incorrect correspondence. This is not a bad result. It is even better when you consider the nature of these false correspondences. The largest group of errors concerns Caucasian skin colors, 16 cases in total. These are largely confused among each other, something that might well occur under a single illuminant as the colors are very similar. In three case there was a confusion of skin colors of different races (Caucasian with Asian or African American). In two cases, Caucasian skin was confused with something totally different (sand, bark, almond). Another group of confusions concerns green leaves. Various leaves, pine needles, and a green pepper are mutually confused. A few white things (tablecloth and sugar) are also mutually confused. A red flower was confused with peach skin, and three red cloths were confused among each other.

Apparently this simple method already yields quite impressive color constancy. You hardly ever commit serious errors in the identification of materials. The reflectance spectra of confused materials turn out to be very similar indeed. Notice that this result was obtained without any spectral estimate whatsoever. For this set of "random" materials, the colors suffice largely to identify a material as you move from a daylight to an incandescent spectrum. This relativizes the magnitude of the color constancy problem.

13.4.2   The "gray world" hypothesis

It is often assumed that the world is gray on average, as in the 18 percent gray chart as used by photographers. This suggests that it might be of interest to compare the averages of the gamuts under the two illuminants. The spatial median of the NCSU colors under daylight illumination is $\{5062, 1858, 750\}$; the eighteen percent gray would be $\{3580, 1810, 1052\}$, whereas these colors under light bulb illumination yield values of $\{7797, 1626, 236\}$ and $\{5446, 1505, 324\}$ respectively.

Thus, the "gray world" hypothesis is not particularly successful in this case. Of course, there is little reason to believe in the "gray world" hypothesis to begin with, at least, little reason for it to hold in any precise sense. It is a default assumption. If there is no reason to assume the world has any particular overall hue, then perhaps Occam's razor suggests "average gray".

The gray world hypothesis is not needed for the problem of color constancy. You can simply use centroids to normalize the colors. I divided every color by 1/0.18 times the centroid. In figure 13.4, I show a plot of the remaining differences in color space. Apparently this simple method does an excellent job.

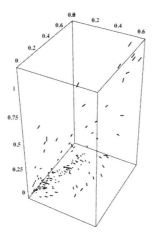

Figure 13.4 A comparison of the object colors after normalization by the centroids of the gamuts. The line segments indicate the shift in color space due to the change of illuminant.

In this case I find only four confusions, thus color constancy is almost perfect. These are two cases in which Caucasian skin was confused with Caucasian skin and two cases in which rocks were confused with rocks. The reflectances of the confused cases are almost identical, and thus these cases are hopeless to begin with. Apparently the simple centroid method lets you identify objects correctly when you switch from daylight to artificial illumination.

13.4.3   Estimation of the illuminant spectra

How well can you estimate the illuminant spectra from the colors? This is an important problem in its own right. It is closely connected to the problem of color constancy.

In order to approach this problem, I start by constructing the lattice of (color) dominance. For any given color I construct the inverted spectrum cone with vertex at that color, and I find all colors that are in the interior of the inverted cone. These are the colors that are dominated by the fiducial color. I repeat this for all colors in the gamut and end up with the "top points" of the gamut, the colors that are dominated by no other color.

For both the daylight and the artificial illumination gamuts I find four top colors, and these turn out to be the same four-tuples.

   75 White T-shirt,
   144 Sugar (White),
   158 Fabric—White,
   163 Tablecloth—White,

that is to say, by this procedure I have automatically found the white colors of the gamut. I use these to find estimates of the white points by looking for colors that just dominate the top points. It is a simple linear programming problem (see figure 13.5). I estimate the white point for daylight illumination as $\{17446, 8562, 4812\}$ and for the artificial illumination as $\{24697, 6657, 1387\}$.

These estimates are quite reasonable, except for the intensities. In both cases the intensity is about 85 percent of the actual intensity. This makes sense because the method always underestimates and (in particular) because "white objects" rarely reflect more than 90 percent. The ratios of the estimated white points are $1.42 : 0.78 : 0.29$, close to the ratio of the colors of the actual illuminants, which are $1.52 : 0.83 : 0.31$.

You can use these estimated white points to normalize the gamut in order to solve the constancy problem. It works very well. Only the Caucasian skin colors are confused among each other.

The illuminant spectra can be estimated from the white point estimates by finding the standard spectrum that is characterized by spectral level, spectral slope and spectral curvature. Results are plotted in figure 13.6. Apparently you can estimate these illuminant spectra quite well. Thus a set of object colors like these in the NCSU data base (a mere 170 colors, many repeated) suffices to gain an excellent estimate of the illuminant.

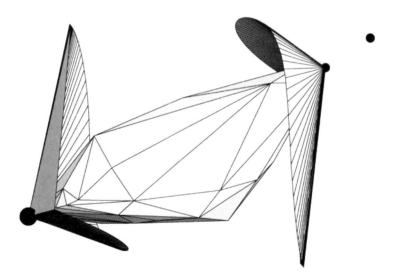

Figure 13.5 The estimation of the white point for the daylight colors. I show the spectrum cone and the inverted spectrum cone at the estimated white point. The actual white point is also indicated. The colors in the gamut are indicated by their convex hull.

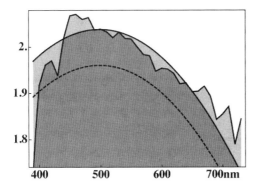 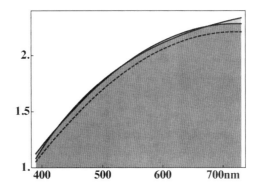

Figure 13.6 Estimated spectra for the two illuminants. The spectral radiant power density is an a logarithmic scale. The main difference with the vertical spectrum is a shift, that is to say, the radiant power has been underestimated. The estimate of spectral shape is quite good. (The dashed curves are the estimates, and the real spectra are compared with shifted copies.)

13.4.4 Estimates of the object reflectance spectra

How well can you estimate object reflectance spectra? This is evidently an important problem since reflectance spectra are *material properties*. This is a key issue in philosophical debate. Are color mere figments of mind, or do they correspond to something real? In the philosophical literature, one often finds the notion that colors do not correspond to anything in the physical world.

You can estimate reflectance spectra of colors if you know the illuminant spectrum. I have already shown that you can estimate illuminance spectra for sufficiently articulated scenes pretty well. In this section I will use the actual illuminant. This will show you the problems involved in the estimation of spectral reflectances per se. In real life the errors in the illuminant estimate and the spectral reflectance estimates are always compounded of course.

Given a color, you can estimate the spectral reflectances in many different ways. A popular method is to find the principal components[6] of the object reflectance spectra. This requires that you have these in the first place. Here we have them, and I show the principal component analysis in figure 13.7.

The singular values fall off very fast, which suggests that just a few might do a good job in describing any spectrum. The first three principal components show (roughly) a uniform spectrum, a monotonic gradient, and a strongly curved spectrum. This is indeed typical.

For most collections of spectra, the first three principal components look like this, which raises the question why you should bother with principal components analysis in the first

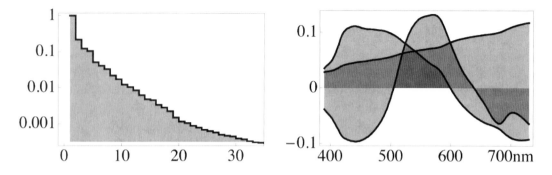

Figure 13.7 A singular values decomposition of the NCSU reflectance spectra yields the singular values plotted at left on a 10-log scale. Evidently, the values fall of very fast and a few dimensions might do. At right, a plot of the first three principal components.

place. You might as well use a level, gradient, and curvature triplet as I did in the illuminant estimation.[7] This would solve the problem of how to get your hands on a large set of spectral reflectance instances, too.

Anyway, once you have the initial three principal components, you can describe any spectrum as a linear combination of those. The coefficients in the combination can be calculated from the color coordinates, so here you have a method to estimate spectral reflectances.

In figure 13.8 I illustrate some results. On the whole, the method doesn't do that badly. A major problem of this method is that it is fully *linear*. Thus, there is no guarantee that the values of the spectral reflectance will turn out to be in the range $(0, 1)$. Indeed, in the example you see several instances where the spectral reflectance has to be clipped because it turned out negative. This is a general drawback of any linear method.

A much more sophisticated method is to use the Ostwald color theory. In Ostwald's description any color is naturally assigned a spectral reflectance.

Given a color, I find the full color of the same hue, and I express the color as a linear combination of the full color and the (typically estimated, here actual) white color. Since both the full color and the white color have fully determined reflectance spectra, I obtain the reflectance spectrum for the color as a linear combination of these. The result is guaranteed to be in range, except in the (very rare) instances that the color is outside the Ostwald double cone.

I show results in figure 13.9. A rare example of an out-of-range case is shown in figure 13.10. For this database and daylight illumination it happens only in a few cases and always in the red. Notice that the white content is negative (-1.8%), though only very

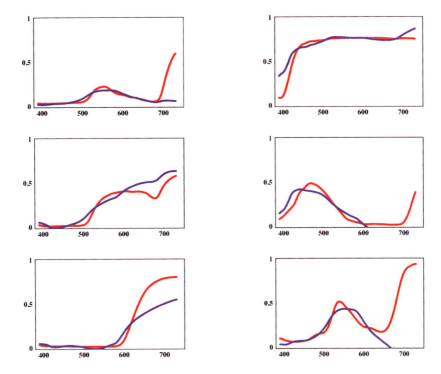

Figure 13.8 Estimated spectral reflectances from color as linear combinations of the first three principal components. Illuminant is daylight. The objects are from left to right, top to bottom: "18: Bush leaf", "31: Daisy — White petals", "32: Daisy — Yellow center", "50: Synthetic cloth blue (with surfing harness)", "51: Synthetic cloth red (with surfing harness)", "62: Green snow hat". Notice that the estimated reflectance becomes negative in certain cases, then the method fails. This is a typical problem with linear methods.

little so. In these cases you should substitute an optimal color for the full color. This would shift the spectral transition locations and raise the reflectance. This correction will be a minor one in all cases. Almost always, clipping would be acceptable.

The Ostwald method works very well for the objects in the NCSU database. An advantage as compared to the PC method is that no prior knowledge concerning the actual spectral reflectances is necessary[8]. It illustrates my observation that you indeed can "see spectral reflectances" (material properties) in practice, even though this is impossible in theory. It illustrates the considerable usefulness of Ostwald's methods in practice. The violent polemics against the so-called Ostwald's color science were indeed to the point in theory, but rather pointless in practice.[9]

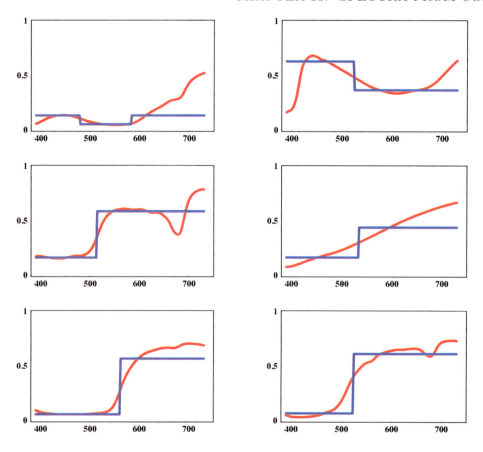

Figure 13.9 Estimated spectral reflectances from color via Ostwald's method. Illuminant is daylight. The objects are from left to right, top to bottom: "35: Small magenta flower (from vine)", "64: Light blue towel", "65: Banana yellow (just turned)", "118: Wood — pine", "142: Carrot", "139: Lemon skin".

The (sometimes extreme) polemics against the so called Ostwald color science were partly due to mutual misunderstanding and to the unfortunate pushing of approximate methods for laboratory use. It is evidently not possible to capture arbitrary spectra via just a few photometric observations using a crude set of filters. Ostwald took his semichrome description (which is quite apt in capturing the major traits of spectra for colorimetric purposes) rather too literally and recommended simply observing the lower and upper reflectance levels through filters roughly tuned to the corresponding spectrum regions in order to get at the color, white, and black contents. The transition wavelengths would

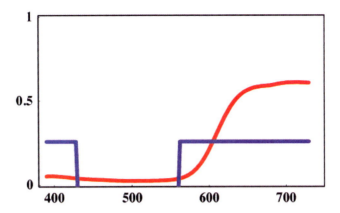

Figure 13.10 A rare example of a color that falls outside of the Ostwald double cone. It is "44 Red flower (Plant had red leaves)" from the NCSU database under daylight illumination. I find a color content 27.4 percent, white content −1.8 percent, black content 74.4 percent. Such cases occur predominantly in the red. Notice that the white content undershoot is so minor as to be irrelevant, though.

simply follow from a comparison of the hue within the color circle. You clearly cannot hope to do spectroscopy this way. Actual spectra may be quite unlike semichromes, especially if their color content is low.

On the other hand, the estimation of the (veiled and attenuated) semichromes definitely gives a semiquantitative handle on the spectra on the basis of color. For instance, this may be all a biological vision system needs to do in order to obtain a useful cue as to the chemical constitution of basic materials.

Exercises

1. Estimate the white point   [Numerical experiment] Set up a scene from a database of object spectral reflectances (or a set of random instances) and an illuminant. "Scene" simply means a set of object colors here. Construct the partial order of dominance, using the inverted spectrum cone and find the top points. From the top points find the white point of minimum luminance. What is the set of all feasible white points?

2. Estimate the illuminant   [Numerical experiment, conceptual] Given a white point, construct a possible illuminant. There are many different methods you might use here. Make sure the illuminant at least has a real spectrum. What is the set of all possible illuminants? What would be a particularly "desirable" instance (think of criteria), and how could you find it?

3. Estimation of spectral reflectances   [Conceptual and numerical experiments] Assume you know the illuminant and you are given object colors. Find the spectral reflectances that correspond to these colors. Think of (and implement) various methods. If you have a possible solution, how would you indicate all possible solutions? Given all possible solutions, what would be the most desirable instance (think of criteria)? How would you find it? Implement the method(s) suggested in this chapter.

4. Principal components   [Elaborate] Collect databases of spectral reflectances as you find them on the Internet and compute the first three principal components of each. Compare the results. (In order to do the comparison, you have to find the angle in the space of all spectral reflectances between the linear subspaces spanned by these PC bases.)

5. Color constancy   [Numerical experiment] Use your data base of object spectral reflectances (or generate a random set) and compute the color gamut for two different illuminants. Now try to find the correspondence between the items in the two gamuts, using the various methods suggested in this chapter. Study the nature of the cases that are confused.

6. Beyond Ostwald   [Research topic] Ostwald's recipe for a spectral reflectance given an object color is by no means the only game in town. The colorimetric data specify an infinite set of spectral descriptions among which (in general) an infinite subset of physically realistic ones. In order to make a choice, you need additional constraints. Ostwald's constraint is only one possibility.[10] For instance, you might try to find solutions that are throughout *smooth* and minimize some measure of variation, say the integral over the second derivative (curvature) squared. Explore such possibilities. (Reading up on "quadratic programming" may help.) How do they compare to the Ostwald solution in the case of grayish or very vivid object colors? Is it worth the additional effort?

7. Edwin Land's "Retinex" [CONCEPUAL] Edwin Land (of Polaroid fame) has devised an interesting algorithm that implements a form of color constancy based on sequences of edges between uniformly colored regions (Retinex theory). Look up the references[11] and discuss the model. How does it differ from the methods discussed in this chapter? Can it be generalized?

8. The "bichromatic model" [FOR PHYSICISTS] Shafer et al.[12] analyzed the beams remitted from glossy dielectric surfaces. They find that the diffuse component has the color of the bulk material, whereas the specular component has the color of the illuminant. Look up the references and discuss the model. How does it differ from the methods discussed in this chapter? Can it be generalized? Can it be used to improve color constancy?

9. Scene interpretation [CONCEPTUAL] Color constancy algorithms typically rely on many prior assumptions that are typically violated in images of actual scenes. Is it easily possibly to "segment" such images into homogeneous regions in order to be able to apply the algorithms per segment? What type of segments do you propose? How do you propose to perform the actual segmentation? What type of problems do you expect to have to cope with? Now look for some literature[13] on the Internet. Are you satisfied with the current state of the art?

10. Spectral reflectance from databases [APPLICATIONS ORIENTED] Suppose you have a database of spectral reflectance factors. You can use this to estimate reflectance factors from RGB colors! Simply select the 5 percent items of the database that are nearest in the RGB cube when illuminated in the standard manner (for instance, average daylight). Take the average of the spectral reflectance factors of this subset and use it as an estimate. This is a simple method that always works. Now explain why this "method" makes no sense. Then explain under what conditions it might be useful.

Chapter Notes

1. This general method was pioneered by Forsyth, D.D. Forsyth, "A novel algorithm for color constancy," *IJCV* 5, pp. 5–36 (1990).

2. The spectral reflectances have to be in the range $[0, 1]$, otherwise they are arbitrary. Thus one needs constraints to arrive at a unique solution. Ostwald's solution is one, but infinitely many others are possible. For instance, it might make sense to go for the smoothest (in some sense) functions. C. van Trigt, "Smoothest reflection functions. I. Definition and main results," *J.Opt.Soc.Am. A* 7, pp. 1891–1904 (1990); —, "Smoothest reflection functions. II. Complete results," *J.Opt.Soc.Am. A* 7, pp. 2208–2222 (1990); —, "Metameric blacks and estimating reflectance," *J.Opt.Soc.Am. A* 11, pp. 1003–1024 (1994).

3. This obviously depends upon your definition of smooth. Ostwald's curves are piecewise constant with at most two discontinuities. In some metrics this counts indeed as very smooth. It all depends on how you weigh gradients and isolated discontinuities.

4. The database can be found at `ftp://ftp.eos.ncsu.edu/pub/spectra/`.

5. On this topic, see G. West, and M. H. Brill, "Necessary and sufficient conditions for von Kries chromatic adaptation to give color constancy," *J.Math.Biol.* 15, pp. 249–258 (1982); M. Brill, and G. West, "Contributions to the theory of invariance of color under the condition of varying illumination," *J.Math.Biol.* 11, pp. 337–350 (1981).

6. D-Y. Tzeng, and R. S. Berns, "A review of principal component analysis and its applications to color technology," *Color Res.Appl.* 2, pp. 84–98 (2005).

7. Indeed, much of the literature on principal components analysis of spectral reflectances makes little sense. It makes good sense for small or very focused databases, but it hardly makes sense if the set of fiducial spectral reflectances is very large. The broader the set, the more the principal components are bound to look as a level, a gradient, and a curvature. You would probably do better to simply accept this and omit the (superfluous) principal components. If the database is very focused, you are bound to find nontrivial results. An example is M. J. Vhrel, and H. J. Trussel, "Color correction using principal components," *Color Res.Appl.* 17, pp. 328–338 (1992). These authors measured samples from the Munsell color atlas and show that these can be approximated quite well using only a small number of principal components. This is indeed to be expected given the extremely focused nature of the data set (how many inks were used in printing the samples?). This study is often cited as showing how *any natural reflectance* can be so approximated, which is of course ludicrous.

8. I think that the literature has made too much of the principal components. People have computed PCs for many data sets. Often these results are then applied to different contexts, throwing much doubt on the basic philosophy (for instance, PCs for the Munsell chips have been applied to reflectances of natural materials). The reason this works (at least somewhat) is obvious when you look at the shape of the PCs. No matter what data set was used, you have a constant, a monotonic slope, and a central extremum, or linear combinations thereof. This is indeed what you would get for a set of random reflectances. Instead of PCs it would make more sense to start with such a simple basis right away, as I did for the case of illuminants. The original data set is largely irrelevant.

9. K. W. F. Kohlrausch, "Bemerkungen zur Ostwald'schen sogenannten Farbentheorie," *Physik.Z.* 22, pp. 402 (1921); J. von Kries, "Physiologische Bemerkungen zur Ostwald'schen Farbenfibel," *Z. Sinnesphysiol.* 50, pp. 117 (1919); E. Schrödinger, "Über Farbenmessung," *Physik.Z.* 26, pp. 349 (1925); C. L. Schäfer, "Über den sogenannten Schwarz- und Weissgehalt von Pigmenten in der Ostwald'schen Farbensystematik," *Physik.Z.* 27, pp. 347 (1926); A. Bernays, "Die Farbenfibel von Wilhelm Ostwald," *Naturwissenschaften* 21, pp. 864 (1933).

10. Van Trigt ibid.

11. E. Land, "The Retinex Theory of Color Vision," *Sci.Am.* 237, pp. 108–128 (1977).

12. S. A. Shafer, "Using Color to Separate Reflection Components," in *Physics Based Vision: Principles and Practice. Color*, edited by G. E. Healey, S. A. Shafer, and L. B. Wolff (Boston: Jones and Bartlett, 1992), pp. 43–51.

13. Ibid.

Part V

RGB colors, Color Gamuts and Images

Chapter 14

The RGB Cube

The RGB colors should be familiar to you. If you (programmatically or via some graphical user interface) cause a colored pixel to appear on your computer monitor, you explicitly or implicitly specify an RGB color (figure 14.1). A generic example is the Mathematica function

    RGBColor[r, g, b]

that specifies an RGB color via three numerical parameters r, g, and b that are supposed to be (real) numbers in the range $[0, 1]$. In the old-fashioned cathode ray tubes[1] each pixel was made up of three luminescent phosphor dots (see figure 14.2), the luminance of which was controlled via digital-to-analog converters[2] setting certain electrode voltages. These again were controlled via the r, g, and b specifications. Thus the Mathematica function invocation caused a certain beam to be emitted whose spectral composition was controlled by these three numbers. The manufacturer set things up such that RGBColor[1,1,1] would produce an achromatic beam ("WHITE"), whereas

RGBColor[1,0,0] would appear "PURE RED",

RGBColor[0,1,0] would appear "PURE GREEN", and

RGBColor[0,0,1] would appear "PURE BLUE".

By suitably setting the parameters r, g, and b you could thus produce "any color," in reality the gamut defined by the linear combinations with fractional (nonnegative) coefficients of the "pure" red, green, and blue beams (figure 14.2). With input devices the situation is analogous, except for the fact that things work the other way around (figure 14.3) and that

About RGB notation

Every computer language or graphics application has its own way to specify RGB colors. These all boil down to the same thing; they are only syntactically different. It is useful to have some concise notation for RGB colors for use in discussions, in text, or just for your own convenience. I will use a notation in this book that I happen to like. I will write $R^{ab}G^{cd}B^{ef}$, where *ab*, *cd*, and *ef* stand for unsigned decimal numbers in the range 00 to 99. I use "99" instead of "100" in order to ensure that all $R^{ab}G^{cd}B^{ef}$ expressions have the same length (that's also why I use the teletype font). The small difference between 99 and 100 is irrelevant in practice. This is useful in tabulations, and the like. If you mainly work with values in the range $[0, 255]$ you may adapt the notation, but I find it (conceptually) more convenient to convert to the 00–99 scale. Thus PURE RED will be $R^{99}G^{00}B^{00}$, PURE GREEN $R^{00}G^{99}B^{00}$, and PURE BLUE $R^{00}G^{00}B^{99}$, whereas $R^{99}G^{75}B^{25}$ is a PALE ORANGE, and so forth. (Of course the same type of notation can also be used in the CMY specifications.)

Figure 14.1 The primary RGB colors, that is to say Red, Green, and Blue. You should know what I mean, at least, if you're a "normal trichromat" as I am.

one deals with metamerism. In all these cases the "RGB" essentially derives from hardware constraints.

In the old-fashioned color TV CRTs a single pixel (picture element) was actually an array of three subpixels, a red, a green, and a blue one. In modern electronic (are there any others left?) cameras a single pixel is typically made up of a square array of four subpixels, two green ones, a red one, and a blue one. Thus the hardware (which is well shielded from the user!) typically implements Schopenhauer's parts of daylight in some way or other.

Typical users are not at all familiar with the phosphor emission spectra, nor need they know whether they use phosphors at all (they are likely to use a liquid crystal display[3] screen nowadays). Thus RGB colors are not strongly related to any colorimetric practice, at least not for typical users. Colorimetry proper is important for the engineer who adjusts the

Figure 14.2 RGB colors as produced on a CRT monitor. Left "white", center "black", right a "gray". The dot pitch is such (typically 72 dots per inch) that the discrete structure is only "just visible" if you look at images intently.

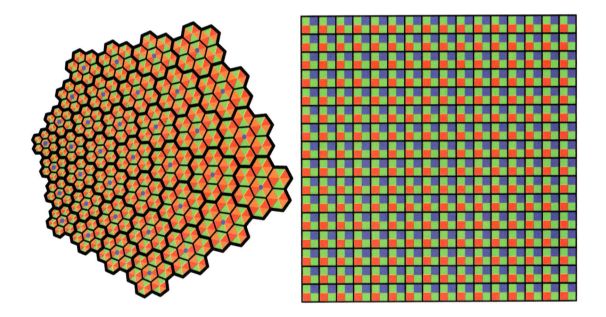

Figure 14.3 Left: The RGB (or LMS for "long, medium, and short wavelength sensitive") receptor mosaic in your retina according to my esteemed advisor Maarten Bouman. Here the inhomogeneous structure of the retina and the relative numerosities of the cones have been taken into account. Notice that the receptors have been grouped into functional units, like the ommatidia of insect eyes, or the Bayer pattern used in CCD chips for consumer cameras. Right: The Bayer pattern used in most CCD chips for consumer cameras. A single pixel is actually a cluster of four subpixels, a long-wavelengths-sensitive one, a short-wavelengths-sensitive one, and two medium-wavelengths-sensitive ones.

RGB colors on display

On a display that emitted beams with radiant powers proportional to the RGB coordinates, the displayed RGB color would cause the same signals to your brain as would the original beam whose color coordinates in the Schopenhauer parts of daylight scheme had these coordinates. Real-life displays are different, though. Without delving into details (there are many) you should at the very least be aware of the gamma correction.

The idea of a correction derives from the time when RGB colors were displayed on CRTs. In vacuum tubes (a CRT is essentially a vacuum tube) the intensity of the electron beam is not proportional with the cathode-grid voltage difference due to the "space charge," a cloud of electrons surrounding the cathode. Since the photon emission by the CRT phosphors is proportional with the electron beam intensity, one felt a correction was needed to linearize the system. For some reason this was called the gamma-correction. This is history, though. Few modern displays are based on CRTs any more.

The main reason for the continued existence of gamma (γ) is a (desirable) attempt to match the monitor images to the subjective scales of generic human users.

On most computers you can prepare a lookup table that sets $\gamma = 1$, which means that you have a linear response. This is perfect for experiments, but not for your daily work, so make sure that you can easily switch between linearized and normal.

What is "normal" depends on your computer. In most systems, images are encoded with a γ of about 0.45 and decoded with a γ of 2.2, whereas the Macintosh system used the corresponding values of 0.55 and 1.8. With the latest operating system Snow Leopard this was (silently) changed to 2.2 (finally, "Apple gave in"). A $\gamma \neq 1$ implies a power law dependency of intensities upon the coordinate values.

I will leave it at this, but if you are interested you might explore the relation between γ and "subjective scales," like that used in the CIE Lab system. Such issues are often important in practice; however, I consider them out of the scope of this book. Simply type "gamma correction" in Google and you'll be in the know soon enough.

monitor, the "power user" who uses the monitor for (say, psychophysical) experiments, or the sophisticated user who needs to synchronize the monitor with artwork, scanners, and printers. Such applications of colorimetry are pretty standard, and I need not specifically cover them in this book (literature abounds[4]). Instead, I will discuss RGB colors *as RGB colors* in this chapter, quite distinct from colorimetry proper.[5] I will highlight the (many) conceptual tangencies with colorimetric concepts, though. This is interesting and much fun as it throws a new light on many colorimetric constructions. In many cases these are but slight complications of the (much simpler, bare bones) RGB structures.

RGB colors need not be introduced as such. They can also be introduced via colorimetry using the Schopenhauer parts of daylight and the notion of coarse grained spectroscopy this induces. Such an introduction provides a strong motivation to study RGB colors as a convenient and very powerful *approximation* of colorimetry. I resist the temptation and treat RGB colors *as such*. However, you are well advised to study the (really very close, much closer than a scan of the literature will suggest to you) relation. It will pay off in terms of both understanding and utility.

14.1   The RGB Colors Defined

Formally, the RGB colors are triples of numbers in the range $[0, 1]$, thus the set of points of the unit cube (see figure 14.4), the body diagonal $R^{00}G^{00}B^{00}$–$R^{99}G^{99}B^{99}$ being special because it contains the "grays" or "monochrome colors". Usually $R^{00}G^{00}B^{00}$ is "black" (or "K") and $R^{99}G^{99}B^{99}$ "white" (or "W").

The RGB cube is often used by designers, since it yields a comprehensive overview of colors. You can buy one or make one yourself (see figure 14.5). It is indeed quite useful to obtain an intuitive overview of the totality of RGB colors. In figure 14.6 I show two other views of the cube, the environments of the white and the black vertices. Evidently

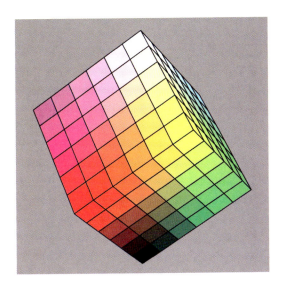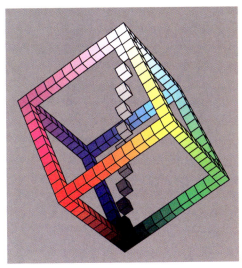

Figure 14.4 The RGB cube. At left, a view of the RGb cube that shows only the exterior. At right, a skeleton view allows you to see the gray axis connecting the black and white vertices.

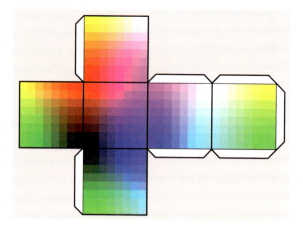

Figure 14.5 Foldout of the RGB cube. You can cut it out, fold it, and glue it into the shape of the RGB cube. Unfortunately, this gives you only the outside. You can construct volumetric cubes from black metal wire and painted beads, but this involves a major undertaking.

the "best colors" are located near the contours in these images. The best colors lie on a polygonal, closed arc that encircles the gray axis. In these figures the contour appears as a (twisted) hexagon, rather than a color "circle."

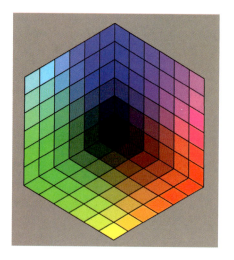 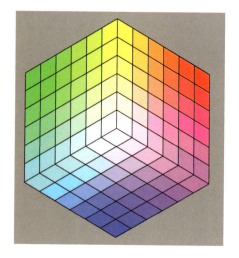

Figure 14.6 The regions around the black and the white vertex of the RGB cube.

It is important to remember that the RGB cube is a *volume*. Since this is difficult to show explicitly in a 2D figure, even in the 3D cardboard models used by designers, it sometimes goes unnoticed. In figure 14.7 I have tried to visualize the internals of the RGB cube.

14.1.1 A special basis

It makes sense to introduce a basis that recognizes the special status of the gray axis, say

$$\mathbf{e_3} = \frac{1}{\sqrt{3}} \{1, 1, 1\},$$

as one of the (normalized) basis vectors. If you want an orthonormal basis (and that makes obvious sense), you need to augment this with two unit vectors orthogonal to the gray axis. You may select (infinitely many choices at your disposal)

$$\mathbf{e_1} = \frac{1}{\sqrt{6}} \{1, 1, -2\},$$

and

$$\mathbf{e_2} = \frac{1}{\sqrt{2}} \{-1, 1, 0\},$$

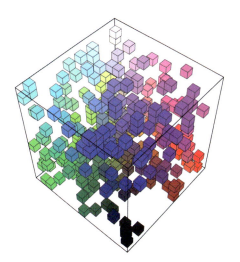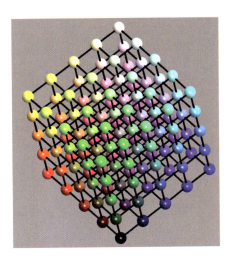

Figure 14.7 Left: A sparsely filled RGB cube suggesting the interior; Right: A "beaded wireframe" model.

as convenient (the orientation is picked in order to obtain a connection to conventional systems), as they are obviously orthogonal to $\mathbf{e_3}$ and to each other, and suitably normalized. Notice that $\mathbf{e_2}$ is in the "red-green," whereas $\mathbf{e_1}$ is in the "yellow-blue" direction. You might say that $\mathbf{e_3}$ spans the achromatic dimension, whereas $\mathbf{e_1}$ spans the "yellow-blue opponent" dimension, $\mathbf{e_2}$ the "red-green opponent" direction. The set $\{\mathbf{e_1}, \mathbf{e_2}, \mathbf{e_3}\}$ is simply an alternative basis to the default $\{1, 0, 0\}$ (red), $\{0, 1, 0\}$ (green), and $\{0, 0, 1\}$ (blue) basis. As you will appreciate soon the opponent basis offers significant intuitive, and conceptual advantages though.[6]

Of course, this procedure introduces "virtual RGB colors" that do not fit in the $R^{dd}G^{dd}B^{dd}$ notation (where $00 < dd < 99$). I discuss this (minor) problem in appendix N.1.

Since the red-green and yellow-blue opponent dimensions carry the hue information, you may parameterize the hues through an angle φ, such that the color

$$\alpha(\cos \varphi \, \mathbf{e_1} + \sin \varphi \, \mathbf{e_2}) + \beta \mathbf{e_3},$$

has a hue given by the parameter φ, whereas the parameter α measures the amount of color ($\alpha = 0$ clearly specifies a gray) and β the amount of achromatic color. The angle φ can be varied on $\{-\pi, +\pi\}$, thus the hues naturally fall on a color circle. The parameters α and β cannot be varied at will, for the RGB components necessarily fall in the range $[0, 1]$. You have a number of constraints.[7] For any value of φ there exist unique values for α and β such that the value of α is a maximum under the given constraints. These colors are the equivalents of the *full colors* (section 7.3.1), they are as different from gray as possible for any specified hue.

The full colors locus (see figure 14.8) is a hexagonal, polygonal arc, a concatenation of edges of the RGB cube. The sequence visits the vertices Y, G, C, B, M and RYGCBMR in cyclical order. Notice that the full color locus is not a planar curve, but goes "up" (toward W) and "down" (toward K). The "up" vertices are the secondary colors C, M, Y, whereas the "down" vertices are the *primary* colors R, G, B. This makes sense because each *secondary* color dominates a pair of primary colors (thus C dominates G and B, M dominates R and B, Y dominates R and G) and thus is brighter than these.

14.1.2   The color circle

Notice that I have defined the hue angle φ such that $\varphi = 0$ corresponds to yellow, whereas an increase produces a greenish cast. Thus both the phase and the sense (orientation) of the color circle are well defined. As the hue angle increases you meet the cardinal colors yellow, green cyan, blue, magenta, and red in that sequence. Of course yellow is next after red, completing the round trip $\varphi = 0 \ldots 2\pi$. The color circle is illustrated in figure 14.9.

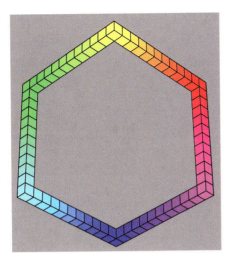

Figure 14.8 The full color locus, a twisted hexagon.

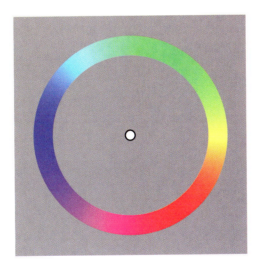

Figure 14.9 The color circle.

Since there exists a well-defined (unique) "full color" for any hue, you can specify any RGB color in terms of that full color (that is φ), its color content, white content, and black content. For any color must lie in the convex hull (a triangle) of its corresponding full color, white and black. Color, white, and black content lie in the range $[0, 1]$ and add to unity.

Thus you obtain a perfect Ostwald-type color specification.

The Ostwald specifications correspond in a 1–1 fashion to the RGB specifications, yet they are much more intuitive to use.

The full colors are necessarily of one of the types

$$\{\alpha, 1, 0\} \quad \text{Green to Yellow,}$$
$$\{0, 1, \alpha\} \quad \text{Green to Cyan,}$$
$$\{0, \alpha, 1\} \quad \text{Blue to Cyan,}$$
$$\{\alpha, 0, 1\} \quad \text{Blue to Magenta,}$$
$$\{1, 0, \alpha\} \quad \text{Red to Magenta,}$$
$$\{1, \alpha, 0\} \quad \text{Red to Yellow,}$$

where α proceeds from 0 to 1. This is indeed formally evident from the nature of the algebraic constraints, and also geometrically evident because the full colors have to lie on those edges of the RGB cube that are not connected to the gray axis at K or W. These edges form a polygonal closed arc, topologically equivalent to the color circle. In order to assign hues "Ostwald style," that is to say, in accord with Ostwald's principle of internal symmetry, you should substitute arc length along this polygonal arc for the hue angle. The total arc length is six, and using the same phase and orientation you obtain:

$$0 \ (\text{or} \quad 0°) \quad R^{99}G^{99}B^{00}: \quad \text{Yellow (Y),}$$
$$1 \ (\text{or} \ 60°) \quad R^{00}G^{99}B^{00}: \quad \text{Green (G),}$$
$$2 \ (\text{or} \ 120°) \quad R^{00}G^{99}B^{99}: \quad \text{Cyan (C),}$$
$$3 \ (\text{or} \ 180°) \quad R^{00}G^{00}B^{99}: \quad \text{Blue (B),}$$
$$4 \ (\text{or} \ 240°) \quad R^{99}G^{00}B^{99}: \quad \text{Magenta (M),}$$
$$5 \ (\text{or} \ 300°) \quad R^{99}G^{00}B^{00}: \quad \text{Red (R),}$$
$$6 \ (\text{or} \ 360°) \quad R^{99}G^{99}B^{00}: \quad \text{Yellow again,}$$

thus arc length $6 = 0$ or hue angle $360° = 0°$. From here on I will use the arc length, though hue angle works just as well in practice. The arc length is much more natural though, the hue angle really being "forced" on the RGB structure. It is also simpler to convert between arc length and full color then between hue angle and full color since there is no simple way to *normalize* the color obtained by hue angle, whereas no normalization is needed when the full color is obtained from arc length. Notice that the hue angle runs over an interval $2\pi = 6.28\ldots$ whereas the total arc length is 6. Thus you need to scale if you mix the two.

The "cardinal colors" are illustrated in figure 14.10. They are indeed special, because they occupy the colored (as opposed to achromatic) vertices of the color cube. They can be

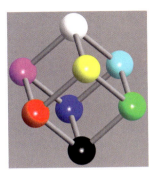

Figure 14.10 The cardinal colors. These are the primary colors R, G, and B and the secondary colors C, M, and Y. They occur in the sequence YGCBMRY.... All most colorful RGB colors are obtained through interpolation of two adjacent cardinal colors.

divided into the "primary" cardinal colors R, G, and B, which are adjacent to K, and the "secondary" cardinal colors C, M, and Y which are adjacent to W. The secondary colors are the sums of two primary colors, whereas the primary colors are indivisible. W may be considered the "tertiary" cardinal color as it is made up of all three primary colors.

14.1.3   The lattice of domination

If you add K ("nothing") to the primary, secondary, and tertiary colors, you obtain a *lattice* structure[8] with W as the lowest upper bound and K as the greatest lower bound. The Hasse diagram is shown in figure 14.11. Notice that the diagram looks like a projection of a cube. With some imagination, you can view the RGB cube itself as a continuous Hasse diagram of the lattice defined by the *domination* relation (see figure 14.12).

> *If color A (say $\{r_A, g_A, b_A\}$) and color B (say $\{r_B, g_B, b_B\}$) are such that $r_A \geq r_B$, $g_A \geq g_B$ and $b_A \geq b_B$, then color A is said to "dominate" color B.*

This is very similar to the notion of domination in color space, or the color solid, except from the fact that the distinction between *spectral* domination and *color* domination does not apply to the RGB colors, since there is no notion of "spectrum" at all.

Notice that color A necessarily *equals* color B in the case that both A dominates B and B dominates A. Notice also that the relation of domination is not defined for all color pairs. For instance, R does not dominate B, nor does B dominate R. The colors thus form a "poset" (partially ordered set) under the relation of domination. If A dominates B, then there exists an RGB color C (say) such that

$$A = B + C.$$

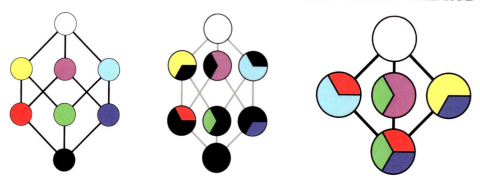

Figure 14.11 Left and center: The Hasse diagram of the RGB colors regarded as a lattice; Right: Another way to show the hierarchical structure is to display the basic partitions of white. You either split white into a secondary and a primary RGB color, or into the three primary RGB colors. Thus you see graphically that cyan is "minus red," magenta is "minus green," and yellow is "minus blue".

Thus A is obtained by *adding* something to B. If you believe (as I do) that a color cannot be made to look *darker* by *adding* to it, then you must also accept that when A dominates B it cannot be that B appears *brighter* than A. Thus the relation of domination goes some way towards the construction of a brightness function.

Because it is so important in applications, there is a lot of literature on how to "translate" RGB colors into monochrome (black and white) (figure 14.13). The key application occurred in the time many people looked at (color) TV transmissions using monochrome receivers. The "standard" rule of thumb is that "luminance" (the RGB way) is 60 percent green, 30 percent red and a meagre 10 percent blue (see appendix N.2). In practice

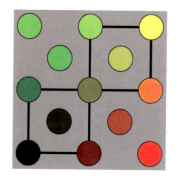
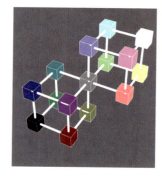

Figure 14.12 The partial order of domination.

Figure 14.13 The "standard" translation of RGB color into monochrome.

you easily accept different recipes as reasonable though, although a change of luminance formula certainly implies a change of key.

14.2 RGB Colors in the Ostwald Description

Suppose you have any RGB color, say, for concreteness, $R^{80}G^{60}B^{30}$. What does this color look like? It isn't so easy, since the color contains red, green, and blue components simultaneously. You might guess that it would be somewhat reddish, perhaps, because the first coordinate (0.8) is largest. In the first panel of figure 14.14, I illustrate what this color looks like and also its RGB components. The first bar simply shows the color. In the second bar, the color has been broken up into R, G, and B components. A better way to do that is illustrated in the next panel of figure 14.14. Here the three bars form a cluster showing the R, G, and B components separately. This is useful because the total bar height now denotes the maximum possible amount. Thus, only 80 percent of the red bar is occupied by red, while the remainder is black (nothing). For the green bar only 60 percent is occupied by green, and for the blue bar only 30 percent.

So what does this color look like? Here the Ostwald method works wonders. Notice that I can add at most 0.2 times W (that is $R^{99}G^{99}B^{99}$), for

$$\{0.8, 0.6, 0.3\} + 0.2W = \{1, 0.8, 0.5\},$$

and I can't add any more white because the first coordinate "has reached the ceiling." Since the color apparently "*lacks* 0.2 W", I will say that the color "*contains* 0.2 K", black

Figure 14.14 The analysis of an RGB color in terms of bar diagrams. From left to right: 1. A color and its RGB-components; 2. Left, the color shown as RGB parts; at right I have identified the common white part. In a similar manner you may identify the common black parts; 3. The color and its decomposition in white, black and "full color" parts. The full color is simply the color minus its white and black parts. 4. The full color and its composition in terms two analog parts, here a red "primary" and a fraction of a yellow "secondary color."

being the opposite of white, that is to say $R^{00}G^{00}B^{00}$. Similarly, I cannot *subtract* more than $0.3\,W$ from the color, for this would yield

$$\{0.8, 0.6, 0.3\} - 0.3\,W = \{0.5, 0.3, 0\},$$

and I can't subtract more white because the last coordinate "has reached the floor". Thus I will say that the color "*contains* $0.3\,W$". This is graphically illustrated in the right side of the second panel of figure 14.14.

Thus the color $R^{80}G^{60}B^{30}$ has 20 percent black content, 30 percent white content, leaving $100 - 30 - 20 = 50$ percent "color content", the "*color*" being $C = R^{99}G^{60}B^{00}$, because

$$0.5\,C + 0.3\,W + 0.2\,K = \{0.8, 0.6, 0.3\},$$

that is indeed the given color. Notice that the "color" you end up with is necessarily a *full color*, thus is located on the color circle. This is graphically illustrated in third panel of figure 14.14. Everything below it is occupied by white in all three (RGB) bars, and everything above it is occupied by black in all three (RGB) bars.

It is now immediately evident that the 30 percent "blue" present in the RGB color $R^{80}G^{60}B^{30}$ will be "neutralized," because part of the white content of the color. The hue is just that of its full color, that is to say $R^{99}G^{60}B^{00}$. This full color is located in between yellow $R^{99}G^{99}B^{00}$ and red $R^{99}G^{00}B^{00}$, indeed

$$(1 - 0.6)\,R + 0.6\,Y = 0.4\,R + 0.6\,Y.$$

Since the color is more yellow than red, it may be denoted "reddish-yellow". In this case the equal mixture between yellow and red actually has a name, "orange"; thus an alternative

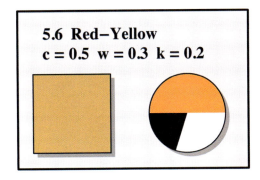

Figure 14.15 The Ostwald denotation of the RGB color analyzed in the previous figure.

name for the hue would be "orange with a yellowish tinge". The full color is illustrated in the last panel of figure 14.14. The full color is mainly yellow, with some red mixed in. It is evidently located at 40 percent on the way from the yellow to the red cardinal color on the color circle.

Since Y occurs at arc-length $0 = 6$ and red at arc-length 5, the hue is located at arc-length $6 - 0.4 = 5.6$. By now I have fully analyzed the color "Ostwald style": It is 50 percent full color 5.6 (orange with a yellowish tinge), with 30 percent white and 20 percent black content. In figure 14.15 I illustrate the resulting "Ostwald chip".

Notice that there exist no "non-Ostwaldian" colors in RGB space.

It is instructive to consider the decomposition of a RGB image in terms of its Ostwald components (figure 14.16). Notice that the white content looks roughly like a monochrome rendering. Notice the difference between the white and black content images, for example, the way shadows come out. The image in which all full colors have been replaced by red is especially striking. Although by no means a "colored image" (its colors are contained in a plane in color space that contains the achromatic axis), it looks "colored" in a weird way.

Notice that

> *every full color is a unique binary mixture of a certain primary with a certain secondary color.*

Not all such combinations occur, for instance, combinations of green with yellow and cyan, occur whereas combination of green with magenta do not. The latter combinations do not occur because these pairs of colors "destroy" each other. This tallies very well with your perceptual experiences, you can easily imagine a "YELLOWISH GREEN" or a "CYANISH GREEN" (common speach is such that you would probably say "bluish green" in the latter case, but the intended "blue" is actually cyan, it is the "blue" from the kindergarten recipe "blue

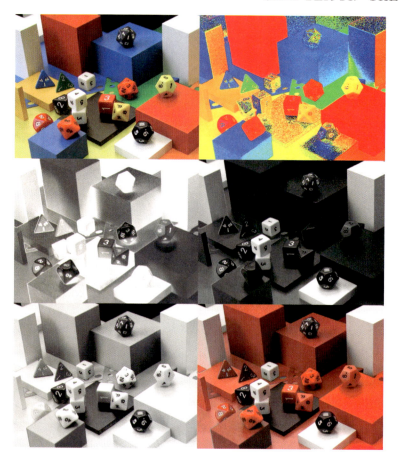

Figure 14.16 The Ostwald decomposition of a RGB image. From left to right, top to bottom you have: RGB image, its full color components, the color content (black is more), the white content (white is more), the black content (black is more) and finally the picture in which all full colors have been replaced by red. This yields a more vivid impression of the color content of the image.

and yellow make green"), but not a "PURPLISH GREEN" (I use "purple" as synonymous with "magenta"). Colors that can join forces to yield a full color together are frequently called "analogous colors" by artists.

All this can be nicely illustrated with the diagrams shown in figures 14.17–14.19. This diagram of figure 14.17 should be cyclically read, that is, its right boundary should be thought as glued to its left boundary. This representation thus graphically proves that the full color gamut is a topological circle. The "color circle" is not an ad hoc phenomenological

Figure 14.17 Diagram showing the full colors as unique binary mixtures of a primary/secondary pair of colors. The diagram should be read as periodically repeated or bent into a cylindrical strip and glued together at the red. The diagram thus also graphically illustrates that the full colors form a topological circle.

Figure 14.18 Diagram showing the full colors as unique binary mixtures of a primary/secondary pair of colors. The diagram thus also graphically illustrates that the full colors form a topological circle.

construct; it is forced upon you by the very structure of the RGB cube. This diagram is rather similar to one proposed by Hering (figure 14.20) in order to illustrate his four color theory (in response to Helmholtz's three color theory). The difference is that Hering leaves out the cyan and magenta. Thus the diagram 14.17 can be thought of as illustrating a six color theory in Hering's sense. Hering argues (convincingly in my opinion) for the primacy of yellow as an elementary, unitary experience. In my *experience* a PURE YELLOW contains neither RED nor GREEN, even though you may *know* that yellow can be obtained by additive mixture of red and green. In the mixture the components are absent; it is only "YELLOW" that you "see".

Hering did not extend the argument to cyan and magenta (purple). However, I fail to see RED and BLUE in a PURE PURPLE, just as I fail to see RED and GREEN in a PURE YELLOW.

Figure 14.19 "Red" content is not limited to the red cardinal color, it is found in yellow and magenta too. It stops at the green and blue and is totally absent from cyan (left). Similar reasonings apply to the "green content" and "blue content". On the right, a figure after the painter Paul Klee of Bauhaus fame. Notice that Klee (as many painters do) thinks in terms of red, yellow, and blue, though.

Of course, this is all introspection rather than science and thus highly speculative. But for reasons of a consistent formalism I think the unitary nature of purple and cyan is almost forced on you. The way colorists use colors and talk about them also suggests that.

When you peruse the full (periodic) set of full colors it is striking (at least to me) that you see an alternation of "brightness". Thus yellow is (very) bright, green is less so (as compared to yellow), cyan is clearly brighter than green, blue is (very) dark, magenta is much brighter than blue, and red dark (although perhaps less so than blue), certainly darker than either magenta or yellow. Thus the sequence YGCBMR is (in terms of brightness) something like ↑↓↑↓↑↓. This is clearly reflected in the six color theory for in figure 14.17 the vertical may double as brightness, but it cannot very easily (at least I see not how) be incorporated in the Hering four color theory. It is of course very difficult to decide on the issue of whether "there is no such a color as CYAN" (CYAN being a mixture of GREEN and BLUE in which the components are easily identified), or whether "there is no such a color as PURPLE" (PURPLE being a mixture of RED and BLUE in which the components are easily identified). But it is—in principle—just as difficult to decide on the issue of whether "there is no such a color as YELLOW" (YELLOW being a mixture of RED and GREEN in which the components are easily identified). What if you happen to meet someone who tells you that he sees YELLOW as distinctly composed of RED and GREEN? If you don't agree, does this show a lack of judgment on your part? Could you conceivably convince the person that he or she is "wrong" about the matter? What rational arguments could you use? You are evidently on slippery ground here. Could the problem be solved by convention? Then what would be the scientific content of such a convention?

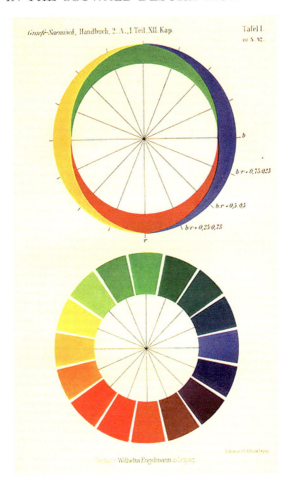

Figure 14.20 Hering's color circle illustrating his four color theory. In this theory both cyan and magenta are not "primary" colors, but are binary mixtures. It is of course a matter of opinion whether you can "see blue and red in magenta." I confess I cannot, but probably Hering could convince himself of that. It may be that Hering's scheme fits very well in our common language use, but I doubt whether it accounts all that well for our experiences. After all, there are many facts of my experience that I feel hard put to formulate in words. Experience has greater scope than language.

The difference also implies different mensuration of the color circle (figure 14.21). In figure 14.22 I have drawn color circles according to Hering and according to the scheme I have advocated here. In the latter (six color) scheme the scale is "well tempered" according

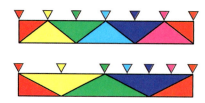

Figure 14.21 A comparison of the four color and the six color theory. If you believe that the primary colors should be equally spaced in the color circle, then the cyan and purple (which are not primary in the four color theory) have to be interpolated. Thus—apart from the issue of being "visually binary"—the mensuration of the color circle will be different. You might expect this would decide the issue. But believe it or not, it is hard to decide by eye measure.

Figure 14.22 Color circles according to the Hering's four color theory and according to the six color theory, which yields equal spacing according to Ostwald's principle of internal symmetry. Which one looks more "natural" or "well tempered"? Take your pick.

to Ostwald's principle of internal symmetry, whereas in the former it is "out of tune." Decide for yourself. The fact that I prefer the Ostwald tuning may have to do with my experiences of course, you should consider a majority vote from the population at large here.

The standard RGB color circle, with the colors YGCBMRY equally distant, automatically satisfies Ostwald's principle of internal symmetry (figure 14.23); thus—according to Ostwald—these cardinal colors should appear about equally spaced along the color circle. To my eye they do, but you should judge for yourself.

With some practice you can easily "eyeball" the Ostwald coordinates, and it is certainly easy to obtain them from the RGB representation, using the method illustrated in figure 14.14. Since the procedure is easily inverted, it is also a simple matter to arrive at the RGB coordinates from the Ostwald representation. The Ostwald representation[9] is the intuitive one, whereas your computer typically enjoys the RGB coordinates. As long as the human interface remains as primitive as it is, you have to put up with that and do frequent conversions in your head. It pays to become adept at that.

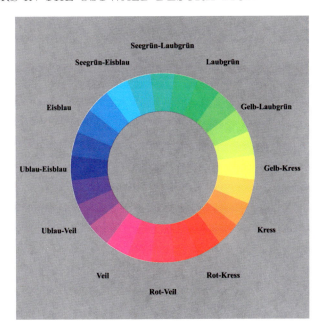

Figure 14.23 The Ostwald principle of internal symmetry fits the YGCBMRY color circle.

The notions of "black content" and "white content" are very intuitive and can can easily be eyeballed after a little practice. It helps to study some examples, they are easily generated on a computer screen (see figures 14.24 and 14.25).

The idea was indeed "natural" enough that it already appeared to Fechner. It may well be that Ostwald has the idea from Fechner, as he was aware of Fechner's work and even visited him once. I cannot find references in Ostwald's work though.

Fechner writes

> Such a veiling triangle as I call it, would represent comprehensively the whole multiplicity of veiled colors corresponding to a specific hue.

and he continues

> Since a chromatic color can be veiled with any color in the black-white series ...each hue can occur in a variety of veiled colors which is much greater ...than the variety of hues themselves.

Again, Fechner writes

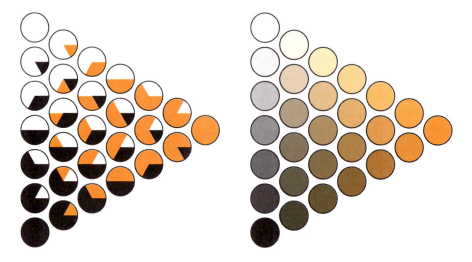

Figure 14.24 An Ostwald page: All combinations of K, W, and some fiducial color occur.

The measure of strength of a complex color, as an opaque paint, for fixed illumination can be changed by mixing black or white to it. In the former case the result is a partial removal of all simple rays equally, in the latter case an addition of a fraction of all simple rays equally. In both cases the colorfulness diminishes. Mixture with black makes the color darker, finally black and less colorful; admixture of white makes the color brighter, washed out, finally white and gains in terms of brightness but loses in terms of colorfulness, for dark, desaturated colors with an increase, for colorful and bright colors with a loss of forcefulness.

Notice how Fechner has already most of Ostwald's major insights and how he thinks of (de)composition in exactly the terms I have proposed in this chapter.

That the Ostwald-specified colors are easily generated on a computer screen is to be considered a major advantage of the Ostwald color atlas. It is a *conceptual* atlas, so you can easily program one on your computer; you don't have to lay out a few hundred dollars to buy an impressive and no doubt beautiful but inconvenient physical specimen.[10]

The various linear color series contained in the Ostwald atlas are often used by designers and architects. The most obvious ones are the "tints" (mixtures of full colors with white) and the "shades" (mixtures of full colors with black). The "mixture with black" appears very "natural" in this context, despite the fact that several well known "systems" for the characterization of RGB colors do not even explicitly recognize the existence of black. Both

Figure 14.25 An Ostwald double page. Here an Ostwald page is combined with the page of its complementary color.

intuitively and formally, black and white play similar (though of different polarity) roles. For this reason it is surprising that the "color pickers" that are provided in about any computer application that deals with graphics in some or other are still based on various—typically not very intuitive—systems of color denotation (see appendix N.4). Apart from the linear series, designers often extract "harmonic chords" from the atlas, typically triples of colors at mutually symmetrical positions.

There appears to be no end to the systems that have been proposed to let the user specify or understand the RGB colors easily. The consensus is apparently that the RGB coordinates (or CMYK coordinates) as such are not very user-friendly. I would agree with that, but then, neither are most of the "systems." In terms of both ease of use (algebraically and algorithmically, I mean) and intuitiveness, the Ostwald system is far ahead of the systems in common use today. I will not enter into a discussion of the various systems and the conversions between them. It is easy enough to find such facts in the countless books on computer graphics.[11] A minimum background is offered in appendix N.3.

Notice that the full colors of the RGB cube are indeed the analogues of the full colors of colorimetry. But then, they are simultaneously the analogues of the ultimate colors, and thus of the monochromatic colors. In the RGB system you cannot distinguish between these. Whereas in colorimetry the spectrum double cone contains the Schrödinger color solid (section 7.3), which again contains the Schopenhauer RGB crate (chapter 9), in the RGB system these have all collapsed into the RGB cube. As you have seen, the Schopenhauer RGB crate exhausts not *all*, but *most* of the object colors. In many respects the Schopenhauer RGB crate can be used much as you would the RGB cube. Indeed, much of what I say here concerning the RGB colors can easily be applied to the Schopenhauer RGB crate of colorimetry.

14.2.1   The "spectrum" and the RGB colors

One important difference between the colorimetric structures and the RGB space is that there exists no natural notion of "spectrum" and thus "spectrum locus" (section 4.5) in the RGB case. Indeed, the three coordinates of an RGB color play equivalent roles, it is not that "B comes before G comes before R." The color circle is indeed naturally *closed*, there is no notion of "spectrum limits." This implies also that the Goethe edge colors (chapter 6) play no particular role in RGB space. In figure 14.26 I show continuous sequences between K and W via the cardinal colors such that you first reach a "primary" cardinal color (R, G, or B) when departing from K and then a "secondary" cardinal color (C, M, or Y) before arriving at W. The sequences travel via the edges of the RGB cube. Notice that the

Figure 14.26 The black-to-white scales via the cardinal colors.

leftmost scale happens to be the "temperature scale," that is, the warm Goethe edge color sequence, whereas the rightmost is the cool Goethe edge color sequence. The other scales have no obvious relevance in colorimetry. They occur here because of the lack of a notion of spectrum limits.

The analogue of the chromaticity diagram (section 4.7) is perhaps the "RGB color triangle" (see figure 14.27). This is indeed a "chromaticity diagram" in the sense of a projection on a plane orthogonal to the gray axis, from the black point. You can also (the more usual interpretation) regard it as a triangular diagram that shows all mixtures of R, G and B. In the figure I have plotted the brightest mixtures available, thus the center of gravity is tinted white, but there does not exist an established convention here. The location in the triangle simply specifies the ratios r:g:b, but not the absolute values. The triangle is useful as a human interface object, for example, in "color pickers".[12] From a conceptual point of view, it is of much less importance than the RGB cube, though, just as the chromaticity diagram is much less useful than color space.

14.2.2 "Complementary" RGB colors, RGB and CMYK

In RGB space, the notion of complementarity is much simpler (and useful) than in colorimetry proper (section 5.2.1). I define the complementary RGB color to a given RGB color as the color that, when added to the given color, yields white (figure 14.28). This particular notion of complementarity is perhaps better called "supplementarity." It is necessary here because the notion of wavelength does play a role in the world of RGB colors. However,

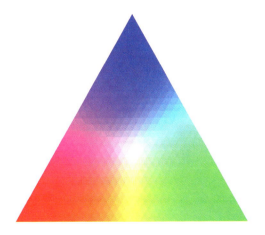

Figure 14.27 The RGB color triangle.

$\mathbf{R^{80}G^{60}B^{30}}$ $\mathbf{C^{20}M^{40}Y^{70}}$

Figure 14.28 The most intuitive way to remember the RGB to CMY conversion is to remember that "cyan is minus red" and so forth. Thus cyan makes up for the lack of red as compared to the maximum amount of red that is present in white.

the notion of supplementarity is also very useful in more general colorimetric settings. It has the advantage that it lends itself to generalization (tetrachromacy and so forth) where the specific notion of complementarity does not.

Thus the complementary of

$$P = \{r, g, b\},$$

is just

$$W - P = \{1 - r, 1 - g, 1 - b\},$$

It is as simple as that.

The complementaries of the primary colors R, G, B are just the secundary colors C, M, Y (in that sequence). Thus

$$R + C = W$$
$$G + M = W$$
$$B + Y = W$$

That is why *cyan* is often denoted "*minus red*," *magenta* as "*minus green*," and *yellow* as "*minus blue*" by practitioners in the printing or photography business. When you print transparent dyes on white paper, you selectively absorb radiation. Thus you "take away from the white." A cyan dye takes away (absorbs) the red; hence it is "minus red." If you overprint the cyan with yellow, the cyan takes away the red, whereas the yellow takes away the blue, leaving you with green. Thus a yellow overprinted over a cyan will appear green. This is what you were taught at kindergarten:

YELLOW *and* BLUE *make (as paints are stirred together)* GREEN.

Of course, the kindergarten teacher should have said "cyan" instead of "blue," but there you are. This introduces the notion of "additive mixture" versus "subtractive mixture". Figures 14.29 and 14.30 show you the basics.

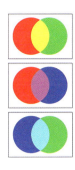
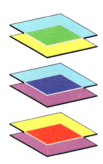

Figure 14.29 "Additive mixture" of beams is done by shining them together on a white screen for instance (left), whereas "subtractive mixture" occurs when you superimpose colored, transparent filters, for instance (right).

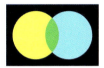

Figure 14.30 "Yellow and blue make green." Left: subtractive mixture of cyan and yellow; center: subtractive mixture of blue and yellow; right: additive mixture of blue and yellow.

If you are mostly occupied with things on your computer screen, you will be interested only in *additive* mixtures (figure 14.31). On a CRT screen you get black for free. It even appears if you switch off your machine.[13] You have to work hard for white, for all three phosphors (red, green, and blue) have to be on in order to produce white. If you do an Ostwald decomposition you describe a color in terms of full color, white, and black. You don't need to care about the black. The white is usually made by having the three phosphors on by the same amount, but this could just as well (perhaps even better) be done by adding a fourth *white* phosphor. The full color is made by increasing the intensities of two of the three colored phosphors. Thus you can do all with just RGB (the usual case), but you might as well use a RGBW system (adding a white phosphor) and use the W to take care of the white content, using (two of the) RGB to produce the (full) color content. The RGBW system actually has some technical advantages, and now and then you hear of patents being filed to cover this innovative discovery.

Notice that "additive mixture" is used in (at least) two different senses. One is straight superposition "$a + b = c$"; the other is an "equal mixture" of the type "$(a + b)/2 = c$." The former sense seems obvious but has the drawback that a color mixed with itself doubles

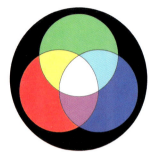
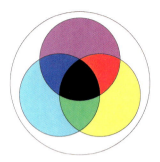

Figure 14.31 Additive and subtractive mixtures.

in intensity. The latter sense takes care of that; a color mixed with itself doesn't change. Notice that this may lead to certain problems, for instance, "YELLOW" is a mixture of RED and GREEN in which "REDDISHNESS" and "GREENISHNESS" mutually cancel out, but an *even* mixture does not at all look YELLOW. It looks BROWN instead (figure 14.32).

If you are mostly occupied with printing transparent dyes on a white paper base, you will be interested in *subtractive* mixtures. On the paper you get white for free; it even appears when you print *nothing* at all. You have to work hard for black, for all three inks (cyan, magenta and yellow, figure 14.33) have to be on the paper in order to produce black. If you do an Ostwald decomposition, you describe a color in terms of full color, white, and black. You don't need to care about the white. The black can be made by having the three inks on the paper by the same amount, but this could just as well (perhaps even better) be done by adding a fourth *black* ink. The full color is made by increasing the intensities of two of the three colored inks. Thus you can do all with just CMY (the usual case in photography), but you might as well use a CMYK system (adding a black ink) and use the K to take care of the black content, using (two of the) CMY to produce the (full) color content. The CMYK system actually has many technical advantages, due to the fact that the inks are not perfect, making it hard to produce a solid black. Thus adding a truly

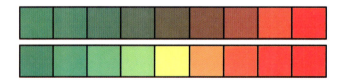

Figure 14.32 Two types of additive mixture of red and green.

Figure 14.33 The secondary colors "cyan," "magenta," and "yellow." You know what I mean if you're trichromatic, as I am.

black ink makes good sense in color printing and this is almost invariably done. The CMY and CMYK systems are usually combined, some of the black content being left to a CMY combination, some of it being applied via the black ink. Printing is something of a black art at the best of times, and a discussion of the various factors that will affect the final result is out of the question here. At least you understand the basics by now. The CMYK and RGBW systems are in many respects very similar of course. In figure 14.34 I show a "subtractive CMY color triangle." You may want to compare this to the additive RGB color triangle shown in figure 14.27.

In figure 14.35 I show the example of a RGB image decomposed into RGB, CMY, and CMYK components.

14.3 Symmetries of the RGB Cube

A cube is a very *symmetrical* object. Many special rotations, reflections and inversions move the cube so as to coincide with itself. The RGB cube has fewer symmetries, though,

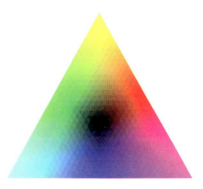

Figure 14.34 A subtractive CMY-color triangle. If you're a printer, this is the way you conceive of colors.

Figure 14.35 An image decomposed into RGB (top), CMY (center), and CMYK (bottom). In the CMYK case the heaviest possible black plate has been extracted. Notice how "thin" the color plates then really are. Most of the printed density comes from the black plate (hence "K" for "black", the "KEY" plate).

because one body diagonal is special. Thus no symmetry of the RGB cube may map the gray axis on another body diagonal, nor is it admissible to interchange the white and the black point. There remain symmetries that interchange the full colors among each other. Such symmetries are interesting because they act like different keys in music. If you apply such a symmetry to a colored image you obtain an image that looks perfectly "natural" (as the negative of an image does not, you may be familiar with this if you have some photographic experience) but appears in a different *key*. It is easy enough to find the symmetries of the RGB cube by exhaustive search among the well-known symmetries of the generic cube. There are two rotations and three reflections (see figures 14.36 and 14.37).

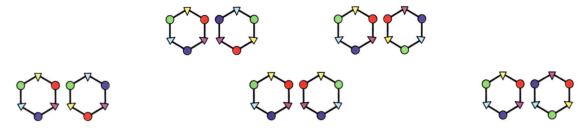

Figure 14.36 Symmetries of the RGB cube.

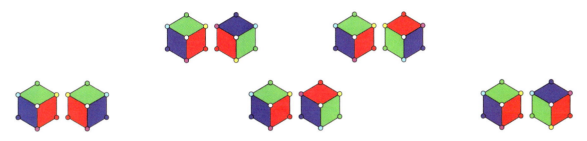

Figure 14.37 Another way to illustrate the symmetries of the RGB cube. This figure should be studied together with the previous figure, both display exactly the same. You probably have a preference, take your pick.

Together with the identity (a trivial symmetry) these form a finite group of "color key transpositions" with multiplication table

	R_1	R_2	M_1	M_2	M_3
R_1	R_2	I	M_2	M_3	M_1
R_2	I	R_1	M_3	M_1	M_2
M_1	M_3	M_2	I	R_2	R_1
M_2	M_1	M_3	R_1	I	R_2
M_3	M_2	M_1	R_2	R_1	I

where I denotes the identity, R_i the rotations, and M_j the mirror reflections. The action of the symmetries on the color circle are illustrated in figure 14.38.

Notice that, since the various full colors have different brightnesses, the symmetries do change the brightness distribution over an image. Although this is indeed fairly obvious on closer comparison, the transposed images tend to look perfectly acceptable as natural images.[14] In figure 14.39 I show this for the case of the two reflections.

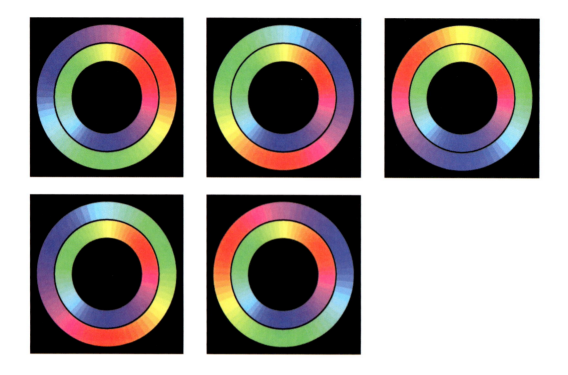

Figure 14.38 Mirror reflection and rotation symmetries of the RGB cube.

Figure 14.39 A natural image and the mirror reflection transpositions of that image.

14.1 Virtual RGB colors

Although most RGB colors you typically encounter (for example, as RGB pixel values in images) are known for sure to be real, "virtual" RGB colors are commonly encountered in image-processing settings. For instance, if you compute the eigenvector of the covariance matrix of the RGB values of all pixels of an image (very informative data, there are often good reasons to do so) you encounter virtual RGB colors naturally.

Here I need to distinguish between "out of gamut" colors and RGB colors proper. Out of gamut depends on the application, but for this discussion I will consider *real* RGB colors to be out of gamut when at least one of the coordinates is outside the $[0, 1]$ range. I will consider *virtual* RGB colors to be out of gamut if at least one of the coordinates has an absolute value exceeding one.

In most cases, out-of-gamut RGB colors are "errors" of some algorithm and should be given special treatment to get in step, for instance (a really Procrustean measure) through clipping.

In this definition there exist three types of virtual RGB colors:

— those with three positive coordinates coincide with the real RGB colors and require no further discussion;

— those with three negative coordinates can be treated as "negative real RGB colors," that is to say, you pull out the sign and make it a special flag. The remaining color is just a normal, real RGB color. Such cases hardly ever appear in any real problem;

— those where the coordinate are of mixed signs are interesting and require further discussion. This type occurs frequently in (especially statistical) applications.

In this appendix I will treat only the (interesting) mixed sign case.

Suppose you have a mixed sign RGB color, say $\{0.8, -0.5, 0.4\}$. The first thing to do is to split it into two real components, one of them flagged negative:

$$\{0.8, -0.5, 0.4\} = \{0.8, 0, 0.3\} - \{0, 0.5, 0\}.$$

Then you scale the colors such that one coordinate is maximal (that is one), and you get

$$\begin{aligned} \{0.8, -0.5, 0.4\} &= 0.8\{1, 0, 0.5\} - 0.5\{0, 1, 0\} \\ &= 0.8(R + 0.5B) - 0.5G, \end{aligned}$$

that is a linear combination of two full colors, in this case a reddish-purple $(R + 0.5G)$ and green (G).

Although this is only a specific example, you easily draw some general conclusions from it:

— You always get two full colors, one of them a primary color.

— These colors are opposite in the color circle, that is to say, typically not complementary, but certainly not analog colors. The full color will be a mixture of the complementary of the primary color and one of the analogous cardinal colors of these.

Thus the virtual RGB colors of this type are all much like opponent colors, and indeed most conveniently treated as such (see figure 14.40).

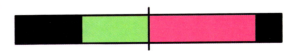

Figure 14.40 Graphical representation of the virtual RGB color $\{0.8, -0.5, 0.4\}$.

This makes it a simple matter to indicate such colors graphically, something that is often desirable if you want to visualize statistical outcomes.

An opponent color is conveniently illustrated as a "split patch" (see figure 14.40). Notice that I have completed the bars with black in order to be able to show the absolute amounts.

14.2 "Luminance" for RGB colors

The standard V_λ curve (section 5.7) doesn't apply to the RGB colors. An alternative that is easily memorized and commonly used (in figure 14.41, I show the alychne) is

$$L(R, G, B) = \frac{1}{10}\left(3R + 6G + B\right).$$

This expression puts very little weight on the blue channel and weights the green channel as double the red channel.

Where do these numbers come from? One way to arrive at such an expression is to take the daylight spectrum, divide it into its Schopenhauer parts, and then to find the luminances of these parts. If I do the calculation, using the second color matching function from the 10° CIE tables, I find

$$L(r, g, b) = \frac{1}{10}(2.1r + 7.3g + 0.6b),$$

which is fairly close to the generally accepted values.

Notice that this puts a rather lopsided structure on the RGB cube, destroying many of its symmetries. The luminances of the primary and secondary colors are (setting white to 10):

Hue	Luminance
Yellow	9
Green	6
Cyan	7
Blue	1
Magenta	4
Red	3

(See figure 14.42.) The sequence is not regular along the color circle; it alternates, following the full color locus. Thus there exist three maxima (the secondary colors) and three minima (the primary colors).

If you use colors in images, their "values" are most important. A "value" is the position of a color on a scale from light to dark, or its "equivalent gray tone". Since each color has a certain value between black and white, you can make a full scale by lightening (mixing with white) and toning down (mixing with black). This results in the scales shown in figure 14.43. Notice that yellow is so bright that almost the only thing you

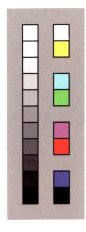

Figure 14.42 The "equivalent gray tones" of the primary and secondary colors.

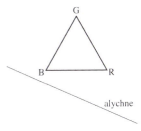

Figure 14.41 The alychne in the GB triangle.

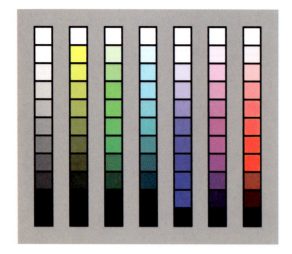

Figure 14.43 Value scales for the primary and secundary colors compared with a regular gray scale.

might do to it is tone it down, whereas blue is so dark that almost the only thing you might do to it is lighten it. Notice that the scales look slightly weird, though. A color is simply never going to look "equivalent" to a gray tone. In practice, photographers using Photoshop use various methods to convert to monochrome. Only the photographer notices, the audience never (figure 14.44).

In practice there is a lot of leeway in using colors to represent values in pictorial rendering, though the rough sequence of values should be respected. In tonal painting (think of Rembrandt), the artist used desaturated colors so the tones are almost from a gray scale, saturated colors being used as typically small accents. In chromatic painting (think of Monet) the tonality is limited to light values; colors are used in highly saturated form. In a monochrome reproduction of a Rembrandt painting most of the structure is retained, though you miss the color accents.

Figure 14.44 Top: An RGB image; Center: The RGB image converted to monochrome via the Photoshop standard method which uses the official RGB luminance functional; Bottom: The RGB image converted to monochrome via the Photoshop "desaturate" method, which does not use the official RGB luminance functional, but gives the three channels equal weights.

In a monochrome reproduction of a Monet, you lose most of what the painting is about and are left with a dull image because there simply is no value structure in place. In order to "translate" the Monet to monochrome, you need to translate chromatic contrast to tonal contrast, that is to say, you have to deviate from the default rendering. This again illustrates the limited value of luminance.

Again, if you do simple technical work, like converting a color photograph to a monochrome one, the "official" luminance functional is not necessarily what you want. Most photo editing applications offer you a variety of choices, and more often than not the "best" rendering (by eye measure of course; the one you *want*) is quite different from the default. The RGB luminance functional is of very limited value, and I find that I rarely use it at all.

In figure 14.44, I show typical examples. Notice that both monochrome renderings are acceptable, though obviously different. Especially notice the relative gray tones of the reds and blues as compared with the greens.

In real applications you would take a good look at the individual color channels and then decide on the "mix" that brings out the best in the monochrome image.

14.3 RGB color denotation "systems"

The "systems" mentioned in this appendix are in current use. There exist several others,[15] but these are the main contenders to fame. They all start with the initial "H," which stands for "Hue," and "S," which stands for "Saturation".

The hue is indicated through the *hue angle* which is periodic with period 2π. The value is most often given in degrees.

The saturation (sometimes described as "vibrancy") ranges from zero to one, or (more conventionally) from 0 percent to 100 percent.

The magnitude of the RGB color, denoted "luminance", "intensity," or "brightness" is in the range zero to one, or (rather common) 0 percent to 100 percent.

All these systems are ad hoc hacks that reinvent and combine classical concepts in a rather hodgepodge manner. It is indeed hard to see any need for these modern inventions. For most technical applications, one will use either RGB or CMYK anyway, whereas for more intuitive specifications the Ostwald system is almost impossible to beat and in addition is simpler than any of the current conventions.

Probably for the very reason that they exist, professional graphics programs usually implement the full sweep of exotic color (really: RGB) denotation systems, one reason why you may want to know about them.

The HSV (or HSB) system

HSL stands for "Hue, Saturation and Luminance." HSV stands for "Hue, Saturation and Value", whereas the "B" in "HSB" stands for "Brightness." The HSV system was proposed in 1978 by Alvy Ray Smith, of computer graphics fame.

HSV color space is often depicted as a cone or a cylinder. The system is supposed to be "intuitive" and is frequently used in the computer graphics community.

The algorithm used to convert RGB to HSV again finds the arc length along the full color locus and scales it to fit the 0–360° range. The "angle" is actually a scaled polygonal arc length, take care.

The "value" is defined as the maximum of the RGB values, thus as one minus the black content. The "saturation" is defined as the maximum minus the minimum, which is simply the Ostwald color content, at least in the "cylindrical coordinate system." In the "conical coordinate system," the saturation is defined as the color content over the sum of the color and the white content.

Notice that the HSV system is similar to the Helmholtz decomposition familiar from classical colorimetry. The system does not explicitly recognize the upper limit for the achromatic color, that is to say, it doesn't recognize the finite range of the black to white grayscale.

The HSL (or HSI) system

HSL stands for "Hue, Saturation and Lightness." Sometimes "Luminance" is used instead of "Lightness."

The I in HSI stands for "Intensity." In this type of models luminance, brightness and intensity tend to be used more or less indiscrimately.

The HSL system is considered to have the geometry of a double cone, thus expressing the fact that both white and black lack true chromaticity.

The algorithm used to convert RGB to HSV actually finds the arc length along the full color locus and scales it to fit the 0–360° range: Thus the "hue angle" of the HSV system should not be confused with the "hue angle" as obtained from a naive interpretation of the cylindrical or conical coordinate systems.

The lightness is defined as the mean of the minimum and the maximum of the RGB values. This can be written as the Ostwald white content plus one-half of the color content.

The saturation is defined as the ratio of the maximum minus the minimum to the maximum plus the minimum of the RGB values in case the lightness is less than one half, and as the ratio of the maximum minus the minimum to two minus the maximum plus the minimum of the RGB values in case the lightness is greater than one half. This equals the color content divided by twice the lightness in the former and the color content divided by twice one minus the lightness in the latter case. It is this part of the definition that makes for the double conical structure.

The HSL system does recognize the finite range of the black to white gray scale and thus fits the structure of the gamut of RGB colors much better than the HSV system does.

14.3.1 The YUV system

The YUV system was defined for PAL television. The Y stands for "luminance," and the UV for "chrominance" components.

The luminance is defined as

$$Y = 0.299R + 0.587G + 0.114B,$$

a discrete approximation to the luminance functional. TV signals invariably have a luminance component, because this is needed for old-fashioned monochrome receivers (the chrominance signals being simply discarded in that case).

The chrominance channel U is the blue-yellow opponent signal

$$U = 0.492(B - Y),$$

and the chrominance channel V is the red-yellow opponent channel

$$V = 0.877(R - Y).$$

14.3.2 The YIQ system

The YUV system was defined for NTSC television. Again, the L stands for "luminance." Of the IQ, the I stands for "in phase," whereas the Q stands for "quadrature." The IQ chrominance signals are merely a "rotated" version of the UV signals discussed earlier. The choice has been made such that the human observer is primarily sensitive to I, thus the Q signal can be transmitted with reduced bandwidth. The definitions used are

$$I = 0.735514(R - Y) - 0.267962(B - Y),$$

and

$$Q = 0.477648(R - Y) + 0.412626(B - Y).$$

14.4 "Color pickers"

Color pickers are little programs that implement user interfaces that let the user select one or more colors by interacting graphically (using a mouse or other picking device) with the computer desktop. They exist in great variety, which makes sense because they are used for a large variety of sometimes very different tasks and subject to often very different constraints.

The simplest color pickers are like the swatch books you see in paint shops. There are a number of colored blotches on the screen and the user clicks on the desired one. Although this simple interface has much to go for it, it has its limitations, especially when the number of available samples is large. It may not be possible to present all options simultaneously, or it may be necessary to show a continuous gradient of colors instead of a discrete set. In the former case you need to let the user page though the various subsets, in the latter case the user cannot indicate a color with absolute precision. In either case you feel a need to order the available gamut in some "natural" way, since this will allow the user to arrive at the destination with much greater ease and thus (in all likelihood) much faster.

The layout of the available gamut on the screen is often based on the coordinate system that the designer of the color picker just happens to use, or just happens to like, or seems indicated by the intended use. For instance, a printer will prefer CMYK over RGB, whereas a computer graphics person will prefer exactly the opposite. But even when the use dictates RGB, it may not be the case that the user finds it convenient to have to indicate a color by its RGB coordinates. Asked for the RGB values of your favorite wine (glass held against normal illuminant) you probably have to think hard and being told that Alice's new blouse has such and

so RGB coordinates probably doesn't help you much either.

Given that RGB space is three-dimensional, it is clear that you will need three independent parameters to characterize a color (figures 14.45–14.51). One of these doubtless has to be hue.

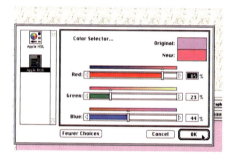

Figure 14.45 A color picker from the Macintosh desktop. This is about the simplest possible; you have sliders for red, green and blue. You can also set the values numerically, and you can monitor the result as a colored patch.

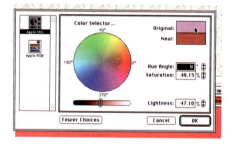

Figure 14.46 Another color picker from the Macintosh desktop. You pick a hue from a Newton-style color disk, then use a slider to change lightness. Lightness, saturation, and hue angle can also be entered numerically.

Figure 14.47 A bizarre color picker from the Macintosh. This one is lets you pick from a (very) finite number of "crayons."

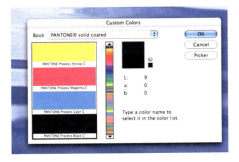

Figure 14.48 A color picker from Adobe. This one lets you pick from a (rather large) but still finite number of Pantone colors. Very primitive, but useful in certain contexts.

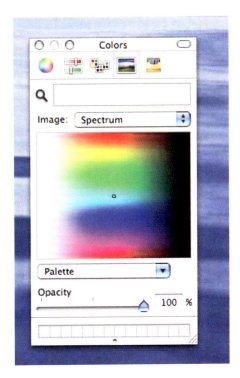

Figure 14.49 A color picker from the Macintosh with a really appalling interface. This example happens to be in particularly bad taste.

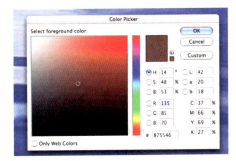

Figure 14.50 A color picker from Adobe with an interface that is easily as appalling as the previous one from Apple (figure 14.49). Bad!

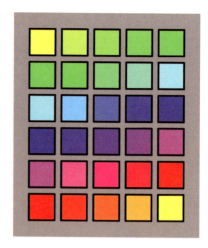

Figure 14.52 It is easy to (visually) interpolate between "analogous colors" (the rows in this array). Using a linear slider to pick a hue is best done on an analog color range. The switch between the six different ranges of analog colors is best done through clicking in the color circle.

Figure 14.51 A color picker from the Painter drawing program. This one is rather sophisticated and intuitive; you have essentially an Ostwald color atlas. What a pity the designers judged it necessary to add HSV-sliders instead of the (much more intuitive) Ostwald coordinates. It reveals a certain lack of either good taste or sound judgment.

The hue is largely independent of the properties a color has. If you have the layout of the color circle in your head then hue will also mean location to you. Thus, you would expect the large majority of color pickers to have an interface for hue based upon the color circle. Re-

markably, that is not the case. Only a few color pickers explicitly use the color circle. One reason might well be that computer interface elements are usually preprogrammed in the window system and are based on horizontal and/or vertical linear scales (think of the familiar sliders, for instance). Thus, circular scales are a programmer's headache. Linear scales are fine if you confine them to subranges of hue that are "naturally" linear, the ranges of "analogous colors" (figure 14.52). I have never seen such an obvious idea implemented though.

The remaining parameters are usually saturation and intensity or entities derived thereof. These are geometrically represented as points on linear segments of finite length; thus the conventional sliders make great interface elements. A problem is that this assumes that human observer find it natural to "see" the (even rough) value of the saturation, and so forth, which is not at all the case. When using such parameters it is always necessary to show the current setting as a colored patch. Then the user fiddles the sliders until that patch looks good. Although this technique works, it is evidently not optimal because badly tuned to the natural workings of the user's mind.

What is needed is to have the user set parameters that make visual sense. It should be obvious to the user which slider to use and which way to shift it, rather than to twiddle things more or less at random till the final patch looks good. In cases where the user becomes very experienced, this is no problem. Some people can "think" RGB or CMYK or hue, saturation, and intensity. They are convinced that setting the sliders is the fastest way.[16] For less experienced people, things are not so obvious. Probably the best interface uses the Ostwald color, white, and black content. In some color pickers this is implemented by letting the user click in the color, white, and black content triangle, a method that works very well. If you look at any number of programs with color pickers you will soon discover that most color picker designs are of a rather different nature. Most are tuned to the convenience of the programmer, rather than the user. Some are so hard to use that I often prefer to just type in RGB values and look at the end result patch to check whether I did the right thing. Pathetic.

Exercises

1. Render the RGB cube   [COMPUTER GRAPHICS] Write code to render the RGB cube from various views. Experiment with skeleton representations. Try to render the interior in various ways, for instance, by using planar cuts or sparsely filled volumes.

If you are a graphics person you will probably program a fly through. It is easy enough to think of all kinds of cool embellishments.

2. The lattice of domination   [NUMERICAL] Write an algorithm that decides for an ordered pair of colors whether the first dominates the second, whether the second dominates the first, or whether the two colors are unrelated. Generate a random gamut and find the partial order for it. This takes some understanding of algorithmics, though you don't need to be a professor.

3. Ostwald color picker   [PROGRAMMING EXERCISE] Design (and implement) a color picker based on the Ostwald representation. Also do the reverse, when you indicate a color in an image the Ostwald representation should be indicated in some intuitive manner. Color pickers being what they are, you might be able to sell your implementations!

4. The CMYK representation   [CONCEPTUAL] Write an RGB to CMYK conversion algorithm where you can indicate how much weight should be placed on the black plate generation.

5. Symmetries   [PROGRAMMING EXERCISE] Write an algorithm that applies the various symmetries of the RGB cube to images. Discuss the results on a set of fiducial images, for instance, tonal (Rembrandt-like) or chromatic (Monet-like) paintings, or low-key and high-key color photographs.

6. Structure of RGB space   [FOR PURISTS] Ideally, "RGB space" should be discussed without any reference to either regular colorimetry or optics. "Spectrum" and "wavelength" do not belong. Cleanse this chapter from such alien terms and come up with a purified theory of RGB space. (I did a few concessions at various places for the sake of didactic exposition, sorry to let you down!)

7. Engineering problems [FOR THE APPLICATIONS ORIENTED] I have concentrated on conceptual matters and ignored numerous topics of engineering interest. For a start you may want to look for the definitions of the "sRGB-standard," or the techniques of "gamut mapping." (The Internet is a great source of material, simply use Google on "sRGB" and "gamut mapping".) Try to relate these engineering topics to this chapter. To analyze the structure and purported advantages of the sRGB-standard into some depth is very instructive, even apart from the fact that it will be of considerable use to you.

Chapter Notes

1. Cathode Ray Tube is generally abbreviated to CRT.
2. Digital-to-analog converter is generally abbreviated to DAC.
3. Liquid crystal display is usually abbreviated LCD.
4. See, for instance, R. S. Berns, *Billmeyer and Saltzman's Principles of Color Technology*, (New York: Wiley-Interscience, 2000).
5. An attractive book that is completely dedicated to the RGB cubical system is H. Küppers, *Das Grundgesetz der Farbenlehre*, (Cologne: Dumont, 1978).
6. Ewald Hering thought in terms of *Gegenfarben*, opponent processes of a "push-pull" type. This is much in the tradition of Goethe, although Hering added many speculative features of a physiological nature.
7. You have the constraints

$$(\alpha \geq 0) \wedge (\beta \geq 0)$$

$$0 \leq \frac{\beta}{\sqrt{3}} + \alpha(\frac{\cos\varphi}{\sqrt{6}} - \frac{\sin\varphi}{\sqrt{2}}) \leq 1$$

$$0 \leq \frac{\beta}{\sqrt{3}} + \alpha(\frac{\cos\varphi}{\sqrt{6}} + \frac{\sin\varphi}{\sqrt{2}}) \leq 1$$

$$0 \leq \frac{\beta}{\sqrt{3}} - \sqrt{\frac{2}{3}}\alpha\cos\varphi \leq 1.$$

For any value of φ there exist unique values for α and β such that the value of α is a maximum under the given constraints.
8. A "lattice" is a partially ordered set with a least upper bound and a greatest lower bound for each pair of its points.
9. Basic references by Ostwald himself are: W. Ostwald, *Die Farbenfibel*, (Leipzig: Unesma, 1916); —, *Der Farbatlas*, (Leipzig: Unesma, 1917).
10. You may want to look at the web-page `http://usa.gretagmacbethstore.com/`. The *Munsell Book of Color*, *Glossy Edition* was priced at $567.50 when I looked in March 2006. I bought one for my laboratory, but considered it too pricey to put on my bookshelves at home. The blurb says "*Munsell Book of Color* provides the ultimate flexibility in designing bright vibrant color palettes. It s based on A. H. Munsell's original master atlas of color and includes two volumes with over 1600 removable glossy Munsell color chips". So you really get something for your money. These atlases simply look gorgeous!
11. For instance, look at the standard text by J. D. Foley, A. van Dam, S. K. Feiner, and J. F. Hughes, *Computer Graphics: Principles and Practice*, (Reading, MA: Addison-Wesley, 1990).
12. It is often convenient to use "triangular coordinates" or "barycentric coordinates." If you have a triangle defined by the vertices $\mathbf{v_{1,2,3}}$, you can indicate any point $\mathbf{p}$ as

$$\mathbf{p} = \mu_1\mathbf{v_1} + \mu_2\mathbf{v_2} + \mu_3\mathbf{v_3},$$

where

$$\mu_1 + \mu_2 + \mu_3 = 1.$$

The coordinates run from zero to one for the interior of the triangle. Such a representation has many uses in colorimetric settings; it is also very convenient for purposes of computer graphics. The barycentric coordinates were introduced by Möbius in 1827. See H. S. M. Coxeter, "Barycentric Coordinates", §13.7 (pp. 216–21) in H. S. M. Coxeter, *Introduction to Geometry*, (New York: Wiley, 1969). Since the barycentric coordinates are a form of homogeneous coordinates, they are extremely useful in many settings.
13. This is actually a hairy topic. When you watch TV, you will notice that (if the TV set is in tune) you get very good whites and blacks, even (as is usual) with your room lights on. Now switch off your TV set. You will notice that the screen is not at all black, but a light grayish tone. With the TV switched off, the screen is only illuminated by the room lights. When the TV set is on, any point of the screen must send more radiation to your eye than when the set is switched off. So where do the "rich blacks" come from? Notice that this is a "problem" only if you fall for the "photometer fallacy."
14. This implies, of course, that some symmetries are more likely to be interesting than others. The major factor is that blue is a relatively very dark, yellow a very light color. Taking this seriously kills off all symmetries except for the reflection in the yellow-blue plane. An interesting empirical study is L. D. Griffin, "Similarity of Psychological and Physical Colour Space Shown by Symmetry Analysis," *Color Res.Appl.* 26, pp. 151–157 (2001).
15. If you need to know, a good source is available on the Internet (use Google to find it): it is Danny Pascale's document, "A Review of RGB Color Spaces ... from xyY to RGB."
16. Some, of course, will hold that using sliders is for sissies and will insist on typing in numerical values from the keyboard. These are the real men. Cojones!

Chapter 15

Extended Color Space

When you say "foliage is green" what could you possibly mean? An analysis of photographs of foliage shows that it is not a single color green that is applied as in children's coloring books (figure 15.1). Most "colors" you see are actually more or less extended *regions* of color space rather than mere points[1]. "Foliage is green" is a symbolic, logical expression of language that has no parallel in perceptual experience at all.

A "color gamut" is simply a collection of colors. In this chapter I will consider RGB colors, but exactly the same discussion applies to the more general colorimetric colors. Gamuts of immediate interest can be very large, for instance, it is now (as of 2005) quite standard to tote a "ten megapixel" electronic camera by way of a cheap sketchbook. An image from such a camera can be considered a gamut if you disregard the particular spatial distribution of the pixels. Why would you want to do that, since the *image* critically

Figure 15.1 Foliage (left). The same image "posterized" (center) shows that the scene is not simply "green" but varicolored. The average color of the scene (right) is indeed a certain shade of green. This is the meaning of "foliage is green."

depends on that spatial distribution? After all, a scrambled jigsaw puzzle is a different object from a "done" jigsaw puzzle. There are plenty of reasons why you might want to regard the pixels as "scrambled" (without determinate spatial locations) though. One example is the *histogram*. Many cameras now let you view histograms as an aid to exposure determination. A histogram is really nothing but a color gamut, ordered by R, G, and B value. It is *very* useful to get the exposure right, much more so than the old-fashioned exposure meters. Thus gamuts are interesting and may well contain many millions of colors. Small gamuts are often of interest too though, like when you use the "eye dropper" sample tool in a program like Photoshop.

If gamuts are very large you will often want to cut them down to size. One method is to describe the gamut statistically. I'll deal with such statistical methods later. Here I will discuss only deterministic methods. One way to cut a gamut down to size is to remove all colors that are somehow "implied" by other colors already in the gamut (figure 15.2). For instance, given a number of colors, I can regard all colors located inside their convex hull as "implied," because they can be obtained by mixture, using only interpolation, no extrapolation. (If extrapolation were allowed any triple of independent colors would "imply" all others. Thus extrapolation is not of any interest.) Thus the subset of colors from the gamut that generate its convex hull imply all colors in the gamut. This might still be a formidable set, but it will almost certainly be much smaller than the "raw gamut."

In figure 15.3 I show an example of a color gamut consisting of fifty colors. For later reference call it "gamut $\mathcal{A}$". In the figure at left I show a square filled with the average color of the gamut. Superimposed on this square I have plotted the various colors of the gamut as little squares at random positions. An image like this yields a good view of what the gamut is like. In the figure at right I have plotted the same gamut, but this time I have

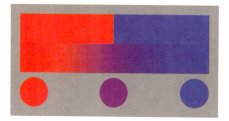

Figure 15.2 On top an image (an edge between two uniform patches) that subtends only a two-color gamut. In the blurred image you have a continuous set of colors including (bottom) magenta, the even mixture of the original red and blue colors. The gamut subtended by the blurred picture is the same as the gamut of the sharp picture, it is spanned by the original red and blue. This illustrates the "intuitiveness" of the notion of "gamut".

Figure 15.3 A color gamut. At left, a square filled with the average color of the gamut on which all points of the gamut are superimposed as little colored squares at random positions. At right, all "implied" colors are omitted. Only the colors that "span" the gamut are displayed.

excluded all "implied colors." There are only twelve colors in the convex hull. For this example I used a two-dimensional subspace, a plane through K in the RGB cube. Thus "color space" is a plane, and I can show simple planar plots of the gamut and its convex hull. You can easily generalize this to the three-dimensional case of course (figure 15.4).

In figure 15.5 I show the gamut with the convex hull at left, and the same gamut "cleaned up" by removal of implied colors at right. The gamut is now "generated" by its "exposed points," the vertices of its convex hull.

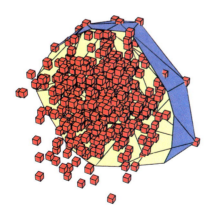
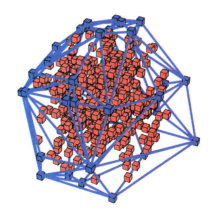

Figure 15.4 The convex hull of a three-dimensional color gamut. Consider an arbitrary linear functional. Generically there will be just one point of the convex hull at which the functional assumes an extremum. This is one reason to call such a point "exposed point". The vertices of these polyhedral convex hulls are such "exposed points".

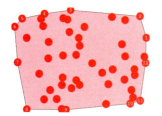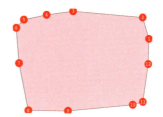

Figure 15.5 The convex hull of the color gamut.

In figure 15.6 I show two other gamuts, $\mathcal{B}$ and $\mathcal{C}$. I can combine these with gamut $\mathcal{A}$ and with each other in various ways. For instance, given two gamuts, I can *join* them. The "join" operation merely amounts to throwing both gamuts into the same bag. After doing this I must find the convex hull for the set union. This will typically contain fewer colors than there are in the union of the generators of each. Thus union is an interesting operation. In figures 15.7 and 15.8 I show results of the join operation. Notice that the join may claim a large volume of "empty space" if the gamuts are remote from each other. On the other hand, if one gamut is contained within the convex hull of another on, it will be annihilated by their join.

The *meet* is an operation on pairs of gamuts that is not unlike a set intersection. The analog of the set intersection is the "meet" of two gamuts. I define the meet as the intersection of the convex hull of the two gamuts, that is either empty or yet another convex hull. That is the best you can do because the *set intersection* of the two gamuts is generically empty, and thus the notion of set intersection is virtually useless in the case of color gamuts.

It is very unlikely that any two gamuts will share the same color, especially in the ideal case where the R,G,B components are specified as real numbers. On the other hand,

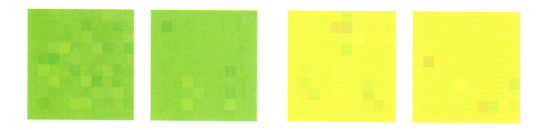

Figure 15.6 The gamuts $\mathcal{B}$ and $\mathcal{C}$.

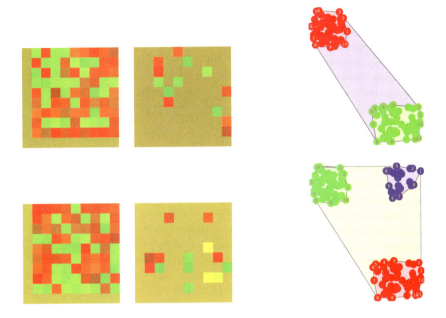

Figure 15.7 The joins of gamuts $\mathcal{A}$ and $\mathcal{B}$ and of $\mathcal{A}$, $\mathcal{B}$ and $\mathcal{C}$.

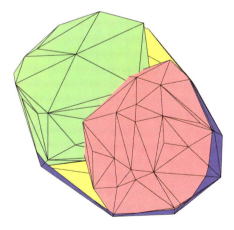

Figure 15.8 The join of gamuts in three dimensions.

the convex hulls are finite volumes, thus there exists a reasonable chance that these might overlap. If not, then the meet is empty. If they do, the meet is a convex region.

In figures 15.9 and 15.10 I illustrate the meet for two gamuts $\mathcal{P}$ and $\mathcal{Q}$. Special cases are the nonoverlap of the convex hulls in case the meet is empty, and the case of one convex hull being included in the other, in which case the meet will annihilate the larger one.

The figures illustrate the more interesting case in which the convex hulls overlap in a nontrivial manner. The generating set of the intersection (which is again a convex hull) contains vertices that are not in the generating sets of the two component convex hulls.

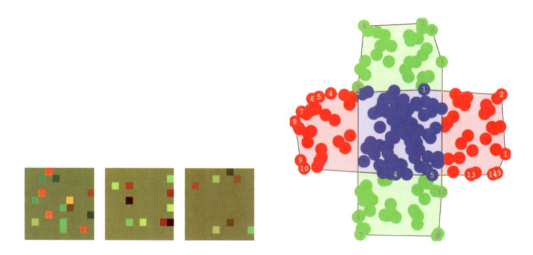

Figure 15.9 The meet of gamuts $\mathcal{P}$ and $\mathcal{Q}$.

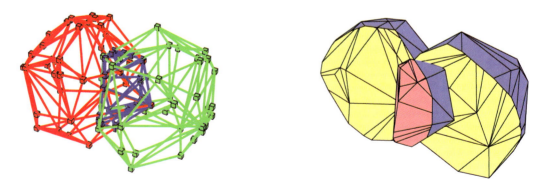

Figure 15.10 The meet of gamuts in three dimensions.

This is perhaps slightly worrisome, for now you represent the meet through colors that are actually in neither gamut. This is a small matter, for when you think about it these definitions obviously make good sense.

What the example of the meet does show, is that the exact generating set that defines a gamut is perhaps "overkill" in the sense that a small perturbation would change that set greatly, whereas the gamut as such would hardly suffer. This is perhaps an indication that a statistical description would often make more sense.

As I will show later, both the convex hull notion of "gamut" and a statistical definition have their uses. It is important to understand when either type makes the best sense of course. I will offer examples later.

15.1   The Notion of Extended RGB Space

In most real life applications a "color" is actually what I have called a "gamut" here. If you analyze a painting you will probably notice that the painter rarely covers large areas with a uniform color. More typically, a colored patch will be made up of a fairly localized gamut. *Fairly localized* it is true, but nevertheless, a *gamut*, not a *color*. The same fact will strike you when you analyze natural scenes. A treetop may be called "green" by the casual observer, but when you segment out the treetop (simply use the lasso selection tool in Photoshop) you will find that the resulting set of pixels have a histogram of more than minimal width. Thus

> "All real life colors" are actually gamuts.

This is a very important observation.

Notice that fairly localized gamuts behave in many respect much like colors. For instance, the join operation is much like the mixing of colors. Such join operations often occur naturally as you change the resolution, by taking distance from an object or squinting. Artists often squint or screw up their eyes in order to see the scene in broader terms. One thing they actually do is to perform various join operations.

I intend to take this seriously, and I will thus refer to fairly localized gamuts as "colors" in some extended sense of what a "color" is. This gains you a lot of extras. The "extended colors" are much richer than the colors themselves. Whereas it is impossible to find a "color" to represent "gold," for instance, it is definitely possible to find an extended color (a gamut) that looks "golden." The colors only live in a three-dimensional space, but the dimensionality of the space of *extended colors* is much higher than that.

In the simplest description you may parameterize "fairly localized" gamuts by their second order statistics, that is the covariance. Then an (extended) color not only has a location in color space (three dimensions), but there is also a volumetric extension (another degree of freedom), a planar attitude (another two degrees of freedom), a direction of elongation (another degree of freedom), and two aspect ratios, a total of three (the location) plus six (the other degrees of freedom) dimensions. The covariance is indeed a three times three symmetric matrix, with

$$\frac{1}{2}3(3+1) = 6,$$

independent components. Let the eigenvalues be σ_1^2, σ_2^2 and σ_3^2, the corresponding eigenvectors $\mathbf{f}_{1,2,3}$, such that $\sigma_1^2 \leq \sigma_2^2 \leq \sigma_3^2$. Then there is a volumetric component of RMS size σ_1, a planary component of RMS size $\sqrt{\sigma_2^2 - \sigma_1^2}$ in the $\mathbf{f_1} \wedge \mathbf{f_2}$-plane, and a linear component of RMS size $\sqrt{\sigma_3^2 - \sigma_2^2}$ in the $\mathbf{f_3}$-direction.

Here are a few examples (figure 15.11). Notice how much more "lively" the extended colors appear as compared to the uniform patches (of the same average color). This is exactly why painters avoid uniform patches and "divide" their touches. Notice that you may

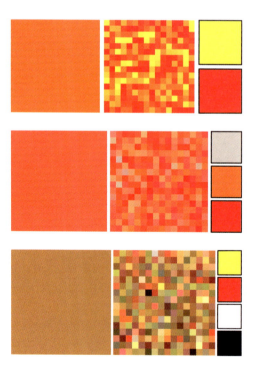

Figure 15.11 Linear, planar, and volumometric extended colors.

need to practice a bit in order to be able to spot "linearly extended," "planary extended," and "volumetrically extended" colors at first blush.

Analysis of generic images (figures 15.12 and 15.13) reveals typical and very interesting structures. You have to define a region of interest (ROI) size that is of reasonable size in order to obtain results that make sense. The ROI should contain a reasonably sized sample in order for the statistics to make sense. In the example I used 8×8 pixel ROIs, and thus the samples contain 64 RGB values. There apparently are regions that are predominantly linear, planar, or volumetric. Notice that the regions where the ROI happens to straddle a contour stand out in the analysis. The reason is obvious: These regions are not homogeneous, thus their statistics are not particularly revealing. There are also homogeneous regions of linear, planar, and volumetric types, though. Linear regions are often caused by shading of curved surfaces; thus you expect their direction to be toward the origin, an expectation that is largely fulfilled in practice, though different directions are clearly present. Planar regions

Figure 15.12 At left, an example image (a photograph taken in Scottsdale, Arizona). At right the type of extended color for 8×8 neighborhoods. The hue has been encoded as R: linear contribution, G: planar contribution, B: volumetric contribution; whereas the maximum of (R, G, B) has been normalized to one.

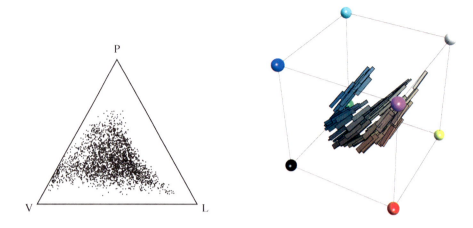

Figure 15.13 At left, the distribution of linear (L), planar (P), and volumometric (V) extended colors in the image of figure 15.12 left. At right the most articulated extended colors (95 percent quantile and over). Notice that they are mainly linear and extended in the achromatic direction.

may be due to a mixture of shading and gloss (diffuse reflection), in which case the planes should be coplanar with the origin. Again, many examples are found, but there are also exceptions. The different spectral content of direct sunlight and diffuse skylight should also produce planar extended colors on curved surfaces, and so forth. There are indeed numerous physical reasons for the generation of extended colors. This is an interesting and potentially important topic that (thus far) has unfortunately drawn hardly any interest from the computer vision community.

Predominantly linear, planar, or volumometric extended colors all occur in a generic image of a natural scene, though combinations appear to be most common. The distribution appears to be not at all uniform, but at present no one has any idea as to the nature of the statistics. The type of scene is no doubt important, but you would expect generic optical processes to imprint their signatures upon the distributions. For the example shown here, the most obviously "extended" colors are mainly of the linear type, and their direction of elongation tends to be the direction toward the white point.

By simply changing the direction of the linear or planar extension, you can change the appearance of a color decisively, *even without changing the average value at all.* I show an example in figure 15.14. All these patches have the same average, a neutral gray. They have been modulated linearly in different directions though. Notice how different they look. Yet when you screw up your eyes or squint, you will notice that these patches are all the same dull gray tone. In the planary extended color patches (figure 15.15), the modulation is in differently oriented planes. All patches have the same average color. In this case the variety looks even richer. In figure 15.16 I show a volumetric, planar, and linear median gray. From a distance the grays look similar, while from nearby they are very different.

The extended color are not just points; they have other spatial properties. Thus the extended colors can stand in relations to each other that simply do not occur for the case of uniform colors. For instance, two linearly extended colors can be colinear, meet at an angle, or cross without meeting; a linear extended color can lie in the plane of a planary extended color, be perpendicular to it, and so forth. The perceptual significance of such relations still has to be investigated.

Simply loosening up a color is something that is widely practiced in the visual arts (figure 15.17). This avoids a "licked appearance" and generally looks more interesting

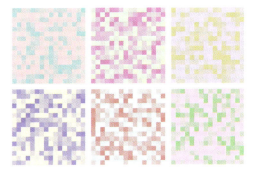

Figure 15.14 A number of "linear grays." All these patches have the same average, a neutral gray. They have been modulated linearly in different directions though.

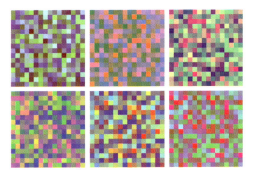

Figure 15.15 A number of "planar colors." All these patches have the same average, a neutral gray. They have been modulated in differently oriented planes though.

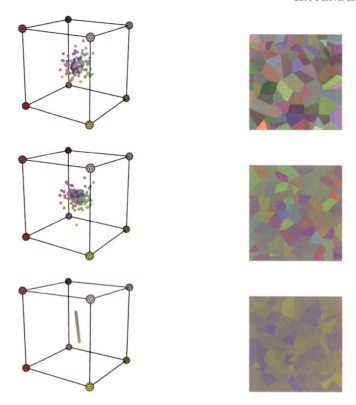

Figure 15.16 From top to bottom, a volumetric extended gray, a planary extended gray and a linearly extended gray.

than a painting in uniform patches. In more sophisticated schemes you split a color into components, thus achieving something that is much like a "chord" in music. This was done in pointillism and divisionism, though by no means in a unitary manner (figure 15.18). A painter like Georges Seurat assigns meanings to the various components of the chords,[2] the meaning deriving from observation of nature (sunlight orange, skylight blue, vegetation proper color green, and so forth) as well as an observation of subjective visual effects (the blue "calls forth" a yellow color by contrast and so forth). Such chords can be painted only by giving up spatial resolution, for the colors have to be applied next to each other. It is understood that the spatial resolution of the depiction is less than that of the brush strokes. Yet the shape of the brush strokes is often used to suggest textural properties, for instance the general direction of leaves of grass. This is a highly interesting topic that still awaits much research.

Figure 15.17 Colors that have been "loosened up" to various degrees.

Figure 15.18 Detail from Seurat's "La Parade".

Splitting a uniform color into a "chord" could be done in many ways. All that is constrained is the average, for the color should look the same as seen through your eyelashes. The splitting could be used to introduce volumometric, planar, or linear extended colors; in the latter two cases you have the orientation as an additional tuning tool. The chord can be constrained in many other ways; for instance, you might decide to use only colors from a limited palette and so forth. This is a huge area open to experimentation in rendering.

The spatial distribution of color is very important. When the microstructures can be parsed in terms of surface properties you see much more than just a dithered average hue (see figure 15.19). In the example, the WHITE blotches are read as specularities and make the whole surface look "GLOSSY RED," whereas the DARKISH blobs are read as "seeds" and are

Figure 15.19 Three red patches, one articulated (top), one locally scrambled (bottom left), one uniform (bottom right), that have precisely the same spatial average. The original looks solid, is glossy and has non–trivial texture. The scrambled one looks solid and shiny, but has lost most of the typical texture. The flat one looks indeed "flat", not solid, is textureless and matte instead of glossy.

discounted from the overall RED. The average looks DULL (not GLOSSY) and of a decidedly less vivid RED. Dithering or breaking up the uniform color would no doubt improve the rendering. However, the fine detail structure of high lights and dark seeds introduces *meaningful* structure, different from mere random dither. Whether such structure is used or not depends upon the observer and the degree of attention devoted to this red blob. In many cases mere dither would perhaps be just as realistic as renderings go.

Exercises

1. Representing gamuts   [NUMERICAL EXPERIMENTS] Study representations of color gamuts that you generate with some random generator. Try statistical descriptions via the covariance ellipsoids, representation via convex hulls, and RGB bounding boxes. Compare volumes, fraction of the gamut that is represented, and complexity of the representation. Use the representations to generate new instances of the gamut and compare.

2. Actual gamuts   [EXPERIMENTAL INVESTIGATION] Select "uniform" patches of (digital) photographs, for example, "sand", "lawn", "foliage","water", and "sky", and construct the gamuts. How large are these gamuts? Are they obviously structured or oriented? Do they overlap? Try to relate these structures to their physical causes.

3. Extended colors   [EXPERIMENTAL INVESTIGATION] Generate instances of "extended colors". (An example might be "GOLD" color.) Study these in the various representations. What happens when you "mix" such extended colors in various ways?

4. Blurring versus scrambling   [EXPERIMENTAL INVESTIGATION] Study the difference between "blurring" and "local scrambling". Vary the amount of blurring, the range of local scrambling, and the extent (pixels, for instance) of the areas that are moved in the scrambling. Compare cases where blurring and local scrambling degrade spatial resolution to the same degree.

5. Local resampling   [EXPERIMENTAL INVESTIGATION] Consider "local resampling": you simply replace pixel values by samples from distributions obtained by sampling the immediate neighborhood of the pixel. Vary the diameter of the sampling neighborhood. Either pick a random instance from the shuffled bag of local samples, or fit some standard probability density to the samples. How does this compare with local scrambling or a median filter?

6. Emulate Seurat   [TAKES ARTISTIC HORSE SENSE] Given a straight photograph, produce a rendering in Seurat's style. To make this a more appealing exercise you may start with a photograph that emulates the scene seen in a well known Seurat painting. Is it possible to arrive at reasonable results at all? Try whether you can fool friends with the "discovery of a lost Seurat". Is this possible at all with people you hold in some esteem?

Chapter Notes

1. Notice the two scales involved here. The *inner scale* is the resolution, that is the scale of the pixels in an image, or the scale of the cones in your retina. The *outer scale* is the scale at which you see the colors. Typically the outer scale *much* exceeds the inner scale. The outer scale is like a "region of interest" (ROI) that may be picked at will. In painting the inner scale is set by the *touches* that are the individual brush strokes, whereas the outer scale is the scale of the smallest "glyphs" (iconically meaningful elements). In a Seurat painting the outer scale *far* exceeds the inner scale. Yet you are not supposed to stand back so much as to equalize inner and outer scales (that is to say, "optically mix" the touches). You are supposed to enjoy the disparity between the inner and outer scales. Seurat's "colors" are "extended colors" in the sense of this chapter.

2. W. I. Homer, *Seurat and the Science of Painting*, (Cambridge, MA: MIT Press, 1964).

Chapter 16

The RGB Structure of Images

The word "image" is used for multifarious ontologically distinct entities and thus is bound to lead to endless confusion. This is indeed what happens, as you will soon discover when you peruse the literature. An eye-opener might be to type "image" in Google and to classify the various meanings implied in the responses. You are going to prove your true grit in the course of this investigation.

That "color" adds unique information to images is obvious when you compare (colored) abstract figures with their monochrome renderings, using the standard "translation" (figure 16.1). Even in real scenes color adds information (figure 16.2), although this goes often unrecognized in say, landscapes since you know that the grass is unlikely to be BLUE and the sky GREEN. This is also the reason that the "standard translation", which is (roughly) $I = 0.3R + 0.6G + 0.1B$ is rarely used by visual artists. The photographer will use the Photoshop "channel mixer" in order to tweak the grays of the monochrome to his or her liking. The results can be strikingly different according to whim (figure 16.3).

Colors are optical properties of visual space (here I mean the two dimensional manifold often called "visual field"). Even with identical geometrical structure, images can be strikingly different if the colors change (figure 16.4). It is not that there are no interactions

Figure 16.1 Colors are the optical qualities of visual space that cannot obtained from a monochrome rendering.

Figure 16.2 You can easily parse the monochrome rendering, though it is hard to guess at the colors of the various objects.

between the geometry and the colors in your experience though, there are and they can be very striking. In this chapter I will largely ignore the geometrical structure, and I will also ignore the interactions, since these properly belong to psychology rather than colorimetry.

In this part I will use "image" mainly for the type of object often referred to as "straight photograph of some scene" and presented to you in numerical form, say an "uncompressed" or (semi)"raw" image file such as you might obtain from a generic digital camera or a scanned piece of photographic film exposed in what is now generally known as "an analog camera."[1] For our purposes, this may be formalized as a rectangular array (or "matrix") of "pixel values," where each pixel value is a small, ordered set of numbers[2]. In practice, the array will be of the order of a thousand by a thousand and a pixel will be just one ("monochrome") or three ("color") numbers. The numbers will be in the range of 0 to 2^n, where n is the number of (significant) "bits," typically at least 8 ($256 = 2^8$) and at most 16 ($65,536 = 2^{16}$). Such images can be regarded as the result of many simultaneous measurements of the irradiance at the focal plane of a lens. These measurements are the recorded responses of photosensitive elements, whose sensitive areas are distributed over the focal plane in some kind of pattern. The photosensitive elements are characterized by a number of crucial parameters, such as their efficiency in yielding electrons for photon absorptions, the probability of photons of given energy and direction to be absorbed, the activity in the absence of radiation, the capacity to deal with a high photon flux, and so forth. The various properties of the camera (its mechanics, its optics, its electronics, and so on) and the photodetector are important and the influence on the eventual image complicated enough that one might write a book on it.

Figure 16.3 Three different monochrome renderings of color pictures. Which one is "right"? How much of the color can you guess from a single monochrome image? Context helps. What color do you think the face of my daughter is, GREEN? I don't think so.

Figure 16.4 Same geometry, different colors. I took these photographs from the same location (my hotel room window that day) at different times.

It is not my intention to deal with engineering details here. I will simply assume an "ideal" camera, that is, one that

- has perfect optics (no aberrations, flare, . . .);

- samples all photons arriving at the focal plane and has photocells that report pure photon flux;

- packages the raw photon flux data in the image file without any fancy algorithms interfering with that.

Such cameras are nonexistent, of course, though equipment improves every year.[3] I assume that the camera has so many pixels that I can assume the resolution to be effectively infinite, that is to say, I get all the resolution that I ask for, and in practice I will always look at the image at a resolution much lower than what the image can provide. Expensive modern cameras indeed yield higher resolution than the eye can use for reasonably sized presentations. It is not unrealistic to expect that in the near future you will indeed have this overdose of resolution.[4] That is nice, because then the discrete pixelation becomes irrelevant and for all practical purposes the image becomes a *continuous* object. Something similar goes for the number of bits, the "dynamic range." I will consider the dynamic range to be essentially *infinite*. Modern cameras are still far from that ideal, although "raw" files typically have a bit depth of 12–16, from which the user picks an 8-bit subrange. Thus the idealization is not totally unrealistic, perhaps.[5]

These idealizations are often useful because they allow me to completely abstract from the hardware. This is often the natural setting for what may be called "image science." Of course you often deal with *actual* images, in which case you are in the *engineering* department of image science. In that case images become almost arbitrarily complicated, it depends on how far you intend to pursue the various calibration and sampling issues.

16.1 The Histogram

The simplest investigation of the distribution of colors in a image would involve the overall statistics of RGB values. This is an extremely crude measure, but very popular nevertheless, first of all because it is simple (users at least vaguely understand it), and because the overall statistics roughly determine the "overall impression" you obtain as you glance at an image. Here are some common uses:

- Many digital cameras display histograms of pixel values, this is a great help in determining an effective exposure (there is no such a thing as a "correct" exposure) if the user knows what to aim at. Thus the feature is mainly present on "professional" equipment, the common user would only be confused by it. Unfortunately, at present few cameras display separate R, G, and B histograms,

- Most Photoshop users are familiar with histograms and use them in the "LEVELS" adjustment, which amounts to the minimum requirement for a tolerable print;

- TV sets are adjusted to produce a reasonable overall impression (often called "natural" impression);

- The printer at the photo shop adjusts the overall impression of your prints such as to look "natural." Thus a *pink* wedding dress is likely to be printed *white*, and so forth. There is no way the operator at the printer could know the actual color of the dress and generic wedding dresses being white. That's how it will appear in print, though this will cause a greenish cast on the faces. In case the operator notices a choice has to be made,

- Exposure meters and color temperature meters as used in photography also yield overall measures of the scene in front of the photographer. If you always use the filters suggested by the color readings, your sunset and noon photographs will come out globally the same hue, an effect that is actually appreciated by some very *careful* workers but hopefully should not attract you. I consider it to be in rather doubtful taste.

Thus the *global statistics* of images is certainly something to worry about. Having said that, I hasten to add that overall measures are essentially worthless in images composed of very different local regions. In such cases you will need to check the regions and *make a choice*. The overall statistics are not particularly informative in such cases.

In figure 16.5 I show the "mandrill image" often used in examples in the image processing literature. It is very colorful (figure 16.6) and will do fine as an example.

The first thing to do is to consider the color channels separately, as you would probably do in Photoshop as a start for essentially anything.

The red, green, and blue channels are shown in monochrome in figure 16.7. It is an interesting exercise to compare these images in some detail. Compare the eyes, nose, cheeks, eyes, and fur in the three channels. The nose is very red, which shows up as almost white in the red channel and almost black (except for gloss) in the blue channel, and so forth. The distribution of image intensities in the channels are the red, green, and blue histograms. They are shown in figure 16.8. In this case the histograms are almost uniform, except for a bimodality due to gloss. That the high-intensity mode is likely due to gloss is clear because it appears in all three histograms, and thus has to be roughly achromatic. It is typical for gloss to be roughly achromatic, or better, to reflect the color of the main source, which is itself typically fairly achromatic.

Figure 16.5 The "mandrill image".

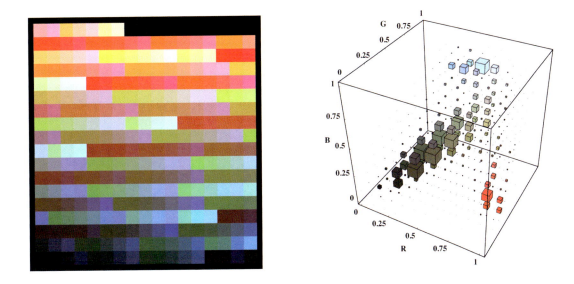

Figure 16.6 Left, a "palette" showing all colors present in the mandrill image in some (lexicographical) sorted order, on the right the density of colors in the RGB cube. Notice that the mandrill image—though very colorful—is by no means composed of "all RGB colors."

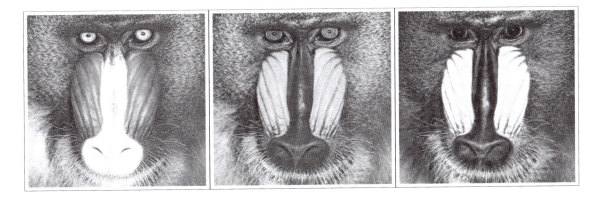

Figure 16.7 The red, green, and blue channels of the mandrill image.

A representation that is even coarser than the histograms looks at the ordinal relations $R \gtrless G$ and $G \gtrless B$. There exist clearly four distinct categories. An analysis of the relevant physics suggests that the boundaries between different categories are likely to coincide with boundaries between different material properties.[6] This categorization is shown in figure 16.9 for the case of the mandrill image. Notice that the specularity is evidently misclassified.

A useful indicator of the "extendedness" of local color is the rank of the gamut in small environment. It is found by considering the eigenvalues of the covariance matrix for the gamut. From this you obtain measures for the linear, planar and volumetric extendedness of the gamut. The result depends upon the size of the subimage of course (see figures 16.10 and 16.11), but it can be found even for very small subimages. A surface of uniform

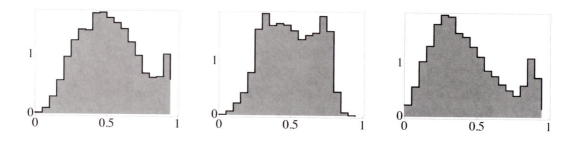

Figure 16.8 The histograms for the red, green, and blue channels of the mandrill image.

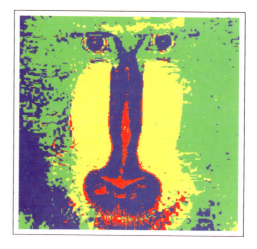

Figure 16.9 The RGB color categories as defined by Rubin and Richards for the mandrill image.

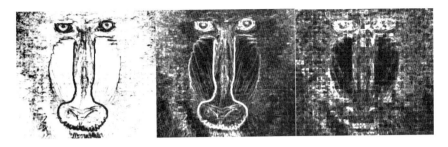

Figure 16.10 The rank of 3×3 subimages.

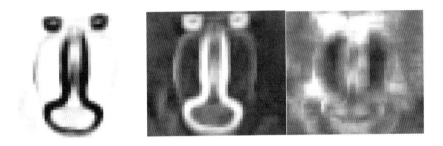

Figure 16.11 The rank of 20×20 subimages.

reflectance, illuminated by a single source, will generate a linear extension due to shading. When illuminated with two, spectrally distinct sources, it will generate a planar extension. A dielectric colored object with a glossy surface will also generate a planar extension because the Fresnel reflection from the surface will be in the color of the source, whereas the diffusely scattered beam from the bulk material will be colored by the pigment in the material. Of course, transitions between different materials are also likely to generate planar extensions. This is very visible in the case of the mandrill image. Notice also the square lattice appearing in the distribution of volumetric extension. It is evidently due to JPEG artifacts in this image.[7]

The histograms reveal very little concerning relations between the RGB channels. Yet this is what you have to put up with, even with fancy digital cameras. A better statistic is to consider the covariance matrix. In order to do that you must prepare flat lists of all pixel values in the red, green, and blue channels. Of course, these lists need to be in sync. These can be *very* long lists, for instance, for a (fairly typical) 1000×1000 pixel image the length would be a million. Let RGB be a matrix that has these lists as columns, that is, $RGB = \{R, G, B\}$ (for the example considered above a $3 \times 1,000,000$ matrix). Then the covariance matrix for the mandrill image is

$$C = \frac{1}{n} RGB^T.RGB = \begin{pmatrix} 0.61 & 0.59 & 0.53 \\ 0.75 & -0.22 & -0.62 \\ 0.25 & -0.78 & 0.57 \end{pmatrix},$$

where n ($n = 512^2 = 262,144$ for the mandrill image) is the length of the lists. It is evident that *the red, green, and blue channels are significantly correlated*. This need not surprise you, since the mandrill is easily recognized in all channels. The square roots of the eigenvalues are in the ratio 1:0.25:0.09; thus the eigenvalues are rather different in magnitude. The first eigenvalue belongs to an eigenvector $\{0.61, 0.59, 0.53\}$, almost an equal mixture of the three channels; it is a slightly brownish, light gray color, the "overall" color of the image. The eigenvectors are illustrated in figure 16.12.

The second eigenvector is $\{0.75, -0.22, -0.62\}$, which can be decomposed as

$$\{0.75, 0, 0\} - \{0, 0.22, 0.62\},$$

that is a red minus a blue. Likewise, the third eigenvector, which is $\{0.25, -0.78, 0.57\}$, can be decomposed as

$$\{0.25, 0, 0.57\} - \{0, 0.78, 0\},$$

that is a purple minus a green. The eigenvector can be displayed as RGB colors as shown in figure 16.12. In this basis the channels are completely de-correlated. Notice that this is much like a Hering-type "opponent" basis. Notice also how much of the image is already "explained" by the first component, that is, a monochrome rendering.

In figure 16.13 I show the "amount of variance explained" by each opponent channel. Clearly the first, monochrome channel takes the brunt. The truly "opponent" channels explain only very little of the variance. These channels are nevertheless important. In

  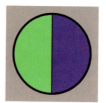 

Figure 16.12 The eigenvectors of the covariance matrix of the mandrill image. The first one is similar to the achromatic beam, the other two are "opponent channels." The rightmost figure illustrates the overall structure.

Figure 16.13 The projection on the eigenvectors of the covariance matrix of the mandrill image.

figure 16.14 I show the contribution of the first, and the first plus the second channel (adding the third channel simply restores the original image of course). Notice that the monochrome rendering, though it does an excellent job in rendering the scene, completely fails to reveal the dramatic coloring of the nose and cheeks. The sum of the first and second channels takes care of much of the image, but notice that the colors of the eyes are way off. The third channel is crucial for that.

The basis of eigenvectors of the covariance matrix is not unlike the basis of RGB space I used throughout the chapter on RGB colors (chapter 14). Thus if you project the image on $\{1, 1, 1\}$, $\{1, -1, 0\}$, and $\{1, 1, -2\}$ (mutually orthogonal directions), a "Hering opponent basis" (section 14.2), you obtain a monochrome and two dichromatic images that are not

Figure 16.14 The contribution of the first, and the sum of the first and the second channel. Adding the third channel restores the original.

unlike the empirical principal components basis for this image (see figure 16.15). The images give some rough notion of what it would be like to have monochromatic or dichromatic color vision. Infinite variations could be thought of.

Thus far I have considered the overall histograms and covariance, as well as some spatial distributions. A topic of considerable interest is the distribution in color space. The histograms and covariance matrix give little of a cue here because they are mere overall measures. You also need the correlation with the distribution in the image. After all, you destroy an image if you scramble its pixels.

If you find the Ostwald decomposition (section 14.2) of each pixel you can find how much of a given full color exists in the image. Plotted as a polar histogram you obtain figure 16.16. The spatial distribution of the full colors over the image is shown in figure 16.17. Notice that this polar histogram yields completely novel information; this is not something that is at all evident from the RGB histograms. The red mode corresponding to the nose and the blue mode corresponding to the cheeks are immediately evident. The yellows must be due to the fur and the eyes. Notice how few hues are really present in this image. A painter would need only a few pigments. The spatial distribution does not give a particularly good insight because the color extent of the full colors is not evident. Scaling with the color content helps a lot to bring things into proper perspective.

A formal modal analysis of the polar hue histogram typically reveals many modes, though most of them are really minor and best considered noise. Plotting the locations of the modes with their weights yields a more balanced picture (figure 16.18). The mandrill image has only a few significant modes (figure 16.16). You could indeed paint it with a very limited palette.

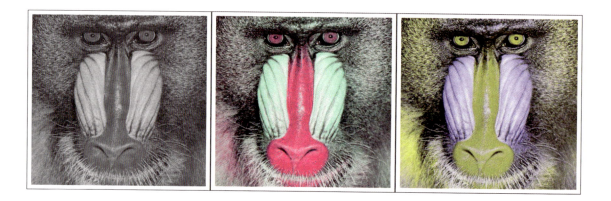

Figure 16.15 The projections on $\{1, 1, 1\}$, $\{1, -1, 0\}$, and $\{1, 1, -2\}$ of the mandrill image.

Figure 16.16 The distribution of hues for the total image over the color circle. Here I used the discretization induced by the cardinal colors of the color circle, this is not a true modal analysis.

In figure 16.19 I show the distributions of black, white, and color content over the image. The color content is visualized by changing all full colors to red. Notice how much more informative these "channels" are as compared with the RGB channels.

You can also draw global histograms for the black, white, and color content. These are shown in figure 16.20. It is particularly interesting to see that the majority of pixels is not very strongly colored at all.

It is of much interest to plot the pixels in the Ostwald color solid (section 14.2) because the 3D density is *so much* more informative than the mere R, G, and B histograms. In the case of RGB colors you may simply plot in the RGB cube. The RGB cube is like the spectrum double cone in that the color solid is tangent to it.

In figure 16.21 I show the resulting "cloud." This representation repeats the information revealed by the polar histogram, but improves on that because it also reveals the white and black contents of the pixels. The polar diagram combined with the 3D "cloud" representa-

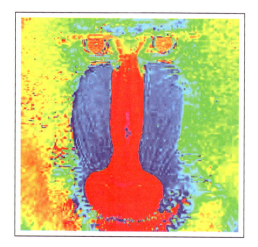 

Figure 16.17 The distribution of full colors over the image. The left image looks weird because even mostly achromatic colors are represented by their full color. Notice that this representation looks "flattish" and that it seems to contain only a few different hues. In the right image the color content has been taken into account, this image looks dark, but more "natural."

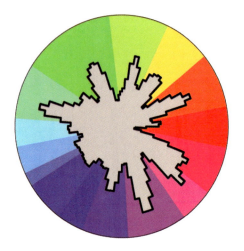 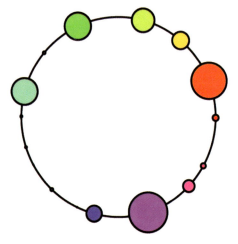

Figure 16.18 A modal analysis of a polar histogram reveals many modes, though only a few major ones. This is better represented in the figure at right which shows the locations of the modes with their weights. There exist only a few very significant modes. (This is just an example of modal analysis, not of the mandrill image.)

Figure 16.19 The distribution of the black, white, and color content over image. The color content distribution is suggested by replacing all full colors by a single (red) color. You may need some initial exercise, but such representations soon come to "look natural."

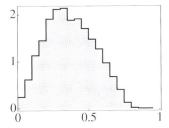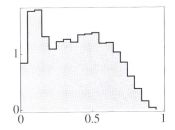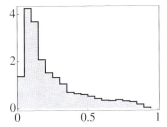

Figure 16.20 The black, white and color content histograms for the total image. Notice how these histograms are rather more informative than the RGB histograms.

tion yield perhaps the best, intuitively useful, handles on the global RGB structure of the image.

Although I have shown results for a specific example, much of what I said applies to the majority of images of natural scenes.

Histogram-based methods are among the simplest, but often surprisingly effective, tools in image analysis. They become especially useful when results can be interpreted in terms of general optical processes that go on in the scene.

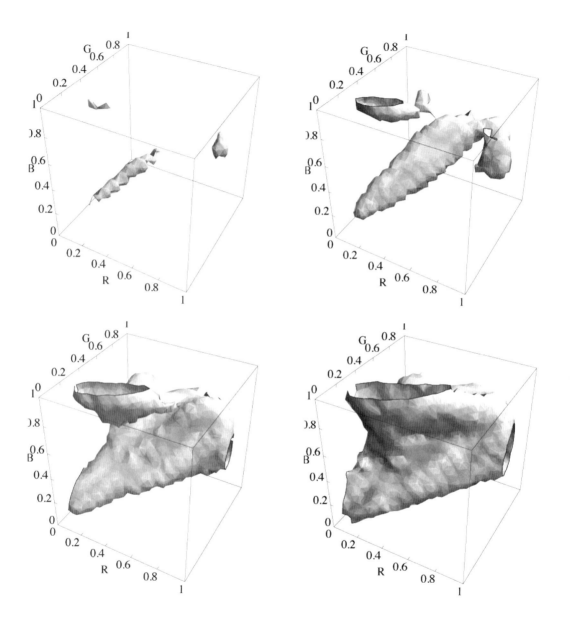

Figure 16.21 The pixel distribution in the RGB cube. The levels are of 1/2, 1/8, 1/32 and 1/128 times the maximum.

This chapter is evidently different in character from the rest of this book. I almost singularly concentrated on a single example, showing many results of a numerical character that have no meaning except in the context of the specific example! Although apologies are certainly needed to some extent, I still think that this method was the best way to introduce you to these (essentially very simple) image analysis techniques.[8]

This type of image analysis has few pretensions, yet is most useful in many contexts. Several of the analyses I introduced in this chapter should be available as a matter of course in various software packages dealing with images.

However, at least on the short term, you'll have to roll your own. Notice that there is lots of room for improvement and elaboration too.

Exercises

1. Histograms [EXPERIMENTS] Take some image and find the histograms for the red, green, and blue channels. Try to identify the origin of the various modes. Do a covariance analysis and move to a basis in which the three channels are uncorrelated. Again, construct the histograms and interpret the modes.

2. Polar hue histogram [TAKES GRAPHICS PROGRAMMING] Construct the polar hue histogram of images. Don't forget to take the color content into account.

3. The air light [EXPERIMENT] Take the image of a landscape with distance running from nearby to the horizon. Estimate the black point for various positions in the image, and thus the amount of "air light." How does the amount of air light depend upon the distance in the scene? Can you "lift the airlight" and render the scene as it would look in vacuum?

4. Material categories [CONCEPTUAL] The Rubin-Richards segmentation is designed to split the image into regions that might correspond to different material categories. How reasonable is this claim? Can you think of a better method? To what extent do you think such a goal can be reached at all?

5. Rank [CONCEPTUAL, FOR PHYSICISTS] A gamut is a set of color vectors. Devise a robust estimate for the "rank" of such a collection, that is the degree to which the vectors span a one-, two- or three-dimensional manifold. What are the physical factors that cause a gamut to be of a certain rank?

6. Modal analysis of hue [ELABORATE] Implement an algorithm that picks out the major modes from the hue histogram. Run it on images from paintings and try to decide on the nature of the palette used by the painter.

7. Variations in instances encountered in the public domain [ELABORATE BUT EASY] Use Google to find a dozen or more versions of a single (obviously popular) image, for instance, a well known painting (typing "bedroom in arles" in Google gave me dozens of copies of Vincent van Gogh's well known painting of his bedroom in Arles, for example). Compare the modal analysis of hues on these instances. What type of judgments would you dare to pass on a work if you had only access to a *single* instance, based on this experience?

Chapter Notes

1. It is rare to deal with *really* "raw" files. Most cameras and scanners perform various operations on the raw photoelectron counts already on the chip. A "raw" file is simply what you get from your electronics at the earliest point where you can tap it. It will rarely be something simple in optical terms. Of course this is worrisome. In this chapter I skip most of such (important) technicalities.

2. The so-called Bayer pattern is fairly standard. It is a repeated pattern of short, medium, and shortwave photoelectric sensors.

3. The modern CCD cameras are already much better than film-based cameras. Very good calibration is obtained in professional applications (astronomy, remote sensing). No doubt prosumer cameras will profit.

4. The spatial resolution and pixel counts of modern cameras already exceed the complexity that typical image processing algorithms can handle. The major bottleneck is the dynamic range. There is no reason not to expect significant progress in this area though.

5. There exists an engineering community dedicated to the manipulation of images at the pixel level, often known as "pixel fucking" by those who believe they enjoy a broader view.

6. J. M. Rubin and W. A. Richards, "Color vision: representing material categories," *MIT AI Memo* 764, pp. 1–38 (1984).

7. JPEG stands for "Joint Photographic Experts Group." The group issued a standard for image compression in 1992, which was accepted as ISO 10918–1 in 1994. It is a lossy compression scheme that coarsely quantizes the spatial spectrum of square (8×8) pixels subimages in a regular tiling individually. Due to a fast cosine transform method, such compression can be done on the fly when writing images to a storage medium. The compression is not done in RGB, but in YCbCr space (intensity-chrominance space like in the PAL television transmission). The compression artifacts show the characteristic square tiling pattern.

8. Notice that although "image processing" is a established field, there does not exist a community claiming to specialize in "image analysis". Yet this is an important endeavor with numerous applications. No need for me to apologize.

Part VI

Beyond Colorimetry

Chapter 17

How Color Is Generated

In this chapter I consider various ways in which "colors" are generated in the generic human biotope. Since there exist many texts on the topic, this chapter can be a relatively short one. The content is differently oriented from the standard accounts though. The reason is that the standard accounts are typically compiled by physicists and chemists with a keen interest in their respective fields but perhaps less interest in the biological or ecological perspective. As a result, the physicist is likely to stress colors due to dispersion of refractive index (the prismatic spectrum), diffraction and interference (the rainbow, colors in oil films), Mie scattering (atmospheric colors, ruby glasses (gold sols), various photoelectric interactions in solids, and so forth, and the chemist to stress colors due to photoelectric effects on the molecular level (metal ionic colors, organic dyes, and so forth). Fascinating as such accounts may be, they hardly address the color effects of the generic human biotope.[1] Here I consider a piece of white paper or a clot of mud as optically just as intrinsically interesting as the rainbow and indeed rather *more* interesting (because involved in your interactions with the physical world) from an ecological perspective.

17.1   Radiometry

Radiometry is a bookkeeping routine, a technique to keep track of where the radiation goes. According to your interests, it allows you to keep track of the numerosity of photons, of the radiant power or of the number of rays. Differences are slight, though in these three cases your perspective on the physics is rather different. In some cases, radiometry can be viewed as an approximation of more fundamental descriptions. For instance, when considering radiant power you can draw on Maxwell's electromagnetic theory and seek for tangencies with the theory of coherence in wave fields, the Poynting vector, or the energy-momentum tensor.[2] From the perspective of geometrical optics, radiometry is merely counting rays[3] and there are no underlying theories to draw on. For most problems, the geometrical optics perspective is just fine. The only aspect missing is that of speed. "Photons" are thought of as particles that travel at certain speeds. The number of photons contained in a certain space-time volume evidently depends on these speeds.[4] In the case of geometrical optics, you can simulate this when you consider ray density as dependent on the refractive index of the medium. In most cases of ecological interest, radiometry is applied to radiation propagating through the atmosphere, which is for most practical purposes a constant refractive index roughly equal to that of the vacuum. Consequently the geometrical optics perspective serves just fine.

17.1.1   The étendue and the radiance

From geometrical optics courses, you are probably familiar with the notion of beams. These are bundles of rays with such a structure that through each point of space there passes a finite number (typically, just one) of rays. Although important in the theory of lenses, mirrors, and so forth, this notion of "beams" is of no interest to radiometry. In order to keep things sorted out, I will denote these "beams" as "pencils"[5] and I will redefine the notion of "beam".

When I say "ray," I generally mean a directed straight line. A Euclidian straight line thus represents two possible rays, one for each direction. In special settings it may be necessary to consider directed polygonal arcs that are only piecewise straight. An instance is the case of reflection or refraction at surfaces dividing regions of different refractive index. It may even be necessary to consider arbitrarily curved rays in cases where the refractive index is a smoothly varying function of position in inhomogeneous media. In radiometry, you typically consider things from the infinitesimal perspective and piece things together by integration. In the infinitesimal domain, rays are generic straight lines.

Whereas a "pencil" has at most a finite number of rays passing through any point of a volume (figure 17.1), a "beam" has an *infinite* number of rays passing through any point of a volume. Consider a single point of space. There may be rays passing through it in any direction. Consider a single direction. There may be rays of that direction at any point of space. Sometimes this type of structure is called "diffuse beam" (people who use this term usually will say "beam" in the case of a "pencil"). Don't be confused.

There is some structure to pencils that is due to the physics and finds its "explanation" in the theory of geometrical optics as the limit of wave optics for vanishing wavelength. The "rays" of geometrical optics are thought of as the normals of the "wavefronts" or "isophasal surfaces" of a wave field. Such "wavefronts" are generically *surfaces* (the phase singularities are curves and points that are rare and far apart); thus, the "rays" cannot act

Figure 17.1 A parallel and a diverging "pencil." Notice that just one ray passes through any point of space. This is the defining property of a "pencil."

independently but must always be such that there exists an orthotomic surface to them (figure 17.2). A differential geometric analysis shows that infinitesimal pencils must be "elliptic", implying that such a pencil has two "focal lines" such that each ray passes through both focal lines. At a focal line the pencil is squashed such that the ray density explodes (figure 17.3). That explains why the foci are also called "caustics," this derives from the theory of burning mirrors (originally developed for military purposes by the ancient Greeks).

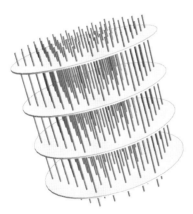
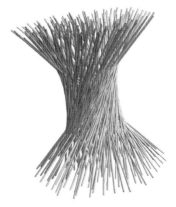

Figure 17.2 Pencils occurring in optics ("elliptic") always admit of "orthotomic surfaces" ("wave fronts"). A "twisted" (hyperbolic) structure can be produced with a bunch of toothpicks, but doesn't describe optical pencils.

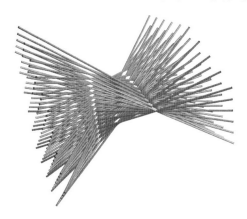

Figure 17.3 Mutually skew caustic lines form in an elliptic ("astigmatic") pencil. This is the generic case for an infinitesimal pencil, you have two caustic lines. In general each ray is tangent two a pair of caustic surfaces. At a caustic the ray density is infinite: This is the principle of burning mirrors.

The ellipticity of geometrical optical pencils was discovered empirically in eigthteenth and nineteenth centuries, even before the full development of wave optics. The differential geometry divides pencils into elliptic and hyperbolic ones. Only the elliptic ones figure in optics, the hyperbolic ones can be realized in various other, non-optical ways. For instance, if you twist a bundle of toothpicks (figure 17.2) you obtain a hyperbolic pencil that indeed fails to show foci.

Beams differ essentially from pencils in that *infinitely many* rays pass through each point of space (figure 17.4). Thus, beams are much more complicated objects than pencils. However, they are also much more realistic and radiometry deals only with beams, never with pencils. The reason is simple enough; pencils are unfit to transport any radiant power. They are only figments of the mind. The real world is described through beams.

In order to set up a suitable bookkeeping scheme for rays, it is most convenient to start with *infinitesimal* beams. Infinitesimal beams are limited in both space and direction. Think of a *slender* beam, for instance, a beam that passes through a small aperture, and of a *highly directional* beam, for instance, a beam that passes through a sequence of two infinitesimal apertures spaced at some finite mutual distance (figures 17.5 and 17.6). Instead of physically isolating such a beam, you will more often isolate it conceptually. Such a beam is characterized by a certain area element (its cross-section) of area dA and (unit) normal $\mathbf{n}$, as well as a certain direction, which is the unit vector $\mathbf{r}$ and angular spread given by the solid angle $d\Omega$. It is convenient to define a vectorial area element $d\mathbf{A} = \mathbf{n}\, dA$ and a vectorial solid angle element $d\mathbf{\Omega} = \mathbf{r}\, d\Omega$. The vector area element specifies both the

Figure 17.4 A "beam" is different from a pencil in that infinitely many rays (not all of them shown) pass through any point of space.

area and the spatial attitude, while the vector solid angle specifies both the direction and angular spread. The inner product of these quantities is a volume element of the phase space of rays. You have

$$d\mathcal{E} = d\mathbf{A} \cdot d\mathbf{\Omega}.$$

The "throughput" or "étendue" of the beam[6] is thus simply the "phase volume" element, that is, the cross-section times the angular spread.

In many cases you isolate a slender, directional beam through a pair of apertures. You obtain

$$d\mathcal{E} = \frac{(d\mathbf{A_1} \cdot d_{12})(d\mathbf{A_2} \cdot d_{12})}{d_{12}^2} = d\mathbf{A_1} \cdot d\mathbf{\Omega_2} = d\mathbf{A_2} \cdot d\mathbf{\Omega_1},$$

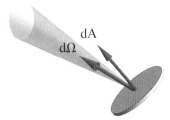

Figure 17.5 An infinitesimal beam defined by way of a vector-area $d\mathbf{A}$ and vector-solid angle $d\mathbf{\Omega}$.

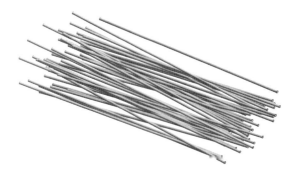

Figure 17.6 Some random rays of the infinitesimal beam defined in figure 17.5.

where d_{12} denotes the direction, d_{12} the mutual spacing, $d\mathbf{A_{1,2}}$ the vector areas and $d\mathbf{\Omega_{1,2}}$ the vector solid angles subtended by the second aperture as seen from the first and vice versa. Notice the pretty symmetry. It makes perfect sense, for the étendue is the same for beams in either direction.

In radiometric calculations it is most convenient to always use infinitesimal vector areas and solid angles.

The étendue of a finite (not infinitesimal) beam is defined through integration over infinitesimal parts

$$\mathcal{E}_{\text{beam}} = \int_{\text{beam}} d\mathcal{E},$$

exploiting the additivity of mutually incoherent beams.

Consider a simple example, the étendue of a hole of area A in a planar screen. By integration[7] you have $\mathcal{E}_{\text{beam}} = \pi A$. This is the étendue of a *very* diffuse beam, since all directions pass through the hole. Notice that you don't get the area times the solid angle of a half-space $(2\pi A)$, as might be naively expected. You integrate over the vector solid angle projected on the normal to the aperture, not solid angle per se. This is an expression that finds frequent application in radiometric problems.

The étendue is a *capacity*, much like the volume of a bucket. You can fill it with rays. The total number of rays that pass through the beam is known as the "radiant flux" Φ. For an infinitesimal beam, I define

$$dN = \frac{d\Phi}{d\mathcal{E}},$$

where dN is the "radiance" of the beam. The flux is the radiance times the étendue, and thus the radiance is a *density*. I'll refer to it as the "density of rays."

The radiance is easily the most important radiometric entity. It is measured as the "number of rays" per area and per solid angle. In the photon picture you would measure the number of photons per second and in physical optics the radiant power instead of the number of rays. It makes little difference.

The radiance is a function of both position and direction, thus a function

$$N(\mathbf{r}, \mathbf{d}) : \mathbb{R}^3 \times \mathbb{S}^2 \mapsto \mathbb{R}^+.$$

At any given location you have a radiance in all directions, much like a fisheye photograph. Apparently the radiance can be understood as an infinite filing cabinet of such fisheye photographs, one for any location. The radiance can be a very complicated function indeed. The only constraints are that it is nonnegative throughout and that the radiance is conserved along a ray, that is to say,

$$\boldsymbol{\nabla}_{\boldsymbol{r}} N(\mathbf{r}, \mathbf{d}) \cdot \mathbf{d} = 0.$$

Radiance times the étendue of the photoreceptors and eye media and optics is the physical correlate of your retinal excitation. It is all important in vision.

17.1.2 The irradiance

Consider a surface that intercepts a beam. At each point of the surface consider an infinitesimal area and integrate the radiance to obtain the flux in the infinitesimal beam. This flux, divided by the (infinitesimal) area is called the *irradiance* of the surface at that point. Irradiance is measured as the number of rays striking the surface per unit area irrespective their direction. (See figure 17.7.) For a finite surface the irradiance will typically vary with location over the surface. This is often important. Think of a projection screen, for instance.

For vision the irradiance at surfaces in a scene is only of *indirect* importance. Only if the surface scatters some of the incident rays towards the eye will you become visually aware of the irradiation of the surface. But irradiance often figures in the intermediate steps in radiometric calculations.

17.1.3 Diffuse beams and the light field

I use "diffuse beam" in an informal sense as a beam with large étendue, thus with wide areal and angular spreads as opposed to a "collimated beam" with very small étendue. Of course, it will depend on the application whether you consider an angular spread "wide" or "narrow". In typical circumstances you would consider the beam of radiation delivered by the sun (angular spread about 0.5°) as "collimated" and the beam delivered by a heavily

Figure 17.7 Lambert's cosine law is illustrated here for a Lambertian surface. In this case the radiance scattered to your eye is simply proportional with the cosine of the inclination of the beam (here 0°, 30°, 60°, and 90°). In generic scenes the direction of the beam will be fixed, whereas the surface orientation will change (over a curved surface, for instance). In such a case, the varying irradiance is known as "shading".

overcast sky (angular spread about 180°) as "diffuse". Whether a beam is to be considered collimated or diffuse is often judged by the quality of cast shadows, for instance, the sharpness of the edge of the shadow of your body on the ground. The sun casts a hard shadow, whereas overcast sky light is often considered "shadowless illumination" by photographers.

The most extreme case of a diffuse beam is encountered in cases of "polar white out" in which the radiance is the same regardless of direction. In the psychology of perception this is known as a Ganzfeld, in the physics laboratory such beams are most commonly produced in "integrating spheres" ("Ulbricht spheres" in continental Europe[8]). In such a beam objects with a white Lambertian surface become literally invisible, regardless of their shape, and so do perfectly reflecting surfaces. Fortunately, Ganzfelds are rare in generic human environments. On the other hand, very diffuse beams (as from the overcast sky) are quite common. Collimated beams are rare, except for the sun's beam. Otherwise, some artificially illuminated scenes are about the only examples. Even in sunlight the parts that are in shadow (not directly illuminated by the sun) are not totally dark. They are illuminated by very diffuse beams of scattered radiation from the environment.[9]

For many purposes, the structure of diffuse beams can be summarized through the first few terms of a development of the radiance into spherical harmonics $Y_{lm}(\vartheta, \varphi)$. The coefficient of the first term (that is Y_{00}) describes a scalar, the "volume density of radiation." In the ray description, this is the total length of all rays per unit volume. In a Ganzfeld, the volume density yields a full description of the beam.

The next three terms (coefficients of $\{Y_{1,-1}, Y_{1,0}, Y_{1,+1}\}$) transform as a vector. It is the so-called net flux vector $\mathbf{F}(\mathbf{r})$. Its significance is that for a surface element $d\mathbf{A}$ the inner product $\mathbf{F} \cdot d\mathbf{A}$ yields the net flux that passes the area element. In general rays will pass the area element in either direction, so you need to count the rays with proper sign. If $\mathbf{F} \cdot d\mathbf{A} = 0$, this means that as many rays pass the area element in one direction as they do in the other direction. You can construct tubular surfaces such that $\mathbf{F} \cdot d\mathbf{A} = 0$ pertains all over the surface of the tube (figure 17.8). Then *the net flux of radiation evidently propagates through the tube*. Different from the generally straight rays, such tubes are typically curved (they can even be closed). The field lines of the net flux vector field thus parcellate space into flux tubes and define the net transport of radiation. This is a great way to visualize important properties of the structure of a diffuse beam.

The following five terms of the series, that are the coefficients of

$$\{Y_{2,-2}, Y_{2,-1}, Y_{2,0}, Y_{2,+1}, Y_{2,+2}\},$$

transform as a symmetric tensor of trace zero. It is possible to construct a geometrical interpretation that renders this "quadrupole" part visually intuitive, although this is rarely done. The quadrupole contribution is a combination of an illumination set up with two sources at opposite directions or a "ring light" surrounding the scene.

A simple diffuse beam that is often used as a model is the beam emitted by an infinite planar Lambertian radiator. It is a (very approximate) model of the heavily overcast sky. For such a beam, the net flux vector field is a uniform, constant field. The flux propagates via straight tubes at right angles to the luminous plane. It is just as important as the uniform, collimated beam is. I will refer to it as the "hemispherical diffuse beam."

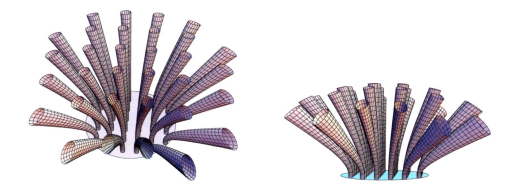

Figure 17.8 "Light tubes" for a uniform, Lambertian, circular disc source. The radiation is "transported by way of the tubes" in the sense that there is no net transport through the surfaces of the tubes. Notice that the tubes—different from the rays—are curved.

17.1.4   Diffuse beams and the structure of scenes

The "scene" may conceptually be divided into "sources," "objects," and the "medium."
In many cases the "medium" can be taken as the vacuum and simply ignored. It is the
medium through which the radiation propagates, it merely enables the existence of rays. In
such cases the medium is simply empty space. If the medium is dispersive or turbid, this
has to be taken into account. In the generic terrestrial environment this typically means
the effect of "atmospheric perspective." I'll treat that later and will ignore the medium
here.

With "sources" I mean what are often known as "primary radiators," the sun or artificial
sources such as candles, fires, and fluorescent tubes. All "objects" in the scene interact with
the beam, typically by scattering radiation. Thus, they act as sources too and are often
called "secondary radiators" for this reason.[10]

Generically, "objects" are opaque volumes bounded by surfaces that scatter impinging
radiation into all directions. The simplest model would involve Lambertian surfaces. This
is of the utmost importance because it enables "vision" as you know it. The major optical
effects that make "vision" possible are:

— *Objects are opaque. Thus, you can only see their front sides, and when one object is
in front of another one it renders the more remote one invisible;*

— *Objects scatter radiation into all directions. Thus, an object is visible from any point
that is located such that rays have a free path between the object and the eye;*

— *If sufficient primary radiation is available (daytime) anything in an open scene will
be visible (irradiated), even in places that are not directly irradiated by the primary
source(s).*

These are the basic facts of (optical) life. Any one of these important facts is not *necessary*
in the sense that you could easily set up physically possible worlds in which they would
become of minor importance. Vision as you know it is bound to break down in such worlds.
Vision is made possible through the generic structure of the human biotope. Of course, it
also (at least partially) *determines* the biotope.

Since objects scatter radiation, anything in a scene optically interacts with anything
else. These contributions are often known as "interreflections" or "reflexes." You can write
a balance equation for this (merely doing the bookkeeping for the rays). It is a Fredholm
integral equation of the first kind in which the integration is over the surfaces of all objects
that make up the scene. Such an equation is far too complicated for you so much as to
think of solving it in any realistic case. Approximate numerical solutions are provided by ray
tracing algorithms. In practice you break up the scene into small parts, and you treat the

parts as mutually independent. You pay for this by having to assume a different beam for every part. The parts can be simple enough that you can understand them from canonical optical models. "Vision" no doubt is based on similar principles. Most complexity is simply ignored,[11] and the remainder is broken down into digestible chunks.

17.2 The Illuminant

The primary radiator is characterized by the geometry of its beam and by its spectral composition.[12]. Most of the radiators of ecological interest are approximately black body radiators with a spectral composition that is approximately described by Planck's formula and can be parameterized by a *temperature*. (See figure 17.9.)

The most important example is the sun, which has an approximately Planckean spectrum for a temperature of 5700°K.[13] Incandescent sources are also approximately Planckean radiators but for much lower temperatures (2400–3000°K). Candlelight is even closer to 1500°K.

Radiation from the sun has to pass the atmosphere before it irradiates the scene in front of you, and in the course of propagating through the air loses some spectral content in a specific way. As a result, the daylight spectrum looks rather rough. The roughness

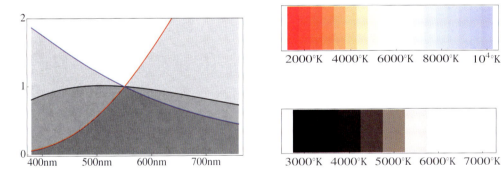

Figure 17.9 Left: Relative spectral power computed from Planck's radiation law for the temperature of a light bulb (2500°K, red), the solar photosphere (5700°K, black), and 10,000°K (blue); Right: Colors of "black body radiators" at various temperatures. These RGB colors were computed via the Schopenhauer crate for a reference illuminant that was a black body radiator at a temperature of 5700° (about the temperature of the photosphere of the sun). At top the true colors, at bottom they have been normalized to (RGB) full colors. For temperatures tending to infinity, the colors tend to a standard bluish color; for (ecologically) "reasonable" temperatures you obtain the standard sequence from red to blue (at the reference temperature you have white of course).

makes hardly any difference because of the low resolution of the human "trichromatic spec-troscopy," though. For all practical purposes, the spectrum can be treated as Planckean.

Of the artificial sources it are mainly the gas discharges and fluorescences that frequently present you with markedly non-Planckean spectra. I will ignore them here. There is plenty of technical literature if you are interested. (The ecological relevance of such sources is slight, even in the build environment.)

17.3   The Medium

For the generic human biotope (the terrestrial environment), the medium is invariably *air*. Clean air is very transparent and for most purposes (seeing objects at close distances) is not different from the vacuum. Thicker masses of clean air scatter sufficiently that the effect becomes noticeable. Rays may be scattered out of the beam, thus decreasing the radiance through extinction. Rays may also be absorbed out of the beam, thus decreasing the radiance through absorption.[14] (See Figure 17.10.) Scattering also *adds* rays to the "scattered beam," which thus increases in radiance. This is known as "air light."

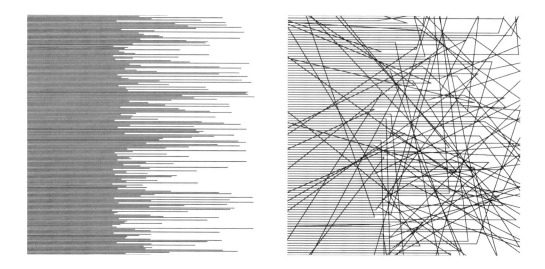

Figure 17.10 Absorption and extinction should not be confused. In these figures a collimated beam enters from the left. In the case of absorption (left), the rays "end" at each "absorption event." In the case of extinction, the rays are scattered and suddenly change direction at an "scattering event." In both cases the result is an attenuation of the collimated beam. But in the case of absorption the rays are lost, in the case of scattering they show up in a diffuse, scattered beam.

In clean air the scattering is due to density fluctuations at the scale of a wavelength of the radiation. Density fluctuations lead to refractive index fluctuations, or turbidity. From a simple dimensional analysis, it follows that the scattering cross-section is proportional to the inverse fourth power of the wavelength, so-called Rayleigh scattering. The result is the blue color of the clear sky, it is Rayleigh scattered sunlight.

Since the extinction is much stronger for short wavelength (Rayleigh scattering) the transmitted beam becomes skewed toward the long wavelengths, an effect that becomes dramatically visible in the red color of the setting sun.

For horizontal air columns (viewing distant objects near the horizon), the beam scattered by the distant object suffers an extinction due to the fact that rays are scattered out of the line of view. At the same time, the beam that arrives at the eye will contain sunlight that has not been scattered by the distant object, but by the air, into the line of view. This so-called air light is bluish and is added to the (attenuated) beam from the distant object, thus decreasing contrast. This is the reason for the bluish color and low contrast of distant objects (figures 17.11 and 17.12). A simple analysis reveals that the effect is exponential with distance.[15]

If the air contains water droplets or dust particles the scattering cross-section becomes much larger. Moreover, the scattering changes from Rayleigh scattering to "Mie scattering"[16] (scattering by arbitrarily large dielectric spheres). This generally implies that the inverse fourth power relationship becomes much less pronounced. Such effects are clearly

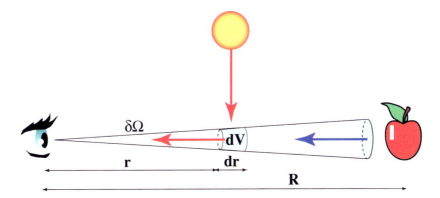

Figure 17.11 An object (red apple) is seen through an air column of length R. Consider a slender cone of solid angle $\delta\Omega$. A volume element dV of thickness dr at distance r absorbs the radiation coming in from the right and scatters sunlight to the eye (the "air light"). A simple integration over the length of the beam yields Koschmieder's law: The contrast at which the eye sees the apple decreases exponentially with distance.

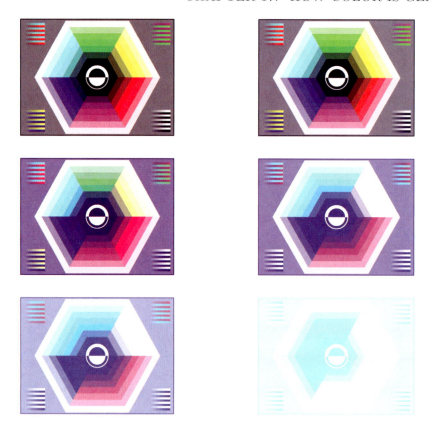

Figure 17.12 A color target seen through a clean atmosphere for various distances turns bluer and less contrasty as is moves farther and farther away. It eventually assumes the (uniform) color of the horizon sky, which is the color of outer space (black) seen through a very (effectively infinite) horizontal layer of air. The eye receives only "air light" in this case, which is sunlight scattered in the layer.

visible, even for short distances, in cases of fog, sandstorms, and the like. If the scatterers are very numerous, single scattering gives rise to multiple scattering and the medium becomes effectively opaque[17] (white clouds).

17.4   Surfaces

Surfaces are by far the most important entities for human vision. Almost all objects of immediate interest for survival are opaque and bounded by surfaces that scatter the radiation impinging upon them. What arrives at the eye are rays due to surface scattering. The patterns of such scattered rays are the bases for the visual "perception" of various important object properties, such as distance, spatial attitude, and physicochemical composition.

17.4.1 Shading, vignetting, reflexes

The basic fact of optical life is that typical objects are opaque. If you see their frontal side, their backsides are hidden; if one object is in front of another, the more distant one is hidden from sight because it is occluded.

This works not only for "visual rays," but just as well for "illuminating rays." One side of objects is illuminated by the sun, the other side not, because it is part of the "body shadow". Parts of objects may not be illuminated by the sun because of distant objects that cast a shadow on them.

These effects can also occur partially if the light source is an extended one. Thus, depressions are often dark because regions in their interior see only part of the source. The effects are particularly important if the main source is very diffuse, like the overcast sky. Such effects are often confused with shading, which occurs also for collimated beams, but it is of a different nature and often known as "vignetting."

Notice that objects not only take away illuminating rays, but that they also scatter rays into other directions, thus becoming secondary light sources themselves. In a typical scene, *everything* acts as a source and nothing is independent of anything else. In this respect, the "light field" is much like Indra's net.[18]

The optical interactions in a generic scene are so complicated that nobody can take all these interactions into account. The only way to have some inkling of what is going on is to have a basic understanding of the various possibilities. Thus you conceptually divide the optical interactions into weakly interacting, almost autonomous parts, and you try to understand these individually. This divide-and-conquer attitude is the only way to deal with real life objects and retain your sanity. This is how artists deal with the complications of the light field, and it is also how illumination engineers have to deal with it.

I will simply discuss some of the major factors in a summary way here.

Rays

For these discussions I need to specify somewhat more precisely what is meant by a ray. Given a cameralike optical system, a ray can be defined as a slender beam with the entrance pupil of the objective (or "lens," but typically a photographic objective will be composed of many lenses) as base area and a solid angle defined by the area of a photosensitive element as seen from the center of the back nodal point. The direction of the beam is defined by the camera geometry in the usual manner. In certain cases it may be more useful to substitute the area of the image of a point source for the area of a photosensitive element. It depends upon the sampling density in the focal plane. The idea remains the same.

Thus a ray is not just a straight line, but a (slender) beam that contains an infinity of lines. Thus it is quite okay for a ray to split, as it has to do in the case of Fresnel refraction/reflection at a planar interface, or in the case of scattering by a rough surface. I will also consider the merging of rays as when you see a superposition of a reflection and a transparent image in a window pane. In both geometrical and wave optics you can develop methods to handle such "thick" rays effectively.

Since the rays are indeed thick, it is conceivable that an object would fit into a ray. In such a case, the object would not be "resolved."[19] This depends on distance. Standing up, you can just about resolve a grain of sand at your feet, at ten times your body length you resolve just about a pebble, and at a hundred times your body length you just about resolve head-sized rocks. This is important, for it implies that "material properties" depend upon distance. A "property" is due to an average over what can just be resolved. For a distant tree that may involve many leaves. Such a tree is made up of "foliage stuff." What counts as a "material property" depends on the resolution at the given viewing distance. This changes the standard account from physical optics decisively.

A ray that arrives at the entrance pupil may have a very complicated history. Its spectral composition will depend on all optical interactions it encountered on its way. Each time the radiant power spectrum is multiplied by the bidirectional reflectance distribution function of the current surface. This is important for the discussion of color.

A pencil of rays carries spatial information. For instance, if the pencil was scattered from a surface just before hitting your eye, the pencil writes an "image" of that surface on your retina. Apart from issues of resolution, this image is topologically equivalent to the pattern of scattering on the actual surface. The rays in the pencil will have various histories, which are largely "erased" by the final surface scattering. Almost the only trace of these "histories" that remains is in the spectral composition, although you only receive a (geometrical rather than arithmetic) average over the various histories of rays that merged (in the surface scattering process) into the ray that finally hit your eye.

Occlusion

Occlusion is due to the fact that most objects are opaque, that is to say, when a ray hits their outer boundary it loses its identity and becomes virtually useless as a source of information concerning its source. The "source" could be either another object, or an actual "primary" source. The fate of a ray on its way from a primary source toward its final destination, your eye, can be arbitrarily complicated. It may have been scattered by many surfaces.

What you eventually see is due to the final surface scattering, though. The result is that you cannot see surfaces that cannot scatter rays directly toward your eye. Such surfaces are occluded by objects in front of them.

Occlusion is defined relative to a point of view. This need not be your eye. It could be the sun for instance. Objects for which the sun is occluded (that "cannot see the sun") are not irradiated, but are in the cast shadow of the occluder. The occluder can be arbitrarily distant as in an eclipse. More generally, you can think of a Boolean function of pairs of points that would be TRUE if the points could mutually see each other and FALSE otherwise. The complexity of this function for a natural scene is mind-boggling. Yet it is a most important function, since it sums up the relevant geometry. Together with the material properties and source distribution, it determines the radiance.

Remember that the radiance describes all there is to see from any view point in any direction. The radiance is just the plenoptic function.[20]

Shading

With shading one indicates the fact that surfaces receive different irradiances depending on their spatial attitudes, even in cases where the source is not occluded.

The paradigmatic case is that of a planar surface element facing an impinging uniform collimated beam. Sunlight is a good enough approximation in most cases. The irradiance of the surface is proportional to the cosine of the inclination of the (inward) surface normal with respect to the beam direction. This is known as Lambert's cosine law.

The cosine law of shading works just as well for diffuse beams, as long as the surface element can "see" all of the (extended) source. In such cases you can always construct an equivalent collimated beam.

Vignetting

Although Lambert's cosine law works for diffuse beams if the surface can "see" all of the extended source, this is often of little use. For really extended sources, such as the overcast sky, it is almost certain for a surface element to *not* see part of the source. For an extended object with a boundary surface that contains surface elements of many different spatial attitudes, this means that almost any point of the surface is essentially irradiated by its unique (effective) source. In such cases, shading in the proper sense is not the major source of irradiance variations. One says that these irradiance variations are due to vignetting.

In typical cases, vignetting defines the effective source (the part of the source as seen from the surface element). This effective source can be replaced with an equivalent collimated source. Then Lambert's cosine law is used to find the irradiance. Thus the irradiance is always due to a mixture of vignetting and shading.

Vignetting is tightly bound up with surface shape. For instance, a deep depression in a surface is likely to receive less irradiance than a high protrusion, quite independent of the precise source distribution. Thus, the eye sockets in a face are generally darker than the nose or cheekbones.

Reflexes

Painters refer to a "reflex" in cases where an object is clearly seen to act as a secondary radiator that happens to "throw a reflex" on a nearby object. Such cases are valued because they allow the painter to "pull the scene together" by painting mutual dependences of otherwise disjunct entities.

Reflexes are the visually more apparent effects of the general distribution of radiant power over a scene. (Figure 17.13.) You can set up a balance equation for radiant power; it is a linear equation. The fact that any surface element in a scene receives (directly or indirectly) radiation from essentially any other surface element is expressed as a Fredholm differential equation of the first kind. Such equations can be neither explicitly written down nor solved in any interesting situation. Approximate solutions are numerically obtained via ray tracing methods in computer graphics.

Because of the linearity, you may find representations in which the elements (linear combinations of surface elements) are independent. Such "modes" are characterized by

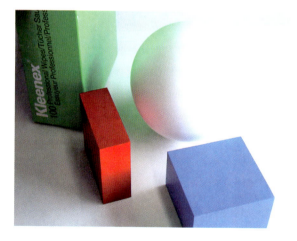

Figure 17.13 Notice the reflexes of the surrounding objects on the white sphere.

(pseudo) "reflectances."[21] The modes are descriptions of the scene geometry (the occlusion function plays an important role) that are especially fit for radiometry. Modes with particularly high pseudo-reflectances often dominate, with the result that the irradiance distribution over the scene may be dominated by the geometry, rather than the precise source distribution. Examples of this can be detected in many paintings.

Roughness texture

It should not be thought that the irradiance suffices to describe the radiometric structure. The actual source distribution matters if the BRDF significantly differs from the Lambertian one. Moreover, most surfaces are rough on scales that can be resolved, leading to *texture*. You can often see this in rough plastered walls, which rarely appear uniformly white. The wall texture depends critically on the irradiation. It is quite different from "wallpaper texture" in that respect.

In the rough structure of a surface, you encounter vignetting, shading, and reflexes on a small scale. Thus the formal description of roughness induced texture is rather complicated. It is important in colorimetry because it makes that you have to deal with local gamuts, or "extended colors," rather than colors of uniform patches.

17.4.2 Fresnel reflection and refraction

In the simplest case you deal with perfectly transparent dielectric media (air, water, glass, and so on).

When a ray passes a planar interface between two such media with different refractive indexes, it changes direction. The change is described by Snell's formula

$$n_i \sin \vartheta_i = n_t \sin \vartheta_t,$$

where n_i denotes the refractive index at the side of incidence and n_t the refractive index at the side of transmission. The angles $\vartheta_{i,t}$ measure the inclination of the ray with the normal at the interface.

Because of the refraction, the étendue of the beam changes at the interface. Of course this implies that the radiance changes, too.[22] In a "mechanical" interpretation this is understood because the refractive index governs the *speed* of the "light particles."

In cases of total internal reflection there is no transmitted ray if the angle of incidence exceeds the so called "critical angle ϑ_c."[23]

For unpolarized radiation the reflectance depends on the direction of incidence.[24] In the case of normal incidence, the simple limiting form is

$$R_\perp = \left(\frac{n_i - n_t}{n_i + n_t} \right)^2,$$

whereas for $\vartheta > \vartheta_c$ you have $R(\vartheta) = 1$ (total internal reflection). The transmission is simply $T = 1 - R$ in all cases. These are known as Fresnel's formulas.

The reflectance depends strongly upon the refractive index and on the angle of incidence (figure 17.14). With some experience you can estimate the refractive index of a substance from its gloss. For instance, diamond is very "glossy" (high refractive index) as compared to glass (intermediate refactive index), whereas a "glassy gloss" is much glossier than a "fatty gloss" (a fat, such as butter, has a low refractive index). Notice that all media have a high reflectance for almost raking illumination, though.

When the media are not transparent but also absorb radiation, the refractive index gains an imaginary component.[25] For normal incidence from air to an absorbing medium you have in the limit for very high absorption κ

$$\lim_{\kappa \to \infty} R = 1,$$

that is, a perfect mirror. This is often seen in highly absorbing dielectra. For instance, dried red ink looks specular greenish. An extreme case are the metals that indeed closely approximate perfect mirrors.

17.4.3 The bidirectional reflectance distribution function

Real world "surfaces" are indeed surfaces, but at the same time *not* surfaces, that is to say, very unlike mathematical "(smooth) surfaces." Just to drive the idea home, consider the "surface" of a treetop. From nearby you see twigs, leaves, and air, while from a distance the "treetop" is an opaque solid (for instance, many oaks have roughly hemispherical treetops),

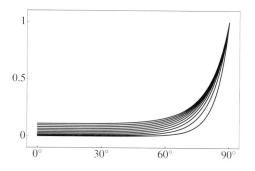 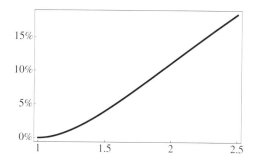

Figure 17.14 Left: The reflectance of the interface of a clear medium with air for various angles of incidence. The curves are for refractive indexes that about span the scale of substances you might encounter in real life. Right: The reflectance for normal incidence for the interface of a clear medium with air.

apparently bounded by some surface or other. Evidently, *scale* is a key issue here. For the physicist and chemist, the only scale of relevance here is the *molecular scale*. On this scale the wavelength of radiation in the visual region is huge, thus *any* "pure" material is homogeneous. In those sciences one only considers (true) planar interfaces; thus everything is simple. Fresnel's equations tell all there is to tell.

In real life the molecular scale is irrelevant. If the molecular scale indeed counts, then the bulk matter is perfectly homogeneous. The simplest kind of phenomenological physics is all you need. But such cases are very rare. In real life you have nontrivial structure on many scales, from the molecular scale to that of many miles (as far as you can see). Moreover, the "scale" is not *absolute* as in physics, where you gauge scale against the molecular dimension or the wavelength of radiation, but *relative*—that is, scale is measured with respect to visual resolution, and thus depends (upon more) on viewing distance.

It makes sense to consider a rough distinction like the following:

the submicro-scale is the scale at which bulk properties of the stuff are determined. This may be the molecular scale, but usually it isn't. Think of milk. Its bulk properties are determined by oil droplets (themselves invisible) in an aqueous medium, or paper, its bulk properties determined by clear cellulose fibers in air;

the micro-scale is the scale at which surface properties below the level of resolution interact with the incident radiation. Think of a distant treetop where you are just about unable to make out the leaves. The optical interaction of the radiation at the scale of the leaves determines the "surface properties";

the macro-scale is the scale just above resolution. Modulations at this scale cause visual *texture*. Texture is in between a "surface property" and such a thing as "surface shape";

the mega-scale is the scale at which the surface manifests itself as "surface," that is, as having a well-defined spatial attitude (tangent plane), curvature, and so forth.

In the "surface approximation" an optical system is considered as a system with a well-defined tangent plane, and such that a ray that impinges on the surface is either absorbed or scattered *from the same point* into some arbitrary direction. This is different from a "thick" optical system in which a ray can enter at one point and exit at another point (after frolicking about in the internal structure). The "surface approximation" applies only to a given scale. At finer scales the "surface" is likely to be a "thick" optical system. The surface of a treetop serves again as a perfect example.

The scattering in the surface approximation is conventionally described via the bidirectional reflectance distribution function (BRDF). The BRDF is the ratio of the scattered radiance in the exit direction to the irradiance caused by an incident collimated beam from the direction of incidence. The BRDF is thus a function of two directions. These directions are measured with respect to the surface. For an isotropic surface, it is sufficient to specify the angles subtended with the surface normal and the angle between the incident and the scattered beams (the so-called phase angle, a term that derives from the lunar cycle). For a general, anisotropic surface the directions of incidence and scattering must also be referred to a fiducial direction in the plane of the surface. Thus, you require three parameters in the isotropic and four parameters in the general case.

The BRDF is a phenomenological description, and there are hardly any constraints on its form. The only constraints are

— the BRDF is nonnegative throughout;

— the BRDF is "conservative," such that the total scattered radiant flux does not exceed the total incident radiant flux;

— the BRDF is invariant against permutation of the directions of incidence and scattering.

The latter constraint is known as "Helmholtz reciprocity" and is due to the fact that geometrical optics is invariant against the reversal of rays.

The simplest BRDF is the white Lambertian BRDF. It is constant, and has a value of $1/\pi$. The incident flux is the area times the irradiance. The scattered flux is $1/\pi$ (the BRDF) times π times the area (the étendue of the scattered beam); thus, all of the radiant power of the incident beam ends up in the scattered beam. People say that "the albedo is one." A gray Lambertian BRDF of albedo α is α/π. The albedo (literally "whiteness") measures the fraction of the incident radiation that is remitted by the surface.

Many different forms of BRDF are encountered in nature (see figures 17.15 and 17.16). Most naturally occurring BRDFs can be approximated by the superposition of a small number of "scattering lobes." The important scattering lobes are

the diffuse lobe is a broad mode in the direction of the surface normal. The Lambertian BRDF is the generic example, but the diffuse lobe is typically somewhat narrower than the Lambertian one. Most surfaces scatter rather less radiant power at large angles with the normal than the Lambertian lobe would suggest. The diffuse lobe is usually due to scattering from within the bulk matter (as in paper), but may also be due to surface roughness (as in brushed metal);

the backscatter lobe is a broad mode in the opposite direction of the incident beam. Thus the radiation is returned toward the direction from which it came. Backscattering is common in the case of rough surfaces such as plaster, grass, and so forth.

Figure 17.15 The three spheres have very different BRDFs. Notice that you see the environment "mirrored" in one sphere, and somewhat visible in another one (though *very* blurred) and that the white sphere merely reveals the overall direction of the light field.

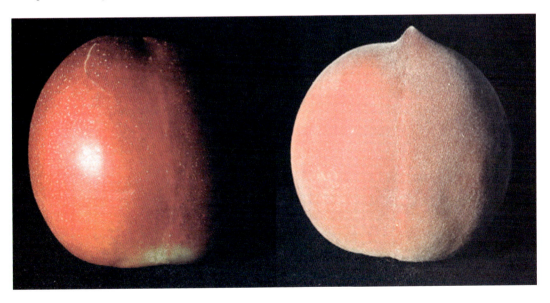

Figure 17.16 These two fruits have roughly the same color but very different BRDFs.

The explanation is a simple geometrical one. If a ray can enter a crevasse, then the odds are that the only way out is by the direction from which it entered. Many other effects may contribute to backscattering. Examples are the "cat's eye" retroreflectors used in traffic markings;

the specular lobe is a mode that is roughly in the direction of the mirror reflection. This lobe may be narrow or broad as the case may be. Often it is composed of a (sometimes large) number of slightly offset directional lobes. Examples occur in the case of orange peels The "specularity" is usually composed of a number of smaller specularities;

the asperity lobe is the lobe due to "asperity scattering." Asperity scattering is usually due to tiny hairs at right angles to the surface (possibly below resolution), but powder or actual asperities have much the same effect. Obvious examples are peaches, velvet, and so forth.

Although this by no means exhausts the possibilities, the large majority of naturally occurring surfaces can be described reasonably well through some combination of these basic processes.

17.4.4   Volume scattering and absorption

As a ray enters bulk matter it typically suffers many scattering events within the volume until it emerges from the surface again. That is to say, if it does emerge at all, for often the ray will be absorbed before it has a change to escape. The rays that do escape from the surface tend to be heavily "shuffled" and have fully forgotten their past when they emerge. The direction of exit is probabilistic and the probability density is determined by the nature of the interface and hardly depends upon the direction of incidence.

The important phenomena are the interface, which determines the processes of entrance into and exit from the bulk matter, and the scattering within the bulk of the stuff. The main dichotomy (figure 17.17) is in these cases of

a cloud of scatterers in air Examples include paper, which consists of cellulose fibers in air and many woven fabrics and felts. For white paper or fabrics, the scatterers are perfectly clear and colorless fibers. They act as scatterers because their refractive index differs from that of the air.

a cloud of scatterers in an optically dense medium Examples include milk, which consists of oil droplets in water, and oil paint, which consists of pigment grains in a dried oil. The oil droplets are perfectly clear and colorless, and so are the pigment

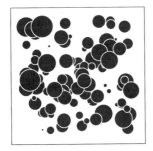 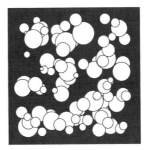

Figure 17.17 Optically dense particles in air (water droplets) and bubbles (air) in an optically dense medium (water) have much the same effect. Both (in sufficiently dense concentrations) lead to "white stuff" (clouds, foam on beer).

grains in a white paint. They act as scatterers because their refractive index differs from that of the oil or water. In "white water" you have the opposite case: the scatterers (air bubbles) have a *lower* index than the medium (water).

The difference is in the interface with the air. In the former case there is no interface, but in the latter case you have to reckon with Fresnel reflection and refraction on entrance and—even more importantly—total internal reflection on the exit of the beam. The precise nature of the processes at the interface depends on the nature of the surface and may be influenced by a variety of surface treatment (for example, think of the effect of varnishing).

If the medium is selectively absorbing and/or the scatterers have a spectrally selective scattering cross-section, you obtain substances that are "colored in the bulk." If Fresnel reflection at the interface is important, you will obtain a specular lobe that has roughly the color of the illuminant and a diffuse lobe that shows the bulk color. This is often referred to as "the dichromatic model," and you will indeed come across cases regularly as you study distributions in RGB space due to the illumination of glossy dielectric surfaces.[26]

17.4.5 Absorption

A transparent medium may show selective spectral absorption and thus shows color when you look through it at a light source behind it. A glass of red wine is an example. In order to judge the color you hold your glass against a light source (for instance, a candle in a romantic setting).

The color you see obviously depends upon the absorption characteristics, but it also depends very strongly upon the thickness of the layer. The radiance of the transmitted beam decreases exponentially with a wavelength depending characteristic length. This is known as Beer's Law. As a consequence, the chromaticity of the transmitted beam depends upon the thickness of the layer in a very nonlinear way. As you change the thickness of the layer, you thus see strong and sometimes unexpected changes of chromaticity (figure 17.18).

An example is shown in greater detail in figures 17.19 and 17.20. In figure 17.19, I show the Ostwald white, black, and color contents as a function of the thickness of the layer. Notice that the hue also changes. For thin layers it starts out orange-yellow, whereas for very thick layers it moves toward pure red. This example was computed in RGB. It is easy to show that in this case you always end up at one of the primary colors as the layer becomes very thick. For a true spectral calculation you likewise show that, in the limit for very thick layers, the color becomes monochromatic. In the chromaticity diagram (figure 17.20) the trajectory (chromaticity as a function of layer thickness) is markedly curved. These orbits always move from the achromatic point to one of the vertices. The color content is obviously low for either very thin layers (they look white) or very thick layers (they look black). Somewhere on the orbit the color content reaches a maximum (figures 17.19 and 17.20 right); thus there is a "best" thickness if you want to produce a colored transparent layer (for instance, a varnish, a colored window pane, a nail polish, and so forth).

Figure 17.18 Changing the layer thickness may change the color of an absorbing layer in unexpected ways. This is easily understood by straightforward application of Beer's law, but you need to know the precise absorption spectrum for that.

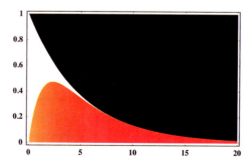

Figure 17.19 Ostwald decomposition of the color of a beam passed by a certain absorbing medium as a function of the thickness of the absorbing layer.

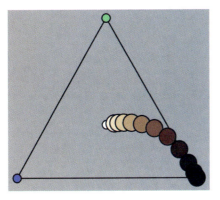 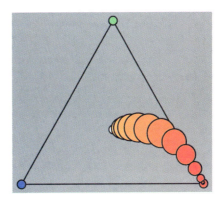

Figure 17.20 The data from figure 17.19 plotted in the chromaticity diagrams. In the figure at right the color content is shown by pure hue and by size; at left, the actual colors are printed.

17.4.6 The colors of thin layers

The colors of thin layers are very important in daily life. Many important things are actually composed of thin layers; think of paper, clothes, and the like. Other important things have been expressly covered with particular thin layers, think of paints, varnishes, and so forth. Even chunks of bulk matter can often be treated as being covered with thin layers because the optical interactions are confined to only a small depth below the surface. In a great many cases this optically important outer "skin" differs in (optically) important ways from the bulk matter, due to weathering, oxidation, wetting, and so forth. Thus the "colors of thin layers" are everywhere (figure 17.21).

A major distinction is that of "optically thick" thin layers and "optically thin" thin layers. Notice the oddness of these expressions. A piece of paper is generally considered a "thin" layer. Here the measure is simply the human scale. Yet the paper may be said to be

Figure 17.21 As you gradually add milk to black coffee, the color changes from black over brown to creamy white. Such effects can easily be described via the theory explained here. These "coffee colors" were calculated through the expressions explained in the text.

"optically thick." This is the case when the paper fully hides any substrate onto which you happen to put it. Thus "optically thick" is synonymous with "good hiding power." This "hiding power" is measured by the degree to which the contrast of a white-black transition in a substrate is attenuated when you put the layer over it. Good hiding power is desirable for paints, paper (you don't want to be able to read through the page of a book), underwear, and so forth. Minimal hiding power is desirable in other cases, for instance, in varnishes and sexy lingerie. A varnish is often used to color a substrate (for instance, in violins) but it is not supposed to hide it.

Thus the color of a varnish must be due to absorption (Beer's law). Hiding power is due to scattering. In order to obtain good hiding power in a thin layer, the scattering has to be very strong. This requires a large difference in the refractive indexes of the medium and the scattering particles. Anything can scatter when embedded in a medium of very different refractive index. For instance, air can become an efficient scatterer when present as a small bubble in water. In many cases of practical interest the scattering particles have an index that is higher than that of the medium, for instance, in a white oil paint the oil has a fairly high refractive index and you need scatterers of the highest index you can find. Titanium-dioxide crystals are commonly used, their refractive index being extremely high.[27]

It makes a huge difference how you happen to look at the layer. By way of a demonstration, I added droplets of milk and red ink to a glass of clear water (figures 17.22 and 17.23). The clear oil droplets in the milk act as efficient scatterers. In high concentrations (the stuff you drink) milk is almost opaque and very white (the oil droplets are almost pure

Figure 17.22 Left: A glass of water into which droplets of milk are added is shown against a light background. (Glass on the right has been stirred.) Right: A glass of water into which droplets of milk are added is frontally illuminated and shown against a dark background. (Glass on the right has been stirred.)

Figure 17.23 A glass of water into which droplets of red ink are added against light and dark backgrounds. (Glass on the right in front of the light background has been stirred.)

scatterers and do not absorb radiation). Red ink, on the other hand, is a clear liquid that does (spectrally selectively) absorb radiation, but does not appreciably scatter radiation. If the milk is placed against a light background, it doesn't look "milky" at all. The reason is the strong extinction due to the scattering that attenuates the beam that travels from the background to your eye. As a result, the milk looks like "mud." A dark background and frontal illumination make the milk look "milky." The reason is again evident, in this case you see the radiation that is scattered by the milk.

The case of the red ink is very different. Seen against a dark background and frontally illuminated, the red ink is all but invisible. The reason is that the ink doesn't scatter radiation, so nothing arrives at your eye. In this case a light background saves the day in that the red ink looks exactly as expected, that is to say, very red. The reason is the absorption of all but the long wavelengths from the beam that travels from the background to your eye. This gives the ink something to work on.

Simple as these observations are, they are very basic and important. The lesson is simple enough. A thin layer of scatterers has only any visual effect as seen against a dark background and when frontally illuminated, whereas a thin absorbing layer has only visual effect when back lighted or frontally illuminated and seen against a white background. If the scattering layer becomes very thick, its apparent color will move toward white. If the absorbing layer becomes very thick, its apparent color will move toward black.

Such rules of thumb are known to any photographer of materials and can be traced to the implementation of colored layers throughout the living world.

It is rather complicated to treat the physics of turbid media in an exact manner. Mostly one uses a variety of approximations. In the case of thin layers, the "two flux" model is commonly used. You treat the diffuse beam as composed of two parts, one beam traveling "up" though the layer, the other beam traveling "down." As they propagate, the beams are attenuated through absorption and extinction. In the case of extinction, a ray is scattered out of the beam. In this case the ray turns up into the other beam; thus the beams also *gain* rays as they propagate.

The bookkeeping is easy enough. You write two coupled first-order ordinary differential equations to describe the balance of radiant power.[28] The equations contain the location in the layer as parameter and depend on a pair of constants K and S.

The coefficient K denotes the absorption per unit thickness of the layer, whereas the coefficient S similarly denotes the scattering power. A layer is also characterized by its thickness d, often occurring in the combination Sd known as the "turbidity" of the layer.

This simple model was proposed for the description of planetary atmospheres by Schuster in 1909. In the 1940s the model was popularized by Kubelka and Munk, who proposed standard solutions in the form of hyperbolic functions.[29] The model is generally known as the Kubelka-Munk two-flux model. The model is sufficient to handle many problems involving thin layers in a quantitative manner. It is commonly used in the paper and paint industries. Various variations may be found in the literature, but for most purposes the Kubelka-Munk model suffices.

For an "infinitely thick" layer, in practice a layer that is opaque, you measure a reflectance R_∞. It can be shown that

$$\frac{K}{S} = \frac{(R_\infty - 1)^2}{2R_\infty},$$

which is a pure material property. In its dependence upon wavelength, it is known as the "spectral signature" of the material.

The "turbidity" Sd can be measured[30] by finding the reflectance R of the layer on a white background and comparing it with the reflectance of an infinitely thick layer R_∞. The turbidity is apparently large when $R \approx R_\infty$, because this means that the layer "hides" the white substrate.

Thus all phenomenological material properties, the turbidity and the spectral signature, have simple operational meanings. In terms of these the reflectance of a turbid layer on any substrate of reflectance R_g can be calculated[31] The ratio R_0/R_∞ is called the "opacity." Here R_∞ is the reflectance of an infinitely thick layer, whereas R_0 is the reflectance of the layer on a black background ($R_g = 0$). The opacity is also called "hiding power" and is evidently important if you want to hide something from sight. Thus paint layers, book paper, and underwear depend upon high opacity.

This theory suffices to describe the colors of many thin layers to some reasonable approximation. Applications include the colors of eyes, (many) flowers and (some) feathers (figure 17.24).

17.4.7 White stuff

What is common to polar bears, paper, underwear, foam on beer, teeth, clouds, hair of old people, and sandy beaches, apart from the fact that they are all *white*? In physics and chemistry texts on materials you won't find chapters on "white matter."

What these materials hold in common is that they are all turbid and that scattering far exceeds absorption. They are turbid layers with $K \approx 0$, $Sd \gg 1$ in the Kubelka-Munk description.

The actual mechanisms responsible for the high turbidity are widely different for these materials. Yet they look very similar, and they do so for very similar optical reasons. This illustrates the fact that "ecological optics" is not simply a subset of "optics." In the optics of materials, the standard textbooks distinguish dielectrics and conductors (metals). In a textbook on "ecological optics" (no such texts are available right now) you would certainly add a chapter on "white stuff," though.

17.5 Photonic Structures

So called structural colors (recently often referred to as "photonic colors") are due to multilocal, coherent interactions with radiation on scales of the order of a wavelength. Thus, the simple "physical colors," such as the colors of the rainbow (local, essentially single scattering) are not "structural." The simplest examples in anorganic nature include the colors of opals, which are periodic structures of silica spheres. Structural colors are common in the animal kingdom, especially in insects, birds, and fishes.

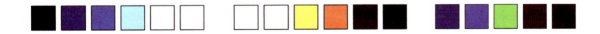

Figure 17.24 Blue eyes are due to a teneous layer of Rayleigh scatterers on a dark (melanin) ground. Yellow flowers are due to an absorbing layer (carotene) in front of a light background (cells with air spaces yield a white petal). A green parrot feather is a combination of both Rayleigh scattering and absorption (in front of the scattering layer). In these series I show the influence of layer thickness.

Photonic band gap structures imply nanotechnology. Such periodic nano structures of surprising complexity are revealed through electron microscopy of bird feathers, fish scales, and insect scales (butterfly wings) and carapaces. Some of the most striking colors to be found in nature are of this type. The BRDF of these materials is very non-Lambertian; thus the colors may depend strongly on the illumination and viewing geometry, leading to iridescence or metallic sheen.[32] Such colors are not at all represented in the conventional color atlases, and the standard theory of object colors is insufficient as a description.

17.6 Inhomogeneous Substances

Many important "materials" from daily life are not recognized as such (as "materials" that is) in the textbooks of physics. Examples include wood, granite, milk, blood, bone, marble, and sand. For the physicists, these are "composite materials" or "inhomogeneous mixtures." Yet these are definitely "bulk materials" from the perspective of ecological physics, neither composites nor mixtures.

The point is that the physicists treat bulk matter on the atomic or molecular scale. The human observer treats bulk matter on a scale that is even coarser than that of visual resolution. Materials like blood, milk, or bone are evidently "homogeneous bulk matter," because you need a microscope in order to see their composite structure. But materials like wood, sand, or marble clearly reveal some structure when closely scrutinized. Yet they are easily recognized in any reasonable (that is clearly visible) size. Thus they count as "homogeneous bulk materials" because they "have no parts." As you cut them in half either side is like the original chunk. Here you escape the sorites paradox[33] because you simply cannot cut things into arbitrarily small pieces.

The case of granite is subtly different. Of course, you see at first sight (no scrutiny needed here) that the stone is composed of various minerals, easily distinguished by their colors. Yet you don't doubt that a variety of sculptures or gravestones could be cut from a large block, though. If you cut a block of granite into two, you obtain two pieces of granite. In that sense, granite is no doubt a "homogeneous bulk material," and that is indeed how it is marketed. However, in the case of a coarse granite, you may doubt the "homogeneity" of pieces that are small, but still easily handled, pebble-size, say. In a small piece, the granite become a composite because the constituents assume an individual existence. "Here is the white-bluish quartz crystal, there the pink feldspar, here the silvery mica flake," and so forth. In a large block the constituents have no individual existence; when you talk of them you use the plural. In "the feldspar in this granite has a particularly attractive hue," you talk about all the visibly feldspar crystals, not a particular one. Thus granite is a homogeneous material in large chunks, a composite in small pieces. This is indeed a common enough phenomenon. Many building materials, fabrics and printed papers fall in this category.

Parts become individuals if you can count them, and the human condition is such that you (visually) lose count after about five to eight. If you see about seven parts, then the block of stuff will contain about $7^{(3/2)} \approx 20$ distinct grains. Thus a chunk that contains over a hundred grains (in the volume) remains "homogeneous" on cutting it into two parts, whereas a piece containing two dozen grains will in all likelihood yield two pieces that would be considered "composites." (See figure 17.25.)

This is important in the case of color. You can certainly speak of the "color of a granite" (although it will be an extended rather than a simple color), but a small piece is made up of parts of different colors. Notice that you don't necessarily need hammer and chisel to produce pieces. You can do the same (with less effort) by directing your visual attention.

17.7 Why Color Vision Works as Well as It Does

Why does color vision works as well as it does (for us humans, I mean)? After all it only offers a glimpse of a very narrow part of the electromagnetic spectrum (less than an octave!) and even that glimpse is not very detailed since the visual region is effectively sampled in only three bins (think of the Schopenhauer parts of daylight representation). As spectroscopic analysis systems go, the visual system is hardly worth a second look. Yet I'm sure most people are quite happy with their visual systems. I certainly am. Just think of a meadow with various wild flowers in bloom—who would be ready to give up that experience? Apparently this lousy attempt at spectroscopy serves us very well.

Figure 17.25 The object at left is evidently inhomogeneous (put together from two parts, each part of a different "stuff"), the object at right is a structured, but "homogeneous" substance (made of a single "stuff"). The center object is an ambiguous case.

There obviously are various natural constraints that serve to roughly determine bounds on a useful "visual range" of the electromagnetic spectrum. The temperature of our main illuminant, the sun, and the filtering by the earth's atmosphere yield bounds that are somewhat broader, but indeed include the visual range. Of the radiation "skipped" by humans, the ultraviolet part interacts destructively with our biological tissues, and the infrared part fails to be very useful because of our body temperature. Yet insects exploit the near ultraviolet and pit vipers the infrared. But the use of these spectral ranges is unlike that of simply enjoying the scene in front of you. Insects mainly use the ultraviolet for navigational purposes, detecting the polarization pattern over large expanses of sky, whereas pit vipers use heat radation to locate (resolution doesn't allow them to "see" detail) warm-blooded animals (mosty small rodents) in the cold nights of the desert. The color vision system of humans is neatly tuned to the range of chemical binding energies. That is to say, the spectral analysis allows inferences concerning chemical constitution and change. Just think of the distinction between unripe, ripe, and rotten apples, for instance.

A major factor in rendering the human color vision system as useful as it is, is that typical illuminant (daylight) spectra are fairly smooth and typical reflection spectra somewhat articulated, but not too much. The most vivid colors that appear to us are the Ostwald full colors, the Goethe edge colors, and the Schrödinger optimal colors. (Of course, the former two are special cases of the latter one.) These colors are due to well-determined spectra. Because the spectral reflectance factor is limited to the range zero to one, it is not possible to define a highly articulated (many high-amplitude wriggles) spectrum that has a vivid color. If something has a vivid color, it *must* have a simple spectrum, close to that of an optimal color. So this is why the system works as well as it does:

— Naturally occurring spectra are often not too different from optimal colors (if they are, they are likely to look grayish);

— There is a one-to-one relation between optimal color hues and reflectance spectra.

The first fact means that you indeed encounter many vivid colors at all in the world; the second means that such vivid colors are actually quite good indicators of spectra. Thus vivid colors indicate physical properties. Color vision is really a form of "coarse grained spectroscopy"!

These facts of ecological physics are typically ignored or downplayed in discussions of color vision. For some reason, the emphasis tends to be on metamerism and the idea that "color" is not a certain indicator of spectral structure (in many such discussions the author actually says "wavelength"). This misses the boat entirely. Metamerism is a fact, but so is the fact that the spectra of the most vivid colors are quite well determined. Metamerism makes that one can say very little concerning the spectral articulation of grayish colors. In principle, these spectra could be almost arbitrarily complicated. But such cases are very rare (though not impossible, rare earth glass powders being a case in point). In real life, the assumption that grayish colors are due to flattish spectra is overwhelmingly likely to apply.

Thus color vision is possible because the world (I mean the generic human environment) being what it is, the system is really an imperfect but quite dependable guide to reflectance spectra of various surfaces, and thus—through another level of indirection—their physical and chemical constitution. The pessimistic notion that colors are "mere mental paint" and have no relation the physical and chemical constitution of things at all is popular in science and (especially) philosophy, but it has no basis in fact.

A simple statistical model of reflectances generates "random telegraph waves," spectra toggling between zero and one. The number of transitions is typically limited, the probability falling off steeply with their number. Suppose the transition locations are uniformly distributed over the visual region. Then it is easy to show that the colors are not at all uniformly distributed over the color solid. Most colors will be white or black (no transitions), many will be Goethe edge colors (one transition), and some will be Schrödinger optimal colors (two transitions), whereas few will fall into the interior of the color solid. This means that most of the vivid dark colors will be red or blue, and most of the vivid light color cyan or yellow, simply because of the geometry of the edge color locus. Thus wavelength nonspecific statistics leads to pronounced hue preferences. The fact that some colors are more frequent than others is most likely the cause of the "focal colors" that have generated tons of literature.[34] Although thus obviously important in discussions of color vision, I fail to find such facts in the standard textbooks.

Exercises

1. The étendue [PARTLY TEDIOUS] Find the étendue of various beams, such as, the diffuse beam emitted by a plane, a circular disc, a spherical surface, and the beam defined by two circular apertures. For applications you will often need the étendue of planar polygonal shapes (think of room windows and so forth).

2. The physical meaning of the étendue [STRICTLY FOR PHYSICISTS] Interpret the étendue as a *phase volume*. This is best done in terms of the photon description. Make sure to incorporate the velocity (this takes care of media of various refractive indexes). Use Liouville's theorem of the conservation of phase volume (familiar from statistical mechanics) to derive the appropriate law of conservation of the étendue along a beam. This serves to define the very meaning of "étendue" as a characteristic descriptor of "beams."

3. Vignetting [SANATORY MEASURES] Consider a Lambertian planar source of infinite extent (coarse model of the overcast sky), illuminating a sphere. Find the irradiance over the surface of the sphere. Do the same for a sphere irradiated with a collimated, uniform beam of infinite extent (model for sun light) and a Ganzeld (model for "ambient light"). In computer graphics the first case is often called "point source at infinity with ambient component." Show that this is formally correct. Explain why it does not make physical sense. Give an example where the "point source at infinity with ambient component" concept leads to erroneous predictions (consider a local perturbation of sphericity).

4. Reflexes [FOR PHYSICISTS] Write down the equation for radiant power balance in the case of a scene made up from Lambertian surfaces of varying albedo. Formally decouple the modes and interpret the result. Apply to the case of the interior (uniform, Lambertian) surface of a sphere.

5. Fresnel reflection [DAILY LIFE EXAMPLE] What is the reflectance of a glass ($n = 1.5$) to air surface, of a windowpane (two parallel surfaces), of a stack of glass plates? How does this change if the illumination is by a diffuse instead of a collimated beam?

6. Beer's Law [SOMEWHAT COMPLICATED] Compute the color of a transparent layer against a light background (glass of red wine) as a function of the thickness of the layer. Plot in color space and chromaticity diagram.

7. Kubelka-Munk two-flux theory   [FOR PHYSICISTS] Consider models of blue eyes (layer of Rayleigh scatterers in front of a dark ground) and marigold flower petals (absorbing layer in front of a white background). What is the influence of layer thickness? A green parrot's feather is green by virtue of a combination of these effects. Set up a model and study it.

8. Statistics of object colors   [PRACTICAL AND CONCEPTUAL INTEREST] Consider "random telegraph wave" models of spectral reflection functions. For instance, let the probability density for a zero-to-one or one-to-zero transition along the wavelength axis be constant. Make sure that you obtain examples with zero, one, two, ... transitions. Study the distribution of the corresponding colors in the Schrödinger color solid (for average daylight, for instance). Show that this distribution is not uniform at all, even though there is no wavelength preference in the statistics of the spectral reflectances. Apparently some colors are more likely than others. Draw (speculative) conclusions, compare with the well known color naming results.

Chapter Notes

1. For instance, K, Nassau, *The Physics and Chemistry of Color: The fifteen causes of color*, (New York: Wiley, 1983) and K. McLaren, *The Colour Science of Dyes and Pigments*, (Bristol: Adam Hilger Ltd, 1983). Books that deal with animal or plant colors are of more interest, for instance H. M. Fox and G. Vevers, *The Nature of Animal Colors*, (London: Sidgwick and Jackson, 1960). On the ecological relevance of coloration, see H. B. Cott, *Adaptive Coloration in Animals*, (London: Methuen & Co. Ltd., 1940). Useful articles are J. B. Hutchings, "Colour and Appearance in Nature. Part I. Colour and Appearance of Photosynthetc Organisms," *Color Res.Appl.* 11, pp. 107–111 (1986); —, "Colour and Appearance in Nature. Part II. Colour and Appearance of Flowering Plants and Animals," *Color Res.Appl.* 11, pp. 112–118 (1986); —, "Colour and Appearance in Nature. Part I. Colour and Appearance of *Homo Sapiens*," *Color Res.Appl.* 11, pp. 119–124 (1986).

2. The "Poynting vector" and the "energy-momentum tensor" describe the flow of electromagnetic energy in spacetime. These entities are only defined in the Maxwell theory, they do not occur in geometrical optics.

3. Of course the manifold of rays is continuous; thus ray counting actually deals with a density of rays.

4. "Photons" describe the interaction of electromagnetic radiation with ponderable matter. It depends upon your perspective whether you consider an electromagnetic wave as a "swarm of photons," there is no pressing need to do so.

5. "Pencil" is indeed an apt term. A painter's pencil is made up of hairs in neat order ("combed," as it were). Since the hairs cannot interpenetrate each other, there is at most one hair at any point of space. The "pencil" of geometrical optics is a similar structure.

6. From the French *étendue géométrique*. English synonyms are throughput, acceptance, light-grasp, collecting power, and $A\Omega$-product.

7. The full derivation is

$$
\begin{aligned}
\mathcal{E}_{\text{beam}} &= \int_{\text{beam}} d\mathcal{E} \\
&= \int_{\text{beam}} d\mathbf{A} \cdot d\mathbf{\Omega} \\
&= \int_{\text{surface}} d\mathbf{A} \int_{\text{directions}} \mathbf{n} \cdot d\mathbf{\Omega} \\
&= A \int_{\text{directions}} \mathbf{n} \cdot d\mathbf{\Omega} \\
&= A \int_{\text{directions}} \mathbf{n} \cdot \mathbf{r} \, d\mathbf{r}
\end{aligned}
$$

$$
\begin{aligned}
&= 2\pi A \int_0^{\pi/2} \cos\vartheta \, \sin\vartheta d\vartheta \\
&= \pi A.
\end{aligned}
$$

8. The integrating sphere is often known after the engineer Richard Ulbricht (1849–1923). R. Ulbricht, *Das Kugelphotometer*, (München, 1920). See H. J. Helwig, "Über praktische Erfahrungen der neuen Messmethode für die Ulbrichtsche Kugel," *Das Licht* 5, pp. 33–34 (1935).

9. Known as "ambient light" in computer graphics.

10. It is not possible, nor useful, to carry the distinction between "primary" and "secondary" radiators through too strictly. Think of "moonlight," for instance. Strictly speaking the moon is a secondary radiator that scatters light from the sun, which is a primary radiator. However, in a night scene the sun's beam only appears as scattered by the moon, and it makes good sense to treat the moon as the primary radiator.

11. In most cases, the "scene in front of you" will have to be segmented into approximately "uniform parts," where the uniformity refers to the luminous atmosphere. Such a segmentation is by no means trivial and remains a tough problem in machine vision.

12. I ignore coherent sources, polarized sources, and so forth. I also ignore temporal aspects (flickering fluorescent tubes, flash tubes, and so forth).

13. E. R. Dixon, "Spectral distribution of Australian daylight," *J.Opt.Soc.Am.* 68, pp. 437–450 (1978).

14. "Extinction" and "absorption" should not be confused. When a medium scatters radiation out of a beam, that beam becomes attenuated by extinction. The scattered radiation is not lost though, it simply turns up elsewhere.

15. H. Koschmieder, "Theorie der horizontalen Sichtweite," *Beitr.Phys.freien Atm.* 12, pp. 33–53 (1924). The effect of the medium is usually even more pronounced in the submarine than in the terrestrial environment. See P. Emmerson and H. Ross, "Colour constancy with change of viewing distance under water," *Perception* 14, pp. 349–358 (1985).

16. G. Mie, "Beiträge zur Optik trüber Medien, speziell kolloidaler Metallösungen," *Ann.Phys.* 25, pp. 377–445 (1908). See also H. C. van de Hulst, *Light Scattering by Small Particles*, (New York: Dover, 1981).

17. In the case of single scattering, rays either pass the medium unscathed or are scattered out of the beam. This means that you can see sharply through the medium without a problem. In the case of multiple scattering, any ray that reaches you has been scattered several times, and its original direction has been lost. You cannot see distant objects sharply through such a medium. If the scatterers are really dense, the medium becomes translucent or even opaque.

18. "Indra's net" (or "Indra's pearls") metaphorically indicates the Buddhist "independent origination", interpenetration and emptiness. It is found in the scriptures of third century Mahayana. Everything is connected to everything else. The net of the Vedic god Indra is set with multifacetted jewels, each jewel reflecting all the others.

19. This notion was already obvious to Euclid. It is rather less ludicrous than it is often made out to be. In fact, it allows you to derive the modern expressions for the resolving power of telescopes or microscopes in a very simple manner. Of course, the "thickness" of the ray remains as an undefined "constant of nature." In the modern theories this constant is "explained" as the wavelength of radiation in the visual region. See J. J. Koenderink, "Different concepts of "ray" in optics: link between resolving power and radiometry," *Am.J.Phys.* 50, pp. 1012–1015 (1982).

20. E. H. Adelson, and J. R. Bergen, "The Plenoptic Function and the Elements of Early Vision," in *Computational Models of Visual Processing*, edited by M. Landy, and J. A. Movshon (Cambridge, MA: MIT Press, 1991), pp. 3–20.

21. I say "pseudo-reflectances" because such entities are of a formal nature and are less constrained than actual reflectances, for instance, they may have values exceeding unity. A simple example is the integrating sphere (Ulbricht sphere if you are from continental Europe). This is a very simple case because the kernel of the Fredholm equation is just a constant, so the calculation can be done in the margin or in your head. One pseudo-facet is the full area of the sphere with constant weight, the others are any distribution over the sphere with zero mean. If the sphere has reflectance ϱ the pseudo-reflectance of the first pseudo-facet is $\varrho/(1-\varrho)$, which tends to infinity when the true reflectance approaches unity! The physical meaning is simple enough. If you place a source of photons in a sphere with unit albedo the photon density will increase linearly with time. You are "pumping" photons in the sphere whereas none get removed. If the reflectance is somewhat less than unity the photon density increases until absorption removes as many photons per second as the source pumps in.

22. You have

$$\frac{N_1}{n_1^2} = \frac{N_2}{n_2^2},$$

for the change of radiance over an interface of two media of different optical density.

23. The critical angle is given by

$$\vartheta_c = \sin^{-1}\frac{n_t}{n_i} = \sin^{-1}\frac{1}{n} \text{ where } n = \frac{n_i}{n_t}.$$

24. The Fresnel reflectance for an unpolarized beam is

$$R(\vartheta_i) = \frac{1}{2}\left[\frac{\sin^2(\vartheta_i - \vartheta_t)}{\sin^2(\vartheta_i + \vartheta_t)} + \frac{\tan^2(\vartheta_i - \vartheta_t)}{\tan^2(\vartheta_i + \vartheta_t)}\right].$$

The reflection typically causess the reflected beam to be partially polarized.

25. For normal incidence from air to an absorbing medium, you obtain the Fresnel reflection factor

$$R = \|\frac{1 - (n_t + i\kappa_t)}{1 + (n_t + i\kappa_t)}\|^2.$$

The imaginary component captures the absorption.

26. See S. A. Shafer, "Using Color to Separate Reflection Components," *Color Res.Appl.* 10, pp. 210–218 (1985) , and G. J. Klinker, S. A. Shafer, and T. Kanade, "Color image analysis with an intrinsic reflection model," *Proc.Second Int.Conf. Comp.Vis.*, pp. 292–296 (1988).

27. The currently most used pigment for white paints is titanium dioxide (or titania). It occurs naturally and is usually named titanium white or Pigment White 6. Its refractive index is 2.4, which is unusually high. It is much higher than that of lead white, which has historically been the white pigment in common use. White lead or cerussite, which also occurs naturally, is lead carbonate (n=2.07).

28. The equations that describe the balance of radiation densities in the "upward" and "downward" directions are

$$\begin{aligned}\frac{dI}{dx} &= -(K+S)I + SJ, \\ \frac{dJ}{dx} &= -SI + (K+S)J,\end{aligned}$$

where x denotes the distance perpendicular to the layer, I the "up" traveling diffuse flux, and J the "down" traveling diffuse flux. The coefficient K denotes the absorption per unit thickness of the layer whereas the coefficient S similarly denotes the scattering power.

29. A. Schuster, *An Introduction to the Theory of Optics*, (London: Arnold, 1931); P. Kubelka and F. Munk, "Ein Beitrag zur Optik der Farbanstriche," *Z.Techn.Phys.* 12, pp. 593–601 (1931).

30. You have

$$Sd = \frac{R_\infty}{1 - R_\infty^2} \log\left[\frac{(1 - R_\infty)\left(R - \frac{1}{R_\infty}\right)}{(R - R_\infty)\left(1 - \frac{1}{R_\infty}\right)}\right].$$

31. For R you have the lengthy expression

$$\frac{\frac{1}{R_\infty}(R_g - R_\infty) - R_\infty\left(R_g - \frac{1}{R_\infty}\right)e^{Sd\left(\frac{1}{R_\infty} - R_\infty\right)}}{(R_g - R_\infty) - \left(R_g - \frac{1}{R_\infty}\right)e^{Sd\left(\frac{1}{R_\infty} - R_\infty\right)}}.$$

32. P. Vulkusic, and J. R. Sambles, "Photonic structures in biology," *Nature* 424, pp. 852–855 (2003); J. Zi, X. Yu, X. Hu, X Wang, X. Liu, and R. Fu, "Coloration strategies in peacock feathers," *P.N.A.S.* 100, pp. 12576–12578 (2003).

33. The "sorites paradox" is named after the Greek *soros* (heap). It refers to the "heap of sand" (obviously, to me that is the essence of "heapness") that remains a heap of sand when you take away a grain. The "paradox" refers to the dilemma that the "heap" certainly does not count as a heap any more when you are ready to take away the final few grains. Where to draw the line?

34. The notion of "basic color terms" is due to B. Berlin, and P. Kay, *Basic Color Terms: Their universality and evolution*, (Berkeley: Un. California Press, 1969).

Chapter 18

Beyond Colorimetry

I consider colorimetry to be "chapter zero" (or perhaps even "chapter minus one") of any conceivable color science. Whatever you study (scientifically) in the field of color, you will need colorimetry as part of the foundation. It is not possible to pursue color science (as a *science*) in any direction without a fair grasp of what colorimetry is about and what its empirical basis and formal structure are.

18.1 Taking Stock: Where Are You Now?

At this point you know (much) more about colorimetry than the average scientist. It is perhaps time to take stock and consider the vantage point you have reached.

Colorimetry is a form of coarse-grained spectroscopy with a mere three degrees of freedom. It allows you, among other things, to find the equivalent RGB color for any fiducial beam, no matter what the spectrum might be. "Equivalence" means that the RGB color sends exactly the same signals to your brain as the fiducial beam would. This implies, of course, that the RGB color will be *indiscriminable* from the fiducial beam.

Yet the RGB color is not a quale. It becomes one the moment you actually *look* at it and experience it. This is where you step into the domain of psychology. Colorimetry gets you at the threshold but doesn't let you step over it.

The linear and convex formal structure of colorimetry allows you to perform a great many useful calculations that connect optics with the stimulation of your brain. In order to do such calculations you have to put ecological optics and colorimetry in series as it were.

You also have the tools to find your way in the realm of colorimetric colors. It is a rich toolbox of methods, colorimetry (color measurement) is every bit as rich as geometry (earth measurement). Yet it takes you only about as far. This is a point that is rarely understood by most scientists. Geometry is a great science, but it has nothing to say on the topic of space and shape as you experience it—as *qualia*, I mean. Geometry is only chapter zero if you want to describe the experience of seeing the *Venus de Milo* (or, for that matter, your backyard, it makes no conceptual difference), yet it is a necessary prerequisite. Visual space and shape are very much unfinished chapters of psychology. Here science has only scratched the surface, and the toolbox of geometrical methods doesn't serve to force progress. One has to carry on by other—as yet largely undeveloped—means. With colorimetry, it is not essentially different.

18.2   A Broad View

As just I said, it is not possible to pursue color science (as a *science*) in any direction without a fair grasp of what colorimetry is about and what its empirical basis and formal structure are. Unfortunately, instances where this condition has been ignored abound in philosophical, psychological, physiological and medical researches involving color. Perhaps unfortunately, but certainly expectedly, the result is often nonsense.

Colorimetry is so basic that is hardly *about* anything of interest to the sciences. It is not about *qualia* (thus irrelevant to the philosopher), nor about *color attributes* (thus irrelevant to the psychologist), nor about the *brain* (thus irrelevant to the physiologist), not about the *human condition* (hence not important to the physician). But ignoring colorimetry means lacking the very concepts and instruments to talk about color in any scientific context. It is a bit like pursuing anything that involves entities of a spatial nature while ignoring geometry. Geometry per se is hardly of interest to any science, but it would certainly be considered "unprofessional" to pursue such topics without having at least *some* background in concepts such as dimension or length, or without knowing how to apply a ruler (or whatever instrument might be relevant for the task) and so forth. But this is exactly what happens all too frequently in the case of color. Sad to say, it makes for much unprofessional behavior.

If colorimetry is chapter zero of any color science, then evidently there is *much* of scientific interest that lies *beyond colorimetry*. Such is indeed the case.[1] Indeed, most of your interests involving color are likely to lie beyond colorimetry.

Here I will succinctly indicate some fields that may well be considered more or less immediate "extrapolations" of colorimetry. I select only a few examples, because color pops up in so many contexts that it is impossible to be complete.

The literature on color is extremely broad and spread over both the sciences and the humanities. Even if you limit yourself to the sciences there exist numerous fields that are "about color" in some way or other but that could hardly be considered extrapolations of colorimetry in any reasonable sense. Such fields are of importance in their own right, but might be considered *autonomous endeavors*.

Examples are

— the philosophy involving color, which is mainly about qualia, consciousness, and so forth;[2]

— physiology involving color, which is mainly about brain processes of visual areas;[3]

— biophysics and biochemistry of color, which is about the structure of photopigments and the processes of interaction with electromagnetic radiation;

— biology and psychology involving ecological aspects of color;

— genetics involving the heredity of retinal pigments;

— and so forth.

These are large and important fields with their own (huge) literature, journals, conferences, and so forth. I consider these autonomous with respect to colorimetry because they involve either *mechanism* or *behavior*, both independent of colorimetry. Mechanism is independent of colorimetry because colorimetry is *phenomenological* and applies equally well to many different mechanism that you may conceive of, whether brain or machine. The same applies to behavior. It is independent of colorimetry because it might just as well be about *any* physicochemical entity that is perceptually available.

The fields that are indeed obvious extrapolations of colorimetry are image engineering, visual perception, and ecological optics (I'll explain that shortly). I limit both image engineering and visual perception to topics involving color. These fields are very different from each other, visual perception (usually considered a subdiscipline of experimental psychology) being a *science* and image engineering being *engineering* (some would say applied science).

18.2.1 Image engineering

Image engineering involves anything that results in objects or processes that are designed to be looked at by human observers. The processes involve the recording of scenes, the storage and processing of such recordings, and the rendering of images to the human observer. Think of photography, TV, movies, printing, computer monitors, you name it. This is an extremely interesting field, also from a scientific perspective. As an engineering discipline, a lot of the literature tends to be repetitive, ad hoc, extremely focused on trivia (to you), and thus generally boring, but even so there is much to enjoy, and when you skim the literature you often find interesting new angles on things. This book is all the basis you need in order to be able to follow this literature (as it applies to color) if you are not afraid to have to pick up some tidbits of additional knowledge here and there.

Being a kind of engineering largely saves the field from producing the worst nonsense, for engineers are preoccupied with getting things to work (it is hard not to sympathize), and things that work often (but by no means invariably) are based on sound principles. I notice that I find much of interest concerning colorimetry and various extrapolations of it in this literature. For the engineer, colorimetry is evidently of much practical importance. However, since engineers mistrust and generally hate psychology, they tend to rely more on convention and less on actual psychophysics when they design human interfaces, a weakness that is unfortunate and often detracts from the value of otherwise interesting work.

18.2.2   Visual perception

I have had a keen interest in visual perception throughout my professional life, and I have worked in the area; thus I may not be entirely neutral in my remarks.

The study of "visual perception" has many ramifications, sometimes being only minimally related. The field I intend here is perhaps better caught under "visual psychophysics," and "cognitive (visual) science," although I'm not too happy with the "cognitive." After all, vision is presentation (precognitive, something that happens to you; like sneezing, you cannot be held responsible for your presentations) rather than "representation" (thought, something you *do* and can be held responsible for). But psychophysics has connotations of boring measurements under conditions of extreme response reduction that bear little on the topic of vision as understood by the average person. In skimming the vision literature, you will find much that is actually physiology—recently, much that falls roughly under the topic of brain scanning. All this is of much intrinsic interest, to be sure, but has nothing to do with the study of visual perception as intended here.

If you have a background in the exact sciences it will be difficult for you to make sense of the type of literature relevant to visual perception. When I lecture on my work in vision, colleagues from the sciences often remark on the "softness" of the work and even ask me whether I consider it science at all? They tolerate me in their neighborhood only because of my background in physics and mathematics. Yet this is quite unfounded. I am attracted to problems in vision partly because I consider the field conceptually and technically *much* more difficult than physics or mathematics. However, I grant you that this is not apparent from the bulk of the literature.

Nowadays, even many professionals in vision (mostly from a background in experimental psychology) seriously believe that their field is bound to become a true science only because it will be taken over by brain science (via the novel brain-scanning techniques). Brain science is indeed a science in the classical sense, though only if you interpret it as physiology. But few people seriously believe (although some do) that all of visual experience can be reduced to physiology. All this is not a little naïve and mistaken.

There is much to be enjoyed if you take your pick from the vision literature. Important problems abound and still await being attacked.

18.2.3 Ecological optics

Ecological optics is not a recognized field, so I need to do some explanation.

Vision is different for different agents. You and your cat, looking at the garden, see different scenes. Not only does the cat have a different visual system from yours, but it is also—as an agent—*alien* with respect to human agency. No doubt the gardener will also see a different scene than you do (at least if you're not particularly into gardening). Thus, there is also a spectrum of human agents whose vision is different for equivalent optical input. Yet you and the gardener share a "generic human" trait, so the gardener's is not entirely *alien* to yours. You might (given sufficient dedication) even *become* a gardener and develop a "gardener's eye" yourself, but you will never be able to become a cat. Yet you and your cat share a generic animality of the terrestrial kind. Thus you and your cat *do* share at least *some* perspective on the world. Indeed, often you will feel that you actually *understand* what makes your cat tick. Few people experience such *empathy* with spiders or slime molds, though.

The *biotope* is the physical environment as described with respect to the senses, appetites, and interactive potential of that strain of agent. All living species are tuned to a certain biotope, through interaction with the physical world over evolutionary time spans.[4] The "tuning" involves anatomical makeup, physiological structure from the molecular to the organ levels, life style, senses, motorics, and (background) understanding. It changes both the agent and the world. The biotope is entirely meaningful to the agent in the sense that each and every part of it is understood (I mean the term on the gut level, not the cognitive level) in its causal connection to other parts and processes, including one's own body and mind. To the extent that the physical world fails to be meaningful, it is not part of the biotope. Thus, if you're not a physicist, gluons and quarks are definitely not part of your biotope. Most of your biotope is likely to be on the supramolecular level, scientifically speaking. Most of the meaning (what is important to you) is irrelevant to physics and chemistry.

Since the biotope (of a certain strain of agents) is the physical environment from the gut level understanding of the agent, the agent is as much defined by the biotope as the biotope by the agent. The physical world is thoroughly meaningless, whereas

in the biotope, physics and meaning coincide.

This is close to Goethe's understanding that you see "because you want to see" and not merely because you were born with eyes.[5] It is what James Gibson describes as the "affordance structure" of the world.[6] It is also why perception is (much) more like "controlled hallucination" (see appendix R.1) than it is like "inverse optics". "Inverse optics"[7] (if there is such a thing) is certainly unable to compute any meaning. It turns one kind of structure (for example, optical), into another kind of structure (for example, geometrical), both equally meaningless. This is indeed a *necessity* for "inverse optics" to be reckoned among the *sciences*. Few hardcore scientists dare to accept the "perception is controlled hallucination" perspective, though, fearing (and perhaps with some reason) that this would expose them to their fellow scientists as softies, falling for mystical, nonscientific fairy tales.[8] But that perception is "intentional"[9] cannot be understood from any "inverse optics" perspective, meaning is not given prior to the perception-action cycle.

Although this view is thus by no means popular among scientists (it is indeed generally considered mystical or nonscientific), it is the only viable one (do my preferences show through here?). But then, studying the world is not entirely distinct from studying the mind, whereas studying the brain without studying the world has no hope to ever "explain" the mind. It is from such a cheerful perspective that I consider ecological optics. If the concept appears to lack a basis in science, I will work hard to construct such a fundament.

Ecological optics is the study of optics in the setting of a biotope. Such a science largely coincides with the study of vision (of the fiducial species), not vision from a physiological perspective (where vision means little more than "optical irritability"), but vision as a nexus of threads of consciousness.

Thus the study of ecological optics is a necessity in order to be able to obtain an understanding of vision. Since the field as such is not recognized, it is not that easy to find relevant literature.[10] Interesting contributions are scattered through different fields such as (applied) optics, (applied) materials science, experimental psychology, ergonomics, computer vision, image processing, and computer graphics. Especially the field of computer graphics seems to have discovered ecological optics, and it may well be the best starting point if you are interested.

18.1 Visual Perception

"Perception" is a very basic component of experience. Perceptions happen to you, like sneezing. Open your eyes, and there is the scene in front of you. You can't help it. What you know about the scene in front of you is due to the present perception and your memory. Your memory contains accounts of your experiences with the scene in front of you (or similar scenes) at earlier times. The memory is in the present, though; both perceptions and memories exist only NOW, and neither of them is due to conscious efforts on your part. They are *presentations*. Having the perceptions and memories (they can't really be distinguished) you have right now is part of *what you are*. These presentations happen only to you (your friend next to you may have a similar perception and perhaps somewhat similar memories, but you can't be sure), and you can't be held responsible for them. The moment you have perceptions and memories (all the time), you will enter these into your cogitation. You are indeed responsible for the thoughts that arise. The thoughts are also yours, but they don't just happen to you. Thinking is something you *do*, and you are responsible for your actions. Some people would hold you responsible only for thoughts made public through overt communication (such as speech) or physical actions, but this seems somewhat hypocritical to me.

There exists no scientific account of how perceptions arise. Some people identify perceptions with certain brain activities, but this is ludicrous. Perceptions are a prerequisite to any science, including brain science.

Goethe even held that "the phenomena (perceptions) are already the theory (science)." Although not a popular thought nowadays, this view has its merits. Anyway, at this moment you have to accept the fact that no one has even the faintest idea of how to "reduce mind to (natural) science" (don't be fooled by recent accounts of brain scanning and all that), whereas it is a valid question to ask "how science depends on the mind." This puts the study of mind and the sciences in different ball parks. I think it rational and most conducive to progress in our understanding simply to accept that fact. You can accept it irrespective of whether you believe that "mind" and "consciousness" are nonentities that will be explained away by the sciences in due time (as irrelevant epiphenomena), or whether you believe (as I do) that the mental realm is ontologically prior to the scientific realm.

Such a stance makes a huge difference in your approach to the topic of perception. The "standard model" (often ascribed to David Marr[11]) is that perceptions are nothing but reconstructions of the scene in front of you on the basis of optical information that is received by your visual system. This is sometimes referred to as inverse optics.[12] In a minor variation, most popular today, perceptions are your best guess of what is the scene in front of you on the basis of memory ("prior knowledge"; notice that I count "abilities" under "memory") and the present optical input.[13] I consider the former theory nonsense and the latter a (mildly interesting) partial insight. My reasons are that you can reconstruct only what was once yours to begin with (and the physical scene in front of you never was) and that physical structure is categorically different from information. Information assumes a definition of signals and their intended meaning. There are no signals in nature—only social agents produce signals—and no meaning either. Any "meaning" is imposed by the perceiver, not by the world.

Perceptions *are* meaning and thus necessarily produced by the perceiver. This means that the perceiver constantly "hallucinates," though the hallucinations are constantly checked against the momentary optical structure (in the world, at the eye, or in the brain, it doesn't matter). Thus perceptions are every bit as "objective" as any scientific theory. Theories in science are free inventions ("hallucinations") that are continually checked against empirical data. The perceiver (as the scientist) actively hunts for structures in perception (measurement) that might falsify the conclusion (perception or theory). I would say that

> science is perception carried on by other means,[14]

where the agent is a cultural community, rather than a single individual.

In order to check perceptions (this goes on largely automatically), the perceiver hunts for certain cues. In many cases this is automatic. Good examples include the "releasers" identified by biologists,[15] though in some cases it may involve conscious actions (as looking for footprints on the beach). Perhaps the best documented family of cues is that of the "depth cues."

The topic of cues became important when Bishop Berkeley[16] convincingly reasoned that inverse optics will never lead to any perception of distance away from the eye ("egocentric distance"). Although Berkeley is frequently ridiculed for his ideas, the main point stuck and has never been refuted.

It is important not to confuse the "cues" with the "data" as intended by the "inverse optics" paradigm. Cues start as mere (meaningless) structures (whereas "data" implies a pre-established meaning or significance) that are promoted to cue status by the observer in the course of checking the current hallucinations against the context of present situational awareness. Thus cues are as much "created" as "found". It follows that an algorithmic "cue detector" is an impossibility. Think of "cues" as similar to the

"clues" in a forensic investigation. A discarded cigarette butt is only meaningless environmental structure, but—in the context of a plot—it may become sufficient evidence to bring someone to the gallows. Notice that the "plot" (a hallucination of the investigator) is crucial in promoting mere meaningless structure to highly significant evidence. No structure is intrinsically "evidence" except in the perception of someone with an appropriate plot. No structure is intrinsically "cue" outside of the mind of an appropriate observer.

In the case of color, the topic of "cues", and its kin, has not been pursued very thoroughly. If you mention "cues for color," experts are likely to laugh. But the notion makes plenty of sense. "Color" clearly has to do with the perception of the material constitution of objects. Like "depth," inverse optics cannot yield solutions to this problem. You have to hallucinate material properties, and you have to check your hallucinations against empirical data, that is to say, you have to hunt for cues. When you "perceive" such color attributes as "METALLICNESS," "WETNESS," "HUE," "GOLDEN," "GLOSSINESS," and so forth you do so on the basis of cues. This evidently indicates the need for a thorough scientific investigation. However, this clearly leads beyond the bounds of my book.

Exercises

1. Philosophy of color [CONCEPTUAL] The Internet is a rich source of papers on philosophical aspects of color. Collect papers on a few subtopics that defend conflicting views and comment on them. Do they get at least the colorimetric aspects right?

2. Psychophysics of color [CONCEPTUAL] You will be able to find a bewildering variety of psychophysical papers on the general topic of color. Collect some on the topic of "color appearance" and comment on them. How well are the experiments controlled from a colorimetric point of view?

3. Inverse optics [CONCEPTUAL] Much of "machine vision" is concerned with inverse optics. Collect papers on some topics that bear on color and comment on them. Are colorimetric methods used, and if so, used correctly? Is colorimetry relevant to these topics at all?

4. Odds and ends [CONCEPTUAL] If you have the mindset to enjoy such things, look for the (many!) weird applications of "color science" that you encounter on the Internet. Try to figure out whether such things stand a chance of ever becoming "science" (in the true sense of the word—like pornography, I mean: you know it when you see it). If not, do they still hold some intrinsic interest (for example, in the sense of aesthetics, after all music is not science either)?

5. Ecological optics [CONCEPTUAL] Prepare a "Table of Contents" for a book on "Ecological Optics" as you would like to see it. How would this be different if you had to do it not just for the human observer, but for the whole animal kingdom? What if you had to do it for all living nature, including plants? How about arbitrary agents (such as Martians, and so forth)?

6. Cues for color [CONCEPTUAL] Consider the concept of "cues for color". Does it make sense? Interpret color in the broad sense of "surface quality" and/or "surface property". Prepare a list of examples. Would you feel confident to write a formal proposal for a scientific project on the topic?

Chapter Notes

1. You may obtain some notion of the often bizarre topics that bear on "color" by taking a look at F. Birren, *Color: A Survey in Words and Pictures from Ancient Mysticism to Modern Science*, (Secaucus, NJ: Citadel Press, 1963). See also R. Steiner, *Das Wesen der Farben*, (Dornach: Rudolf Steiner Verlag, 1980). More down to earth, and interesting, is Ch. A. Riley II, *Color Codes. Modern Theories of Color in Philosophy, Painting and Architecture, Literature, Music and Psychology*, (Hanover: U.P. of New England, 1995).

2. For an introduction, see D. R. Hilbert, *Color and Color Perception, a Study in Anthropocentric Realism*, (Menlo Park CA: CSLI/SRI Int., 1959).

3. A good introduction is S. Zeki, *A Vision of the Brain*, (Oxford: Blackwell, 1993).

4. Jakob von Uexküll (1864–1944) may have been the most influential biologist to pursue this topic. Some of his better known works are J. von Uexküll, *Umwelt und Innenwelt der Tiere*, (Berlin: Springer, 1909); ——, and G. Kriszat, *Streifzüge durch die Umwelten von Tieren und Menschen: Ein Bilderbuch unsichtbarer Welten (Sammlung: Verständliche Wissenschaft, Bd. 21.)*, (Berlin: Springer, 1934).

5. Well known is "Parabase":

> Wär nicht das Auge sonnenhaft,
> Die Sonne könnt es nie erblicken;
> Läg nicht in uns des Gottes eigne Kraft,
> Wie könnt uns Göttliches entzücken?

A translation that remains fairly close to the meaning, but fails to capture rhyme and rhythm of the original might be:

> If the eye weren't sun-like,
> How could it ever see it;
> If we hadn't godlike power,
> How could the divine turn us on?

and "Epirrhema":

> Müsset im Naturbetrachten
> Immer eins wie alles achten;
> Nichts ist drinnen, nichts ist draußen:
> Denn was innen das ist außen.
> So ergreifet ohne Säumnis
> Heilig öffentlich Geheimnis.
>
> ———
>
> Freuet euch des wahren Scheins,
> Euch des ernsten Spieles:
> Kein Lebendiges ist ein Eins,
> Immer ist's ein Vieles.

It is hard to capture rhyme and rhythm of the original here, this at least captures the meaning:

> In observing nature always
> Attend to this, and everything;
> Nothing's inside, nothing outside:
> For the inside is the outside.
> So be quick to grasp
> The sacred public secret.
>
> ———
>
> Rejoice in true appearance,
> And in earnest play:
> No being is a unit,
> It's always an arrangement.

6. Gibson's "affordances" merely reflect the fact that in the biotope meaning and structure coincide.
   J. J. Gibson, *The Ecological Approach to Visual Perception*, (Boston: Houghton Mifflin, 1979).

7. The term "inverse optics" is due to T. Poggio, "Low-level Vision as Inverse Optics," in *Proc.Symp. Comp. Models of Hearing and Vision*, edited by M. Rauk (Tallin: Acad.Sci. Estonian S.S.R., 1984), pp. 23–127.

8. It is often held that "controlled hallucination" implies that perception cannot be *objective*. Such is not the case. "Controlled" implies that the "hallucinations" are constantly tested against the available optical structure. The observer hunts for structures that might conceivably be at odds with perceptions and, if necessary, adjusts the perceptions accordingly. This is not different from the "scientific method" in that theories are free inventions of the mind (hallucinations) but are constantly confronted with empirical data. The theory helps to hunt for possibly conflicting data. Experimenting at random (often called "butterfly collecting") would be far less effective. In this view perception works from the inside out (the perceiver probes for conflicting evidence) rather than from the outside in (the perceiver computes perceptions from optical data). In my view the latter (mainstream!) view is actually incoherent. What would happen if you applied this notion to science?

9. "Intentionality" is to be understood in the sense that it is usually used in philosophical context, "referring to the world." Perceptions are indeed "about the world," which is why they cannot be prior to experience and interaction. Anything prior to experience can only have formal meaning, no meaning in the sense of "affordance".

10. A rare example is J. N. Lythgoe, *The ecology of vision*, (Oxford: Clarendon, 1979); however, it is focused on the submarine rather than the terrestrial environment.

11. D. Marr, *Vision: A Computational Investigation into the Human Representation and Processing of Visual Information*, (New York: Freeman, 1982).

12. Ibid.

13. R. D. Rosenkrantz, "Inference, Method, and Decision: Towards a Bayesian Philosophy of Science," *J.Am.Stat.Assn.* 74, pp. 740 (1979).

14. This of course echoes Carl von Clausewitz *"Krieg ist die Fortsetzung der Politik unter Einbeziehung anderer Mittel"* (war is politics carried on by other means). —, *Vom Kriege*, (Berlin: Ferdinand Dümmler, 1833-1834).

15. K. Lorenz, *The Foundations of Ethology*, (New York: Springer, 1981).

16. G. Berkeley, *An Essay Towards a new Theory of Vision*, (Dublin: Aaron Rhames, 1709).

Index

A

A4 format, 26
aberrations, 630
Abney
    law, 422
    shift, 431, 432
Abney, W., 431, 457
Abramowitz, M., 210
absolute color judgments, 19
absorbed photon, 78
absorbing dielectric media, 668
absorption
    of radiation, 62, 660, 672, 673
    spectra of cones, 125
abstract figures, 627
Académie des Sciences, 33
achromacy, 288
achromatic
    axis, 223, 246, 258, 271, 306
    beam, 135, 136, 138–140, 143, 156, 162, 191,
        194, 218, 223, 241–243, 246, 268, 284,
        285, 381, 481, 563
    color, 20, 138, 189, 258, 481
    component, 38, 190, 191, 278
    content, 296
    direction, 144, 187
    hue, 138
    line, 224
    point, 195, 223, 227, 234, 242, 258, 268, 310,
        370, 432, 541

action
    of an operator, 89
    spectra, 95
actual, 75
    composition, 75
    images, 631
addition of
    beams, 68
    linear functionals, 86
    vectors, 83
additive
    color mixture, 19
    inverse, 68–70
    mixture of beams, 43, 588
additivity law, 65
Adelson, E. H., 687
adjoint, 362
affine
    differential invariants, 112
    hyperplanes, 339
    hyperspaces, 72, 90, 101, 336
    invariance, 413
    invariant, 112, 262, 268
    line, 89
    space, 107, 132
    subspaces, 89
    transformations, 242
affinely invariant distance measures, 413
affordances, 694
Afriat, S. N., 412
agents, 693
Aguado, L., 458
air, 660, 661, 667
    light, 524, 661

J

O

T

U

Z